CYTOKINES
Stress and Immunity

CYTOKINES
Stress and Immunity

edited by
Nicholas P. Plotnikoff
Robert E. Faith
Anthony J. Murgo
Robert A. Good

CRC Press

Boca Raton Boston London New York Washington, D.C.

Acquiring Editor: Liz Covello
Project Editor: Maggie Mogck
Marketing Manager: Becky McEldowney
Cover design: Jonathan Pennell

Library of Congress Cataloging-in-Publication Data

Cytokines: stress and immunity / edited by Nicholas P. Plotnikoff . . . [et al.].
    p. cm.
    Includes bibliographical references and index.
    ISBN 0-8493-3150-1 (alk. paper)
    1. Cytokines.   2. Stress (Physiology).   3. Stress (Psychology).   4. Immunity.
 5. Psychoneuroimmunology.   I. Plotnikoff, Nicholas P.
QR185.8.C95C996   1998
616.07'9—dc21
                                                                                       98-34473
                                                                                           CIP

This book contains information obtained from authentic and highly regarded sources. Reprinted material is quoted with permission, and sources are indicated. A wide variety of references are listed. Reasonable efforts have been made to publish reliable data and information, but the author and the publisher cannot assume responsibility for the validity of all materials or for the consequences of their use.

Neither this book nor any part may be reproduced or transmitted in any form or by any means, electronic or mechanical, including photocopying, microfilming, and recording, or by any information storage or retrieval system, without prior permission in writing from the publisher.

All rights reserved. Authorization to photocopy items for internal or personal use, or the personal or internal use of specific clients, may be granted by CRC Press LLC, provided that $.50 per page photocopied is paid directly to Copyright Clearance Center, 222 Rosewood Drive, Danvers, MA 01923 USA. The fee code for users of the Transactional Reporting Service is ISBN 0-8493-3150-1/99/$0.00+$.50. The fee is subject to change without notice. For organizations that have been granted a photocopy license by the CCC, a separate system of payment has been arranged.

The consent of CRC Press LLC does not extend to copying for general distribution, for promotion, for creating new works, or for resale. Specific permission must be obtained in writing from CRC Press LLC for such copying.

Direct all inquiries to CRC Press LLC, 2000 Corporate Blvd., N.W., Boca Raton, Florida 33431.

Trademark Notice: Product or corporate names may be trademarks or registered trademarks, and are only used for identification and explanation, without intent to infringe.

© 1999 by CRC Press LLC

No claim to original U.S. Government works
International Standard Book Number 0-8493-3150-1
Library of Congress Card Number 98-34473
Printed in the United States of America 1 2 3 4 5 6 7 8 9 0
Printed on acid-free paper

Preface

It has not been long since the immune system was considered to be an independent system designated to stand alone in defense of the body from infections. This view was fostered by the fact that many immune functions can be induced *in vitro* and contribute to the appearance that these functions are independent of other bodily systems. Increasing knowledge in recent years of the immune system and its functioning has made it clear that the immune system does not stand alone, but is profoundly affected by other organ systems, especially the central nervous and neuroendocrine systems, and in turn affect the functioning of these systems as well. The three systems can now be considered to be a kind of super system. In many ways this makes sense and seems right. The more our knowledge of the systems increases, the more it becomes apparent that these three networks are intimately tied together as a major set of networks or super network. Elements of the three systems share a common origin in the neuroectoderm and also share controlling feedback loops. All three systems produce a number of cytokines, hormones, or neuropeptides in common, and cells of all three systems appear to share receptors for some of these molecules. This super system performs the critically important function of many of the homeostatic controls.

The information contained in this book extend our knowledge of the close relationships and multifaceted interactions of the three components of the super system, exemplifying how each exerts checks and controls on the others. The book provides new insights into the potential role of the increasing number of cytokines in stress responses and neuroendocrine-immune reactions. During the last decade, substantial advances have been made in understanding the nature of these relationships and the interactions they imply. A recurrent theme in this volume is the role played by cytokines and neurohormones in mediating bidirectional communications between the immune, central nervous, and neuroendocrine systems. Many previously elusive factors have now been identified and well characterized. The synthesis and secretion of neurohormones by lymphocytes and the production of cytokines by cells of the neuroendocrine system are documented and described. Thus, there exists in the mammalian body a complex, yet unified, neuroendocrine-immune network which is undoubtedly very much involved in the responses to stress.

Improved knowledge of tumor biology and tumor host relationships has led to a better understanding of how neuroendocrine and immune factors mediate the effects of stress on tumor growth and development. Conversely, nothing is more a stimulus of physical and psychological stress in humans than a malignant tumor. Cytokines and neuroendocrine factors appear to be responsible for many of the remote effects of cancer on the host, e.g., pain, anorexia, weight loss, cachexia, fever, and depression. Several chapters in this book provide insights into the nature of interactions between cancer and neuroendocrine and immune systems, and address clinical implications of these interactions.

Our earlier reviews of research into clinical depression revealed mechanisms involved and reactions generated by stress, including correlates such as reduction in mitogen-induced blastogenesis and other measures of immune function. Our present effort represents an attempt to bring together the approaches in investigating changes in cytokine levels that result from clinical depression or other forms of adaptations to stress. If stress can result in induced parameters of immune function, the mechanisms involved could operate at the cytokine level. Central inhibitory mechanisms via the autonomic nervous systems, as well as the endocrine system, may account for alterations of cytokine responses and functions and activities. Thus, cancer and AIDS patients exhibit reduced interleukin-2 and γ interferon levels possibly related, at least in part, to clinical depression or other manifestations of the stress to which the patients are exposed.

New findings in clinical research are presented by the several contributors to this book. They represent neuroimmunological challenges for investigation of clinical manifestations of stress. These findings may help clarify some of the complexities of clinical stress as they relate to infectious diseases, including AIDS and even cancer.

Editors

Nicholas P. Plotnikoff, Ph.D., is a professor of Pharmacology at the College of Pharmacy, University of Illinois/Chicago, as well as the College of Medicine and the Graduate College. His principal research interests are in the areas of psychoneuroimmunology with special emphasis on stress hormones and cytokines (enkephalins-endorphins). His current research has been focused on the clinical effects of methionine enkaphalin in the treatment of cancer and AIDS patients. In 1991, together with Anthony Murgo, Robert Faith, and Joseph Wybran, he helped organize an update of the state-of-the-art in *Stress and Immunity* (CRC Press LLC).

Robert E. Faith, D.V.M., Ph.D., is the Director of the Center for Comparative Medicine at Baylor College of Medicine. He received his D.V.M. from Texas A&M University in 1968, his Ph.D. in immunology from the University of Florida in 1979, and his residency training in Laboratory Animal Medicine at UF from 1968 to 1971. From 1974 to 1978, Dr. Faith was a Staff Fellow and Senior Staff Fellow at the National Institute of Environmental Health Sciences. He was Director of the Biomedical Research Center at Oral Roberts University from 1978 to 1984, and Director of the Animal Care Facilities at the University of Houston from 1984 to 1990. He joined Baylor College of Medicine in 1990.

Anthony J. Murgo, M.D., M.S., FACP, a medical oncologist and hematologist, is currently a senior investigator at the National Cancer Institute (NCI) in the Cancer Therapy Evaluation Program (CTEP). He also is a Clinical Professor of Medicine at the Uniformed Services University of the Health Sciences in Bethesda, MD. Dr. Murgo received his M.D. and M.S. in Immunology in 1975 from the State University of New York. He did his medical residency training at Maimonides Medical Center in Brooklyn and a fellowship in Hematologic Oncology at Memorial Sloan-Kettering Cancer Center in New York. He was on the faculty of Medicine at the Oral Roberts University School of Medicine from 1980 to 1983. He later joined the faculty of West Virginia University School of Medicine where he was Professor of Medicine in the Section of Hematology/Oncology. Prior to joining NCI, Dr. Murgo was a medical officer at the Food and Drug Administration in the division of Oncologic Drug Products from 1989 to 1996.

Robert A. Good, Ph.D., M.D., D.Sc., FACP, received his graduate degrees from the University of Minnesota in the early 1950s. He is presently Physician-in-Chief at All Children's Hospital in St. Petersburg, FL, as well as Distinguished Research Professor at the University of Florida College of Medicine. His remarkable career includes President and Director of the Sloan Kettering Institute for Cancer Research in New York, professorships in Pediatrics, Pathology, and Medicine at Cornell

University, head of cancer research at the Oklahoma Medical Research Foundation, and head of research at Memorial Hospital in New York. He has distinguished himself with major contributions to both clinical and basic immunology. His research achievements include discovery of the function of the thymus and definition of the two distinct, major cellular components of the immune systems. Good is the recipient of 13 honorary doctorate degrees, over 100 major citations and awards, and has written/edited/published 50 books and book chapters. He has been president of the American Association of Immunologists, Association of American Pathologists, Central Society for Clinical Research, American Society for Clinical Investigation, the Reticuloendothelial Society, and Society for Experimental Biology and Medicine. In 1978, he was recognized by Citation Index as the most cited scientist in the world over a 10-year period.

Note: This book was co-edited by Dr. Anthony J. Murgo in his private capacity. No official endorsement by the National Cancer Institute or the Uniformed Services University of the Health Sciences is intended or should be inferred.

Table of Contents

Chapter 1
Cytokines: Stress and Immunity — An Overview 1
Ronit Weizman and Hanna Bessler

Chapter 2
Neuroendocrine Regulation of the Immune Process 17
George Mastorakos. Christoph Bamberger, and George P. Chrousos

Chapter 3
Immune and Clinical Correlates of Psychological Stress-Induced Production of
Interferon-γ and Interleukin-10 in Humans 39
*Michael Maes, Cai Song, Aihua Lin, Raf De Jongh, An Van Gastel,
Gunter Kenis, Eugene Bosmans, Ingrid De Meester, Hugo Neels,
Aleksandar Janca, Simon Scharpé, and Ronald S. Smith*

Chapter 4
Depression and Immunity: The Role of Cytokines 51
Ziad Kronfol

Chapter 5
Bacterial Infections Due to Emerging Pathogens: Cytokines and Neuro-AIDS ... 61
*Vito Covelli, Angela Bruna Maffione, Giuseppe Giuliani, Salvatore Pece,
and Emilio Jirillo*

Chapter 6
The Anticytokine Peptide, α-MSH, in Infectious and Inflammatory Disorders 69
Anna Catania, Lorena Airaghi, and James M. Lipton

Chapter 7
Methionine Enkephalin in the Treatment of AIDS-Related Complex 77
Bernard Bihari and Nicholas P. Plotnikoff

Chapter 8
Central Nervous System Toxicities of Cytokine Therapy 93
Robert C. Turowski and Pierre L. Triozzi

Chapter 9
Stress and Inflammation 115
Paul H. Black and Ari S. Berman

Chapter 10
Neuropeptides, Cytokines, and Cancer — Interrelationships 133
Anthony J. Murgo, Robert E. Faith, and Nicholas P. Plotnikoff

Chapter 11
Cytokines, Stress Hormones, and Immune Function 161
Robert E. Faith, Nicholas P. Plotnikoff, and Anthony J. Murgo

Chapter 12
Bidirectional Communication between the Immune and Neuroendocrine
Systems 173
Douglas A. Weigent and J. Edwin Blalock

Chapter 13
The Thymus–Pituitary Axis: A Paradigm to Study Immunoneuroendocrine
Connectivity in Normal and Stress Conditions 187
Wilson Savino and Eduardo Arzt

Chapter 14
New Insights into the Hypothalamic Control of FSH and LH by
Cytokines and Nitric Oxide 205
*S. M. McCann, M. Kimura, A. Walczewska, S. Karanth, V. Rettori,
and W. H. Yu*

Chapter 15
Interferons and the Central Nervous System 221
Nachum Dafny

Chapter 16
Opioid Systems: Cytokines and Immunity 233
Hemendra N. Bhargava

Chapter 17
Opioid Growth Factor, [MET^5]-Enkephalin, and the Etiology, Pathogenesis,
and Treatment of Gastrointestinal Cancer 245
Ian S. Zagon, Jill P. Smith, and Patricia J. McLaughlin

Chapter 18
Control of Pain in Peripheral Tissues by Cytokines and Neuropeptides 261
Michael Schäfer and Christoph Stein

Chapter 19
Interleukins and Immunocyte β-Endorpin 271
Paola Sacerdote

Chapter 20
Opioids, Opioid Receptors, and the Immune System 281
Ricardo Gomez-Flores and Richard J. Weber

Chapter 21
Methionine Enkephalin-Induced Immune Modulation and Cytokines 315
Rebecca Bowden-Elkes and Steven C. Specter

Chapter 22
The Immune System and the Hypothalamus–Pituitary–Adrenal (HPA) Axis 325
P. Falaschi, A. Martocchia, A. Proietti, and R. D'Urso

Index 339

Contributors

Lorena Airaghi
IRCCS Ospedale Maggiore
Milano, Italy

Eduardo Arzt
University of Buenos Aires
Buenos Aries, Argentina

Christoph Bamberger
University of Hamburg
Hamburg, Germany

Ari S. Berman
Boston University School of Medicine
Boston, Massachusetts

Hanna Bessler
Rabin Medical Center
Petah Tiqwa, Isreal

Hemendra N. Bhargava
University of Illinois
Chicago, Illinois

Bernard Bihari
SUNY Health Sciences Center
Brooklyn, New York

Paul H. Black
Boston University
School of Medicine
Boston, Massachusetts

J. Edwin Blalock
University of Alabama
Birmingham, Alabama

Eugene Bosmans
University of Antwerp
Antwerp, Belgium

Rebecca Bowden-Elkes
University of South Florida
College of Medicine
Tampa, Florida

Anna Catania
IRCCS Ospedale Maggiore
Milano, Italy

George P. Chrousos
National Institute of Health/NICHHD
Bethesda, Maryland

Vito Covelli
University of Napoli
Napoli, Italy

Nachum Dafny
University of Texas
Medical School
Houston, Texas

Raf De Jongh
Algemeen Ziekenhuis, St. Jan
Genk, Belgium

Ingrid De Meester
University of Antwerp
Antwerp, Belgium

R. D'Urso
University La Sapienza
Roma, Italy

Robert E. Faith
Baylor College of Medicine
Houston, Texas

P. Falaschi
University La Sapienza
Roma, Italy

Giuseppe Giuliani
University of Bari
Bari, Italy

Ricardo Gomez-Flores
University of Illinois
College of Medicine
Peoria, Illinois

Aleksandar Janca
World Health Organization
Geneva, Switzerland

Emilio Jirillo
University of Bari
Bari, Italy

S. Karanth
Louisiana State University
Baton Rouge, Louisiana

Gunter Kenis
Algemeen Ziekenhuis, St. Jan
Genk, Belgium

M. Kimura
Institute of Medical and Dental
  Engineering
Toyko, Japan

Ziad Kronfol
University of Michigan
Medical Center
Ann Arbor, Michigan

Aihua Lin
University of Antwerp
Antwerp, Belgium

James M. Lipton
University of Texas
Southwestern Medical Center
Dallas, Texas

Michael Maes
University of Antwerp
Antwerp, Belgium

Angela Bruna Maffione
University of Bari
Bari, Italy

A. Martocchia
University La Sapienza
Roma, Italy

George Mastorakos
University of Athens
Athens, Greece

S. M. McCann
Louisiana State University
Baton Rouge, Louisiana

Patricia J. McLaughlin
Penn State University
College of Medicine
Hershey, Pennsylvania

Anthony J. Murgo
National Cancer Institute
Rockville, Maryland

Hugo Neels
University of Antwerp
Antwerp, Belgium

Salvatore Pece
University of Bari
Bari, Italy

Nicholas P. Plotnikoff
University of Illinois
College of Pharmacy/College of Medicine
Chicago, Illinois

A. Proietti
University La Sapienza
Roma, Italy

V. Rettori
Cientificas y Technicas-Serrano
Buenos Aires, Argentina

Paola Sacerdote
University of Milano
Milano, Italy

Wilson Savino
Institute Oswaldo Cruz
Rio de Janerio, Brazil

Michael Schäfer
Ludwig-Maximilians
University of Müchen
München , Germany

Simon Scharpé
University of Antwerp
Antwerp, Belgium

Jill P. Smith
Penn State University
College of Medicine
Hershey, Pennsylvania

Ronald Smith
Sierra Pacific Seminars
Gilroy, California

Steven C. Specter
University of South Florida
College of Medicine
Tampa, Florida

Cai Song
University of Antwerp
Antwerp, Belgium

Christoph Stein
Johns Hopkins University
Baltimore, Maryland

Pierre L. Triozzi
Ohio State University
Compreshensive Cancer Center
Columbus, Ohio

Robert C. Turowski
Ohio State University
Comprehensive Cancer Center
Columbus, Ohio

An Van Gastel
University of Antwerp
Antwerp, Belgium

A. Walczewska
Medical University of Lodz
Lindleya, Poland

Richard J. Weber
University of Illinois
College of Medicine
Peoria, Illinois

Douglas A. Weigent
University of Alabama
Birmingham, Alabama

Ronit Weizman
Tel Aviv Community Health
Tel Aviv, Isreal

W. H. Yu
Louisiana State University
Baton Rouge, Louisiana

Ian S. Zagon
Penn State University
College of Medicine
Hershey, Pennsylvania

1 Cytokines: Stress and Immunity — An Overview

Ronit Weizman and Hanna Bessler

CONTENTS

1.1 Introduction 1
1.2 Cytokines Involved in Inflammation and the Host Immune Response 2
1.3 Physiologic Aspects of Stress 3
    1.3.1 Hormones Involved in the Stress Response 3
    1.3.2 Immunoregulatory Activity of Stress Hormones 3
        1.3.2.1 ACTH and Opioid Peptides 3
        1.3.2.2 Prolactin and Growth Hormone 4
    1.3.3 Cytokines and the Neuroendocrine Response to Stress 4
        1.3.3.1 Interleukin-1 5
        1.3.3.2 Interleukin-2 5
        1.3.3.3 Interleukin-3 5
        1.3.3.4 Interleukin-4 6
        1.3.3.5 Interleukin-6 6
        1.3.3.6 Other Cytokines 6
1.4 Cytokine Production in Stress-Related Psychiatric Disorders 7
    1.4.1 Obsessive-Compulsive Disorder 7
    1.4.2 Panic Disorder 8
    1.4.3 Generalized Anxiety Disorder 8
    1.4.4 Post-Traumatic Stress Disorder (PTSD) 8
    1.4.5 Anorexia Nervosa 8
1.5 Conclusion 9
References 9

1.1 INTRODUCTION

Selye in 1935[1] was the first to examine the impact of emotional states on immune system function. He demonstrated that exposure to stressors is associated with a substantial decrease in the number of circulating lymphocytes and in the weight of the thymus and lymph nodes, especially in the cortical region. The underlying biological mechanisms, however, as well as the relevance of such changes to clinical conditions, remained unclear. Over the next 50 years, innovative studies in neuroscience, immunology, and endocrinology led to the emergence of the new field of psychoneuroimmunology.[2-6] Several investigators studying the link between stress and disease found that alterations occur in the resistance to viral and bacterial infections following environmental stress (for review see References 7 to 9). Ramirez et al.[10] demonstrated an association between severe life stressors and the first recurrence of breast cancer following surgery, and in an animal study Ben-Eliyahu et al.[11] noted that stress increases the metastatic spread of mammary tumors.

Current research in psychoneuroimmunology is geared to a more comprehensive understanding of the complex, bidirectional interaction between the central nervous system and the immune system and the role of emotions and hormones in this interaction, using more sophisticated methodological approaches in animals and humans and a wide range of behavioral, cellular, and molecular parameters. Cumulative data indicate that neurohormones regulate the activity of immunocompetent cells, and that cytokines secreted by the immunocompetent cells modulate neuroendocrine function. During stress, the extent to which the relevant neurohormones and neuropeptides are selectively recruited determines the impact of the specific stress on the immune function. The immunoregulatory effects of the same neurohormone or neuropeptide can be either facilitatory or inhibitory, depending on the target immune cell. This chapter focuses on the effect of stress on cytokine production and release.

1.2 CYTOKINES INVOLVED IN INFLAMMATION AND THE HOST IMMUNE RESPONSE

The initiation and maintenance of the immune response requires a complex interplay of humoral and cellular mediators. These include the cytokines which are soluble polypeptide signaling proteins that generate and augment the immune response by recruiting and activating the necessary cells. Cytokines are produced by many cells in response to stimuli; some have many targets and multiple functions, and others have a unique or synergistic biologic effect when combined with other cytokines. They are involved in all levels of the immune reaction, including recognition, differentiation, and cell proliferation.

Cytokines may be arbitrarily classified into interferons, lymphokines, monokines, hematopoietic growth factors, chemokines, and other cytokines.[12] *Interferons* (IFN)[13,14] (α, β, γ types) exert an antiviral activity, stimulate major histocompatibility complex antigen expression, and activate phagocyte killing of bacteria, fungi, parasites, and tumor cells, in addition to other activities. *Lymphokines*, originally thought to be produced only by lymphocytes, are now known to be products of nonlymphoid cells as well.[15,16] They include interleukin (IL)-2 to IL-6, IL-9, IL-10, IL-13, IL-14, low molecular weight B-cell growth factor, oncostatin M, and lymphotoxin or tumor necrosis factor (TNF)-β. In addition to their action on T lymphocytes, B lymphocytes, or both, and on other cells of the immune system, some lymphokines affect various stages of hematopoiesis and have diverse effects on other cell types. *Monokines*[15,17-19] are produced predominantly by mononuclear phagocytes, but may also be secreted by other cell types. The monokines that affect immune response are IL-1α and β, IL-1 receptor antagonist (IL-1ra), IL-12, IL-15, and TNF-α. *Hematopoietic growth factors*[15,20] stimulate the proliferation and differentiation of blood cells. They are further classified as colony-stimulating factors, erythropoietin, and stem cell factors. *Chemokines*[21-24] stimulate leukocyte chemotaxis, but are often involved in leukocyte activation as well. The list of chemokines includes interferon-inducible protein 10, IL-8, lymphotactin, melanoma growth-stimulating activity, monocyte chemotactic protein-1/monocyte chemotactic and activating factor, macrophage inflammatory protein-1α and -1β, and RANTES. *Other cytokines* are those that are difficult to place in the above categories because they are produced by cells other than lymphocytes or mononuclear phagocytes (e.g., IL-7).[25] Their biologic action is mainly observed in concert with other cytokines (e.g., IL-11 and leukemia inhibitory factor)[26,27] and they may both stimulate and suppress different kinds of inflammatory responses (e.g., platelet-derived growth factor and transforming growth factor-β).[28,29]

1.3 PHYSIOLOGIC ASPECTS OF STRESS

1.3.1 HORMONES INVOLVED IN THE STRESS RESPONSE

Stressors are stimuli that interfere with the organism's homeostasis. They may be physical/metabolic (e.g., surgery, cold, starvation) or psychological (e.g., fear, maternal separation, restraint). Emotional factors are common to both types. Stressors differ in controllability, novelty, predictability, anticipation, intensity, duration, and frequency. The response to an environmental stressor is associated with activation of the sympathoadrenomedullary system and the hypothalamic–pituitary–adrenal (HPA) axis. Activation of the sympathoadrenomedullary system causes increased secretion of norepinephrine and epinephrine, as well as neuropeptides such as *enkephalin* and *β-endorphin*. The stress-induced HPA activation is secondary to an increase in hypothalamic *corticotropin-releasing hormone* (CRH) and *pituitary adrenocorticotropin hormone* (ACTH), which elevate circulatory adrenal glucocorticoids. On the central nervous system level, stress enhances the activity of the brain noradrenergic neurons which, together with the stimulation of vasopressin release, directly activate the pituitary to release ACTH. Other hormones involved in coping with stress are *prolactin* and *growth hormone*. Severe acute or chronic stress may inhibit the hypothalamic–pituitary–gonadal (HPG) and hypothalamic–pituitary–thyroid (HPT) axes (for review see Reference 30). It is noteworthy that the neuroendocrine response is not a general, nonspecific phenomenon; it is modulated by the type of stressor and the organism's previous experience and behavioral control over the stressor, i.e., the ability to alter its onset, duration, intensity, pattern, or point of termination.[31]

1.3.2 IMMNOREGULATORY ACTIVITY OF STRESS HORMONES

1.3.2.1 ACTH and Opioid Peptides

Both ACTH and β-endorphin are produced in the pituitary from the polyprotein precursor proopiomelanocortin (POMC).[32] ACTH inhibits antibody synthesis by lymphocytes, especially in response to proteins,[33] while it amplifies the proliferation of activated B cells.[34] β-endorphin modulates the function of lymphocytes, natural killer (NK) cells, macrophages, and mast cells.[33,35,36] *Methionine-enkephalin*, a pentapeptide endorphin, like β-endorphin, can enhance the natural cytotoxicity of lymphocytes and macrophages toward tumor-cell and T-cell IL-2 production.[37-39] Since methionine-enkephalin is produced by immunocompetent cells and enhances immune function by binding to specific receptors, some authors classify it as a cytokine.[39] The assumption that ACTH and the endorphin-like peptides play an immunoregulatory role was supported by the finding of Smith and Blalock[40] that they are synthesized by lymphocytes infected with Newcastle disease virus.

Similar to IL-2 and several interferons, the POMC- and proenkephalin-A-derived opioids have a biphasic activity: immunostimulation at low concentrations and immunosuppression at high concentrations.

1.3.2.2 Prolactin and Growth Hormone

Both *prolactin* and *growth hormone* possess significant immunomodulatory properties.[41-43] In one study of hypophysectomized rats, prolactin or growth hormone replacement therapy restored immunocompetence, whereas other pituitary hormones had no effect.[44] Receptors for these hormones have been identified on lymphocytes and, like most interleukin receptors, they are part of the hematopoietic growth factor receptor family.[45] Prolactin, which is released from a human B lymphoblastoid cell line variant as well as from murine splenocytes,[46,47] facilitates lymphocyte growth and function.[48] Growth hormone (somatotropin) affects lymphocyte function via specific membranal receptors.[49] Its effect on immunological responses is achieved in an autocrine or paracrine manner, since it is synthesized by both lymphocytes and macrophages.[50] Growth hormone administration can augment T-cell cytolytic activity, NK cell activity, antibody synthesis, and TNF production.[51,52] Weigent et al.[53] suggested that it is probably essential for interleukin-induced lymphocyte proliferation, since antisense knockout of the growth hormone gene in lymphocytes inhibits T- and B-cell proliferation.

1.3.3 CYTOKINES AND THE NEUROENDOCRINE RESPONSE TO STRESS

Most endocrine glands contain lymphoid cells which release cytokines that directly modulate neuroendocrine activity.[54] The release of the immunoregulatory cytokines IL-1 to IL-6, TNF-α and IFN-γ is sensitive to glucocorticoid inhibition. This indicates the presence of a negative feedback mechanism between the HPA axis and the immune system. At the same time, corticosteroids act to attenuate the signal transduction of cytokine receptor stimulation, as demonstrated by the T-cell responses to IL-2 activation.[55] The resulting bidirectional regulatory loop between the HPA axis and the immune system serves as an adaptive defense mechanism to inhibit overactivity of hormones, cytokines, and inflammatory agents during stress.

Specifically, cytokines IL-1 to IL-6, IFN, and TNF have been found to play a significant role in the regulation of pituitary hormone secretion. IL-1, IL-2, and IL-6 are known to stimulate ACTH secretion[56,57] and, on the pituitary cellular level,

IL-1 also stimulates the release of growth hormone, luteinizing hormone, and thyroid-stimulating hormone, and inhibits prolactin secretion.[58] In addition, IL-1, and also IL-2, stimulate hypothalamic CRH secretion.[59]

Many studies have used physical stress models to determine the impact of stress on alterations in the production of cytokines, and results have been relatively consistent over the range of models employed. The relevant findings are described below.

1.3.3.1 Interleukin-1

An increase in IL-1 release was demonstrated in Sprague-Dawley rats 2 h after loss of 30% of blood volume[60] and IL-1-like activity was observed following burn injury.[61] This increase in IL-1 release may be part of the accelerated acute-phase reaction associated with the stress-induced hypermetabolic state. Thus, IL-1 may play a major role in the induction of acute-phase reactant synthesis.[62,63] An increase in serum ACTH and cortisol as well as in IL-1α and IL-2 was observed in humans following acute stress of angioplasty. A similar immunoendocrine pattern was obtained in healthy humans in response to CRH administration, indicating that CRH may play a major role in neuroendocrine-immune response to acute stress.[64] The stress-induced increase in IL-1 release may affect brain function, leading to activation of the HPA axis, fever and depression. Since the synthesis and action of cytokines in the periphery and brain are under the regulatory control of glucocorticoids,[65] the IL-1-related activation of the HPA axis may be ascribed to activation of the noradrenergic and indolaminergic systems.[66] The stress-induced IL-1 activation of the HPA axis is probably modulated by IL-6.[67]

1.3.3.2 Interleukin-2

In contrast to IL-1, IL-2 production is usually suppressed in response to hemorrhage, burns, and surgery.[68,69] The diminished ability of the peripheral blood mononuclear cells (PBMC) to produce IL-2 following accident injury is in direct correlation with the severity of the injury.[70] Furthermore, exhaustive stress leads to a suppression of phytohemagglutinin (PHA)-induced IL-2 release and consequent transient alterations in immune system function;[71,72] the psychological stress of training and competition may be an additive factor. Finally, emotional stressors may also have an impact. A decrease in the mitogenic response of lymphocytes to PHA was detected in unemployed subjects in the first year of unemployment (but not later),[73] and reduced IL-2 receptor expression but with no interference with IL-2 production was noted in students under examination stress.[74] It is possible that stress is associated with activation of suppressor T lymphocytes which inhibit both mitogen-induced proliferation of lymphocytes as well as synthesis of cytokines.[75]

1.3.3.3 Interleukin-3

Together with the suppression of IL-2, concanavalin A (con A)-stimulated IL-3 and IL-5 production by the splenocytes is reduced soon after hemorrhagic stress in mice; levels return to normal range after 24 h. These findings may indicate a stress-induced impairment in T-cell function as well as inhibition of T-helper (Th) cell activity.[76] Data on the effect of stress on IL-3 are still sparse.

1.3.3.4 Interleukin-4

While surgery suppresses IL-2 production, it activates Th cell type 2 (Th2) activity which, in turn, stimulates IL-4 production. In a study by Decker et al.[77] cholecystectomy increased PHA-induced IL-4 production in freshly isolated PBMC. Moreover, conventional cholecystectomy caused a greater increase than the less stressful laparoscopic cholecystectomy.

1.3.3.5 Interleukin-6

Hisano et al.[78] found that IL-6 levels rose in humans after thoracoabdominal surgery and decreased thereafter. Restraint stress in rats also increased mRNA for IL-6 in the midbrain, but not in the hypothalamus.[79] In a related study, mice exposed to influenza viral infection placed under restraint stress showed enhanced IL-6 production in the regional lymph nodes but diminished production in the splenocytes. Both alterations were inhibited by administration of the type II glucocorticoid receptor antagonist RU-486.[80] Thus, IL-6 production plays a role in both IL-1 activation of the HPA axis, as noted above, and in stress-induced activation of the HPA axis.[67,81]

1.3.3.6 Other Cytokines

Bacterial lipopolysaccharide-induced IL-12 production is inhibited in a dose-dependent fashion and in physiologically relevant concentrations by dexamethasone, norepinephrine, and epinephrine.[82] This suppressive activity is antagonized by RU-486 and the β-adrenoreceptor antagonist propranolol. By contrast, IL-10 production is stimulated by the catecholamines in a propranolol-sensitive manner, but not by dexamethasone. IL-12 is a key inducer of the differentiation of uncommitted Th cells toward the Th1 phenotype which regulates cellular immunity; IL-10 inhibits Th1 functions and potentiates the Th2-regulated response (humoral immunity). Thus, since stress is associated with increased secretion of glucocorticoids and catecholamines it may also affect IL-12 and IL-10 secretion, with a selective suppression of Th1 and a shift toward Th2. This change in the Th1/Th2 balance may be involved in the stress-induced susceptibility to infectious, autoimmune, allergic, and neoplastic diseases.[82]

IFN-γ is involved in amplifying the acute-phase response and in modifying T- and B-cell function. Stress has an apparently immunomodulatory effect. Mice and humans show a decreased production of IFN-γ following injury[83] and an increased con A-stimulated splenocyte production of IFN-γ (studied in mice only) following hemorrhagic stress.[76] In humans, exhaustive exercise stress decreased both con A- and lipopolysaccharide-induced release of IFN-γ.[71]

TNF plays a major role in the response to septic shock and the induction of fever, hypotension, and acute-phase proteins.[75] Weinstock et al.[71] found that lipopolysaccharide-induced release of TNF-α, like IL-1, IL-6, and IFN-γ, was suppressed 1 h after exhaustive exercise stress.

IL-13 can effectively downregulate the traumatic stress-induced elevation of monocyte-macrophage production of IL-1β, IL-6, IL-8, and TNF-α. Thus, IL-13 may be a modifier of acute states of traumatic stress-induced host defense deficiency.[84]

There are not enough data on the impact of stress on the remaining cytokines, especially the hematopoietic growth factors and the chemokines.

1.4. CYTOKINE PRODUCTION IN STRESS-RELATED PSYCHIATRIC DISORDERS

Depression has been shown to affect cytokine production and release, and thereby immune function, probably via an increase in glucocorticoid production. Depressed patients have increased plasma concentrations of IL-6, soluble IL-6 receptor (sIL-6R) and sIL-2R,[85] and suppressed PBMC synthesis of IL-1β, IL-2, and IL-3; cytokine production normalizes on clinical improvement with antidepressant treatment.[86] Anxiety is frequently complicated by depression, and some antidepressant agents are effective in certain anxiety disorders such as obsessive-compulsive disorder (OCD), panic disorder and, to some extent, post-traumatic stress disorder (PTSD). Data on cytokine production in anxiety disorders are still sparse. The current state of knowledge is summarized in this section.

1.4.1 OBSESSIVE-COMPULSIVE DISORDER

OCD is apparently associated with dysfunction of the central serotonergic system[87] and basal ganglia.[88] There is a body of evidence suggesting a role for the autoimmune mechanisms in some cases of OCD:

1. An increased incidence of OCD has been observed in individuals with Sydenham's chorea.[89]
2. An infection-triggered autoimmune subtype of pediatric OCD and Tourette's syndrome has been described.[90]
3. Serum antibodies for somatostatin and prodynorphin have been detected in OCD patients.[91]

Maes et al.[92] noted that, although plasma levels of IL-1β, IL-6, sIL-6R, and sIL-2R were similar in OCD patients and controls, there was a significant positive correlation between both IL-6 and sIL-6R and the severity of specific compulsive symptoms. Our group recently investigated the IL-1-, IL-2-, and IL-3-like activity (IL-3-LA) induced by PBMC in nondepressed OCD patients before and after tricyclic antidepressant treatment (clomipramine).[93] Since OCD is frequently associated with distress and depression, conditions known to affect cell-mediated immune function, we expected cytokine production to be impaired in these patients; yet no difference was observed between the OCD patients and controls, and the drug treatment did not affect the immunological measures. Therefore, it is possible that compensatory coping mechanisms prevent the long-term modulation of immune function in this chronic disorder. However, these results cannot be generalized to

OCD with comorbid depression. When cerebrospinal fluid cytokines were assessed in patients with childhood-onset OCD, a preponderance of cytokines relevant to cell-mediated immunity was noted, as opposed to cytokines relevant to humoral immunity.[94] This finding suggests that cell-mediated immunity may be involved in the pathophysiology of OCD.

1.4.2 PANIC DISORDER

Plasma IL-1β concentration has been reported to be significantly higher in panic disorder patients than controls, both before and after benzodiazepine treatment.[95,96] In another study, however, a slight alteration in serum IL-2 was noted, but not in serum IL-1α, IL-1β, or sIL-2R.[97] In a study by our group on IL-2 and IL-3 production by the PBMC in nondepressed panic disorder associated with agoraphobia, no differences were found between patients and healthy controls or between patients with and without agoraphobia.[98] A negative correlation was noted between IL-3 production and the severity of anxiety,[98] indicating that IL-3 production may be sensitive to the presence of anxiety and stress.

1.4.3 GENERALIZED ANXIETY DISORDER

La Via et al.,[99] in a study of the relationship between generalized anxiety disorder (GAD) and immunomodulation, found a decreased expression of IL-2R (CD25) on T lymphocytes stimulated with anti-CD3, which correlated with higher stress intrusion scores and a higher number of sick days due to upper respiratory tract infection. The authors concluded that there may be a direct relationship between the level of stress intrusion and immunodepression and between immunodepression and morbidity.

1.4.4 POST-TRAUMATIC STRESS DISORDER (PTSD)

PTSD is a severe and chronic anxiety disorder, often accompanied by depression and other anxiety disorders. Spivak et al.[100] found that circulating levels of IL-1β, but not sIL-2R, were significantly higher in patients with combat-related PTSD than in healthy controls. Serum IL-1β levels correlated with the duration of PTSD symptoms, but not with their severity, and with anxiety and depression; a correlation with cortisol levels was also observed. The authors suggested that the high levels of circulating IL-1β play a role in the increased sensitivity to dexamethasone suppression characteristic of PTSD patients.

1.4.5 ANOREXIA NERVOSA

Anorexia nervosa is an eating disorder manifested by altered food intake, changes in body temperature, suppression of gonadal hormone secretion, metabolic stress, and distortion of body image. It is associated with overactivity of the HPA axis. Increased spontaneous TNF production by the PBMC and decreased PHA-induced IFN-γ levels have been found in anorectic patients in the acute state; weight gain led to normalization of the cytokine level.[101,102] Other studies have demonstrated decreased IL-2 production by the PBMC[103] and elevated serum IL-6 and transforming growth factor -β.[104] Since some of these cytokines affect food intake, mood, and temperature regulation,[105,106] the alterations in cytokine release may contribute to the persistence of the anorexia nervosa symptoms.

1.5 CONCLUSION

Cumulative data indicate that stress and stress-related psychiatric disorders are associated with complex interactions between the nervous and immune systems that are mediated by hormones, cytokines, neuropeptides, and neurotransmitters. The correlation between stress and immunity is not linear, and both increases and decreases in immunocompetence can occur following exposure to different stressors. The cytokines involved in the immune response to stress are pluripotent molecules with the ability to activate the HPA axis, enhance cerebral and peripheral noradrenergic activity, and induce changes in neurobehavioral changes (fever, increased sleep, decreased appetite, and hypomotility) which affect the organism's coping ability. Stress-induced suppression of cytokine production may be geared to prevent overactivity of the immune system and development of autoimmunity. Work in the area of psychiatric disease is just beginning, and further studies are needed for a comprehensive understanding of the implications of direct and indirect stress-induced fluctuations in cytokine levels and their impact on morbidity.

REFERENCES

1. Selye, H., A syndrome produced by diverse noxious agents, *Nature*, 138, 32, 1935.
2. Solomon, G. F. and Moos, R. H., Emotions, immunity, and disease: a speculative theoretical integration, *Arch. Gen. Psych.*, 11, 656, 1964.
3. Stein, M., Keller, S. E., and Schleifer, S. J., Stress and immunomodulation: the role of depression and neuroendocrine function, *J. Immunol.*, 35, 827, 1985.
4. Ader, R., Cohen, N., and Felten, D. L., Brain, behavior, and immunity, *Brain Behav. Immunol.*, 1, 1, 1987.
5. Solomon, G. F., Psychoneuroimmunology: interactions between central nervous system and immune system, *J. Neurosci. Res.*, 18, 1, 1987.
6. Blalock, J. E., A molecular basis for bidirectional communication between the immune and neuroendocrine system, *Physiol. Rev.*, 69, 1, 1989.
7. Irwin, M. R. and Strausbaugh, H., Stress and immunity changes in humans: a biopsychosocial model, in *Psychoimmunology Update*, Gorman, J. M. and Kertzner, R. M., Eds., American Psychiatric Press, Washington, D.C., 1991, 55-80.
8. LaPerriere, A., Antoni, M., Kilmans, N., Ironside, G., Schneiderman, N., and Fletcher, M. A., Psychoimmunology and stress management in HIV-1 infection, *in Psychoimmunology Update*, Gorman, J. M. and Kertzner, R. M., Eds., American Psychiatric Press, Washington, D.C., 1991, 81-113.
9. Koolhaas, J. M. and Bohus, B., Animal models of stress and immunity, in *Stress, the Immune System and Psychiatry*, Leonard, B. E. and Miller, K., Eds., John Wiley & Sons, New York, 1995, 69-83.
10. Ramirez, A. J., Craig, T. K. J., and Watson, J. P., Stress and relapse of breast cancer, *Br. Med. J.*, 298, 291, 1989.

11. Ben-Eliyahu, S., Yirliyar, R., Liebeskind, J. C., Taylor, A. N., and Gale, R. P., Stress increases metastatic spread of a mammary tumor in rats: evidence for mediation by the immune system, *Brain Behav. Immunol.*, 5, 193, 1991.
12. Liles, W. C. and Van Voorhis, W. C., Review: Nomenclature and biologic significance of cytokines involved in inflammation and the host immune response, *J. Infect. Dis.*, 172, 1573, 1995.
13. Baron, S., Tyring, S. K., Fleischmann, W. R., Coppenhaver, D. H., Niesel, D. W., Klimpel, G. R., Stanton, G. J., and Hughes, T. K., The interferons: mechanisms of action and clinical applications, *J. Am. Med. Assoc.*, 266, 1375, 1991.
14. Farrar, M. A. and Schreiber, R. D., The molecular cell biology of interferon-γ and its receptor, *Annu. Rev. Immunol.*, 11, 571, 1993.
15. Abbas, A. K., Lichtman, A. H., and Pober, J. S., *Cellular and Molecular Immunology*, 2nd ed., W. B. Saunders, Philadelphia, 1994, 239-260.
16. Defrance, T., Carayon, P., Billian, G., Guillemot, J. C., Minty, A., Caput, D., and Farrara, P., Interleukin 13 is a B stimulating factor, *J. Exp. Med.*, 179, 135, 1994.
17. Dinarello, C. A., Interleukin-1 and interleukin-1 antagonism, *Blood*, 77, 1627, 1991.
18. Burton, J. D., Bamford, R. N., Peters, C., Grant, A. J., Kurys, G., Goldman, C. K., Brennan, J., Roessler, E., and Waldmann, T. A., A lymphokine, provisionally designated interleukin T, that stimulates T-cell proliferation and the induction of lymphokine-activated killer cells, *Proc. Natl. Acad. Sci. U.S.A.*, 91, 4935, 1994.
19. Trinchieri, G., Interleukin-12: a cytokine produced by antigen-presenting cells with immunoregulatory functions in the generation of T-helper cells type 1 and cytotoxic lymphocytes, *Blood*, 84, 4008, 1994.
20. Dale, D.C., Liles, W. C., Summer, W. R., and Nelson, S., Review: Granulocyte-colony-stimulating factor — role and relationships in infectious diseases, *J. Infect. Dis.*, 172, 1061, 1995.
21. Miller, M. D. and Krangel, M. S., Biology and biochemistry of the chemokines: a family of chemotactic and inflammatory cytokines, *Crit. Rev. Immunol.*, 12, 17, 1992.
22. Taub, D. D., Lloyd, A. R., Conlon, K., Wang, J. M., Ortaldo, J. R., Harada, A., Matushima, K., Kelvin, D. J., and Oppenheim, J. J., Recombinant human interferon inducible protein 10 is a chemoattractant for human monocytes and T lymphocytes and promotes T cell adhesion to endothelial cells, *J. Exp. Med.*, 177, 1809, 1993.
23. Schall, T. J., Biology of RANTES/SIS cytokine family, *Cytokine*, 3, 165, 1991.
24. Mukaida, N., Harada, A., Yasumoto, K., and Matsushima, K., Properties of pro-inflammatory cell type-specific leukocyte chemotactic cytokines, interleukin-8 (IL-8) and monocyte chemotactic and activating factor (MCAF), *Microbiol. Immunol.*, 36, 773, 1992.
25. Appasamy, P. M., Interleukin-7: biology and potential clinical applications, *Cancer Invest.*, 11, 487, 1993.
26. Kawashima, I. and Takiguchi, Y., Interleukin 11: a novel stroma-derived cytokine, *Prog. Growth Factor Res.*, 4, 191, 1992.
27. Metcalf, D., Leukemia inhibitory factor — a puzzling polyfunctional regulator, *Growth Factors*, 7, 169, 1992.
28. Heldin, C. H., Structural and functional studies on platelet-derived growth factor, *EMBO J.*, 11, 4251, 1992.
29. McCartney-Francis, N. L. and Wahl, S. M., Transforming growth factor β: a matter of life and death, *J. Leukocyte Biol.*, 55, 401, 1994.
30. Martin, J. B. and Reichlin, S., *Clinical Neuroendocrinology*, F. A. Davis, Philadelphia, 1987, 669-694.

31. Maier, S. F., Ryan, S. M., Barksdale, C. M., and Kalin, N. H., Stressor controllability and the pituitary-adrenal system, *Behav. Neurosci.*, 100, 669, 1986.
32. Douglas, J., Givelli, O., and Herbert, E., Polyprotein gene expression: generation of diversity of neuroendocrine peptides, *Annu. Rev. Biochem.*, 53, 665, 1984.
33. Johnson, H. M., Downs, M. O., and Pontzer, C. H., Neuroendocrine peptide hormone regulation of immunity, in *Neuroimmunoendocrinology*, 2nd ed., Blalock, J. E., Ed., Karger, Basel, Switzerland, 1992, 49-83.
34. Alvarez-Mon, M., Kehrl, H. H., and Fauci, A. S., A potential role for adrenocorticotropin in regulating human B lymphocyte functions, *J. Immunol.*, 135, 3823, 1985.
35. Gilmore, W. and Weiner, L. P., β-endorphin enhances interleukin (IL-2) production in murine lymphocytes, *J. Neuroimmunol.*, 18, 125, 1988.
36. Sibinga, N. E. S. and Goldstein, A., Opioid peptides and opioid receptors in cells of the immune system, *Annu. Rev. Immunol.*, 6, 219, 1988.
37. Faith, R. E., Laing, H. G., Murgo, A. J., and Plotnikoff, N. P., Neuroimmunomodulation with enkephalins: enhancement of human natural killer (NK) cell activity *in vitro*, *Clin. Immunol. Immunopathol.*, 31, 412, 1984.
38. Oleson, D. R. and Johnson, D. R., Regulation of human natural cytotoxicity by enkephalins and selective opiate agonists, *Brain Behav. Immunol.*, 2, 171, 1988.
39. Plotnikoff, N. P., Faith, R. E., Murgo, A. J., Huberman R. B., and Good, R. A., Short analytical review. Methionine enkephalin: a new cytokine-human study, *Clin. Immunol. Immunopathol.*, 82, 93, 1997.
40. Smith, E. M. and Blalock, J. E., Human lymphocyte production of corticotropin- and endorphin-like substances: association with leukocyte interferon, *Proc. Natl. Acad. Sci. U.S.A.*, 78, 7530, 1981.
41. Bernton, E. W., Prolactin and immune host defense, *Prog. Neuroendocrinoimmunol.*, 2, 21, 1989.
42. Cross, R. J. and Roszman, T. L., Neuroendocrine modulation of immune function: The role of prolactin, *Prog. Neuroendocrinoimmunol.*, 2, 17, 1989.
43. Kelley, K. W., Growth hormones, lymphocytes and macrophages, *Biochem. Pharmacol.*, 38, 705, 1989.
44. Nagy, E., Berczi, I., and Friesen, H. G., Regulation of immunity in rats by lactogenic and growth hormones, *Acta Endocrinol.*, 102, 351, 1983.
45. Kelley, P. A., Djian, J., Postel-Vinay, M. C., and Edery, M., The prolactin/growth hormone receptor family, *Endocrine Rev.*, 12, 235, 1991.
46. DiMattia, G. E., Gellersen, B., Bohnet, H. G., and Friesen, H. G., A human B-lymphoblastoid cell line produces prolactin, *Endocrinology*, 122, 2508, 1988.
47. Montgomery, D. W., Zukoksi, C. F., Shah, G. N., Backlery, A. R., Pacholczyk, T., and Russell, D. H., Concanavalin A-stimulated murine splenocytes produce a factor with prolactin-like bioactivity and immunoreactivity, *Biochem. Biophys. Res. Commun.*, 145, 692, 1987.
48. Clevenger, C. V., Russell, T. H., Appasamy, P. M., and Prystowsky, M. B., Regulation in interleukin-2 driven T-lymphocyte proliferation by prolactin, *Proc. Natl. Acad. Sci. U.S.A.*, 87, 6460, 1990.
49. Kiess, W. and Butenandt, O., Specific growth hormone receptors on human peripheral mononuclear cells: reexpression, identification and characterization, *J. Clin. Endocrinol. Metab.*, 60, 740, 1985.
50. Weigent, D. A. and Blalock, J. E., The production of growth hormone by subpopulations of rat mononuclear leukocytes, *Cell. Immunol.*, 135, 55, 1991.

51. Davila, D. R., Brief, S., Simon, J., Hammer, R. E., Brinster, R. L., and Kelley, K. W., Role of growth hormone in regulating T-dependent immune events in aged nude and transgenic rodents, *J. Neurosci. Res.*, 18, 108, 1987.
52. Kelley, K. W., The role of growth hormone in modulation of the immune response, *Ann. NY Acad. Sci.*, 594, 95, 1990.
53. Weigent, D. A., Blalock, J. E., and LeBoeuf, R. D., An antisense oligodeoxynucleotide to growth hormone messenger ribonucleic acid inhibits lymphocyte proliferation, *Endocrinology*, 128, 2053, 1991.
54. Smith, E. M., Hormonal activities of cytokines, in *Neuroimmunoendocrinology*, 2nd ed., Blalock, J. E., Ed., Karger, Basel, Switzerland, 1992, 154-169.
55. Paliogianni, R., Ahuja, S. S., Balow, J. P., and Boumpas, D. T., Novel mechanism for inhibition of human T cells by glucocorticoids, *J. Immunol.*, 151, 4081, 1993.
56. Woloski, B. M. R. N. J., Smith, E. M., Meyer, W. J., Fuller, G. M., and Blalock, J. E., Corticotropin-releasing activity of monokines, *Science*, 230, 1035, 1985.
57. Breard, J., Costa, O., and Kordon, C., Organization and functional relevance of neuroimmune networks, in *Stress, the Immune System and Psychiatry*, Leonard, B. E. and Miller, K., Eds., John Wiley & Sons, Chichester, U.K., 1995, 17-45.
58. Bernton, E., Beach, J. E., Holaday, J. W., Smallridge, R. C., and Fein, H. C., Release of multiple hormones by a direct action of interleukin-1 on pituitary cells, *Science*, 238, 519, 1989.
59. Gonzalez, M. C., Riedel, M., Rettori, V., Yu, W. H., and McCann, S. M., Effect of recombinant human β-interferon on the release of anterior pituitary hormones, *Prog. Neuroendocrinoimmunol.*, 3, 49, 1990.
60. Abraham, E., Richmond, N. J., and Chang, Y.-H., Effects of hemorrhage on interleukin-1 production, *Circ. Shock*, 25, 33, 1988.
61. Saunder, D. N., Sample, J., and Truscotte, D., Stimulation of muscle protein degradation by murine and human epidermal cytokines: Relationship to thermal injury, *J. Invest. Dermatol.*, 87, 711, 1986.
62. Dinarello, C. A., Interleukin-1 as mediator of the acute-phase response, *Surv. Immunol. Res.*, 3, 29, 1984.
63. Stahl, W. M., Acute phase protein response to tissue injury, *Crit. Care Med.*, 15, 545, 1987.
64. Schulte, H. M., Bamberger, C. M., Elsen, H., Herrmann, G., Bamberger, A. M., and Barth, J., Systemic interleukin-1α and interleukin-2 secretion in response to acute stress and to corticotropin-releasing hormone in humans, *Eur. J. Clin. Invest.*, 24, 773, 1994.
65. Goujon, E., Laye, E., Parnet, P., and Dantzer, R., Regulation of cytokine gene expression in the central nervous system by glucocorticoids: mechanisms and functional consequences, *Psychoneuroendocrinology*, Suppl. 1, S75, 1997.
66. Dunn, A. J. and Wang, J., Cytokine effects on CNS biogenic amines, *Neuroimmunomodulation*, 2, 319, 1995.
67. Zhou, D., Shanks, N., Riechman, S. E., Liang, R., Kuscov, A. W., and Rabin, B. S., Interleukin-6 modulates interleukin-1 and stress-induced activation of the hypothalamic-pituitary-adrenal axis in male rats, *Neuroendocrinology*, 63, 227, 1996.
68. Abraham, E., Lee, R. J., and Chang, Y.-H., The role of interleukin 2 in hemorrhage-induced abnormalities of lymphocyte proliferation, *Circ. Shock*, 18, 205, 1986.
69. Akiyoshi, T., Koba, F., and Arinaga, S., Impaired production of interleukin-2 after surgery, *Clin. Exp. Immunol.*, 59, 45, 1985.
70. Abraham, E. and Regan, R. F., The effects of hemorrhage and trauma on interleukin-2 production, *Arch. Surg.*, 12, 1341, 1985.

71. Weinstock, C., Konig, D., Harnischmacher, R., Keul, J., Berg, A., and Northoff, H., Effect of exhaustive exercise stress on the cytokine response, *Med. Sci. Sports Exerc.*, 29, 345, 1997.
72. Nieman, D. C., Immune response to heavy exertion, *J. Appl. Physiol.*, 82, 1354, 1997.
73. Arnetz, B. B., Brenner, S. O., Levi, L., Hjelm, R., Petterson, I. L., Wasserman, J., Petrini, B., Enerogh, P., Kallner, A., Kvetnansky, R., and Vigas, M., Neuroendocrine and immunologic effects of unemployment and job insecurity, *Psychother. Psychosom.*, 55, 76, 1991.
74. Glaser, R., Kennedy, S., Lafuse, W. P., Bonneua, R. H., Speicher, C., Hillhouse, J., and Kiecolt-Glaser, J. K., Psychological stress-induced modulation of interleukin-2 receptor gene expression in peripheral blood leukocytes, *Arch. Gen. Psychiatry*, 47, 1707, 1990.
75. Abraham, E., Effect of stress on cytokine production, in *Neuroendocrinology of Stress*, Vol. 1, Jasmin, G. and Cantin, M., Eds., Karger, Basel, Switzerland, 1991, 54-62.
76. Abraham, E. and Freitas, A. A., Hemorrhage produces abnormalities in lymphocyte function and lymphokine generation, *J. Immunol.*, 142, 899, 1989.
77. Decker, D., Schondorf, M., Bidlingmaier, F., Hirner, A., and von Ruecker, A., Surgical stress induces a shift in the type-1/type-2 T-helper cell balance, suggesting down-regulation of antibody-mediated immunity commensurate to the trauma, *Surgery*, 119, 316, 1996.
78. Hisano, S., Sakamoto, K., Ishiko, T. K. H., and Ogawa, M., IL-6 and soluble IL-6 receptor levels change differently after surgery both in the blood and in the operative field, *Cytokine*, 9, 447, 1997.
79. Shizuya, K., Komori, T., Fujiwara, R., Miyahara, S., Ohmori, M., and Nomura, J., The influence of restraint stress on the expression of mRNAs for IL-6 and the IL-6 receptor in the hypothalamus and midbrain of the rat, *Life Sci.*, 61, PL135, 1997.
80. Dobbs, C. M., Feng, N., Beck F. M., and Sheridan, J. F., Neuroendocrine regulation of cytokine production during experimental influenza viral infection: effects of restraint stress-induced elevation in endogenous corticosterone, *J. Immunol.*, 157, 187, 1996.
81. Path, G., Bornstein, S. R., Ehrhart-Bornstein, M., and Scherbaum, W. A., Interleukin-6 and the interleukin-6 receptor in the human adrenal gland: expression and effects on steroidogenesis, *J. Clin. Endocrinol. Metab.*, 82, 2343, 1997.
82. Elenkov, I. J., Papanicolaou, D. A., Wilder, R. L., and Chrousos, G. P., Modulatory effects of glucocorticoids and catecholamines on human interleukin-12 and interleukin-10 production: clinical implications, *Proc. Assoc. Am. Physicians*, 108, 374, 1996.
83. Suzuki, R. and Pollard, R. B., Mechanism for the suppression of γ interferon responsiveness in mice after thermal injury, *J. Immunol.*, 129, 1806, 1982.
84. Kim, C., Schinkel, C., Fuchs, D., Stadler, J., Walz, A., Zedler, S., von Donnersmarck, G. H., and Faist, E., Interleukin-13 effectively down-regulates inflammatory potential during traumatic stress, *Arch. Surg.*, 130, 1330, 1995.
85. Maes, M., Meltzer, H. Y., Bosmans, E., Bergmans, R., Vandoolaeghe, E., Ranjan, R., and Desnyder, R., Increased plasma concentrations of interleukin-6, soluble interleukin-6, soluble interleukin-2 and transferrin receptor in major depression, *J. Affect. Disord.*, 34, 301, 1995.
86. Weizman, R., Laor, N., Podiszewski, E., Notti, I., Djaldetti, M., and Bessler, H., Cytokine production in major depressed patients before and after clomipramine treatment, *Biol. Psychiatry*, 35, 42, 1994.

87. Murphy, D. L., Zohar, J., Benkelfat, C., Pato, M. T., Pigott, T. A., and Insel, T. R., Obsessive-compulsive disorder as a 5-HT-subsystem-related behavioral disorder, *Br. J. Psychiatry*, 155 (Suppl. 8), 15, 1989.
88. Baxter, L. R., Rhelps, R., Mazziotti, J., Guze, B. H., Schwartz, J. M., and Selin, C. E., Local cerebral glucose metabolic rates of obsessive-compulsive disorder compared to unipolar depression and normal controls, *Arch. Gen. Psychiatry*, 44, 211, 1987.
89. Swedo, S. E., Rapoport, J. L., Cheslow, D. L., Leonard, H. L., Ayoub, E. M., Hosier, D. M., and Wald, E. R., High prevalence of obsessive-compulsive symptoms in patients with Sydenham's chorea, *Am. J. Psychiatry*, 146, 245, 1989.
90. Allen, A. J., Leonard, H. L., and Swedo, E., Case study: New infection-triggered autoimmune subtype of pediatric OCD and Tourett's syndrome, *J. Am. Acad. Child. Adolesc. Psychiatry*, 34, 307, 1995.
91. Roy, B. F., Benkelfat, C., Hill, J. L., Pierce, P. F., Dauphin, M. M., Kelly, T. M., Sunderland, T., Weinberger, D. R., and Breslin, N., Serum antibody for somatostatin-14 and prodynorphin 209-240 in patients with obsessive-compulsive disorder, schizophrenia, Alzheimer's disease, multiple sclerosis and advanced HIV infection, *Biol. Psychiatry*, 35, 335, 1994.
92. Maes, M., Meltzer, H. Y., and Bosmans, E., Psychoimmune investigation in obsessive-compulsive disorder: assays of plasma transferrin, IL-2 and IL-6 receptor, and IL-1 β and IL-6 concentrations, *Neuropsychobiology*, 30, 57, 1994.
93. Weizman, R., Laor, N., Barber, Y., Hermesh, H., Notti, I., Djaldetti, M., and Bessler, H., Cytokine production in obsessive-compulsive disorder, *Biol. Psychiatry*, 40, 908, 1996.
94. Mittleman, B. B., Castellanos, F. X., Jacobsen, L. K., Rapoport, J. L., Swedo, S. E., and Shearer, G. M., Cerebrospinal fluid cytokines in pediatric neuropsychiatric disease, *J. Immunol.*, 159, 2994, 1997.
95. Brambilla, F., Bellodi, L., Perna, G., Garberi, A., Petraglia, R., Panerai, A., and Sacerdote, P., Psychoimmunoendocrine aspects of panic disorder, *Neuropsychobiology*, 26, 12, 1992.
96. Brambilla, F., Bellodi, L., Perna, G., Bertani, A., Panerai, A., and Sacerdote, P., Plasma interleukin-1β concentrations in panic disorder, *Psychiatry Res.*, 54, 135, 1994.
97. Rapaport, M. H. and Stein, M. B., Serum cytokine and soluble interleukin-2 receptors in patients with panic disorder, *Anxiety*, 1, 22, 1994.
98. Weizman, R., Laor, N., Wiener, Z., Wolmer, L., and Bessler, H., Cytokine production in panic disorder patients, 1998.
99. La Via, M. F., Munno, I., Lydiard, R. B., Workman, E. W., Hubbard, J. R., Michel, Y., and Paulling, E., The influence of stress intrusion on immunodepression in generalized anxiety disorder patients and controls, *Psychosom. Med.*, 58, 138, 1996.
100. Spivak, B., Shohat, B., Mester, R., Avraham, S., Gil-Ad, I., Bleich, A., Valevski, A., and Weizman, A., Elevated levels of serum interleukin-1β in combat-related posttraumatic disorder, *Biol. Psychiatry*, 42, 345, 1997.
101. Shattner, A., Steinbock, M., Tepper, R., Schonfeld, A., Vaisman, N., and Hahn, T., Tumor necrosis factor production and cell-mediated immunity in anorexia nervosa, *Clin. Exp. Immunol.*, 79, 62, 1990.
102. Shattner, A., Tepper, R., Steinbock, M., Hahn, T., and Schonfeld, A., TNF, interferon-γ and cell-mediated cytoxicity in anorexia nervosa; effect of refeeding, *J. Clin. Lab. Immunol.*, 32, 183, 1990.
103. Bessler, H., Karp, L., Notti, I., Apter, A., Tyano, S., Djaldetti, M., and Weizman, R., Cytokine production in anorexia nervosa, *Clin. Neuropharmacol.*, 16, 237, 1993.

104. Pomeroy, C., Eckert, E., Hu, S., Eiken, B., Mentink, M., Crosby, R. D., and Chao, C. C., Role of interleukin-6 and transforming growth factor-β in anorexia nervosa, *Biol. Psychiatry*, 36, 836, 1994.
105. Holden, R. J. and Pakula, I. S., The role of tumor necrosis factor-α in the pathogenesis of anorexia and bulimia nervosa, cancer cachexia and obesity, *Med. Hypothesis*, 47, 423, 1996.
106. Janik, J. F., Curti, B. D., Considine, R. V., Rager, H. C., Powers, G. C., Alvord, W. G., Smith, J. W., II, Gause, B. L., and Kopp, W. C., Interleukin-1α increases serum leptin concentrations in humans, *J. Clin. Endocrinol. Metab.*, 82, 3084, 1997.

2 Neuroendocrine Regulation of the Immune Process

George Mastorakos, Christoph Bamberger, and George P. Chrousos

CONTENTS

2.1 Introduction 17
2.2 Inflammatory/Immune Response 18
2.3 Effects of the Hypothalamic-Pituitary-Adrenal Axis on
    Inflammatory/Immune Response 19
2.4 Direct Interactions Between Hypothalamic and Pituitary
    Neuropeptides and the Inflammatory/Immune Response 21
    2.4.1 POMC and Proenkephalin Products 21
    2.4.2 Thyroid-Stimulating Hormone and Related Hormones 21
    2.4.3 Growth Hormone 22
    2.4.4 Prolactin 22
    2.4.5 Direct Effects on "Immune" CRH, AVP, and Sms
          in Peripheral Inflamation 23
2.5 Effects of "Inflammatory" Cytokines on the HPA Axis 26
2.6 Perspectives 29
References 31

2.1 INTRODUCTION

Ancient Greek medicine and common sense accepted a cause and effect association between the state of mind and the ability of the body to fight diseases. Recently, a substantial body of data has been obtained indicating that the brain, by regulating endocrine and peripheral nervous system functions, influences in a major way the immune/inflammatory (I/I) response and, therefore, the defense of the organism.[1,2] Interestingly, this association has a reciprocal component, with products of the I/I reaction influencing the brain along with its endocrine and peripheral nervous system limbs.

The hypothalamic-pituitary-adrenal (HPA) axis, which is activated during stress, represents one of the major pathways by which the brain regulates the I/I process.[1,2]

However, it took almost 400 years from the first description of the adrenal glands (*glandulae renis incumbentes*) by Eustachius in A.D. 1563, and 2000 years since the annunciation of four (*rubor, et tumor, cum calore et dolore*) of the five (*et functio laesa*) cardinal signs of inflammation by Celsus in the first century until the two were put together by Hans Selye in 1936.[3,4] Selye noted that rats exposed to diverse stressors simultaneously developed enlargement of the adrenal glands and shrinkage of their thymuses and lymph nodes and suggested that a cause and effect relationship might be present between the two, with former secreting a hormone that suppressed the latter.[4]

Cells of the immune carry receptors for a number of hormones, neuropeptides, and neurotransmitters.[5] Changes in neuroendocrine and/or autonomic activities may therefore lead to modulation of the responses of these cells. In addition, immune cells produce a number of hormones and neuropeptides, such as corticotropin-releasing hormone (CRH), corticotropin (ACTH), β-endorphin, growth hormone (GH), somatostatin (Sms), prolactin (PRL), and thyroid-stimulating hormone (TSH), which probably act locally as autacoids during both the early and late stages of the I/I process. As part of the bidirectional communication between the brain and the I/I reaction, the inflammatory cytokines, specifically tumor necrosis factor-α (TFNα) and interleukins (IL) 1 and 6, may act as endocrine hormones of the immune system, produced at distant sites and acting upon the central components of the HPA axis and the sympathetic system.[1,2]

2.2 INFAMMATORY/IMMUNE RESPONSE

The immune system exerts its surveillance-defense functions constantly and, mostly, unconsciously for the individual.[6] The inflammatory reaction is part of the response of immune system to septic or nonseptic aggressors. This reaction is microscopically characterized at the local level by dilatation and increased permeability of microvessels, leading to both increased blood flow and exudation of plasma, and migration and accumulation of leukocytes into the inflammatory area. The result of this is to destroy or wall off attracting microorganism or the destroyed tissue.

The cellular components of the inflammatory reaction consist of circulating nonlymphoid leucocytes and lymphocytes and local immune accessory cells. Non-lymphoid leucocytes include the monocytes/macrophages, neutrophils, basophils, and eosinophils. Local immune accessory cells include the endothelial cells, tissue fibroblasts, resident macrophages and macrophage-related cells such as liver Kupffer cells, type A synovial lining cells and CNS glia cells, as well as the basophil-related mast cells.[7,8]

Many substances secreted locally in the inflammatory area by the above cells act as auto- or paracrine regulators and/or mediators of the inflammatory reactions, as well as endocrine messengers between the inflammatory process and other systems such as the CNS, HPA axis, peripheral nervous system, etc. These substances include the vasoactive amines, histamine and serotonin, the kallikrein/kinin system, the Hageman and other clotting factors, the fibrinolytic system, several components of the complement system, as well as eosinophil and platelet activators. They also include cytokines, such as TFN-α and -β, interferons α, β, and γ, the interleukins 1 through 14 as well as their binding proteins and natural antagonists, and many lipid and glucolipid products triglyceride metabolism, such as several active products of arachidonic acid, including the endoperoxides, the thromboxanes, prostacyclin, and leukotrienes, as well as platelet activating factor. In addition, active oxygen radicals, nitric oxide, and lysosomal constituents like neutral proteases participate in the inflammatory reaction.[2]

2.3 EFFECTS OF THE HYPOTHALAMIC-PITUITARY-ADRENAL AXIS ON INFLAMMATORY/IMMUNE RESPONSE (FIGURE 2.1)

Corticotropin-releasing hormone (CRH) is a 41-amino-acid neuropeptide secreted by parvocellular neurons of the paraventricular nucleus of the hypothalamus into the hypophyseal portal system. Parvocellular neurons of this nucleus secrete not only CRH but also vasopressin (AVP), another potent stimulator of ACTH secretion, which acts synergistically with CRH and which also is secreted by magnocellular neuron terminals of the posterior pituitary. ACTH is transported through the systemic circulation to the cortex of the adrenal glands, where it stimulates the synthesis and release of cortisol and adrenal androgens. There is negative feedback regulation of the HPA axis exerted by cortisol on both CRH and ACTH secretion, and to a lesser extent on parvocellular AVP secretion.

Cortisol, the final effector of the HPA axis, as well as synthetic glucocorticoids, have multiple and profound effects on the I/I process.[9] They affect both the production and traffic of leukocytes, causing lymphopenia, monocytopenia, eosinopenia, as well as neutrophilia.[8-10] Also, they affect the function of these cells by decreasing neutrophil exodus to inflammatory sites, T-lymphocyte proliferation and cytotoxicity, as well as monocyte chemotaxis and bactericidal activity. Furthermore, the synthesis and action of numerous mediators of inflammation are inhibited by glucocorticoids. For example, prostanoid synthesis is inhibited by glucocorticoids — a well-known major action of this hormones. Among the genes inhibited by glucocorticoids are those of group II phospholipase A2, cyclooxygenase 2, 15-lipoxygenase, and nitric oxide synthetase 2, whose suppression of transcription and/or translation is associated with decreased production of arachidonic acid, prostanoids, platelet activating factor, and nitric oxide. Glucocorticoids also inhibit the production of many cytokines and may cause resistance of the cells to the effects of pro-inflammatory molecules. In addition, glucocorticoids inhibits the expression of adhesion and adhesion receptor molecules on the surface of immune and other cells[11] and potentiate the acute-phase reaction induced by cytokines, primarily IL-6.[12-13]

The anti-inflammatory effects of glucocortcoids are exerted by a receptor-mediated mechanism, as we have shown using a rat air-pouch model of acute inflammation.[14] In this experimental model an air-pouch is created on the neck of the animal into which carrageenin, a seaweed polysaccharide, is injected. This produces a local inflammatory response which can be quantified in terms of the volume and leukocyte concentration of the inflammatory exudate in the pouch. When different groups of

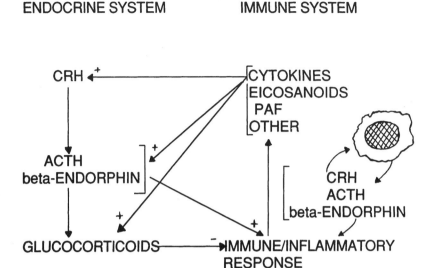

FIGURE 2.1 Interactions between the HPA axis (left) and the main mediators of the I/I response (cytokines, eicosanoids, PAF) (right). Cytokines stimulate hypothalamic CRH production, as well as pituitary ACTH and adrenal glucocorticoid secretion. The ACTH and β-endorphin, both metabolic products of proopiomelanocortin (POMC), generally enhance the I/I response. The accessory immune cells secrete and are regulated by peripheral neuropeptides such as CRH, ACTH, and β-endorphin.

animals were pretreated with an anti-inflammatory dose of dexamethasone (0.1 mg) and received increasing doses of RU 486, a glucocorticoid receptor-specific antagonist, the anti-inflammatory effects of the glucocorticoid were antagonized in a dose-depended fashion. Interestingly, when intact, nondexamethasone-treated animals were given increasing doses of RU 486, an enhancement of the inflammatory response was observed, suggesting that endogenous glucocorticoids normally harness inflammation.[14]

The human glucocorticoid receptor is a 777-amino-acid cytoplasmic protein, which in the ligand-unbound inactive stage is complexed as a heterooligomer with two molecules of heat shock protein (hsp) 90 and one molecule each of hsp 70, immunophilin of the FK 506/rapamycin-binding type, and hsp 24.[8,10,15] Of these, hsp 90 is crucial for allowing proper ligand binding and further activation of the receptor. The glucocorticoid-receptor complex translocates into the nucleus, where it binds on specific sequences of DNA present in to the regulatory regions of glucocorticoid-responsive genes (glucocorticoid-responsive elements, GRE) through the middle portion of glucocorticoid receptor, the so-called DNA-binding domain.[10,16-18] There, the glucocorticoid-receptor complex modulates transcription either directly or through the mediation of other factors (adaptors) by altering the stability of the initiation complex. Glucocorticoids, indirectly by altering the transcription and translation rate of intermediate proteins, may also modulate the stability of messenger RNA and the translation rates of various mediators of inflammation.

In addition, glucocorticoids may modify the transcription rates of inflammation-specific genes by interaction of the glucocorticoid-receptor complex with transcription factors such as c-jun and NF-kB that regulate the promoters of such genes.[8,10]

2.4 DIRECT INTERACTIONS BETWEEN HYPOTHALAMIC AND PITUITARY NEUROPEPTIDES AND THE INFLAMMATORY/IMMUNE RESPONSE

2.4.1 POMC AND PROENKEPHALIN PRODUCTS

The endogenous opioid peptides α-, β-, and γ-endorphin are derived from POMC and are composed of amino acid 61 to 76, 61 to 91, and 61 to 77 of β-lipotropin, respectively. Endorphin receptors similar to those in the brain are present on spleen cells and probably several others types of leukocytes.[5] β-endorphin has been shown to enhance T-cell proliferation and IL-2 production.[19] [Leu]- and [Met]-enkephalin are also endogenous opioid peptides which have their own polyprotein precursors.[20] Interestingly, [Met]-enkephalin is contained within the N-terminal sequence of all three endorphins. α-endorphin, β-endorphin, and/or [Met]-enkephalin can:

1. Enhance the natural cytotoxicity of lymphocytes and macrophages toward tumor cells.[21-25]
2. Enhance or inhibit T-cell mitogenesis; this enhancement might be due to induction of IL-2.[19,26-29]
3. Enhance T-cell rosetting.[30]
4. Stimulate human peripheral blood mononuclear cell chemotaxis.[31,32]
5. Inhibit T-cell chemotactic factor production.[33]
6. Inhibit interferon γ production by cultured human peripheral blood mononuclear cells.[34]
7. Inhibit major histocompatibility class II antigen expression.[35]

Another POMC-derived neuropeptide is α-melanocyte-stimulating hormone (MSH). MSH may be important in the control of fever, the acute-phase response, and inflammation. The antipyretic activity of this neuropeptide was about 25,000-fold greater than that of acetaminophen when given centrally. MSH appears to be a modulator of cytokines-induced responses in the host. Antipyretic doses of MSH inhibit febrile reactions caused by intraventricular injections of crude IL-1 into animals.[36] Moreover, MSH suppressed IL-1-induced neutrophilia and the synthesis of serum amyloid A, another acute-phase protein.

2.4.2 THYROID-STIMULATING HORMONE AND RELATED HORMONES

Initial observations of hypothyroidism in thymectomized or athymic mice suggested a relationship between the thyroid and the immune system.[37] In fact, human lymphocytes stimulated with the T-cell mitogen staphylococcal enterotoxin A (SEA) were shown to produce immunoreactive TSH.[38] Lymphocyte-derived and pituitary TSH were antigenically related and had the same molecular mass and structure.[39,40] In addition to TSH being produced by cells of the immune system, some leukocytes express TSH receptors. Thus, monocytes, NK cells, and mitogen-stimulated B cells, but not phytohemagglutinin (PHA)-stimulated T cells or resting T and B lymphocytes, had binding sites for TSH.[41]

On the other hand, inflammation and sepsis influence thyrotropin secretion, in part through suppressive actions of inflammatory cytokines on the production and secretion of the hypothalamic thyrotropin-releasing hormone (TRH)[42] and through parallel enhancement of somatostatin secretion.[43] The TSH response to TRH is also reduced by TFNα and perhaps by the other inflammatory cytokines.[44] Clearly, the hypothalamic/TRH-pituitary/TSH hyposecretion observed in inflammatory states represents the central component of the "euthyroid sick" syndrome. The homeostatic value of this in patients with nonthyroidal illness is not yet clearly established. However, animal studies with experimental inflammation indicate that hypothyroidism protects infected animals; thus, as compared with animals treated with thyroid hormone, fewer animals die, or death is delayed.[45]

The structural similarity of TSH with FSH, hCG, and LH would suggest that these hormones also should be further investigated as potential regulators of leukocyte function.

2.4.3 Growth Hormone

Growth hormone (GH) receptors are present on lymphocytes; GH-deficient animals have thymic atrophy and are immunodeficient, and this deficiency is reversed by treatment with GH.[46] GH is also responsible for modulating other immunologic activities, such as augmenting cytolytic activity of T cells, antibody synthesis, TNF production, superoxide anion generation from peritoneal macrophages of hypophysectomized mice, and NK activity.[47] Administration of an antisense oligonucleotide of GH mRNA to lymphocytes inhibited T- and B-cell proliferation.[48] Importantly, lymphocyte-derived GH appears to be necessary for interleukin-induced lymphocyte growth, although it is not directly mitogenic.

2.4.4 Prolactin

Prolactin (PRL) seems to have an important role in regulating the immune system. In rats, both hypophysectomy and treatment with the dopamine agonist bromocriptine, a drug that inhibit PRL secretion, inhibit the development of delayed cutaneous hypersensitivity, experimental allergic encephalomyelitis, or adjuvant-induced arthritis, whereas treatment with exogenous PRL reverses these immunosuppressive effects.[49] Bromocriptine, in limited trials in humans, ameliorated several established or putative autoimmune diseases, including psoriasis, iridocyclitis, and iritis.[50] Furthermore, hypoprolactinemia was associated with impaired lymphocyte proliferation and decreased production of macrophage-activating factors by T lymphocytes.[51]

Prolatin binds to specific receptors on several classes of lymphocytes, stimulates the production of cytokines and their secretion by these cells, and is an essential growth factor in one line of lymphoid cells.[52] Cyclosporine, an immunosuppressant drug, also binds to PRL receptors.[53] Prolactin also is shown to be involved in the development of the inflammatory process, evaluated in the rat by carrageenin foot edema and experimental pleurisy, thus confirming the proinflammatory activity of this hormone.[51] The limited effectiveness of bromocriptine in human autoimmune diseases may be due to the fact that the immune system prolactin gene employs a promoter different from that of the anterior pituitary and which may not be regulated by dopamine.[54]

2.4.5 DIRECT EFFECTS OF "IMMUNE" CRH, AVP, AND SMS IN PERIPHERAL INFLAMMATION

Since CRH is the main stimulator of POMC production from the pituitary,[55-57] and certain POMC products (like ACTH and β-endorphin) can directly affect the I/I response, we hypothesized that CRH itself might be directly involved with the I/I response. The putative role of CRH in the I/I response was further suggested by the presence of CRH-specific binding sites in human lymphocytes and macrophages[58-61] and by the CRH-stimulated production of POMC-derived peptides (ACTH and β-endorphin) by leukocytes.[62-65] By employing the rat air-pouch model of acute aseptic chemical inflammation and performing immunoneutralization studies *in vivo* with a highly specific anti-CRH polyclonal antiserum, we demonstrated significant suppression of the inflammatory response by this antiserum, suggesting that CRH had a positive role in inflammation.[66] The decrease in inflammation was similar to the one caused by TNF immunoneutralization that was used as a positive control, since the latter is a well-known mediator of the I/I response. The effects of a combination of anti-CRH and anti-TNFα were not additive, indicating that the two antisera might interfere with a common aspect of the inflammatory response.

Immunoreactive CRH was localized in the inflammatory tissue by immunohistochemistry with a biotin-avidin/horseradish peroxidase system and rabbit polyclonal, highly specific anti-human/rat CRH, affinity-purified by adsorption to synthetic rat/human CRH antiserum. Immunoreactive CRH was found in the cytoplasm of immune accessory cells like macrophages, endothelial cells surrounding vessels, and tissue fibroblasts[66] (Figure 2.2). The mobility of this "immune" CRH was similar, by HPLC, to that of r/hCRH 1-41, the form produced by the rat and human hypothalamus as well as the human placenta.[67-70] We also demonstrated presence of CRH on peripheral inflammatory sites in other animal models of both acute and chronic inflammation. Immunoreactive CRH was present in inflammatory cells of rat joint tissue with streptococcal cell wall- and adjuvant-induced arthritis.[71,72] CRH mRNA was present in the inflamed synovia from arthritic rat joints expressed specific CRH binding sites. Furthermore, irCRH was seen in the synovial lining cell layers and blood vessels from the joints of patients with rheumatoid arthritis and osteoarthritis, whereas high levels of CRH immunoreactivity were found in the synovial fluids of the former patients.[72] IrCRH was also present in immune accessory cells from uveitic retinas and corpora vitrea from Lewis rats with experimentally induced autoimmune uveitis.[73]

CRH may have a peripheral, primarily activating, role on the immune system. The mechanisms of the CRH-mediated component of the I/I response are still

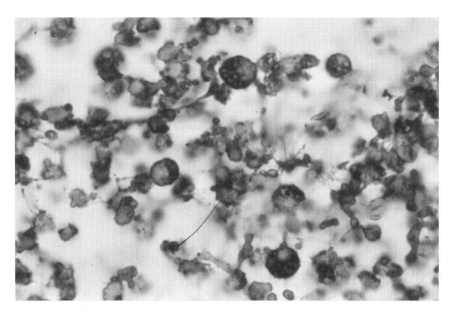

FIGURE 2.2 Magnification of the inflammatory granuloma from the animal pouch stained for irCRH immunoreactivity. Positive staining is revealed as black-dark green spots, whereas a light green staining is used as a nonspecific counterstaining to reveal the tissue architecture. The specific irCRH staining is seen in the cytoplasm of macrophages and tissue fibroblasts.

unclear. Mediation by local POMC gene products with known proinflammatory activity and/or by inflammatory cytokines are potential mechanisms. It seems that while "central" CRH participates in the systemic endocrine inhibition of the I/I reaction, "local" or "peripheral" immune CRH may participate in an auto/paracrine stimulation of inflammation. A possible source of "immune" CRH and other neuropeptides thought to be important in inflammation (such as substance P and Sms) other than the accessory immune cells, may be the primary afferent (sensory) nerves and the sympathetic postganglionic neurons. CRH and substance P are depleted in the rat spinal cord and dorsal root ganglia in response to capsaicin, which is toxic for the sensory afferent fibers.[74] Also, irCRH is present in the interomediolateral sympathetic column as well as the ganglia of the sympathetic chain and, therefore, could contribute to the inflammatory process through the sympathetic postganglionic fibers.[75,76]

In addition to CRH, some other neuropeptides have also been localized at the site of inflammation. We have demonstrated their presence by employing the same animal air-pouch model and looking for immunoreactivity at the inflammatory site by immunocytochemistry. Thus, we localized β-endorphin, substance P, a neuropeptide known to participate in the I/I process, and stomatostatin, known for its immunosuppressive activity *in vitro* and for its inhibitory actions upon other neuropeptide secretion, including that of substance P.[77] Furthermore, we demonstrated that somatulin, a potent long-acting Sms analogue, when administered to animals with the acute carrageenin-induced inflammation model, resulted in a reduction of both the volume and leukocyte concentration of the inflammatory exudate (Figure 2.3). The

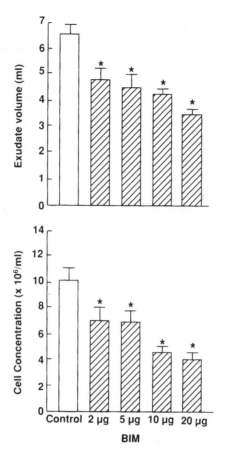

FIGURE 2.3 Effects of systemic administration of Sms analogue BIM 23014 on inflammation. The volume and the leukocytic concentration of the exudate were reduced significantly after administration of the analogue in a dose-dependent fashion. *Significant difference at 95% compared with control levels (open bars). (From Karalis, K., Mastorakos, G., Sano, H., Wilder, R., and Chrousos, G.P., *Endocrinology,* 136(9), 1995. With permission).

inhibition of inflammation was similar to that produced by indomethacin, a known nonsteroidal anti-inflammatory agent. In corroboration of these, immunohistochemical evaluation of the levels of local inflammatory mediators such as irTNF-α, irsubstance P, and irCRH was inhibited significantly by the Sms analogue treatment.[77]

In addition, when this model of acute aseptic inflammation was applied to dexamethasone-pretreated animals, irCRH disappeared 7 h after the induction of inflammation, in parallel with the reduction of the parameters of local inflammation.[78,79] A similar inhibition of the local expression of other neuropeptides such as TNF and substance P was also found after dexamethasone pretreatment of the animals. Interestingly, dexamethasone pretreatment did not seem to affect expression of irSms in the inflammatory tissue and rather increased its presence. Therefore, we could postulate that part of the anti-inflammatory action of glucocorticoids is also mediated by locally secreted Sms. It should be noted that Sms has been shown to significantly inhibit Molt-4 lymphoblast proliferation and PHA stimulation of human T lymphocytes.[80] Other immune responses, such as SEA-stimulated IFNγ secretion,[81] endotoxin-induced leukocytosis,[82] and colony-stimulating activity release[83] are also inhibited by Sms. These functional data suggest that these responsive tissues express receptors for Sms. Indeed, several cells of the immune system express receptors for this hormone.[84,85]

Arginine vasopressin, another major regulator of the HPA axis, has been also shown to participate directly in the peripheral inflammatory response (Figure 2.4). We have demonstrated by employing the animal air-pouch carrageenin-induced inflammation model that AVP immunoneutralization significantly decreased the parameters of inflammation.[86] Lymphocytes possess specific receptors for AVP, which seems to play an important role in positive regulation of IFNγ production by providing a helper signal[87,88] mediated by the helper T-cell lymphokine IL-2.[90]

2.5 EFFECTS OF THE "INFLAMMATORY" CYTOKINES ON THE HPA AXIS (FIGURE 2.4)

During infection, autoimmune processes, or trauma a complex cascade of events ensues that is characterized by fever, circulation of cytokines, and alterations of acute phase proteins in plasma that are important to initiate, propagate, and terminate host defense mechanisms. In addition, it has been known for several decades that activation of the HPA axis occurs in parallel. Recently, it became apparent that several mediators of inflammation play a major role in this phenomenon, which has been extensively studied in models employing lipopolisaccharide (LPS), a component of the cell walls of Gram-negative bacteria. Among all the cytokines, three (TNF, IL-1, and IL-6) are responsible for most of the stimulation of the HPA axis associated with the I/I response. These three cytokines are produced at inflammatory sites and elsewhere in response to inflammation. TNF, which has a tumoricidal activity and is responsible for cachexia, is the first to appear in the inflammatory cascade of the events and stimulates both IL-1 and IL-6; similarly, IL-1 stimulates both TNF and IL-6.[90-92] In contrast, IL-6, which participates in a major fashion in the acute-phase reaction, inhibits the secretion of both the other cytokines (Figure 2.5). Recent *in vivo* and *in vitro* studies have demonstrated that IL-1 and TNF can synergistically stimulate IL-6 production. All three cytokines, on the other hand, stimulate their own production in an auto/paracrine way.[12,92,93]

All three inflammatory cytokines have been shown to activate the HPA axis *in vivo*, alone or in synergy with each other.[94-99] This effect can be blocked significantly with CRH-neutralizing antibodies, glucocorticoids, and prostanoid synthesis inhibitors. Indeed TNF peaked approximately 1 h after LPS administration and rapidly declined afterwards, whereas IL-1 and IL-6 levels peaked somewhat later (2 to 4 h) and were sustained longer. Pretreatment of LPS-stimulated rats with anti-IL-6 antibody completely blocked the synergistic induction of ACTH secretion after coinjec-

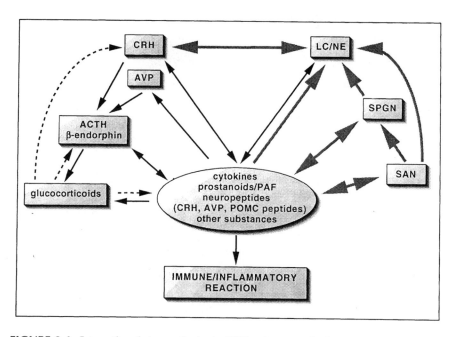

FIGURE 2.4 Interactions between the brain, HPA axis, sympathetic system, peripheral nervous system, and the I/I response. The inflammatory cytokines TFNα, IL-1, and IL-6 acutely stimulate the hypothalamic CRH and AVP neurons to secrete CRH and AVP, and more chronically the pituitary corticotroph to secrete ACTH and the adrenal cortex to produce glucocorticoids. The same mediators may stimulate the central catecholaminergic system (LC/NE) and the sympathetic postganglionic nerves (SPGN). Glucocorticoids directly inhibit peripheral immune target tissues, while CRH, AVP, ACTH, and β-endorphin play primarily immunopotentiating or pro-inflammatory roles. These neuropeptides are produced locally by sensory afferent fibers (SAN), sympathetic postganglionic nerves, and immune or immune accessory cells, and act as autacoids. Thin lines: humoral route, stimulation; interrupted lines: humoral route, inhibition; thick lines: neural route, stimulation. (Modified from Chrousos, G. P., *N. Engl. J. Med.*, 332, 1351, 1995.)

tion of IL-1 and TNF.[99] When administered to humans, both IL-1 and TNF have significant toxicity including fever, general malaise, and hypotension at the doses needed to activate the HPA axis. Recently, we demonstrated that IL-6, with its ability to inhibit the two other inflammatory cytokines and its modest toxicity in experimental animals, was a potent stimulator of the HPA axis in human, causing an impressively marked and prolonged elevation of plasma ACTH and cortisol when administered either s.c. or i.v.[101,102] The levels of ACTH and cortisol attained after stimulation with IL-6 were well above those observed with maximal stimulatory doses of CRH, suggesting that parvocellular AVP and other ACTH secretagogues were also stimulated by this cytokine (Figure 2.6). In a dose-response study, maximal levels of ACTH were seen at doses at which no peripheral AVP levels were increased. At higher doses, however, IL-6 stimulated peripheral elevations of AVP, indicating that this cytokine might also be able to activate magnocellular AVP-secreting neu-

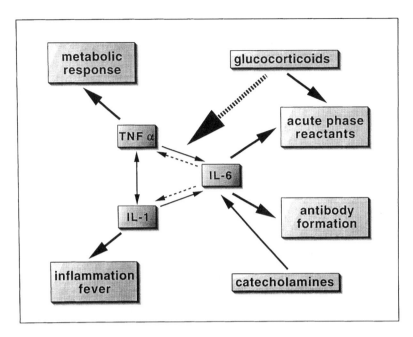

FIGURE 2.5 Interactions between the inflammatory cytokines and their stimulatory effects on CRH and AVP neurons during the I/I response. The brain capillary endothelium and glia cells may participate in a local cascade leading to activation of the hypothalamic neurons. (Modified from Chrousos, G. P., *N. Engl. J. Med.*, 332, 1351, 1995.)

rons. This suggested that Il-6 might be involved in the genesis of the syndrome of inappropriate secretion of antidiuretic hormone (SIADH), which is observed in the course of infectious of inflammatory diseases, or during trauma.

Preliminary data showed that IL-6 levels were increased in plasma in patients with early, untreated rheumatoid arthritis and had a circadian rhythm with a zenith around 07:00 a.m., about 1 to 1.5 h before the zenith of plasma ACTH, suggesting that in patients with inflammatory disease IL-6 levels correlate with the activity of the axis.[102] IL-6 and TFNα were undetectable during 24-h studies after IL-6 administration in humans,[101] confirming that IL-6 stimulates the HPA axis independently of circulating levels of IL-6 and TFNα.

The route of access of the inflammatory cytokines to the CRH- and AVP-secreting neurons is not, as yet, clearly elucidated, given that the cellular bodies of both are within the blood-brain barrier.[2,94,103] It has been suggested that they may act on nerve terminals of these neurons at the median eminence through the fenestrated endothelia of this circumventricular organ (Figure 2.7). Alternatively, they could stimulate nonfenestrated endothelia and glia cells in a cascade-like fashion, or they might cross the barrier with the assistance of a transport system. Also, the inflammatory cytokines might stimulate the hypothalamic CRH/AVP neurons indirectly by first stimulating ascending catecholaminergic neurons of the area posterma.[104]

In addition to their hypothalamic effects, the inflammatory cytokines can apparently stimulate directly pituitary ACTH and adrenal cortisol secretion.[94,95,103,105-108]

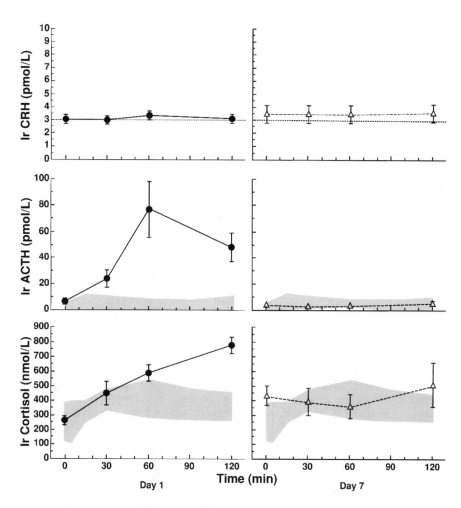

FIGURE 2.6 Mean ± SE plasma immunoreactive CRH (top), immunoreactive ACTH (middle), and immunoreactive cortisol (bottom) responses to a s.c. bolus injection of 30 μg/kg IL-6 in five cancer patients with good performance status on the first and seventh days of IL-6 administration. The gray areas represent the mean ± SD hormone responses of healthy normal volunteers to a standard bolus dose (1 μg/kg) of oCRH and were included for comparison.

This may be related to the chronicity of the elevation of the inflammatory cytokines or may be a dose-related phenomenon. It is noteworthy that IL-1 and IL-6 are themselves produced in the anterior pituitary and adrenal glands, where they may have auto/paracrine effects.[94,109,110]

2.6 PERSPECTIVES

In conclusion, the reaction of the organism to an immune system aggressor results in interrelated changes of both central neuroendocrine functions and the peripheral

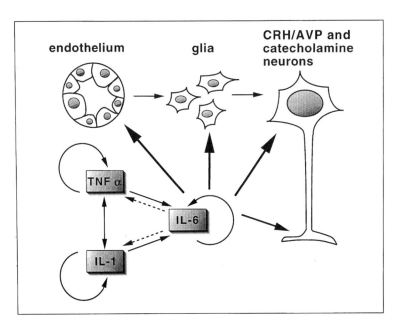

FIGURE 2.7 Interactions between the inflammatory cytokines and the effects of glucocorticoids and cetecholamines. Most of the production of all three cytokines is inhibited by glucocorticoids. Catecholamines, on the other hand, stimulate IL-6 which, by inhibiting the other two cytokines, stimulating glucocorticoids, and inducing the acute-phase response, ultimately participate in the control of inflammation. Solid lines: humoral route, stimulation; interrupted lines: humoral route, inhibition. (Modified from Chrousos, G. P., *N. Engl. J. Med.*, 332, 1351, 1995.)

I/I response. These interrelations may be disturbed and lead to development of disease or susceptibility to disease.

An illustrative paradigm is that of the Lewis rat strain its near-histocompatible, highly inbred Fischer rat strain.[9] The former has clearly decreased HPA axis responsiveness to inflammatory stimuli, whereas the latter has slightly increased HPA axis responsiveness to the same stimuli. Further characterization of the defect in the Lewis rat localized it in the hypothalamic CRH neuron, which was globally defective in its response to all stimulatory neurotransmitters.[111] Interestingly, Lewis rats are prone to develop a host of inflammatory diseases, including a rheumatoid arthritis-like syndrome in response to a streptococcal cell-wall peptidoglycan, autoimmune uveitis in response to an epitope of the interphotoreceptor retinol-binding protein, and encephalomyelitis in response to myelin basic protein. In contrast, Fischer rats are resistant to the development of similar syndromes. Other similar animal paradigms include the obese strain of chickens that develop spontaneous autoimmune thyroiditis and have a defective response of their HPA axis to inflammatory cytokines; the lupus-prone MRL/Mp-1pr/1pr and (NZBx NZW) FL mice, the nonobese diabetic (NOD) mouse and the BB rat, which also have abnormal interrelations between inflammatory cytokines and the HPA axis. A human homologue of the inherent, probably hypothalamic defect of the Lewis rat might be found in a subgroup of patients with rheumatoid arthritis, who demonstrate low or normal ACTH and cortisol concentrations during active disease and poor responses to the inflammatory mediators that are normally elevated during major surgery.[112]

From a pathophysiologic perspective, the above information suggests that other human autoimmune/inflammatory diseases should be examined for presence of disturbances in the neuroendocrine control of the I/I reaction. From a therapeutic perspective, the above information suggests that proper systemic or compartmentalized use of glucocorticoids and/or agents that potentiate their actions, remains a fully rational option for the therapy of autoimmune/inflammatory diseases. Use of inflammatory cytokine or proinflammatory neuropeptide antagonists — such as those for CRH, AVP, or substance P, alone or in combination — might be added in the future as another therapeutic option for the systemic or local control of inflammatory diseases. In addition, of inflammation in subgroups of patients with autoimmune/inflammatory diseases might be useful in the prevention and therapy of such diseases.

REFERENCES

1. Chrousos, G. P., The hypothalamic-pituitary-adrenal axis and the immune/inflammatory reaction, *N. Engl. J. Med.*, 332, 1351, 1995.
2. Reichlin, S., Neuroendocrine-immune interactions (review article), *N. Engl. J. Med.*, 329, 1246, 1993.
3. Gordon, B. L., *Medicine Throughout Antiquity*, F. A. Davis, Philadelphia, 1949, Chap. 27,
4. Selye, H., Stress syndrome. A syndrome produced by diverse noxious agents, *Nature (London)*, 138, 32, 1936.
5. Johnson, H. M., Neuroendocrine peptide hormone regulation of immunity, in *Neuroimmunoendocrinology*, 2nd rev. ed., Blalock, J. E., Ed., Karger, Basel, 52, 49, 1992.
6. Gallin, J. I., Goldstein, I. M., and Snyderman, R., Overview, in *Inflammation. Basic Principles and Clinical Correlates*, Gallin, J. I., Goldstein, I. M., Snyderman, R., Eds., Raven Press, New York, 1988, 1.
7. Paul, W. E. and Seder, R. A., Lymphocyte responses and cytokines, *Cell*, 76, 241, 1994.
8. Boumpas, D. T., Chrousos, G. P., Wilder, R. I., Cupps, T. R., and Balow, J. E., Glococorticoid therapy for immune-mediated diseases: basic and clinical correlates, *Ann. Intern. Med.*, 119, 1198, 1993.
9. Chrousos, G. P. and Gold, P. W., The concepts of stress and stress system disorders. Overview of physical and behavioral homeostasis, *JAMA*, 267, 1244, 1992.
10. Chrousos, G. P., Detera, Wadleigh, S. D., and Karl, M., Syndromes of glucocorticoid resistance, *Ann. Intern. Med.*, 119, 1113, 1993.
11. Crostein, B. N., Kimmel, S. C., Levin, R. I., Martiniuk, F., and Weissmann, G., A mechanism for the anti-inflammatory effects of corticosteroids: the glucocorticoid receptor regulates leukocyte adhesion to endothelial cells and expression of endothelial-leukocyte adhesion molecule 1 and intercellular adhesion molecule 1, *Proc. Natl. Acad. Sci. U.S.A.*, 89, 9991, 1992.

12. Hirano, T., Akira, S., Taga, T., and Kishimoto, T., Biological and clinical aspects of interleukin 6, *Immunol. Today,* 11, 443, 1990.
13. Brasier, A. R., Ron, D., Tate, J. E., and Habener, J. F., Synergistic enhansons located within an acute phase responsive enhanser modulate glucocorticoid induction of angiotensinogen gene transcription, *Mol. Endocrinol.,* 4, 1921, 1990.
14. Laue, L., Kawai, S., Brandon, D. D., et, al., Receptor-mediated effects of glucocorticoids on inflammation: enhancement of the inflammatory response with a glucocorticoid antagonist, *J. Steroid. Biochem.,* 29, 591, 1988.
15. Smith, D. F. and Toft, D. O., Steroid receptors and their associated proteins, *Mol. Endocrinol.,* 7, 4, 1993.
16. Evans, R. M., The steroid and thyroid receptor superfamily, *Science,* 240, 889, 1988.
17. Guiochon-Mantel, A. and Milgrom, E., Cytoplasmic-nuclear trafficking of steroid hormone receptors, *Trends. Endo. Metab.,* 4, 322, 1993.
18. Truss, M. and Beato, M., Steroid hormone receptors: interaction with deoxyribonucleic acid and transcription factors, *Endo. Rev.,* 14, 459, 1993.
19. Gilmore, W. and Weiner, L. P., β-Endrophin enhances interleukin (IL-2) production in murine lymphocytes, *J. Neuroimmunol.,* 18, 125, 1988.
20. Udenfriend, S. and Kilpatrick, D. L., Biochemistry of the enkephalins and enkephalin-containing peptides, *Arch. Biochem. Biophys.,* 221, 309, 1983.
21. Kay, N. E., Morley, J. E., and Allen, J. I., Interaction between endogenous opioids and IL-2 on PHA-stimulated human lymphocytes, *Immunology,* 70, 485, 1990.
22. Faith, R. E., Liang, H. J., Murgo, A. J., and Plotnikoff, N. P., Neuroimmunomodulation with enkephalins: enhancement of human natural killer (NK) cell activity *in vitro, Clin. Immunol. Immunopathol.,* 31, 412, 1984.
23. Foster, J. S. and Moore, R. N., Dynorphin and related opioid peptides enhance tumoricidal activity mediated by murine peritoneal macrophages, *J. Leukocyte Biol.,* 42, 171, 1987.
24. Oleson, D. R. and Johnson, D. R., Regulation of human natural cytotoxicity by enkephalins and selective opiate agonists, *Brain Behav. Immunol.,* 2, 171, 1988.
25. Carr, D. J., De Costa, B. R., Jacobson, A. E., Rice, K. C., and Blalock, J. E., Corticotropin-releasing hormone augments natural killer cell activity through a naloxone-sensitive pathway, *J. Neuroimmunol.,* 28, 53, 1990.
26. Gilman, S. C., Schwartz, J. M., Minler, R. J., Bloom, F. E., and Feldman, J. D., β-endorphin enhances lymphocyte proliferative responses, *Proc. Natl. Acad. Sci. U.S.A.,* 79, 4226, 1982.
27. McCain, H. W., Lamster, I. B., Bozzone, J. M., and Grbic, J. T., β-endorphin modulates human immune activity via non-opiate receptor mechanisms, *Life Sci.,* 31, 1619, 1982.
28. Gilmore, W., Moloney, M., and Beristein, T., The enhancement of polyclonal T cell proliferation by β-endorphin, *Brain Res. Bull.,* 24, 687, 1990.
29. Hemmick, L. and Bidlack, J. M., β-endorphin stimulates rat T lymphocyte proliferation, *J. Neuroimmunol.,* 29, 239, 1990.
30. Miller, G. C., Murgo, A. J., and Plotnikoff, N. P., Enkephalins: enhancement of active T-cell rosettes from lymphoma patients, *Clin. Immunol. Immunopathol.,* 26, 446, 1983.
31. Van Epps, D. E. and Saland L., β-endorphin and [Met] enkephalin stimulate human peripheral blood mononuclear cell chemotaxis, *J. Immunol.,* 132, 3046, 1984.
32. Ruff, M., Schiffmann, E., Terranova, V., and Pert, C. B., Neuropeptides are chemoattractants for human tumor cells and monocytes: a possible mechanism for metastasis, *Clin. Immunol. Immunopathol.,* 37, 387, 1985.

33. Brown, S. L. and Van Epps, D. E., Supression of T lympocyte chemotactic factor production by the opioid peptides β-endorphin and [Met] enkephalin, *J. Immunol.*, 134, 3384, 1985.
34. Peterson, P. K., Sharp, B., Gekker, G., Brummitt, C., and Keane, W. F., Opioid-mediated suppression of interferon-γ production by cultured peripheral blood mononuclear cells, *J. Clin. Invest.*, 80, 824, 1987.
35. Morgano, A., Setti, M., Pierri, I., Barabino, A., Lotti, G., and Indiveri, F., Expression of HLA-class II antigens and proliferative capacity in autologous mixed lymphocyte reactions of human T lymphocytes exposed *in vitro* to α-endorphin, *Behav. Immunol.*, 3, 214, 1989.
36. Lipton, J. M., Modulation of host defense by neuropeptide α-MSH. *Yale J. Biol. Med.*, 63, 173, 1990.
37. Pierpaoli, W., *Psychoneuroimmunology*, Academic Press, New York, 575, 1981.
38. Smith, E. M., Phan, M., Kruger, T. E., Coppenhaver, D., and Blalock, J. E., Human lymphocyte production of immunoreactive thyrotropin, *Proc. Natl. Acad. Sci. U.S.A.*, 80, 6010, 1983.
39. Harbour, D. V., Kruger, T. E., Coppenhaver, D., Smith, E. M., and Meyer, W. J., Differential expression and regulation of thyrotropin (TSH) in T cell lines, *Mol. Cell. Endocrinol.*, 64, 229, 1989.
40. Kruger, T. E., Smith, L. R., Harbour, D. V., and Blalock, J. E., Thyrotrophin: an endogenous regulator of the *in vitro* immune response, *J. Immunol.*, 142, 744, 1989.
41. Coutelier, J.-P., Kerhl, J. H., Bellur, S. S., Kohn, L. D., Notkins, A. L., and Prabhakar, B. S., Binding and functional effects of thyroid stimulating hormone on human immune cells, *J. Clin. Immunol.*, 10, 204, 1990.
42. Lechan, R. M., Update on thyroid releasing hormone, *Thyroid Today*, 16 (1), 1993.
43. Scarborough, D. E., Lee, S. L., Dinarello, C. A., and Reichlin, S., Interleukin-1 stimulates somatostatin biosynthesis in primary cultures of fetal rat brain, *Endocrinology*, 124, 549, 1989.
44. Pang, X. P., Hershman, J. M., Mirell, C. J., and Peraky, E. A., Impairment of hypothalamic-pituitary-thyroid function in rats treated with human recombinant tumor necrosis factor-α (cachectin), *Endocrinology*, 125, 76, 1989.
45. Reichlin, S. and Glaser, R. J., Thyroid function in experimental streptococcal pneumonia in the rat, *J. Exp. Med.*, 107, 219, 1958.
46. Weigent, D. A. and Blalock, J. E., Growth hormone and the immune system, *Progr. Neuroendocrinoimmunol.*, 3, 231, 1990.
47. Kelley, K. W., The role of growth hormone in modulation of the immune response, *Ann. N.Y. Acad. Sci.*, 594, 95, 1990.
48. Weigent, D. A., Blalock, J. E., and Le Boeuf, R. D., An antisense oligodeoxynucleotide to growth hormone messenger ribonucleic acid inhibits lymphocyte proliferation, *Endocrinology*, 128, 2053, 1991.
49. Nagy, E., Berczi, I., Wren, G. E., Asa, S. L., and Kovacs, K., Immunomodulation by bromocriptine, *Immunopharmacology*, 6, 231, 1983.
50. Reichlin, S., Endocrine-immune interaction, in *Endocrinology*, 3rd ed., DeGroot, L., Ed., W.B. Saunders, Philadelphia, 1998.
51. di Carlo, R. and Meli, R., Interactions between prolactin and immune processes, *Acta Neurol. (Napoli)*, 13, 380, 1991.
52. Kelley, K. W., Arkins, S., and Li, Y. M., Growth hormone, prolactin, and insulin-like growth factors: new jobs for old players, *Brain Behav. Immunol.*, 6, 317, 1992.

53. Russell, D. H., New aspects of prolactin and immunity: a lymphocyte-derived prolactin-like product and nuclear protein kinase C activation, *Trends Pharmacol. Sci.,* 10, 40, 1989.
54. Berwaer, M., Martial, J. A., and Davis, J. R. E., Characterization of an up-stream promoter directing extrapituitary expression of the human prolactin gene, *Mol. Endocrinol.,* 8, 635, 1994.
55. Saffran, M., Schally, A. V., and Benfey, B. G., Stimulation of the release of corticotropin from the adenohypophysis by a neurohypophysial factor, *Endocrinology,* 57, 439, 1995.
56. Guillemin, R. and Rosenberg, B., Humoral hypothalamic control of anterior pituitary: a study with combined tissue cultures, *Endocrinology,* 57, 599, 1995.
57. Vale, W., Spiess, J., Rivier, C., and Rivier, J., Characterization of a 41-residue ovine hypothalamic peptide that stimulates secretion of corticotropin and β-endorphin, *Science,* 213, 1394, 1981.
58. Dave, J. R., Eiden, L. E., and Eskay, R. L., Corticotropin-releasing factor binding to peripheral tissue and activation of the adenylate cyclase-adenosine 3', 5'-monophosphate system, *Endocrinology,* 116, 2152, 1985.
59. Webster, E. L. and De Sousa, E., Corticotropin-releasing factor receptors in mouse spleen: identification, autoradiographic localization and regulation by divalent cations and guanine nucleotides, *Endocrinology,* 122, 609, 1988.
60. Singh, V. K. and Fudenberg, H. H., Binding of [^{125}I] corticotropin releasing factor to blood immunocytes and its reduction in Altzeimer's disease, *Immunol. Lett.,* 18, 5, 1988.
61. Webster, E. L., Tracey, D. E., Jutila, M. A., Wolfe, S. A., Jr., and De Souza, E. B., Corticotropin-releasing factor receptors in mouse sleen: identification of receptor-bearing cells as resident macrophages, *Endocrinology,* 127, 440, 1990.
62. Smith, E. M., Morrill, A. C., Meyer, W. M., and Blalock, M. E., Corticotropin-releasing factor induction of leukocyte- derived immunoreactive ACTH and endorphins, *Nature,* 321, 881, 1986.
63. Kavelaars, A., Berkenbosch, F., Croiset, G., Ballieux, R. E., and Heijnen, C. J., Induction of β-endorphin secretion by lymphocytes after subcutaneous administration of corticotropin-releasing factor, *Endocrinology,* 126, 759, 1990.
64. Hargreaves, K. M., Costello, A. H., and Joris, J. L., Release from inflamed tissue of a substance with properties similar to corticotropin-releasing factor, *Neuroendocrinology,* 49, 476, 1989.
65. Kavelaars, A., Ballieux, R. E., and Heinjen, C. J., The role of IL-1 in the corticotropin-releasing factor and arginine-vasopressin-induced secretion of immunoreactive β-endorphin by human peripheral blood mononuclear cells, *J. Immunol.,* 142, 2338, 1989.
66. Karalis, K., Sano, H., Redwine, J., Listwak, S., Wilder, R. L., and Chrousos, G. P., Autocrine or paracrine inflammatory actions of corticotropin-releasing hormone in vivo, *Science,* 254, 421, 1991.
67. Calogero, A. E., Bernardini, R., Margioris, A. N., et al., Effects of serotonergic agonists and antagonists on corticotropin-releasing hormone secretion by explanted rat hypothalami, *Peptides (Elmsford),* 10, 189, 1989.
68. Shibahara, S. Y., Morimoto, Y., Furutani, Y., et al., Isolation and sequence analysis of the human corticotropin-releasing factor precursor gene, *EMBO J.,* 2, 775, 1983.

69. Suda, T., Tomori, N., Tozawa, F., et al., Immunoreactive corticotropin and corticotropin-releasing factor in human hypothalamus, adrenal, lung cancer and pheochromocytoma, *J. Clin. Endocrinol. Metab.,* 58, 919, 1984.
70. Grino, M., Chrousos, G. P., and Magrioris, A. N., The corticotropin-releasing hormone gene is expressed in human placenta, *Biochem. Biophys. Res. Commun.,* 148, 1208, 1987.
71. Crofford, L. J., Sano, H., Karalis, K., et al., Local secretion of corticotropin-releasing hormone in the joints of Lewis rats with inflammatory arthritis, *J. Clin. Invest.,* 90, 2555, 1992.
72. Crofford, L. J., Sano, H., Karalis, K., et al. Corticotropin-releasing hormone in synovial fluids and tissues of patients with rheumatoid arhtritis and osteoarthritis, *J. Immunol.,* 151, 1, 1993.
73. Mastorakos, G., Silver, P. B., Bouzas, E. A., Caspi, R. R., Chan, C. C., and Chrousos, G. P., Immunoreactive corticotropin-releasing hormone in experimental uveitis, *Invest. Ophthalmol. Vis. Sc.,* 33, 933, 1992.
74. Scofitsch, G., Hamill, G. S., and Jacobowitz, D. M., Capsaicin depletes corticotropin-releasing factor-like immunoreactive neurons in the rat spinal cord and medulla oblongata, *Neuroendocrinology,* 38, 514, 1984.
75. Merchenthaler, I., Hynes, M. A., Vigh, S., Schally, A. V., and Petrusz, P., Immonucytochemical localization of corticotropin releasing factor (CRF) in the rat spinal cord, *Brain Res.,* 275, 373, 1983.
76. Udelsman, R., Harwood, J. P., Millan, M. A., et al., Functional corticotropin-releasing factor receptors in the primate peripheral sympathetic nervous system, *Nature (London),* 319, 147, 1986.
77. Karalis, K., Mastorakos, G., Chrousos, G. P., and Tolis, G., Somatostatin analogues suppress the inflammatory reaction *in vivo, J. Clin. Invest.,* 93, 2000, 1994.
78. Karalis, K., Mastorakos, G., Sano, H., Redwine, J., and Wilder, R., Neuropeptide and cytokine interplay in the regulation of the inflammatory response in a rat experimental model, Paper presented at 74th Annual Meeting of the Endocrine Society, San Antonio, Texas, 1992.
79. Karalis, K., Mastorakos, G., Sano, H., Wilder, R., and Chrousos, G. P., Somatostatin may participate in the anti-inflammatory actions of glucocorticoids, *Endocrinology,* 136(9), 4133, 1995.
80. Payan, D. G., Hess, C. A., and Goetzl, E. J., Inhibition by somatostatin of the proliferation of T-lymphocytes and Molt-4 lymphoblasts, *Cell. Immunol.,* 84, 433, 1984.
81. Muscettola, M. and Grasso, G., Somatostatin and vasoactive intestinal peptide reduce interferon-γ production by human peripheral blood mononuclear cells, *Immunobiology,* 180, 419, 1990.
82. Wagner, M., Hengst, K., Zierden, E., and Gerlach, U., Investigations of the antiproliferative effect of somatostatin in man and rats, *Metab. Clin. Exp.,* 27, 1381, 1978.
83. Hinterberger, W., Cenry, C., Kinast, H., Pointer, H., and Trag, K. H., Somatostatin reduces the release of colony-stimulating activity (CSA) from PHA-activated mouse spleen lymphocytes, *Experentia,* 34, 860, 1977.
84. Sreedharan, S. P., Kodama, K. T., Peterson, K. E., and Goetzl, E. J., Distinct subsets of somatostatin receptors on cultured human lymphocytes, *J. Biol. Chem.,* 264, 949, 1989.

85. Nakamura, H., Koike, K. T., Hiruma, K., Sate, T., Tomioka, H., and Yoshida, S., Identification of lymphoid cell lines bearing receptors for somatostatin, *Immunology*, 62, 655, 1987.
86. Patchev, V. K., Mastorakos, G., Brady, L. S., Redwine, J., Wilder, R. L., and Chrousos, G. P., Increased arginine vasopressin secretion may participate in the enhanced susceptibility of Lewis rats to inflammatory disease, *Neuroendocrinology*, 58, 106, 1993.
87. Johnson, H. M., Farrar, W. L., and Torres, B. A., Vasopressin replacement of interleukin-2 requirement in γ interferon production: lymphokine activity of a neuroendocrine hormone, *J. Immunol.*, 129, 983, 1982.
88. Johnson, H. M. and Torres, B. A., Regulation of lymphokine production by arginine vasopressin and oxytocin: modulation of lymphocyte function by neurohypophyseal hormones, *J. Immunol.*, 135, 773, 1985.
89. Torres, B. A., Farrar, W. L., and Johnson, H. M., Interleukin-2 regulates immune interferon (IFNγ) production by normal and suppressor cell cultures, *J. Immunol.*, 128, 2217, 1982.
90. Hess, D. G., Tracey, K. J., Fong, Y., et al., Cytokine appearance in human endotoxemia and primate bacteremia, *Surg. Gynecol. Obstert.*, 166, 147, 1988.
91. Van Deventer, S. J., Buller, H. R., Cate, J. W., Aarden, L. A., Hack, C. E., and Sturk, S., Experimental endotoxemia in humans: analysis of cytokine release and coagulation, fibrinolytic, and complement pathways, *Blood*, 76, 2520, 1990.
92. Akira, S., Hirano, T., Taga, T., and Kishimoto, T., Biology of multifunctional cytokines: IL-6 and related molecules (IL-1 and TNF), *FASEB J.*, 4, 2860, 1990.
93. Ulich, T. R., Guo, K. Z., Remick, D., Castillo, J., and Yin, S. M., Endotoxin-induced cytokine gene expression *in vivo*. III. IL-6 mRNA and serum protein expression and the *in vivo* hematologic effects of IL-6, *J. Immunol.*, 146, 2316, 1991.
94. Imura, H., Fukata, J., and Mori, T., Cytokines and endocrine function: an interaction between the immune and neuroendocrine systems, *Clin. Endocrinol.*, 35, 107, 1991.
95. Bernardini, R., Kamilaris, T. C., Calogero, A. E., Johnson, E. O., Gold, P. W., and Chrousos, G. P., Interactions between tumor necrosis factor α, hypothalamic corticitropin-releasing hormone and adrenocorticotropin secretion in the rat, *Endocrinology*, 126, 2876, 1990.
96. Sapolsky, R., Rivier, C., Yamamoto, G., Plotsky, P., and Vale, W., Interleukin-1 stimulates the secretion of hypothalamic corticotropin releasing factor, *Science*, 238, 522, 1987.
97. Naitoh, Y., Fukata, J., Tominaga, T., et al., Interleukin-6 stimulates the secretion of adrenocorticotropin hormone in conscious, free-moving rats, *Biochem. Biophys. Res. Commun.*, 155, 1459, 1988.
98. Perlstein, R. S., Mougey, E. H., Jackson, W. E., and Neta, R., Interleukin-1 and interleukin-6 act synergistically to stimulate the release of adrenocorticotropic hormone *in vivo*, *Lymphok. Cytok. Res.*, 10, 141, 1991.
99. Perlstein, R. S., Whitnall, M. H., Abrams, J. S., Mougey, E. H., and Neta, R., Synergistic roles of interleukin-6, interleukin-1, and tumor necrosis factor in adrenocorticotropin response to bacterial lipopolysaccharide *in vivo*, *Endocrinology*, 132, 946, 1993.
100. Mastorakos, G., Chrousos, G. P., and Weber, J., Recombinant inteleukin-6 activates the hypothalamic-pituitary-adrenal axis in humans, *J. Clin. Endocrinol. Metab.*, 27, 1690, 1993.
101. Mastorakos, G., Weber, J. S., Magiakou, M. A., Gunn, H., and Chrousos, G. P., Hypothalamic-pituitary adrenal axis activation and stimulation of systemic vasopressin

secretion by recombinant interleukin-6 in humans: potential implications for the syndrome of inappropriate vasopressin secretion, *J. Clin. Endocrinol. Metab.,* 79(4), 934, 1994.
102. Mastorakos, G., Weber, J. S., Magiakou, M. A., et al., Interleukin-6 effects on the human hypothalamic-pituitary-adrenal axis, paper present at 75th Annual Meeting of the Endocrine Society, Las Vegas.
103. Tilders, F. J., De Rijk, R. H., Van Dam, A. M., Vincent, V. A., Schotanus, K., and Persoons, J. H., Activation of the hypothalamic-pituitary-adrenal axis by bacterial endotoxins: routes and intermediate signals, *Psychoneuroendocrinology,* 19, 209, 1994.
104. Matta, S., Singh, J., Newton, R., and Sharp, B. M., The adrenocorticotropin response to interleukin-1β instilled into the rat median eminence depends on the local release of catecholamines, *Endocrimology,* 127, 2175, 1990.
105. Uehara, A., Gillis, S., and Arimura, A., Effects of interleukin-1 on hormone release from normal rat pituitary cells in primary culture, *Neuroendocrinology,* 45, 343, 1987.
106. Spangelo, B. L., Judd, A. M., Isakson, P. C., and McLeod, R. M., Interleukin-6 stimulates anterior pituitary hormone release *in vitro, Endocrinology,* 125, 575, 1989.
107. Roh, M. S., Drazenovich, K. A., Barbose, J. J., Dinarello, C. A., and Cobb, C. F., Direct stimulation of the adrenal cortex by interleukin-6, *Surgery,* 102, 140, 1987.
108. Salas, M. A., Evans, S. W., Levell, M. J., and Whicher, J. T., Interleukin-6 and ACTH act synergistically to stimulate the release of corticosterone from adrenal gland cells, *Clin. Exp. Immunol.,* 79, 470, 1990.
109. Vankelecom, H., Carmeliet, P., Van Damme, J., Billiau, A., and Denef, C., Production of interleukin-6 by folliculo-stellate cells of the anterior pituitary gland in a histiotypic cell aggregate culture system, *Neuroendocrinology,* 49, 102, 1989.
110. Schultzberg, M., Anderson, C., Undin, A., Troye-Blomberg, M., Svenson, S. B., and Bartfai, T., Interleukin-1 in adrenal chromaffin cells, *Neuroscience,* 30, 805, 1989.
111. Calogero, A., Sternberg, E., Bagdy, G, et al., Neurotransmitter-induced hypothalamic-pituitary-adrenal axis responsiveness is defective in inflammatory disease-susceptible Lewis rats: *in vivo* and *in vitro* studies suggesting globally defective hypothalamic CRH secretion, *Neuroendocrinology,* 55, 600, 1992.
112. Chikanza, I. C., Petrou, P., Chrousos, G. P., Kingsley, G., and Panayi, G. S., Defective hypothalamic response to immune/inflammatory stimuli in patients with rheumatoid arthritis, *Arthritis Rheum.,* 35, 1281, 1992.

3 Immune and Clinical Correlates of Psychological Stress-Induced Production of Interferon-γ and Interleukin-10 in Humans

Michael Maes, Cai Song, Aihua Lin, Raf De Jongh, An Van Gastel, Gunter Kenis, Eugene Bosmans, Ingrid De Meester, Hugo Neels, Aleksandar Janca, Simon Scharpé, and Ronald S. Smith

CONTENTS

Abstract 39
3.1 Introduction 40
3.2 Subjects and Methods 41
    3.2.1 Subjects 41
    3.2.2 Methods 41
3.3 Statistics 42
3.4 Results 43
3.5 Discussion 43
Acknowledgment 46
References 47

ABSTRACT

Recently, we have shown that academic examination stress in university students significantly increases the stimulated production of proinflammatory cytokines, such as interferon-γ (IFNγ), interleukin-6 (IL-6), and tumor necrosis factor-α (TNFα), and that of the negative immunoregulatory cytokine IL-10. The present study examines the immune and clinical characteristics of psychological stress-induced increases in IFNγ or IL-10 in these students. This study reports that there are two different profiles of cytokine production to psychological stress, i.e., a first with a predominant IFNγ vs. IL-10 response (labeled IFNγ reactors) and a second with a predominant IL-10 vs. IFNγ response (labeled IL-10 reactors). IFNγ reactors, but not IL-10 reactors, show a significantly stress-induced increase in anxiety and depressive feelings. IFNγ reactors, but not IL-10 reactors, show a significantly stress-induced increase in the stimulated production of TNFα and IL-6 and an increase in serum IL-1 receptor antagonist, soluble CD8, and serum IgA, IgG and IgM, as well as suppression of serum CC16 (Clara Cell Protein). IL-10 reactors, but not IFNγ reactors, have a significantly stress-induced increase in the number of peripheral blood lymphocytes and $CD4^+$ and $CD8^+$ T cells. A lower baseline $CD4^+/CD8^+$ T-cell ratio is a predictor for a subsequent stress-induced IFNγ response. The findings suggest that the response to psychological stress in humans entails two different profiles of cytokine production, i.e., either a predominant proinflammatory response or a predominant negative immunoregulatory cytokine response. A stress-induced proinflammatory response is associated with significantly greater stress-induced anxiety and depressive feelings and may be causally related to a stress-induced decrease in serum CC16.

3.1 INTRODUCTION

There is now evidence that the principal messengers responsible for the bidirectional communications between the immune system and the central nervous system (CNS) following stress comprise glucocorticoid hormones and catecholamines, as well as proinflammatory cytokines. Internal stressors such as injury, infections, immune responses, inflammation, necrosis, and toxins induce a stress response characterized by proinflammatory cytokine-induced (1) activation of the hypothalamic-pituitary-adrenal (HPA) axis, with increased glucocorticoid hormone concentrations; (2) increased catecholamine neurotransmission in selected brain areas; and (3) sickness-behavior: sleep disturbances, anorexia, weight loss, psychomotor retardation, anergy, and loss of interest.[1-6] External stressors, that is psychological stress, also is known to induce hyperactivity of the HPA axis and to increase catecholaminergic turnover in (experimental) animals and humans.[7,8] The stress-induced activation of the HPA axis and catecholaminergic turnover is related to stress-induced activation of corticotropin releasing hormone (CRH).[9,10]

There is now also some evidence that external stressors may induce the production of proinflammatory cytokines in experimental animals and humans. Thus, in the rodent, psychological stressors (e.g., immobilization or open field stress) may increase the production of IL-6 and IL-1 mRNA expression in the hypothalamus.[11-14] In humans, no effects of psychological stress on plasma IL-1 or IL-6 concentrations could be found, although a significant and positive correlation between plasma IL-6 and serum C-reactive protein (CRP) concentrations was found in the stress condition.[15] Recently, we reported that academic examination stress in biomedical students significantly increased the stimulated production of proinflammatory cytokines such as IL-6, tumor necrosis factor-α (TNFα), and interferon-γ (IFNγ), and of the negative immunoregulatory cytokine IL-10.[16] Students with a high stress perception had a significantly higher production of TNFα, IFNγ, and IL-6 than students with a low stress perception during the examination period. Moreover, the stress-induced changes in perceived stress were significantly and positively related to the stress-induced production of these proinflammatory cytokines, suggesting that the stress-induced production of these cytokines is sensitive to graded differences in the perception of stressor severity.[16] Finally, in the same study it was shown that subjects with stress-induced anxiety have a significantly higher stress-induced IFNγ and lower IL-10 production than subjects without anxiety.[16]

The aims of the present study were to examine the immune and clinical correlates of stress-induced production of IFNγ or IL-10, i.e., the stress-induced alterations in anxiety or depression ratings as well as the stress-induced alterations in the production of other cytokines, such as IL-6 and TNFα, serum concentrations of CC16 (Clara Cell Protein), soluble CD8 molecule (sCD8), IL-1 receptor antagonist (IL-1RA), serum immunoglobulins IgA, IgG, and IgM, and the number of lymphocytes, CD4$^+$, and CD8$^+$ T cells in the peripheral blood.

3.2 SUBJECTS AND METHODS

3.2.1 SUBJECTS

The subjects were university students attending the second year of medical sciences at the Rijksuniversitair Centrum Antwerpen, University of Antwerp, Antwerp, Belgium. Thirty-eight students participated, ranging in age from 19 to 22 years. There were 13 female students not taking contraceptive drugs, 14 female students who had been taking contraceptive drugs (monophasic oestroprogestativa), and 11 male students. All students had normal baseline routine blood tests. They did not suffer from medical disorders and had no past or present history of psychiatric disorders. None had ever taken psychotropic medications. None of the subjects had taken psychostimulants during the study span. There was no tobacco use, except in 6 subjects who did not change their smoking behavior (less than 15 cigarettes daily) during the stress condition.

3.2.2 METHODS

Blood samples for cytokine assays were collected at 9:00 a.m. (±45 min) a few weeks before (mean = 44.5 ± 6.0 days; BASELINE condition), as well as one day before a difficult academic examination during the period January 15, 1995 to February 2, 1995 (STRESS condition). At the same occasions, students completed self-rating instruments aimed at measuring anxiety and depression. The Spielberger State-Trait Anxiety Inventory (STAI) was used to measure the level of state anxiety.[17] The Beck Depression Inventory (BDI) was used to measure cognitive symptoms of depression.[18]

Alterations in cytokine synthesis were studied by stimulating whole blood with phytohemagglutinin (PHA) and lipopolysaccharide (LPS) and analyzing cytokine production in culture supernatant.[19] Whole blood diluted one quarter in RPMI-1640 was employed with or without 1 μg/mL PHA + 5 μg/mL LPS as final concentrations. Samples (400 μL) were pipetted into 24 well-plates prefilled with medium (1200 μL) and incubated for 48 h in a humidified atmosphere at 37°C, 5% CO_2. After incubation, the plates were centrifuged at 1500 rpm for 10 min. Supernatants were taken off under sterile conditions, divided into Eppendorf tubes, and frozen immediately at –75°C until thawed for assay. All cytokines and receptors were quantified by means of ELISA methods (Eurogenetics, Tessenderlo, Belgium) based on appropriate and validated sets of monoclonal antibodies. The intra-assay coefficients of variation were less than 8% for all assays. WBC differentials were determined by means of the Technicon H2 system (Brussels, Belgium). The inter-assay CV values obtained in our laboratory are <2% for the number of lymphocytes. Flow cytometric evaluation of the lymphocyte subsets was carried out on 50 μL of fresh, total blood after NH_4Cl lysis of the RBC, and using the same batches of directly conjugated antisera throughout the study span.

All antisera were purchased from Becton Dickinson (Erenbodegem, Belgium). Samples were assayed after incubation on an Epics XL-MCL Analytical Flow Cytometer (Analasis, Namur, Belgium) using standard software and gating on the lymphocyte cluster defined by light scatter characteristics. IgA, IgG, and IgM were assayed using laser nephelometry (Behring Nephelometer Analyzer, Behringwerke AG, Marburg, Germany). Antiserum was obtained from Dako (Denmark). The analytical intra-assay coefficients of variation were IgA 1.95%, IgG 3.33%, and IgM 0.40%. Serum concentrations of IL-1RA, sCD8, and CC16 were measured by ELISA techniques (Eurogenetics, Tessenderlo, Belgium) based on appropriate and validated sets of monoclonal antibodies. The intra-assay CV values were less than 8% for all variables. Each set of blood (for serum Igs, IL-1RA, CC16, and sCD8) or culture supernatant specimens from an individual subject was assayed in a single run with a single lot number of reagents and consumables employed by a single operator.

3.3 STATISTICS

Repeated measure design analyses of variance (ANOVAs) were used to examine the interindividual variability with (1) effects of different profiles of cytokine production, i.e., increased stress-induced IFNγ vs. increased IL-10 production; and (2) gender or hormonal state (i.e., three groups: males, females without use of oral contraceptives, females with the use of oral contraceptives). Significant time X gender/hormonal state or time X reactor status X gender/sex hormonal state interactions were taken into account. The prime focus of interest were tests on simple effects.[20] A simple effect is defined as the effect of one variable at one level of the other variable. In order to examine the effects of different profiles of cytokine production, subjects were divided into two groups on the basis of their IFNγ and IL-10 production in the STRESS condition, i.e., z-transformed IFNγ – z-transformed IL-10 (labeled the IFNγ/IL-10 ratio) ≥0.00 (labeled IFNγ reactors) vs. <0.00 (labeled IL-10 reactors). Relationships between variables were ascertained by means of Pearson's product moment correlation coefficients, i.e., intraclass correlations pooled over the repeated measurements (BASELINE and STRESS conditions) in the 38 subjects. This method eliminates the interindividual variability and assesses the relationships over time between two sets of variables.

3.4 RESULTS

Table 3.1 shows the measurements of the immune and clinical variables in IFNγ and IL-10 reactors. Tests on simple effects showed significant time effects on the IFNγ/IL-10 ratio, i.e., an increase in IFNγ reactors (n = 15) and a decrease in IL-10 reactors (n = 23). In the STRESS (F = 47, df = 1/72, p = .000002), but not in the BASELINE, condition there was a highly significant difference in the IFNγ/IL-10 ratio between both groups. In IFNγ reactors, but not IL-10 reactors, stress induced a significant increase in IFNγ production. In IL-10 reactors, but not IFNγ reactors, stress induced a significant increase in IL-10 production. In the STRESS condition, there were significant differences in the production of IFNγ (F = 25, df = 1/72, p = .00004) and IL-10 (F = 14, df = 1/72, p = .0006) between IFNγ and IL-10 reactors, respectively.

IFNγ reactors, but not IL-10 reactors, showed a significant stress-induced increase in the STAI and Beck scores. There were significant and positive correlations between the stress-induced changes in the STAI score and IFNγ (R = .40, p = .01) and IL-6 (R = .32, p = .04), but not TNFα (R = .27, NS) or IL-10 (R = .19, NS).

IFNγ reactors, but not IL-10 reactors, showed significant stress-induced increases in the production of TNFα and IL-6 in culture supernatant, as well as in serum IL-1RA, sCD8, IgA, and IgG and IgM concentrations. In the STRESS, but not in the BASELINE, condition, the stimulated production of TNFα was significantly higher in IFNγ than in IL-10 reactors (F = 9.2, df = 1/72, p = 0.004). IFNγ reactors, but not IL-10 reactors, showed a significant stress-induced decrease in serum CC16.

IL-10 reactors, but not IFNγ reactors, had significant stress-induced increases in number of lymphocytes, CD4+, and CD8+ T cells. In the STRESS, but not in the BASELINE, condition, the number of CD4+ T cells was significantly higher in IL-10 than in IFNγ reactors (F = 13, df = 1/72, p = .0009). IFNγ reactors showed a significantly lower CD4+/CD8+ T-cell ratio in the BASELINE (F = 8.9, df = 1/72, p = .004) as well as in the STRESS (F = 8.2, df = 1/72, p = .005) condition than IL-10 reactors. There were significant and positive correlations between the stress-induced changes in the number of lymphocytes and IL-10 (R = .43, p = .007), but not IFNγ (R = .17, NS), IL-6 (R = .22, NS), and TNFα (R = .26, NS). There were significant and positive correlations between the stress-induced changes in the number of CD4+ T lymphocytes and IL-10 (R = .46, p = .003), but not IFNγ (R = −.06, NS), IL-6 (R = .02, NS), and TNFα (R = −.18, NS). There were significant and positive correlations between the stress-induced changes in the number of CD8+ lymphocytes and IL-10 (R = .49, p = .002), but not IFNγ (R = .18, NS), IL-6 (R = .24, NS), and TNFα (R = .30, NS).

3.5 DISCUSSION

This study shows that the response to psychological stress in humans consists of two different profiles of cytokine production, i.e., a first characterized by a higher IFNγ than IL-10 response (labeled IFNγ reactors), and a second characterized by a higher IL-10 than IFNγ response (labeled IL-10 reactors). Phrased differently, the

TABLE 3.1
Immune and Clinical Measurements a Few Weeks Before (BASELINE) as Well as the Day Before a Difficult Academic Examination (STRESS) in 38 Biomedical Students, Divided into INFγ (2) and IL-10 Reactors (1)

Variables	Response	BASELINE	STRESS	F	df	p
IFNγ/IL-10 ratio	1	−0.25 (0.88)	−1.03 (0.87)	6.6	1/36	0.01
	2	0.193 (0.70)	1.38 (1.69)	15.4		0.0006
IFNγ (IU/mL)	1	44 (69)	77 (70)	0.8	1/36	0.6
	2	64 (69)	269 (223)	33.4		0.00002
IL-10 (pg/mL)	1	420 (370)	782 (334)	15.4	1/36	0.0006
	2	307 (211)	381 (329)	0.6		0.6
STAI	1	39.6 (10.0)	45.5 (11.9)	2.6	1/34	0.1
	2	35.3 (9.0)	49.9 (12.9)	17.8		0.0004
Beck	1	6.6 (6.1)	8.3 (6.0)	0.8	1/34	0.6
	2	4.7 (2.5)	9.6 (5.9)	7.4		0.009
TNFα (pg/mL)	1	120 (137)	312 (328)	1.8	1/36	0.2
	2	247 (243)	725 (779)	11.5		0.002
IL-6 (pg/mL)	1	85 (51)	89 (44)	0.1	1/36	0.7
	2	74 (36)	117 (79)	8.4		0.006
sCD8 (U/mL)	1	290 (140)	290 (130)	0.0	1/36	0.9
	2	316 (133)	379 (241)	4.5		0.03
IL-1RA (ng/mL)	1	0.30 (0.11)	0.30 (0.13)	0.0	1/36	0.9
	2	0.40 (0.27)	0.48 (0.35)	9.4		0.004
CC16 (ng/mL)	1	34.8 (10.0)	36.0 (11.4)	0.7	1/36	0.6
	2	36.8 (11.9)	33.5 (8.9)	5.2		0.02
Lymphocytes ($10^3/\mu L$)	1	1.97 (0.70)	2.64 (0.86)	16.0	1/36	0.0005
	2	1.79 (0.36)	2.13 (0.58)	3.9		0.054
CD4+ ($10^3/\mu L$)	1	0.85 (0.30)	1.09 (0.39)	10.4	1/36	0.003
	2	0.70 (0.16)	0.73 (0.21)	0.2		0.7
CD8+ ($10^3/\mu L$)	1	0.57 (0.28)	0.75 (0.30)	11.8	1/36	0.002
	2	0.59 (0.15)	0.69 (0.24)	3.8		0.06
CD4+/CD8+	1	1.63 (0.45)	1.51 (0.44)	4.5	1/36	0.04
	2	1.23 (0.28)	1.13 (0.37)	3.4		0.07
IgA (mg/dL)	1	176 (78)	183 (83)	2.9	1/35	0.09
	2	185 (58)	196 (70)	7.4		0.009
IgG (mg/dL)	1	988 (158)	1022 (164)	3.0	1/35	0.09
	2	1119 (135)	1167 (184)	5.9		0.02
IgM (mg/dL)	1	95 (34)	98 (36)	0.8	1/35	0.6
	2	120 (45)	132 (53)	14.1		0.0009

Note: All results are shown as mean (±SD). Response: (1) subjects with a stress-induced IFNγ/IL-10 ratio <0.0; (2); subjects with a stress-induced IFNγ/IL-10 ratio ≥0.0. F: All results of repeated measure design ANOVAs with study groups (1) and (2) as treatments; listed are the results of tests on simple effects performed on the time series of study groups (1) and (2).

external stress response entails either a predominant proinflammatory cytokine response or a negative immunoregulatory cytokine response. The distinction between IFNγ vs. IL-10 reactors to psychological stress was externally validated by significant differences between both groups in the clinical and immune responsivity to psychological stress.

A second major finding of this study is that IFNγ reactors, but not IL-10 reactors, show significant stress-induced increases in anxiety and depression ratings. Thus, a proinflammatory response and a lower stress-induced production of the negative immunoregulatory cytokine IL-10 appears to be accompanied by increased stress-induced anxiety and depression. This is in accordance with previous reports that administration of interferons, including IFNγ, has neurotoxic effects, such as mood alterations, irritability, fatigue, lassitude, psychomotor retardation, anorexia, and other symptoms reminiscent of depression and anxiety.[21-23] Our results also corroborate previous reports that major depression is characterized by a significantly increased stimulated production of IFNγ[24,25] and a significantly increased secretion of neopterin in CSF, serum, or urine.[24,26-29]

A third finding of this study is that IFNγ reactors, but not IL-10 reactors, to psychological stress show a significantly increased stimulated production of other proinflammatory cytokines, i.e., TNFα and IL-6, and of serum IL-1RA, sCD8 and IgA, IgG and IgM concentrations. The increased production of TNFα and IL-6 in IFNγ reactors may be explained by the effects of IFNγ, boosting cells of the monocyte-macrophage lineage to produce cytokines that will reciprocally activate lymphocytes then produce more IFNγ.[30,31] IL-10, on the other hand, potently inhibits the production of TNFα and IL-6,[32] findings which could explain the lack of stress-induced IL-6 and TNFα responses in IL-10 reactors. The stress-induced stress IL-1RA response in IFNγ reactors may be explained by similar mechanisms. Indeed, IL-1RA is produced and released by activated cells of the monocyte-macrophage lineage in conjunction with IL-1, which secretion is stimulated by IFNγ.[30,33] IL-1RA is an endogenous, pure IL-1R antagonist that inhibits the action of IL-1 by competing for its receptor,[34,35] and reduces the bioactivity of IL-6.[36] By inference, stress-induced IL-1RA serum concentrations in IFNγ reactors could tend to limit the primary proinflammatory response in these subjects. Stress-induced increased serum sCD8 concentrations in IFNγ reactors may also be explained by the biological activities of IFNγ: this cytokine exerts a stimulatory effect on T lymphocytes by augmenting the activity of T cytotoxic $CD8^+$ cells,[37] while sCD8 is secreted by activated $CD8^+$ T cells.[38,39] The stress-induced secretion of IgA, IgM, and IgG in IFNγ reactors may be related to the stress-induced production of IL-6 in these subjects, since it is known that IL-6 stimulates the production of the above immunoglobulins by activated B cells.[40]

The fourth major finding of the present study is that IFNγ reactors, but not IL-10 reactors, showed a significant stress-induced decrease in serum CC16 concentrations. Phrased differently, there is an inverse relationship between the stress-induced decreases in serum CC16 and increases in IFNγ production. CC16 is a natural immunosuppressor and anti-inflammatory secretory protein,[41,42] which acts as an anti-cytokine through suppression of IL-2-related production of IFNγ.[43] Thus, the stress-induced reductions in serum CC16 could be causally related to increased IFNγ production.

Another finding of this study is that there were significant differences in stress-induced changes in lymphocytes (subsets) between IFNγ and IL-10 reactors. Thus, IL-10 reactors, but not IFNγ reactors, showed significant stress-induced increases in the number of lymphocytes, and CD4+ and CD8+ T cells. These findings may reflect the proliferative effects of IL-10 on CD4+ and CD8+ T cells.[32] In this respect, it is interesting to note that the stress-induced changes in IL-10 were significantly and positively related to those in the number of lymphocytes and CD4+ and CD8+ cells, whereas no significant relationships between these lymphocyte subsets and IL-6, TNFα or IFNγ could be found. Finally, IFNγ reactors had a significantly lower CD4+/CD8+ ratio in baseline as well as stress conditions than IL-10 reactors. This lower CD4+/CD8+ ratio is attributable to a lower percentage of CD4+ cells in the IFNγ reactors (results not shown). These results could suggest that subjects with a lower CD4+/CD8+ ratio are more prone to develop an inflammatory response to psychological stress. Previously, we have shown that subjects with a lower homeostatic setpoint of the CD4+/CD8+ ratio have a more distinct immune-inflammatory profile than subjects with a higher ratio.[44]

Our studies report that external stress is accompanied by an increased production of proinflammatory cytokines, such as IL-6, TNFα, and IFNγ, and the negative immunoregulatory cytokine IL-10 and that stress-induced anxiety is accompanied by increased IFNγ, but lower IL-10 production. Therefore, it may be hypothesized that not only internal, but also external stressors are perceived by the immune system and, through secretion of proinflammatory and negative immunoregulatory cytokines, take part in an integrated psychoneuroendocrine homeostatic response which eventually can become harmful, depending on the type of cytokine profile response. The discovery that psychological stress in humans can induce the production of cytokines and can alter the equilibrium between proinflammatory and negative immunoregulatory cytokines in some subjects has important implications for human psychopathology, since there is now extensive evidence of an immune-inflammatory response in major depression, an illness triggered by psychosocial stressors and accompanied by increased stressor perception.[45,46] The etiology of the stress-induced changes in the above cytokines, however, have remained elusive. In the present study, no significant stress-induced changes in plasma cortisol could be found, suggesting that glucocorticoids are not involved (results not shown). There is evidence that CRH has enhancing effects on the production of proinflammatory cytokines such as IL-1, IL-2, and IL-6.[47-49] However, no significant or even suppressive effects of CRH on the stimulated production of IFNγ were found.[50,51]

ACKNOWLEDGMENTS

The research reported here was supported in part by the Clinical Research Center for Mental Health (CRC-MH), Antwerp, Belgium and the Kaplen Investigator Award to Dr. M. Maes (NARSAD).

REFERENCES

1. Dantzer, R. and Mormede, P., Psychoneuroimmunology of stress, in *Stress, the Immune System and Psychiatry*, Leonard, B. E. and Miller, K., Eds., John Wiley & Sons, Chichester, U.K., 1995, 48.
2. Besedovsky, H. O., Psychoneuroimmunology: an overview, in *Psychoneuro-Immunology: Interactions Between Brain, Nervous System, Behavior, Endocrine and Immune Systems*, Schmoll, H.-J., Tewes, U., and Plotnikoff, N. P., Eds., Hogrefe and Huber Publishers, Kirkland, WA, 1991, 13.
3. Plata-Salaman, C. R., Immunoregulators in the nervous system, *Neurosci. Biobehav. Rev.*, 15, 185, 1991.
4. Fukata, J., Imura, H., and Nakao, K., Cytokines as mediators in the regulation of the hypothalamic-pituitary-adrenocortical function, *J. Endocrinol. Invest.*, 16, 141, 1993.
5. Besedovsky, H., del Rey, A., and Sorkin, E., The immune response evokes changes in brain noradrenergic neurons, *Science*, 221, 564, 1983.
6. Blalock, J. E., The syntax of immune-neuroendocrine communication, *Immunol. Today*, 15, 504, 1994.
7. Stratakis, C. A. and Chrousos, G. P., Neuroendocrinology and pathophysiology of the stress system, in *Stress: Basic Mechanisms and Clinical Applications*, Chrousos, G. P., McCarty, R., Pacák, K., Cizza, G., Sternberg, E., Gold, P. W., and Kvetnansky, R., Eds., New York Academy of Sciences, New York, 1995.
8. Kvetnansky, R., Dobrakovoda, M., Jezova, D., Oprsalova Z., Lichardus, B., and Makara, G., Hypothalamic regulation of plasma catecholamine levels during stress: effect of vasopressin and CRF, in *Stress, Neurochemical and Humeral Mechanisms*, Vol. 2, Van Loon, G. R., Kvetnansky, R., McCarty, R., and Axelrod, J., Eds., Gordon and Breach, Newark, NJ, 1989, 549.
9. Song, C., Early, B., and Leonard, B., Behavioral, neurochemical, and immunological responses to CRF administration, in *Stress, Basic Mechanisms and Clinical Implications*, Chrousos, G. P., McCarty, R., Pacak, K., Cizza, G., Sternberg, E., Gold, P. W., and Kvetnansky, R., Eds., New York Academy of Sciences, New York, 1995.
10. Kvetnansky, R., Pacak, K., Fukuhara, K., Viskupic, E., Hiremagalur, B., Nankova, B., Goldstein, D. S., Sabban, E. L., and Kopin, I. J., Sympathoadrenal system in stress: interaction with the hypothalamic-pituitary-adrenocortical system, in *Stress, Basic Mechanisms and Clinical Implications*, Chrousos, G. P., McCarty, R., Pacak, K., Cizza, G., Sternberg, E., Gold, P. W. and Kvetnansky, R., Eds., New York Academy of Sciences, New York, 1995.
11. LeMay, L. G., Vander, A. J., and Kluger, M. J., The effects of psychological stress on plasma interleukin-6 activity in rats, *Physiol. Behav.*, 47, 957-961, 1990.
12. Zhou, D., Kusnecov, A. W., Shurin, M. R., De Paoli, M., and Rabin, B., Exposure to physical and psychological stressors elevates plasma interleukin-6: relationship to the activation of hypothalamic-pituitary-adrenal axis, *Endocrinology*, 133, 2523, 1993.
13. Soszynski, D., Kozak, W., Conn, C. A., Rudolph, K., and Kluger, M. J., β-adrenoceptor antagonists suppress elevation in body temperature and increase plasma IL-6 in rats exposed to open field, *Neuroendocrinology*, 63, 459, 1996.
14. Minami, M., Kuraishi, Y., Yamaguchi, T., Nakai, S., Hirai, Y., and Satoh, M., Immobilization stress induces interleukin-1 β mRNA in the rat hypothalamus, *Neurosci. Lett.*, 25, 254, 1991.

15. Dugué, B., Leppanen, E. A., Teppo, A. M., Fyrquist, F., and Gräsbeck, R., Effects of psychological stress on plasma interleukins-1 and -β and -6, C-reactive protein, tumour necrosis factor α, anti-diuretic hormone and serum cortisol, *Scand. J. Clin. Lab. Invest.*, 56, 555, 1993.
16. Maes, M., Song, C., Lin, A., Gabriels, L., De Jongh, R., Van Gastel, A., Kenis, G., Bosmans, E., DeMeester, I., Benoyt, I., Neels, H., Demedts, P., Janca, A., Scharpé, S., and Smith, R. S., The effects of psychological stress on humans: increased production of proinflammatory cytokines and a Th-1-like response in stress-induced anxiety, *Cytokine*, 10, 313, 1998.
17. Spielberger, C. D., Edwards, C. D., Mantoun, J., and Lushene, R. E., *The State-Trait Inventory*, NFER-Nelson, Windsor, 1987.
18. Beck, A. T., Rial, W. Y., and Rickels, K., Short form of depression inventory: cross validation, *Psychol. Rep.*, 34, 1184, 1974.
19. De Groote, D., Zangerle, P. F., Gevaert, Y., Fassotte, M. F., Beguin, Y., Noizat-Pirenne, F., Pirenne, J., Gathy, R., Lopez, M., Dehart, I., Igot, D., Baudrihaye, M., Delacroix, D., and Franchimont, P., Direct stimulation of cytokines (IL-1β, TNF-α, IL-6, IL-2, IFN-γ and GM-CSF) in whole blood. I. Comparison with isolated PBMC stimulation, *Cytokine*, 4, 239, 1992.
20. Howell, D. C., *Statistical Methods for Psychology*, Duxbury Press, Boston, 1982.
21. Smith, R. S., The macrophage theory of depression, *Med. Hypoth.*, 35, 298, 1991.
22. Gutterman, J. U., Fein, S., Quesada, J., Horning, S. J., Levine, J. F., Alexanian, R., Bernhardt, L., Kramer, M., Spiegel, H., Colburn, W., Trown, P., Merigan, T. and Dziewanowski, Z., Recombinant leukocyte A interferon: pharmacokinetics, single dose tolerance, and biologic effects in cancer patients, *Ann. Intern. Med.*, 96, 549, 1982.
23. Weinberg, S. B., Schulteis, G., Fernando, A. G., Bakhit, C., and Martinez, J. L., Decreased locomotor activity produced by repeated, but not single, administration of murine-recombinant interferon-γ in mice, *Life Sci.*, 42, 1085, 1988.
24. Maes, M., Scharpé, S., Meltzer, H. Y., Okayli, G., Bosmans, E., D'Hondt, P., Vanden Bossche, B., and Cosyns, P., Increased neopterin and interferon-γ secretion and lower availability of L-tryptophan in major depression: further evidence for activation of cell-mediated immunity, *Psychiatr. Res.*, 54, 143, 1994.
25. Seidel, A., Arolt, V., Hunstiger, M., Rink, L., Behnisch, A., and Kirchner, H., Increased CD56+ natural killer cells and related cytokines in major depression, *Clin. Immunol. Immunopathol.*, 78, 83, 1996.
26. Duch, D. S., Woolf, J. H., Nichol, C. A., Jonathan, R. D., and Garbutt, J. C., Urinary excretion of biopterin and neopterin in psychiatry disorders, *Psychiatr. Res.*, 11, 83, 1983.
27. Anderson, D. N., Abou-Saleh, M. T., Collins, J., Hughes, K., Cattell, R. J., Hamon, C. G. B., Blair, J. A. and Dewey, M. E., Pterin metabolism in depression: an extension of the amine hypothesis and possible marker of response to ECT, *Psychol. Med.*, 22, 863, 1992.
28. Dunbar, P. R., Hill, J., Neale, T. J., and Mellsop, G. W., Neopterin measurement provides evidence of altered cell-mediated immunity in patients with depression, but not with schizophrenia, *Psychol. Med.*, 22, 1051, 1992.
29. Bonaccorso, S., Lin, A., Verkerk, R., Van Hunsel, F., Libbrecht, I., Scharpé, S., DeClerck, L., Biondi, M., Janca, A., and Maes, M., Immune markers in fibromyalgia: comparison with major depressed patients and normal volunteers, *J. Affect. Disord.*, 48, 75, 1998.

30. Murray, H. W., Interferon-γ, the activated macrophage, and host defense against microbial challenge, *Ann. Intern. Med.*, 108, 595, 1988.
31. Cheung, D. L., Hart, P. H., Vitti, G. F., Whitty, G. A., and Hamilton, J. A., Contrasting effects of interferon-γ and interleukin-4 on the interleukin-6 activity of stimulated human monocytes, *Immunology*, 71, 70, 1990.
32. Leclerc, C., Interleukine-10, in *Les Cytokines*, Cavaillon, J.-M., Ed., Masson, Paris, 1996, 241.
33. Roux-Lombard, P., Modoux, C., and Dayer, J. M., Production of interleukin-1 (IL-1) and a specific IL-1 inhibitor during human monocyte-macrophage differentiation: influence of GM-CSF, *Cytokine*, 1(1), 45, 1989.
34. Eisenberg, S. P., Evans, R. J., Arend, W. P., Verderber, E., Brewer, M. T., Hannum, C. H., and Thompson, R. C., Primary structure and functional expression from complementary DNA of a human interleukin-1 receptor antagonist, *Nature*, 343, 341, 1990.
35. Hannum, C. H., Wilcox, C. J., Arend, W. P., Joslin, F. G., Dripps, D. J., Heimdal, P. L., Armes, J. G., Sommer, A., Eisenberg, S. P., and Thompson, R. C., Interleukin-1 receptor antagonist activity of a human interleukin-1 inhibitor, *Nature*, 343, 336, 1990.
36. Luheshi, G., Miller, A. J., Brouwer, S., Dascombe, M. J., Rothwell, N. J., and Hopkins, S. J., Interleukin-1 receptor antagonist inhibits endotoxin fever and systemic interleukin-6 induction in the rat, *Am. J. Physiol.*, 270, E91, 1996.
37. Lefevre, F., Charley, B., and La Bonnardiere, C., Interferon γ, in *Les Cytokines*, Cavaillon, J.-M., Ed., Masson, Paris, 1996, 320.
38. Tomkinson, B. E., Brown, M. C., Ip, S. H., Carrabis, S., and Sullivan, J. L., Soluble CD8 during T cell activation, *J. Immunol.*, 142, 2230, 1989.
39. Wijngaard, P. L., Van der Meulen, A., Gmelig Meyling, F. H., De Jonge, N., and Schuurman, H. J., Soluble CD8 and CD25 in serum of patients after heart transplantation, *Clin. Exp. Immunol.*, 97, 505, 1994.
40. Cavaillon, J.-M., Interleukin-6, in *Les Cytokines*, Cavaillon, J.-M., Ed., Masson, Paris, 1996, 184.
41. Wolf, M., Klug, J., Hackenberg, R., Gessler, M., Grzeschik, K. H., Beato, M., and Suske, G., Human CC10, the homologue of rabbit uteroglobin: genomic cloning, chromosomal localization and expression in endometrial cell lines, *Hum. Mol. Genet.*, 1, 371, 1992.
42. Singh, G., Katyal, S. L., Brown, W. E., Kennedy, A. L., Singh, U., and Wong-Chong, M. L., Clara cell 10 kDa protein (CC10): comparison of structure and function to uteroglobin, *Biochim. Biophys. Acta*, 1039, 348, 1990.
43. Dierynck, I., Bernard, A., Roels, H., and De Ley, M., Potent inhibition of both interferon-γ production and biologic activity by the Clara cell protein CC16, *Am. J. Respir. Cell. Mol. Biol.*, 12, 205, 1995.
44. Maes, M., Scharpé, S., Cosyns, P., Cooreman, W., Neels, H., and Wouters, A., Components of biological variation in plasma haptoglobin: relationships to plasma fibrinogen and immune variables, including interleukin-6 and its receptor, *Clin. Chim. Acta*, 239, 23, 1995.
45. Maes, M., Bosmans, E., De Jongh, R., Kenis, G., Vandoolaeghe, E., and Neels, H., Increased serum IL-6 and IL-1 receptor antagonist concentrations in major depression and treatment resistant depression, *Cytokine*, 9, 853, 1997.
46. Maes, M., Scharpé, S., Meltzer, H. Y., et al., Relationships between interleukin-6 activity, acute phase proteins and HPA-axis function in severe depression, *Psychiatr. Res.*, 49, 11, 1993.

47. Singh, V. K. and Leu, S.-J. C., Enhancing effect of corticotropin-releasing neurohormone on the production of interleukin-1 and interleukin-2, *Neurosci. Lett.*, 120, 151, 1990.
48. Pereda, M. P., Sauer, J., Castro, C. P., Finkielman, S., Stalla, G. K., and Holsboer, F., Corticotropin-releasing hormone differentially modulates the interleukin-1 system according to the level of monocyte activation by endotoxin, *Endocrinology*, 136, 5504, 1995.
49. Leu, S.-J. C. and Singh, V. K., Stimulation of interleukin-6 production by corticotropin-releasing factor, *Cell. Immunol.*, 143, 220, 1992.
50. Angioni, S., Petraglia, F., Gallinelli, A., Cossarizza, A., Franceschi, C., Muscettola, M., Genazzani, A. D., Surico, N., and Genazzani, A. R., Corticotropin-releasing hormone modulates cytokines release in cultured human peripheral blood mononuclear cells, *Life Sci.*, 53, 1735, 1993.
51. Perez, L. and Lysle, D. T., Corticotropin-releasing hormone is involved in conditioned stimulus-induced reduction of natural killer cell activity but not in conditioned alterations in cytokine production or proliferative responses, *J. Neuroimmunol.*, 63, 1, 1995.

4 Depression and Immunity: The Role of Cytokines

Ziad Kronfol

CONTENTS

4.1 Introduction 51
4.2 Rationale for Studying Cytokines in Major Depression 52
    4.2.1 Cytokines and Stress 52
    4.2.2 Cytokines and Hypothalamic-Pituitary-Adrenal Axis Activity 53
    4.2.3 Cytokines and Neurovegetative Features of Depression 54
4.3 Cytokines and Major Depression 55
    4.3.1 Plasma Concentrations of Cytokines and Related Soluble Factors in Major Depression 55
    4.3.2 Cytokine Production in Depressed Patients 56
    4.3.3 Effects of Antidepressant Treatment 56
4.4 Conclusion 57
References 58

4.1 INTRODUCTION

Cytokine research has grown very rapidly over the last several years. While initially limited to investigations in the field of immunology, this research is now of interest to a growing number of related fields including oncology, epidemiology, pharmacology, and the neurosciences. This entire volume is a testimony to the shared interest and enthusiasm that immunologists and neuroscientists have in this field.

    The fact that cytokines serve as messengers among both immune cells and nerve cells has sparked the interest and curiosity of investigators in clinical neuroscience. Neurologists are studying possible links between specific cytokines and clinical disorders such as multiple sclerosis and Alzheimer's disease. Psychiatrists are exploring the possibility that schizophrenia and affective disorders also may be associated with altered regulation of cytokine function. This chapter is dedicated to a review of recent developments in cytokine research as they relate to affective disorders. We will start by presenting the rationale for studying cytokines in major depression based on clinical and neuroendocrine features of this disorder. We will then review published investigations assessing the role of cytokines in major depression and also address the issue of cytokines and antidepressant therapy. We will conclude with a general summary on the state of cytokine research in affective disorders and an invitation for more rigorous methodology in future studies.

4.2 RATIONALE FOR STUDYING CYTOKINES IN MAJOR DEPRESSION

Major depression is among the most common psychiatric disorders in the U.S. It is estimated that the lifetime prevalence of this disorder is 20% in women and 10% in men. Stated differently, one of every five women and one of every ten men will suffer a major depression at least once in their lifetime. The price, including healthcare needs, lost productivity, and human suffering is immeasurable; yet little is known about the exact etiology of this disorder and/or pathophysiological mechanism(s). While most investigators have focused their attention, perhaps rightly so, on specific neurotransmitters (e.g., norepinephrine or serotonin), or specific brain regions (e.g., hypothalamus, hippocampus, or prefrontal cerebral cortex), we and few others have put forward the hypothesis that cytokines play a role in the pathophysiology of major depression. First, depression is closely associated with stress and is often portrayed as an exaggerated response to stress on clinical, biological, and even experiential grounds.[1] Second, depression is accompanied by neuroendocrine changes, notably an increase in hypothalamic-pituitary-adrenal (HPA) axis activity.[2] Third, cardinal manifestations of depression include changes in neurovegetative functions such as sleep, appetite, and sex drive.[3] Since there is growing evidence that cytokines are involved in most of these major characteristics of depression, we think a study of cytokine regulation in major depression is warranted. The remainder of this section will therefore be devoted to a review of studies linking cytokines to stress, HPA-axis activity, and neurovegetative functions associated with major depression.

4.2.1 CYTOKINES AND STRESS

The effects of stress on immune function in general, and cytokine production in particular, have been well documented. Weiss and colleagues,[4] using a standard stress paradigm (intermittent tail shocks previously described by Keller et al.)[5] found that the *in vitro* production of both interleukin-2 (IL-2) and of interferon (IFN) by lymphocytes from stressed animals were significantly reduced in comparison to nonstressed animal controls. Furthermore, the expression of IL-2 receptors was diminished in lymphocytes from the animals that had been stressed compared to the nonstressed controls. The effects of stress were still observed when the stressed animals had their adrenals removed prior to the stressor, showing that the stress effects are not necessarily mediated by adrenal hormones. Finally, the same group of investigators showed that, depending on the intensity of the stressors, both enhancement and suppression of cytokine production (in this case IL-2) could be obtained.[4]

In a different line of experiments, Le May and colleagues[6] have examined the effects of psychological stress on plasma interleukin-6 (IL-6) activity in rats. The stressful paradigm here consisted of exposure to an open field and as such did not involve any physical stimulus. Plasma IL-6 activity was significantly more elevated among rats exposed to the open field. There was also a significant positive correlation between changes in plasma IL-6 activity and body temperature. The authors conclude that exposure to psychological stress can elevate the plasma concentration of IL-6, a known mediator of the acute phase response.[6] With regard to the differential effects of stress on specific cytokines, Persoons and colleagues[7] noted that animals exposed to an acute stress paradigm (inescapable mild footshock) had a marked increase in the secretion of IL-1 β and TNF-α from alveolar macrophages in comparison to nonstressed controls. The secretion of IL-6, however, did not differ between the two groups. The results again support the notion that stress can differentially affect cytokine secretion, at least in the case of isolated alveolar macrophages. The authors discuss their findings in light of clinical implications, including the possible contribution to asthma.[7]

With regard to human studies, several investigators have assessed the effects of different forms of human stress on immune function in general and cytokine regulation in particular.[8-11] Several review articles have also addressed this topic.[12] Thus, the effects of such stressors as bereavement,[8] unemployment,[9] marital discord,[10] and caring for an elderly demented relative[11] on various immunological phenomena have recently been investigated. Glaser and colleagues have conducted extensive research in this area. In a series of well-designed experiments,[13,14] they compared various immunological parameters in the same medical students the morning of an important school exam (stressful condition) and the morning of a regular school day (control condition). Compared to the control conditions, stressful situations were associated with an alteration in several immune parameters including NK cell activity, IL-2 production, and IFN production.[13] The same authors showed that the effects of stress could be observed at the mRNA level, suggesting that these effects could be traced back to the level of gene expression.[14]

4.2.2 Cytokines and Hypothalamic-Pituitary-Adrenal Axis Activity

There is also substantial evidence that cytokines are closely associated with hormone secretion, particularly hormones of the HPA axis. Ever since the initial reports showing that IL-1 can initiate the cascade of HPA-axis hormones,[15,16] numerous articles have addressed different aspects of the complex interactions between cytokines and neurohormones.[17,18] Several review articles have also covered this topic.[19,20] Cytokines thus have been shown to stimulate the release of HPA-axis hormones, eventually leading to the secretion of excessive amounts of glucocorticoids. Increased glucocorticoid production then acts as a negative feedback mechanism that tends to suppress an otherwise exaggerated and out-of-control immune or inflammatory reaction brought about by a chain of cytokines. This delicate balance between suppressor factors and enhancer factors, which seems to be operative under optimal physiological conditions, is not always easy to achieve. A breakdown in homeostasis could lead to severe infection or cancer on the one hand, or allergy or auto-immune disorders on the other. It is also now clear that the brain plays an important role in maintaining this balance, and that stressful phenomena can interfere with this process. The clinical implications, as will be discussed later in this chapter, are that emotions and psychiatric disorders can affect immune regulation and thus play a role in the etiology and course of immune-related disorders such as infectious diseases and malignancy.

There is also evidence that under certain conditions immune cells are capable of secreting neurohormones such as adrenocorticotrophic hormone (ACTH), endorphins, and enkephalins.[21] Conversely, cells of the nervous system, particularly glial cells, may be involved in the production of specific cytokines.[22] This obviously raises some interesting questions regarding the proper definition of a hormone and its distinction from a cytokine.

4.2.3 Cytokines and Neurovegetative Features of Depression

We shall now turn to specific neurovegetative functions such as sleep and appetite and review the role of cytokines in the regulation of these phenomena. Sleep and appetite disturbances are cardinal features of major depression, particularly of the melancholic subtype.[3] There is now a good deal of evidence suggesting that specific cytokines are involved in the regulation of sleep. Briefly stated, specific cytokines seem to possess somnogenic properties when injected centrally or systemically to different animal species, suggesting a role in physiological sleep regulation.[23] Cytokines such as IL-1, tumor necrosis factor (TNF), and IFN-α are somnogenic regardless of the route of administration while other cytokines like IL-2, IL-6, and IFN-β lack somnogenic activity.[24] These authors conclude that microbial-cytokine alterations in sleep probably result from an amplification of physiological mechanisms which involve cytokines, several neuropeptides, and neurotransmitters including serotonin, acetylcholine, and nitric oxide, and possibly other factors.[24]

Another neurovegetative function which is often disturbed in major depression, and possibly regulated by cytokine, is appetite. Anorexia is common both to major depression on the one hand and infectious processes and cancer on the other. Recent evidence produced by Plata-Salaman and colleagues[25] indicates that IL-1β, IL-6, IL-8, and TNF-α induce anorexia when administered intracerebroventricularly at doses comparable to pathophysiological concentrations. Sarraf et al.[26] reported that inflammatory cytokines such as IL-1 and TNF that produce anorexia also increase serum leptin levels and leptin mRNA expression in mice, while cytokines not known to induce anorexia, such as IL-2, IL-4, and IL-10, had no effect on leptin level or gene expression. They conclude that leptin levels may be one mechanism by which anorexia is produced during acute inflammation. Perhaps more indicative of a role for cytokines in anorexia are the reports of dysregulated cytokine production in anorexia and bulimia nervosa. Bessler et al.[27] found that IL-2 synthesis was significantly reduced in patients with anorexia nervosa and that this phenomenon was accompanied by an enhanced stimulatory activity of the patients' sera on the production of this cytokine by lymphocytes from healthy subjects.

In summary, depression is associated with stress and is frequently accompanied with HPA-axis abnormalities, particularly hypercortisolemia and escape from suppression by dexamethosone. Cardinal features of depression include neurovegetative disturbances in sleep, appetite, and sex drive. Most, if not all, of these characteristics of major depression have also been associated with a dysregulation in cytokines physiology. Let us now turn to the evidence of cytokine disturbance in major depression.

4.3 CYTOKINES AND MAJOR DEPRESSION

There have been very few published studies devoted to a comprehensive evaluation of cytokine regulation in major depression. This is somewhat surprising in view of the compelling argument (summarized above) linking cytokines to various psychobiological features of depression, including increased HPA axis activity, sleep and appetite disturbances, and similarities between major depression and an exaggerated "stress response." Furthermore, many of the studies in print lack a methodological rigor that would allow for a firm conclusion issue to be addressed.

In reviewing these studies, we will first consider plasma concentrations of various cytokines and related soluble factors in patients with major depression, then focus on studies addressing cytokine production and, finally, review the effects of antidepressant treatment on cytokine regulation.

4.3.1 Plasma Concentrations of Cytokines and Related Soluble Factors in Major Depression

While assessing immune function in patients with major depression, investigators have found various aberrations in cytokines and related soluble factors in the plasma of these patients. We have first noted a significant elevation in the plasma concentrations of complement factors C_3 and C_4 in depressed patients compared to normal controls.[28] Both C_3 and C_4 levels were negatively correlated with lymphocyte mitogenic responses, suggesting the possibility of an association between depression, an acute phase response, and reduced lymphocyte activity. In a related experiment, we compared lymphokine-activated killer (LAK) cell activity in patients with major depression and matched healthy controls.[29] Even when recombinant IL-2 was added to the cultures to generate LAK cells, LAK cell activity was still significantly lower in the depressed group, suggesting that a deficiency in IL-2 production cannot solely explain the defect seen in the lymphocytes of these patients.

With regard to plasma concentrations of cytokines, Maes and colleagues reported significant elevations in IL-1 β and IL-6 in the plasma of depressed patients.[30,31] They also reported significant increases in the plasma concentrations of IL-1 receptor antagonists,[32] soluble IL-2 receptors (sIL-2R), and soluble IL-6 receptors (sIL-6R)[31] in depressed patients compared to controls. They further noted significant elevations in the plasma concentrations of haptoglobin and significant reductions in the plasma concentrations of transferrin in the depressed group.[33] These authors postulate a significant association between HPA-axis hormones, acute phase proteins, cytokine concentrations, and severity of depressive symptoms.[33] Similar findings have been reported by other investigators.[34,35] The clinical significance of these findings in depressed patients, however, remains unclear.

4.3.2 CYTOKINE PRODUCTION IN DEPRESSED PATIENTS

Reports on cytokine production in patients with major depressive disorder are rare indeed. Furthermore, published reports often seem to produce contradictory results. Maes and colleagues compared IL-1β production by PHA-stimulated peripheral blood mononuclear cells (PBMC) in 28 depressed patients and 10 normal controls.[30] They found supernatant IL-1β levels to be significantly higher in the depressed group. They also found a significant positive correlation between IL-1β production and post-dexamethasone suppression test cortisol values, suggesting that IL-1β may be a contributor to HPA-axis hyperactivity in major depression. Maes and colleagues[33] also measured IL-6 production in culture supernatants of PHA-stimulated PBMC in 24 depressed inpatients and 8 healthy control subjects. They found IL-6 activity to be significantly elevated in depressed patients, but only those of the melancholic subtype. Furthermore, the authors found a significant positive correlation between IL-6 production, haptoglobin concentrations, and post-dexamethasone cortisol values. The authors conclude there is an association in depressed patients between elevated IL-6 production, elevated haptoglobin levels, and HPA-axis activity.[33] Weizman and colleagues[36] compared PBMC production of IL-1β, IL-2, and IL-3 -like activity (IL-3-LA) in 10 patients with major depression and 10 matched healthy controls. They noted a significant reduction in the production of all three cytokines in the untreated depressed group. Guidi and co-investigators[37] studied immune regulation in 26 institutionalized aged patients. They found the levels of IL-2, IL-4, and IFN in PHA-stimulated lymphocyte culture supernatants to be significantly lower among the elderly subjects compared to adult controls. They also found a negative association between cytokine production and depression scores. The reason for the discrepancy in results between these studies is not clear. It is worth noting that Maes and colleagues used PHA to stimulate the cells for IL-1 production[30] while Weizman and his group used LPS.[36] Thus, the cells recruited for the different studies may be different. Likewise, the clinical characteristics of the depressed patients may not have been the same. Maes and colleagues studied patients with major depression, dysthymia, and/or adjustment disorder with depressed mood, while Weizman and associates restricted their study group to patients with major depression. More studies with larger, more homogeneous groups using standardized state of the art methodology should be conducted to resolve this issue.

4.3.3 EFFECTS OF ANTIDEPRESSANT TREATMENT

Studies reporting the effects of antidepressant treatment on immune function in general, and cytokine regulation in particular, are even more difficult to find. This is due in part to the difficulties inherent to conducting follow-up studies, major methodological issues such as differentiating between a pure drug (antidepressant) effect and the effect of recovery from depression (i.e., the state vs. trait phenomenon), and the conflicting data regarding the effects of depression as such. In 1985, Albrecht and colleagues[38] reported lack of significant differences in mitogen-induced lymphocyte proliferation between depressed patients and normal controls. They noted, however, that all forms of somatic therapy for depression were associated with a decrease in lymphocyte mitogenic activity. Irwin and colleagues[39] found the opposite to be true with regard to natural killer (NK) cell activity. Depressed patients showed impairment in NK cell activity during the acute phase of depression, but the NK activity returned to normal (control values) following treatment.

With regard to cytokine regulation, Maes and colleagues[31] showed that plasma concentrations of IL-6, sIL-$_6$R, sIL-$_2$R, and Transferrin receptors (TfR) in depressed patients did not significantly differ during the acute phase of the illness and following treatment with antidepressant medications (tricyclic antidepressants or selective serotonin-reuptake inhibitors). These values were also similar to those obtained on drug-free depressed patients in remission. The authors conclude that these soluble immune factors may represent a trait marker for major depression.[31] With regard to cytokine production, Weizman and colleagues[36] found that the secretion of IL-1β and IL-3-LA were reduced in major depression and that treatment with clomipramine, a tricyclic antidepressant medication, significantly increased the production of these cytokines.[36]

Electroconvulsive therapy ECT is another commonly used somatic treatment for depression. After more than 50 years since it was first introduced, ECT remains one of the most powerful and most effective treatments for depression. The effects of ECT on lymphocyte reactivity have only recently been investigated. Data from our own laboratory show that ECT produces significant, though transient, increases in NK cell activity.[40] ECT has also been shown to transiently and significantly increase plasma levels of IL-6.[41] Again, the clinical significance of these findings remains to be determined.

In summary, data regarding the effects of antidepressant treatment on immune function in general, and cytokine regulation in particular, remain very preliminary and inconclusive. One difficulty in interpreting the results is the distinction between possible direct immunological properties of the treatment itself and recovery from depression. Another difficulty is the distinction between the short-term (acute) effects of a given treatment modality and the long-term (chronic) effects of the treatment. Evidently more studies are needed to address these issues.

4.4 CONCLUSION

In conclusion, a great deal of evidence presently supports the notion that mood states, stressful life events, and psychiatric disorders are closely interconnected with the neuroendocrine and immune systems. There is also a growing appreciation for the role that cytokines play in these interactions. Cytokines are secreted by various immune and, possibly, nerve cells. They act on a variety of specialized tissues and organs. Their mechanisms of action are highly complex and tightly regulated. It is therefore no surprise that a number of medical disorders could be linked to cytokine dysregulation. In the case of major depression, although the details remain sketchy, it seems very likely that various cytokines play a role in the pathophysiology of this disorder. The challenge ahead is to translate these ideas into meaningful research protocols and to move from pilot studies to more comprehensive and definitive investigations.

REFERENCES

1. Calabrese, J., Kling, M., and Gold, P., Alterations in immunocompetence during stress, bereavement, and depression: focus on neuroendocrine regulation, *Am. J. Psychiatry*, 144, 1123, 1987.
2. Carroll, B. J., Feinberg, M., Greden, J., Tarika, J., Albala, A., Haskett, R., James, N., Kronfol, Z., Lohr, N., Steiner, M., de Vigne, J. P, and Young, E., A specific laboratory test for the diagnosis of melancholia, *Arch. Gen. Psychiatry*, 38, 15, 1981.
3. APA, *Diagnostic and Statistical Manual of Mental Disorders,* 4th ed., American Psychiatric Association, Washington, D.C., 1994, p 317.
4. Weiss, J. M., Sundar, S. K., Becker, K. J., and Cierpial, M. A., Behavioral and neural influences on cellular immune responses: effects of stress and interleukin-1, *J. Clin. Psychiatry*, 50 (Suppl. 5), 43, 1989.
5. Keller, S., Weiss, J., Schleifer, S., Miller, N., and Stein, M., Suppression of immunity by stress: effects of a graded series of stressors on lymphocyte stimulation in the rat, *Science*, 213, 1397, 1981.
6. LeMay, L. B., Vander, A. J., and Kluger, M. J., The effects of psychological stress on plasma interleukin-6 activity in rats, *Physiol. Behav.*, 47, 957, 1990.
7. Persoons, J. H., Schornagel, K., Breve, J., Berkenbosch, F., and Kraal, G., Acute stress affects cytokines and nitric oxide production by alveolar macrophages differently, *Am. J. Respir. Crit. Care Med.*, 152, 619, 1995.
8. Bartrop, R. W., Luckhurst, E., Lazarus, L., Kiloh, L. G., and Penney, R., Depressed lymphocyte function after bereavement, *Lancet*, 1, 834, 1977
9. Arnetz, B., Wasserman, J., Petrini, B., Brenner, S., Levi, L., Eneroth, P., Salovaara, H., Hjelm, R., Salovarra, L., Theorell, T., and Patterson, I., Immune function in unemployed women, *Psychosom. Med.*, 49, 3, 1987.
10. Kiecolt-Glaser, J. K., Kennedy, S., Malkoff, S., Fisher, L., Speicher, C. E., and Glaser, R., Marital discord and immunity in males, *Psychosom. Med.*, 51, 195, 1989.
11. Kiecolt-Glaser, J. K., Glaser, R., Dyer, C., Shuttleworth, E., Ogrocki, P., and Speicher, C. E., Chronic stress and immunity in family caregivers of Alzheimer's disease victims, *Psychosom. Med.*, 49, 523, 1989.
12. Hillhouse, J. E., Kiecolt-Glaser, J. K., and Glaser, R., Stress-associated modulation of the immune response in humans, in *Stress and Immunity*, Plotnikoff, N., Murgo, A., Faith, R., and Wybran, J., Eds., CRC Press, Boca Raton, FL, 1991, p 3.
13. Glaser, R., Rice, J., Speicher, C. E., Stout, J. C., and Kiecolt-Glaser, J. K., Stress depresses interferon production by leukocytes concomitant with a decrease in NK cell activity, *Behav. Neurosci.,* 100, 675, 1986.
14. Glaser, R., Kennedy, S., Lafuse, W., Bonneau, R., Speicher, C., Hillhouse, J., and Kiecolt-Glaser, J. K., Psychological stress-induced modulation of interleukin-2 receptor gene expression and interleukin-2 production in peripheral blood leukocytes, *Arch. Gen. Psychiatry*, 47, 707, 1990.
15. Sapolsky, R., Rivier, C., Yamamoto, G., Plotsky, P., and Vale, W. W., Interleukin-1 stimulates the secretion of hypothalamic corticotropin-releasing factor, *Science*, 238, 522, 1987.
16. Bernton, E., Beach, J., Holaday, J., Smallridge, R., and Fein, H. G., Release of multiple hormones by a direct action of interleukin-1 in pituitary cells, *Science*, 238, 519, 1987.
17. Wilder, R. l., Neuroendocrine-immune system interactions and autoimmunity, *Annu. Rev. Immunol.*, 13, 307, 1995.

18. Kapcala, L. P., Chautard, T., and Eskay, R. L., The protective role of the hypothalamic-pituitary-adrenal axis against lethality produced by immune, infectious, and inflammatory stress, *Ann. NY Acad. Sci.*, 771, 419, 1995.
19. Bateman, A., Singh, A., Krol, T., and Solomon, S., The immune hypothalamic-pituitary-adrenal axis, *Endocr. Rev.*, 10, 92, 1989.
20. Buzzetti, R., McLoughlin, L., Scara, D., and Rees, L. H., A critical assessment of interactions between the immune system and the hypothalmic-pituitary-adrenal axis, *J. Endocrinol.*, 120, 183, 1989.
21. Smith, E. M. and Blalock, J. E., Human leukocyte production of corticotropin and endorphin-like substances: association with leukocyte interferon. *Proc. Natl. Acad. Sci. U.S.A.*, 78, 7530, 1981.
22. Faggioni, R., Benigni, F., and Ghezzi, P., Proinflammatory cytokines as pathogenetic mediators in the central nervous system: brain-periphery connections, *Neuroimmunomodulation*, 2, 2, 1995.
23. Grazia de Simoni, M., Imeri, L., De Matteo, W., Perego, C., Simard, S., and Terrazzino, S., Sleep regulation: interactions among cytokines and classical neurotransmitters, *Adv. Neuroimmunol.*, 5, 189, 1985.
24. Krueger, J. M. and Majde, J. A., Microbial products and cytokines in sleep and fever regulation, *Crit. Rev. Immunol.*, 14, 355, 1994.
25. Plata-Salaman, C. R., Sonti, G., Borkoski, J. P., and Wilson, C. D., Anorexia induced by chronic central administration of cytokines at estimated pathophysiological concentrations, *Physiol. Behav.*, 60, 867, 1996.
26. Sarraf, P., Frederich, R. C., Turner, E. M., Ma, G., Jaskowiak, N. T., Rivet, D. J., Flier, J. S., Lowell, B. B., Fraker, D. L., and Alexander, H. R., Multiple cytokines and acute inflammation raise mouse leptin levels: potential role in inflammatory anorexia, *J. Exp. Med.*, 185, 171, 1997.
27. Bessler, H., Karp, L., Nutti, I., Apter, A., Tyano, S., Djaldetti, M., and Weizman, R., Cytokine production in anorexia nervosa, *Clin. Neuropharmacol.*, 16, 937, 1993.
28. Kronfol, Z. and House, J. D., Lymphocyte mitogenesis, immunoglobulin and complement levels in depressed patients and normal controls, *Acta Psychiat. Scand.*, 80, 142, 1989.
29. Kronfol, Z., Nair, M. P. N., Goel, K., and Schwartz, S. A., Lymphokine-activated killer cell activity in psychiatric illness, in *Psychiatry and Biological Factors*, Kurstak, E., Ed., Plenum Press, New York, 1991, p 217.
30. Maes, M., Bosmans, E., Meltzer, H., Scharpé, S., and Suy, E., Interleukin-1β: a putative mediator of HPA axis hyperactivity in major depression? *Am. J. Psychiatry*, 150, 1189, 1993.
31. Maes, M., Meltzer, H., Bosmans, E., Bergmans, R., Vandoolaeghe, E., Ranjan, R., and Desnyder, R., Increased plasma concentrations of interleukin-6, soluble interleukin-6 receptor, soluble interleukin-2 receptor and transferrin receptor in major depression, *J. Affect. Disord.*, 34, 301, 1995.
32. Maes, M., Vandoolaeghe, E., Ranjan, R., Bosmans, E., Bergmans, R., and Desnyder, R., Increased serum interleukin-1-receptor antagonist concentrations in major depression. *J. Affect. Disord.*, 36, 29, 1995.
33. Maes, M., Scharpé, S., Meltzer, H., Bosmans, E., Suy, E., Calabrese, J., and Cosyns, P., Relationships between interleukin-6 activity, acute phase proteins and function of the hypothalamic-pituitary-adrenal axis in severe depression, *Psychiatry Res.*, 49, 11, 1993.
34. Joyce, P., Hawkes, C., and Mulder, R., Elevated levels of acute phase plasma proteins in major depression, *Biol. Psychiatry*, 32, 1035, 1992.

35. Kronfol, Z., Singh, V. J., Zhang, Q., Starkman, M., and Schteingart, D. E., Plasma cytokines, acute phase proteins and cortisol in major depression, *Biol. Psychiatry*, 37, 609A, 1995.
36. Weizman, R., Laor, N., Podliszewski, E., Notti, I., Djaldetti, M., and Bessler, H., Cytokine production in major depressed patients before and after clomipramine treatment, *Biol. Psychiatry*, 35, 42, 1994.
37. Guidi, L., Bartolomi, C., Frasca, D., Antico, L., Pili, R., Cursi, F., Tempestar, E., Rumi, C., Menini, E., and Carbonin, P., Impairment of lymphycyte activity in depressed aged subjects, *Mech. Aging Dev.*, 60, 13, 1991.
38. Albrecht., J., Helderman, J. H., Schlesser, M. A., and Rush, A. J., A controlled study of cellular immune function in affective disorders before and during somatic therapy, *Psychiatr. Res.*, 15, 185, 1985.
39. Irwin, M., Lacher, U., and Caldwell, C., Depression and reduced natural killer cytotoxicity: a longitudinal study of depressed patients and control subjects, *Psychol. Med.*, 22, 1045, 1992.
40. Kronfol, Z., Goel, K., Nair, M., and Yu, D., and Brown, M. B., Effects of ECT on natural killer cell activity in patients with major depression, *Biol. Psychiatry*, 41, 29S, 1997.
41. Kronfol, Z., LeMay, L., Nair, M., and Kluger, M., Electroconvulsive therapy increases plasma levels of interleukin-6, *Ann. NY Acad. Sci.*, 594, 463, 1990.

5 Bacterial Infections Due to Emerging Pathogens: Cytokines and Neuro-AIDS

Vito Covelli, Angela Bruna Maffione, Giuseppe Giuliani, Salvatore Pece, and Emilio Jirillo

CONTENTS

5.1 The Nervous System as an Immunological Organ 61
5.2 Emerging Pathogens in AIDS 62
5.3 Neuro-AIDS and the Cytokine Network 64
5.4 Concluding Remarks 65
References 65

5.1 THE NERVOUS SYSTEM AS AN IMMUNOLOGICAL ORGAN

Physiologically, the nervous system has the ability to regulate the immune response via several hormones [e.g., corticotrophin (ACTH), growth hormone (GH), luteinizing hormone (LH)] and certain neuropeptides [e.g., somatostatin (SOM), substance P (SP), vasoactive intestinal peptide (VIP)] which, when released at remote sites, reach immune cells localized within the various organs of the body.[1,2] On the other hand, immune cells possess receptors in order to receive neuroendocrine chemical signals. For instance, murine splenic mononuclear cells exhibit receptors for ACTH with about 3000 high- and 50,000 low-affinity sites per cell.[3] However, ACTH is also found in the immune system following activation with viruses, transformed cells, and bacterial lipopolysaccharides (LPS).[1,2] On the contrary, evidence has been provided that murine splenic macrophages (MO) as well as rat lymphocytes are able to produce ACTH constitutively.[4,5]

Also, endorphins derived from a precursor protein, proopiomelanocortin (POMC), are expressed in leukocytes along with ACTH, and this likely is due to their common derivation from POMC.[6] Many other hormones and neuropeptides (e.g., GH, LH, thyrotropin [TSH], SOM, SP, VIP) have been found in immune cells, thus supporting the idea that the immune system may function as a sensory organ for stimuli not recognized by the central and peripheral nervous systems.[7] For instance, immune cells (monocytes [M], macrophages [MO] and T cells) activated by bacterial, viral and tumor antigens, respectively, release hormones, neuropeptides, and cytokines (CKs) that when conveyed to the nervous system cause a physiologic modification. In this respect, in the course of infection the pattern of release of pituitary hormones is quite altered (e.g., increase in ACTH, prolactin, and GH production, and suppression in LH and TSH release).[7] Evidence has been provided that the altered pattern of hormone secretion is caused by peripheral CKs which act upon the hypothalamus and pituitary gland. These modifications imply penetration of CKs into the central nervous system (CNS) through areas where the blood-brain barrier (BBB) is more vulnerable, such as the median eminence, organum vasculosum, lamina terminalis, and choroid plexus.[8,9] Of note, an active transport system which carries interleukin-1 (IL-1) into the brain has been demonstrated.[9] In this regard, conflicting results have been reported on the ability of IL-1 to act at pituitary or hypothalamic sites, even if the identification of specific receptors on the those areas supports the view that IL-1 acts at both levels.[10-13]

As a further demonstration of the immune functions of the CNS, CKs are also produced by glial cells of the brain following LPS stimulation.[14] For instance, IL-6 is produced in the anterior pituitary by folliculostellate cells acting on brain and/or on pituitary receptors in order to generate certain effects, such as ACTH overproduction.[14,15] Furthermore, in humans IL-1β is produced by neurons located in the paraventricular nucleus, whose axons project to the median eminence.[16] This enables IL-1β molecules to reach the anterior lobe for their subsequent uptake by portal vessels. Finally, IL-1α is present in the dorsolateral preoptic area and anterior hypothalamus and its synthesis is increased by LPS.[14] Taken together, these findings suggest that under physiologic circumstances a bidirectional communication exists between the nervous and immune systems, an imbalance that, however, may lead to severe damage of the neural tissue, as in the case of immune-mediated cerebral pathologies.

In this chapter, emphasis will be placed on the pathogenic mechanisms responsible for neuro-acquired immunodeficiency syndrome (AIDS) with special reference to the role played by proinflammatory CKs released under stimulation of bacterial antigens. In this framework, the pathogenicity of some emerging bacteria isolated with increasing frequency from human immunodeficiency virus (HIV) patients will be outlined.

5.2 EMERGING PATHOGENS IN AIDS

AIDS patients undergo many infectious episodes of bacterial, fungal, parasitic, and viral origin which represent one of the main causes of death in these subjects.[17] With special reference to bacteria, over the past few years emerging organisms have been isolated from AIDS subjects and current studies deal with understanding the pathogenic mechanisms triggered by these agents. In general terms, CKs and other mediators released from immune cells following bacterial stimulation represent detrimental factors for the HIV host, as described in the next sections.

Besides clinical infections due to *Mycobacterium tuberculosis* (MTB), nontuberculous mycobacteria such as *Mycobacterium avium* complex (MAC) provoke disseminated infections in AIDS patients, the severity of which correlate with the degree of immunodepression.[18,19]

It is current opinion that both reduced functional capacities of MO and subversion of the CK network may allow intracellular survival of mycobacteria. In particular, evidence has been provided for a predominant Th2-type response in the course of AIDS, which may account for the release of IL-13 and IL-10.[20] The first CK inhibits IL-12 at the transcriptional level, while the latter plays a suppressive role and activates HIV replication in M and MO.[20] On the other hand, in HIV-infected individuals the diminished production of IL-12 by MO may justify the enhanced replication of intracellular organisms such as MTB and MAC. In fact, IL-12 leads to the release of interferon (IFN)-γ from both α/β and γ/δ T cells and NK cells.[21] In this regard, it is well known that IFN-γ potentiates MO killing activity in cooperation with tumor necrosis factor (TNF)-α.[21,22]

Another emerging pathogens is represented by *Bartonella henselae*, a Gram-negative bacterium which has been considered as the etiologic agent of cat scratch disease (CSD) and bacillary parenchymal angiomatosis (BPA).[23] BPA is a vasculo-proliferative disease whose lesions resemble those of Kaposi's sarcoma and is localized to skin and lymph nodes in AIDS patients. Recent reports emphasize the ability of *B. henselae* to survive and replicate intercellularly in professional phagocytes and *in vitro* data have demonstrated a reduced oxidative metabolism of normal human polymorpho nuclear (PMN) cells following their exposure to *B. henselae* organisms.[24,25] In addition, the demonstration of a low endotoxic potency of *B. henselae* LPS may represent another escape mechanism from the host immune control. It is likely that the chronic persistence of this bacterium into phagocytes, coupled with a low potency of LPS, may give rise to a slow but continuous release of inflammatory CKs in the host.

Rhodococcus equi is a Gram-positive coccobacillus of veterinary importance but has been reported as a pathogen for HIV-infected patients.[26,27] *R. equi* infections provoke severe chronic pyogranulomatous lesions, and systemic manifestations have been described in the immunocompromised host.[27] *R. equi*-killed organisms are potent inducers of proinflammatory CKs such as IL-8, IL-6, and TNF-α from normal human M,[28] while in MO from AIDS patients the killing capacities are profoundly depressed in response to the same bacterium.[29-31] Based on these concepts, it is likely that, analogous to LPS, bacterial products from *R. equi* (e.g., lipoarabinomannan) may cross the BBB in HIV-infected subjects, thus triggering in the CNS a series of pathological events mediated by the cerebral CK network (see Section 5.3).

Many other bacterial infections can occur in AIDS patients and among them *Salmonella*, *Shigella*, *Campylobacter* and *Legionella* organisms are frequently isolated as etiologic agents.

In terms of relationship between bacterial infections and CNS involvement, evidence has been provided that LPS may cause an exaggerated release of inflammatory CKs which pass through the BBB and reach the nervous system. In this respect, LPS is detectable in the plasma of HIV-infected subjects at any stage of the disease and its level very often correlates with an elevated serum amounts of IL-1β and TNF-α in the same patient.[32,33]

5.3 NEURO-AIDS AND THE CYTOKINE NETWORK

The mechanisms by which HIV establishes infection in the brain are still unclear. Undoubtedly, HIV has a neurotropism to astrocytes, microglial cells, and neurons, and these derived brain cells produce toxic cellular factors which could damage both myelin and neurons. However, CNS infection is also facilitated by other factors and, among them, one can include the toxic effect of viral proteins like Nef, Tat, GP 120, and Gp 41, autoimmune disorders, cytotoxicity, and other infections with viruses (e.g., herpes viruses) or with other bacteria.[34] Finally, one should not ignore the ability of HIV to infect endothelial cells, thus gaining access to the brain through the BBB.

Neurological complications represent one of the major clinical manifestations in the course of HIV infection and may occur at any stage of the disease. In the asymptomatic phase, a Guillan-Barrè-like syndrome has been diagnosed, while during the late stage a distal sensor axonal neuropathy may occur as result of the HIV infection and/or of the toxic effects of the antiretroviral therapy.[35,36] Finally, bacterial and viral meningitis are frequently observed, and in some cases it was also possible to isolate HIV from the CSF.[35,36] AIDS dementia complex (ADC) is the most frequent neurological complication of HIV infection. ADC is a subcortical dementia accompanied by cognitive behavioral and motor dysfunctions. Reduced concentration, mental slowing, forgetfulness, leg weakness, tremor, and personality modification are the most precocious symptoms.[37] Late clinical manifestations are represented by global dementia with psychomotor involvement, pyramidal tract alterations, incontinence, and myoclonus.

The subcortical white matter and deep cerebral nuclei are directly affected by HIV and demyelination represents the main pathological finding.[37] HIV is commonly found within perivascular and parenchymal MO and microglial cells as well as in multinucleated giant cells.[37] In this framework, there is evidence that both peripheral and cerebral CKs may play a pathogenic role in the course of ADC.[38]

As indicated in the previous sections, the exaggerated amount of proinflammatory CKs derived from PMN and M/MO is likely dependent on the LPS stimulus, as also is demonstrated by elevated levels of endotoxemia. At the same time, the impaired phagocytic and killing capacities of PMN and M/MO frequently observed in HIV subjects may facilitate penetration by other bacteria in the host, thus increasing the amount of bacterial products (e.g., LPS) and related CKs circulating in biological fluids. In this context, elevated amount of CSF TNF-α were assayed in ADC patients and correlated with the presence of high levels of anti-myelin basic protein (MBP) antibodies, while elevated levels of soluble MBP paralleled the severity of cerebral damage.[39,40] Accordingly, *in vitro* data demonstrated the demyelinating role of TNF-α on oligodendrocytes.[41]

Furthermore, TNF-α has been shown to augment the expression of the HIV in infected CD4+, CD8+, and B cells. In turn, depletion of HIV-infected T and B cells aggravates the immune deficit, allowing further entry of pathogens into the HIV-infected host. On the other hand, it is well known that in human brain several cell types are present such as CD4+ microglial cells, CD4+ M, and astrocytes that are able to release TNF-α, IL-1β, transforming growth factor-β (TGF), and macrophage inflammatory protein (MIP)-1α following HIV and/or LPS stimuli.[41,42] Therefore, according to the proposed pathogenic mechanism, human brain seems to represent a target organ for different mediators of both peripheral and cellular origin, triggered by both HIV and LPS, which may contribute to the development of ADC through a demyelinating process.

5.4 CONCLUDING REMARKS

Despite numerous studies on the CK network in the course of HIV infection, further aspects of phagocytic functions should be considered in the evolution of the disease. In particular, there is evidence that HIV-infected M induces enhanced expression of E-selectin and vascular cell adhesion molecule 1 (VCAM-1) on brain microvascular endothelial cells (BMVEC-1),[43] thus facilitating transendothelial migration of M into brain tissue. Furthermore, local liberation of TGF-β and MIP-1α contributes to the entry of other M into the brain. However, other CKs such as TNF-α and IL-1β enhance expression of HIV genes, E-selectin, and VCAM-1.[43]

It is worth mentioning that mediators of M/MO origin, such as platelet aggregating factor, quinolinate, and eicosanoids, have been detected in CSF from HIV patients with neurological disorders. In this context, it has been demonstrated that in HIV patients production of neurotoxin by M can be detrimental for the CNS, and this neurotoxic activity can be reduced with antagonists of the N-methyl-D-aspartate receptor.[44] This is one of three receptors for glutamate, a neurotransmitter involved in neural potentiation and depression.[44]

Conclusively, M/MO functions and related CK production are in an unstable balance in HIV infection and, therefore, therapeutic correction of these immune responses may reduce clinical complications, leading to a better quality of life and longer survival of the HIV-infected host. In this framework, emphasis should be placed on the capacity of methionine-enkephalin (a new CK) for enhancing immune responses in AIDS patients.[45] This clinical approach further supports the link between the nervous system and immune system.

REFERENCES

1. Blalock, J. E. and Smith, E. M., Human leukocyte interferon: structural and biological relatedness to adrenocorticotropic hormone and endorphins, *Proc. Natl. Acad. Sci. U.S.A.*, 77, 5972, 1980.
2. Smith, E. M. and Blalock, J. E., Human lymphocyte production of ACTH and endorphin-like substances: association with leukocyte interferon, *Proc. Natl. Acad. Sci. U.S.A.*, 78, 7530, 1981.
3. Johnson, H. M., Smith, E. M., Torres, B. A., and Blalock, J. E., Neuroendocrine hormone regulation of *in vitro* antibody production, *Proc. Natl. Acad. Sci. U.S.A.*, 79, 4171, 1982.

4. Lolait, S. J., Lim, A. T. W., Toh, B. H., and Funder, J. W., Immunoreactive β-endorphin in a subpopulation of mouse spleen macrophages, *J. Clin. Invest.*, 73, 277, 1984.
5. Endo, Y., Sakata, T., and Watanabe, S., Identification of proopiomelanocortin-producing cells in the rat pyloric antrum and duodenum by *in situ* mRNA-cDNA hybridization, *Biomed. Res.*, 6, 253, 1985.
6. Herbert, E., Discovery of proopiomelanocortin, cellular polyprotein, *Trends Biochem. Sci.*, 6, 184, 1981.
7. Blalock, J. E., Shared ligands and receptors as a molecular mechanism for communication between the immune and neuroendocrine systems, *Ann. NY Acad. Sci.*, 741, 292, 1994.
8. Hetier, E., Ayala, J., Denèfle, P., Bouseau, A., Rouget, P., Mallat, M., and Prochiantz, A., Brain macrophages synthesize interleukin-1 and interleukin-1 mRNAs *in vitro*, *J. Neurosci. Res.*, 21, 391, 1988.
9. Banks, W. A. and Kastin, A. J., Measurement of transport of cytokines across the blood-brain barrier, in *Neurobiology of Cytokines,* Part A, De Souza, E. B., Ed., Academic Press, San Diego, 1993.
10. Farrar, W., Hill, J., Hael-Bellan, A., and Vinocour, M., The immune logical brain, *Immunol. Rev.*, 100, 361, 1987.
11. Farrar, W. L., Kilian, P. L., Ruff, M. R., Hill, J. M., and Pert, C. B., Visualization and charecterization of interleukin-1 receptors in brain, *J. Immunol.*, 139, 459, 1987.
12. Haour, F. G., Ban, E. M., Milon, G. M., Baran, B., and Fillion, G. M., Brain-interleukin-1 receptors: characterization and modulation after lipopolysaccharide injection, *Prog. Neurobiol. Endocrinol. Immunol.*, 3, 196, 1990.
13. Takau, T., Tracey, D. E., Mitchell, W. M., and De Souza, E. B., Interleukin-1 receptors in mouse brain: characterization and neuronal localization, *Endocrinology*, 127, 3070, 1990.
14. Spangelo, B. L., MacLeod, R. M., and Isakson, P. C., Production of interleukin-6 by anterior pituitary cells *in vitro*, *Endocrinology*, 126, 582, 1990.
15. De Simoni, M. G., Sironi, M., De Luigi, A., Manfridi, A., Mantovani, A., and Ghezzi, P., Intracerebroventricular injection of interleukin-1 induces high circulating levels of interleukin-6, *J. Exp. Med.*, 171, 1773, 1990.
16. Breder, C. D., Dinarello, C., and Saper, C. B., Interleukin-1 immunoreactive innervation of the human hypothalamus, *Science*, 240, 321, 1988.
17. Stanley, S. K. and Fauci, A. S., Acquired immunodeficiency syndrome, in *Clinical Immunology: Principles and Practice*, Rich., R. R., Ed., Mosby-Year Book, St. Louis, 1996, 707.
18. Bloom, B. R. and Murray, C. J. L., Tuberculosis: commentary on remergent killer, *Science*, 257, 1055, 1992.
19. Horsburgh, C. R., Jr., *Mycobacterium avium* complex infection in the acquired immunodeficiency syndrome, *N. Engl. J. Med.*, 324, 1332, 1991.
20. Doherty, T. M., Interleukin-13. A review, *EOS J. Immunol. Immunopharmacol.*, 15, 81, 1995.
21. Mosman, T., Cytokines and immunoregulation, in *Clinical Immunology: Principles and Practice*, Rich, R. R. Ed., Mosby-Year Book, St. Louis, 1996, 217.
22. Flesch, I. E. A., Hess, J. H., Huang, S., Aguet, M., Rothe, J., Bluethmann, H., and Kaufmann, S. H., Early interleukin-12 production by macrophages in response to mycobacterial infection depends on interferon-γ and tumor necrosis factor-α, *J. Exp. Med.*, 181, 1615, 1995.

23. Regnery, R. L., Anderson, B. E., and Clarridge, J. E., III, Characterization of a novel Rochalimaea species, *R. henselae* sp. nov., isolated from blood of a febrile human immunodeficiency virus-positive patient, *J. Clin. Microbiol.*, 30, 265, 1992.
24. Zbinden, R., Hochly, M., and Nadal, D., Intracellular location of *Bartonella henselae* cocultivated with vero cells and used for an indirect fluorescent-antibody test, *Clin. Diagn. Lab. Immunol.*, 2, 693, 1995.
25. Fumarola, D., Pece, S., Fumarulo, R., Petruzzelli, R., Greco, B., Giuliani, G., Maffione, A. B., and Jirillo, E., Downregulation of human polymorphonuclear cell activities exerted by microorganisms belonging to α-2 subgroup of Proteobacteria (*Afipia felis* and *Rochalimaea henselae*), *Immunopharmacol. Immunotoxicol.*, 16, 449, 1994.
26. Harvey, R. L. and Sumstrom, J. C., *Rhodococcus equi* infection in patients with and without human immunodeficiency virus infection, *Rev. Infect. Dis.*, 13, 139, 1991.
27. Verville, D. T., Huycke, M. M., Greenfield, R. A., Fine, D. P., Kuhls, T. L., and Slater, L. N., *Rhodococcus equi* infections of humans: 12 cases and review of the literature, *Medicine*, 73, 119, 1994.
28. Pece, S., Fumarola, D., Giuliani, G., Mastroianni, C. M., Lichtner, M., Vullo, V., Antonaci, S., and Jirillo, E., *In vitro* production of tumor necrosis factor-α, interleukin-6 and interleukin-8 from normal human peripheral blood mononuclear cells stimulated by *Rhodococcus equi*, *Vet. Microbiol.*, 56, 277, 1997.
29. Giuliani, G., Pece, S., Fumarola, D., Caccavo, D., and Jirillo, E., Interaction of *Rhodococcus equi* with human macrophages, *Eur. J. Haematol.*, 60, 337, 1998.
30. Nordmann, P., Ronco, E., and Guenounou, M., Involvement of interferon-γ and tumor necrosis factor-α in host defence against *Rhodococcus equi*, *J. Infect. Dis.*, 167, 1456, 1993.
31. D'Elia, S., Mastroianni, C. M., Lichtner, M., Mengoni, F., Moretti, S., and Vullo, V., Defective production of interferon-γ and tumor necrosis factor-α, by AIDS mononuclear cells after *in vitro* exposure to *Rhodococcus equi*, *Med. Inflamm.* 4, 306, 1995.
32. Jirillo, E., Greco, B., Munno, I., Pellegrino, N. M., Brandonisio, O., Di Venere, A., Riccio, P., De Simone, C., Vullo, V., Mastroianni, C. M., and D'elia, S., Demonstration of an exaggerated serum release of tumor necrosis factor-α and interleukin-1β in HIV-infected patients: a possible correlation with circulating levels of bacterial lipopolysaccharides, in *Cellular and Molecular Aspects of Endotoxin Reactions*, Nowotny, A., Spitlzer, J. J., and Ziegler, E. J., Eds., Elsevier, Amsterdam, 1990, 529.
33. Jirillo, E., Covelli, V., Maffione, A. B., Greco, B., Pece, S., Fumarola, D., Antonaci, S., and De Simone, C., Endotoxins, cytokines, and neuroimmune networks with special reference to HIV infection, *Ann. NY Acad. Sci.*, 741, 174, 1994.
34. Cullen, B. R., Mechanism of action of regulatory proteins encoded by complex retroviruses, *Microbiol. Rev.*, 56, 375, 1992.
35. Collier, A. C., Marra, C., Coombs, R. W., et al., Central nervous system manifestations in human immunodeficiency virus infection without AIDS, *J. Acquir. Immunodefic. Syndr.*, 5, 229, 1992.
36. Epstein, L. G. and Gendelman, H. E., Human immunodeficiency virus type 1 infection on the nervous system: pathogenic mechanisms, *Ann. Neurol.*, 33, 429, 1993.
37. Ho, D. D., Bredesen, D. E., and Vinters, H. U., The acquired immunodeficiency syndrome (AIDS) dementia complex, *Ann. Intern. Med.*, 111, 400, 1989.
38. Pulliam, L., Herndier, B. G., Iang, N. M., and McGrath, M. S., Human immunodeficiency virus-infected macrophages produce soluble factors that cause hystological and neurochemical alterations in cultured human brains, *J. Clin. Invest.*, 87, 503, 1991.

39. Mastroianni, C. M., Liuzzi, G. M., Jirillo, E., Vullo, V., Delia, S., and Riccio, P., Cerebrospinal fluid markers for the monitoring of HIV dementia complex severity: usefulness of anti-myelin basic protein antibody detection, *AIDS,* 5, 464, 1991.
40. Liuzzi, G. M., Mastroianni, C. M., Vullo, V., Jirillo, E., Delia, S., and Riccio, P., Cerebrospinal fluid myelin basic protein as a predictive marker of demyelination in HIV dementia complex, *J. Neuroimmunol.,* 36, 251, 1992.
41. Nottet, H. S. L. and Gendelman, H. E., Revealing the neuroimmune mechanisms for the HIV-1-associated cognitive/motor complex, *Immunol. Today,* 16, 441, 1995.
42. Bradl, M., Immune control of the brain, *Springer Semin. Immunopathol.,* 18, 35, 1996.
43. De Simone, C., Famularo, G., and Cifone, G., HIV-infections and cellular metabolism, *Immunol. Today,* 17, 256, 1996.
44. Lipton, S. A. and Rosenberg, P. A., Mechanisms of disease: excitatory aminoacids as a final common pathway of neurologic disorders, *N. Engl. J. Med.*, 330, 613, 1994.
45. Plotnikoff, N. P., Faith, R. E., Murgo, A. J., Herberman, R. B., and Good, R. A., Methionine enkephalin: a new cytokine-human studies. Short analytical review, *Clin. Immunol. Immunopathol.*, 82, 93, 1997.

6 The Anticytokine Peptide, α-MSH, in Infectious and Inflammatory Disorders

Anna Catania, Lorena Airaghi, and James M. Lipton

CONTENTS

6.1 Introduction 69
6.2 α-MSH and Melanocortin Receptors 69
6.3 Fever and Systemic Inflammation 70
6.4 Experimental and Clinical Arthritis 71
6.5 Reperfusion Injury and Myocardial Infarction 72
6.6 Inflammatory Bowel Disease 73
6.7 HIV Infection 73
6.8 Conclusions 74
References 74

6.1 INTRODUCTION

Proinflammatory cytokines such as interleukin-1 (IL-1) and tumor necrosis factor (TNF) are involved in host responses to inflammatory or infectious stimuli in human and lower animals. The potent effects of these and other cytokines, if not modulated by endogenous actions, would likely be detrimental to the host. It is therefore reasonable to expect that actions of proinflammatory cytokines are normally modulated by endogenous anti-inflammatory mediators. Specific cytokine antagonists, such as IL-1 receptor antagonist, are known to modulate cytokine actions and it appears that a balance between the production of cytokines and cytokine antagonists is important to control host responses.

6.2 α-MSH AND MELANOCORTIN RECEPTORS

A neuroimmunomodulatory peptide, α-melanocyte-stimulating hormone (α-MSH), is believed to modulate production and actions of proinflammatory cytokines. α-MSH is a proopiomelanocortin derivative that shares the 1-13 amino acid sequence with adrenocorticotropic hormone (ACTH).[1] Binding sites for the peptide are widespread within the brain and in peripheral tissues.[2] α-MSH receptors have been characterized and cloned:[3] they are a subset of the guanine nucleotide-binding protein-coupled receptor gene family. At this time, five G-protein-linked receptors (MC-1R through MC-5R) are recognized, all of which react to the peptide including MC-2R, the ACTH receptor on the adrenal glands.[4] Melanocortin receptors occur in several tissues including the brain. MC-1R expression has been found in neutrophils[5] and monocytes[6,7] and in many other cells.[4] MC-5R is likewise expressed in different tissues including the brain[4] and the gut.[8] Melanocortin receptors are coupled to adenylyl cyclase and induce intracellular cAMP.[4] Increases in this mediator are believed to underlie several of the anti-inflammatory effects of melanocortin peptides, including reduced cytokine production and inhibition of chemotaxis.

α-MSH has potent anti-inflammatory influences, partly mediated by inhibition of production/action of proinflammatory cytokines such as TNF-α, IL-1β and IFN γ.[9-11] α-MSH has beneficial influences in the treatment of inflammatory disorders in animal models[12,13] and it reduces production of proinflammatory mediators in human cells during naturally occurring disorders.[14] α-MSH has very low toxicity and is a candidate for treatment of inflammatory disorders in humans.[12,13]

6.3 FEVER AND SYSTEMIC INFLAMMATION

α-MSH inhibits fever caused by endotoxin, endogenous pyrogen, and individual cytokines.[9,14-16] The antipyretic potency of α-MSH is remarkable; with systemic administration it was 20,000 times more potent, on a molar basis, than acetaminophen in reducing endogenous pyrogen-induced fever, and 25,000 times more potent when the peptide was given intracerebroventricularly (i.c.v.).[17] Fever caused in squirrel monkeys by intravenous injection of endotoxin was reduced by α-MSH.[18] Central administration of α-MSH also antagonized fevers induced by injections of recombinant IL-6 and TNF into the brains of rabbits.[9] Therefore, the antipyretic action of this potent peptide is not restricted to a certain pyrogen, but includes major pyrogenic cytokines, and it is observed after both peripheral and central administration of the peptide. The antipyretic message sequence of α-MSH was traced to the COOH-terminal tripeptide KPV.[19,20] Endogenous α-MSH is released within the brain during host challenge and is believed to contribute to physiological control of fever.[21,22] This idea is supported by the observation that immunoneutralization of central α-MSH markedly prolonged IL-1-induced fever.[23]

Inflammation can involve the whole body, and disorders such as endotoxemia, sepsis, septic shock, and adult respiratory distress syndrome (ARDS) are considered systemic inflammation. Therapy for this form of inflammation is still unsatisfactory. It is likely that advances in treatment of this form of inflammation will come from molecules that do not abolish host reactions that are useful to host defense, but from compounds that only "modulate" them. α-MSH was tested in two animal models of systemic inflammation, ARDS, and septic shock.[24] ARDS is marked by increased vascular permeability of the lung to white blood cells (WBC). To test the influence of α-MSH on experimental ARDS, α-MSH was administered i.p. to rats after intratracheal infusion of endotoxin.[24] Bronchoalveolar lavage data showed that α-MSH treatment inhibited endotoxin-induced

WBC migration into the pulmonary tree. Therefore, α-MSH can reduce WBC migration in experimental lung injury.

α-MSH, administered alone or in combination with an inhibitor of Gram-negative bacterial growth, increased survival in a model of peritonitis/endotoxemia/septic shock.[24] Cecal ligation and puncture was performed in female BALB/c mice that were randomly assigned control saline injections, injections of α-MSH (100 µg) or gentamicin sulfate, or both α-MSH and gentamicin. Both α-MSH and gentamicin treatments administered singly improved survival; when given in combination survival was even greater (approximately 75% compared to 12% in control animals). It appears, therefore, that the salutary effects of the peptide and of an antibiotic such as gentamicin can occur together. These findings suggest that the anticytokine α-MSH molecule might be useful for treatment of systemic inflammation, perhaps particularly when used in combination with agents that inhibit bacterial growth.

Research in human subjects determined that α-MSH also is part of the host response during endotoxin challenge in humans.[25] Endotoxin was injected in normal subjects and changes in circulating α-MSH and other aspects of the acute phase response including fever, circulating TNF-α, and ACTH were monitored. Intravenous endotoxin stimulated the release of α-MSH in subjects with marked thermal responses, whereas ACTH increased in all subjects independently of fever magnitude. Therefore it appears that α-MSH increased in the circulation only in those subjects who were particularly sensitive to endotoxin as demonstrated by their high thermal responses. Increase in plasma TNF-α after endotoxin was less in subjects who had a conspicuous increase in α-MSH. This is consistent with *in vitro* experiments which showed that α-MSH inhibits TNF-α release from mitogen-stimulated peripheral blood mononuclear cells.[11]

Although in normal human subjects injected with endotoxin there was an increase in plasma α-MSH, during clinical septic syndrome concentrations of the peptide were reduced.[26] α-MSH was low during the critical phase of the disease and returned to concentrations similar to those of controls after recovery. Therefore, it appears that the peptide is probably released in the circulation early during bacterial invasion, but then, during overt septic syndrome, its production is reduced.

Administration of TNF to animals reproduces many aspects of the clinical septic syndrome.[27] Further, it is clear that this cytokine contributes to the cardiovascular impairment and multiple organ failure that mark septic shock.[27] We investigated influences of α-MSH on TNF production in whole blood samples of patients with septic syndrome. The peptide effectively inhibited endotoxin-induced production of TNF in a dose-related fashion.[28] It appears that, although endogenously produced α-MSH can be beneficial to the host, it may not be released in amounts sufficient to control inflammation in overt disease, and this is the rationale for its administration in pharmacological doses for treatment of disease.

6.4 EXPERIMENTAL AND CLINICAL ARTHRITIS

To investigate the effects of α-MSH in chronic inflammation, we determined whether the neuropeptide can modulate experimental arthritis.[29] This possibility was tested in the most widely used rat model of experimental arthritis, the adjuvant arthritis model. This model, in which immunity to mycobacterial components plays a major part, differs from human rheumatoid arthritis, but certain underlying inflammatory mediators and processes are believed to be shared.

Twice-daily injections of α-MSH markedly inhibited development of adjuvant-induced arthritis. The salutary influence of α-MSH on development of arthritis was reflected in histological studies of affected joints. The general picture of severe inflammatory reaction was the most marked in control animals, particularly in the rear feet. Rats treated with α-MSH showed much less involvement of both soft tissues and joint structures, in keeping with their generally lower arthritis scores. The results indicated that treatment with α-MSH inhibits development of clinical and histological signs of chronic experimental arthritis, and it appears to prevent the weight loss seen in control arthritic animals.

Beneficial results of α-MSH treatment in experimental arthritis prompted us to investigate changes of the peptide in human rheumatoid arthritis. The aim of the study was to determine if α-MSH occurs in synovial fluid of patients with rheumatoid arthritis (RA), juvenile chronic arthritis (JCA), or osteoarthritis.[30] The data showed that α-MSH occurs in the synovial fluid and its concentrations are greater in patients with RA than in those with osteoarthritis. Further, concentration of α-MSH was greater in patients with polyarticular/systemic-onset JCA than in those with pauci-articular disease, that is, in patients with greater joint inflammation. Concentrations of α-MSH were greater in synovial fluid than in plasma in a substantial proportion of patients, suggesting local production of the peptide. Further, it appears that local production of α-MSH is induced particularly in those arthritic joints that have more intense inflammatory reactions. There was a negative correlation between α-MSH and white blood cells in the synovium. This might indicate that α-MSH has a protective influence on migration of neutrophils to inflamed synovial fluid and is consistent with the observation that α-MSH prevents chemotaxis of human neutrophils.[5]

6.5 REPERFUSION INJURY AND MYOCARDIAL INFARCTION

Ischemia and reperfusion cause an inflammatory response with leakage of plasma constituents into the interstitial space and accumulation of polymorphonuclear leukocytes and macrophages at the site of injury.[31] Harmful substances are released by the invading cells, including superoxide anion, hydrogen peroxide, thromboxanes, and leukotrienes.[31] These substances injure further cells already damaged by hypoxia. Thus, the magnitude of tissue injury depends on the degree and duration of the initial ischemic insult and on effects of toxic products released by infiltrating inflammatory cells.

That during myocardial infarction the inflammatory response contributes to cardiac tissue injury is clear from the protection of reperfused myocardium afforded by depletion of circulating neutrophils.[32,33] Further, inhibition by ibuprofen of leukocyte accumulation in ischemic myocardium limits infarct size.[34]

We tested the idea that α-MSH is released during acute myocardial ischemia and reperfusion to counteract effects of proinflammatory cytokines.[35] Whereas concentrations of α-MSH were elevated in early samples of patients with acute myocardial ischemia treated with a thrombolytic agent, they were consistently low in untreated patients. It is reasonable to believe that during myocardial ischemia and reperfusion α-MSH released from the injured myocardium becomes available to reduce inflammation caused by cytokines and other mediators of inflammation. Influences of α-MSH were tested in models of brain reperfusion. The peptide had marked beneficial effects in animals with bilateral vertebral artery occlusion in which brainstem function was monitored via auditory evoked potentials.[36] These results in a model of CNS ischemia-reperfusion injury (marked by local inflammation) suggest a new avenue for preservation of viable nervous tissue in stroke or during intraoperative vascular occlusion. It is likely that α-MSH, which inhibits neutrophil chemotaxis *in vivo* and *in vitro*,[5,37] protects the host from reperfusion injury mainly caused by release of oxygen radicals from inflammatory cells, including neutrophils.

6.6 INFLAMMATORY BOWEL DISEASE

Immunoreactive α-MSH is found in the mucosal barrier of the gastrointestinal tract. It is especially concentrated in the duodenum but also is found in substantial concentrations in the ileum, jejunum, and colon in intact and hypophysectomized rats.[38] The extrapituitary origin of the peptide within these barriers between the host and the external environment is not proof of a host-response modulatory influence of the local peptide; however, with evidence of the anticytokine actions of α-MSH described below, the presence of the peptide in these barriers is consistent with such an influence. Further, the melanocortin receptor MC-5R is expressed in the gut.[8] To test the possibility that α-MSH can be used to control the inflammatory bowel disease, the peptide has been administered in a murine colitis model, the dextran sulfate-induced colitis.[39] In this model, α-MSH reduced the appearance of fecal blood by over 80%, inhibited weight loss, and prevented disintegration of the general condition of the animals. Samples of the lower colon from mice given the peptide showed markedly lower production of TNF-α after stimulation with concanavalin A; the inhibitory effect of α-MSH on production of NO by lower bowel tissue was even greater.

6.7 HIV INFECTION

Proinflammatory cytokines contribute to the progression of HIV infection. Efficient viral replication in T cells and monocytes depends upon activation and differentiation of these cells.[40] IL-1 and TNF activate target cells and promote HIV replication.[41,42] Further, excessive TNF production probably contributes to the cachectic state of AIDS patients. It is clear, therefore, that inhibition of these cytokines could be useful in HIV patients.

Administration of α-MSH could be a way to reduce proinflammatory cytokines in HIV-infected patients. In research on HIV-positive subjects we measured changes of the endogenous molecules and effect of α-MSH on cytokine production by host cells. Patients with HIV infection of CDC group III and IV have elevated concentrations of plasma α-MSH.[43] In related research, we found that elevated concentrations of circulating α-MSH were associated with reduced progression of the disease in HIV-infected subjects (Airaghi et al., unpublished observations). Plasma concentrations of α-MSH were greater in nonprogressors than in progressors; further, association between elevated α-MSH and reduced disease progression was even more pronounced in patients with CD4⁺ T cells <200/µl at baseline.

Production of IL-1 and TNF-α was measured in whole blood samples from HIV-infected patients stimulated with endotoxin in the presence of concentrations of α-MSH.[14] The peptide significantly reduced production of both cytokines. α-MSH likewise reduced production of IL-1β and TNF-α by peripheral blood mononuclear cells stimulated with HIV envelope glycoprotein gp120.[14]

Further, recent observations indicate that α-MSH inhibits HIV expression in chronically and acutely HIV-infected human monocytes (unpublished observations).

6.8 CONCLUSIONS

It is clear that modulation of proinflammatory cytokine production can be useful in several inflammatory disorders in humans. The neuropeptide α-MSH is distinctive relative to other anticytokine molecules in that it combines neural anti-inflammatory influences and direct inhibitory effects on inflammatory cells. The wide range of activity and the very low toxicity make this molecule a candidate for treatment of inflammation in humans.

REFERENCES

1. Eberle, A. N., *The Melanotropins*, Karger, Basel, Switzerland, 1988.
2. Tatro, J. B. and Reichlin, S., Specific receptors for α-melanocyte stimulating hormone are widely distributed in the tissues of rodents, *Endocrinology*, 121, 1900, 1987.
3. Mountjoy, K. G., Robbins, L. S., Mortrud, M. T., and Cone, R. D., The cloning of a family of genes that encode the melanocortin receptors, *Science*, 257, 1248, 1992.
4. Tatro, J. B., Receptor biology of the melanocortins, a family of neuroimmunomodulatory peptides, *Neuroimmunomodulation*, 3, 259, 1996.
5. Catania, A., Rajora, N., Capsoni, F., Minonzio, F., Star, R. A., and Lipton, J. M., The neuropeptide α-MSH has specific receptors on neutrophils and reduces chemotaxis *in vitro*, *Peptides,* 17, 675, 1996.
6. Star, R. A., Rajora, N., Huang, J., Stock, R. C., Catania, A., and Lipton, J. M., Evidence of autocrine modulation of macrophage nitric oxide synthase by α-melanocyte-stimulating hormone, *Proc. Natl. Acad. Sci. U.S.A.*, 92, 8016, 1995.
7. Rajora, N., Ceriani, G., Catania, A., Star, R. A., Murphy, M. T., and Lipton, J. M., α-MSH production, receptors and influence on neopterin in a human monocyte/macrophage cell line, *J. Leukocyte Biol.*, 59, 248, 1996.
8. Gantz, I., Konda, Y., Tashiro, T., Shimoto, Y., Miwa, H., Munzert, G., Watson, S. J., DelValle, J., and Yamada, T., Molecular cloning of a novel melanocortin receptor, *J. Biol. Chem.*, 268, 8246, 1993.

9. Martin, L. W., Catania, A., Hiltz, M. E., and Lipton, J. M., Neuropeptide α-MSH antagonizes IL-6- and TNF-induced fever, *Peptides,* 12, 297, 1991.
10. Taylor, A. W., Streilein, J. W., and Cousins, S. W., α-Melanocyte stimulating hormone inhibits IFN-γ-dependent T-cell and macrophage participation in immunogenic inflammation, *J. Immunol.,* 150, 129A, 1993.
11. Lipton, J. M., Catania, A., Rajora, N., Ceriani, G., Star, R. A., and Boccoli, G., Modulation of inflammatory effects of cytokines by the neuropeptide α-MSH, in *Proinflammatory and Antiinflammatory Peptides, Lung Biology in Health and Disease,* Said, S., Ed., Marcel Dekker, New York, 1998, 112, 281.
12. Catania, A. and Lipton, J. M., α-Melanocyte stimulating hormone in the modulation of host reactions, *Endocr. Rev.,* 14, 564, 1993.
13. Lipton, J. M. and Catania, A., Antiinflammatory influence of the neuroimmunomodulator α-MSH, *Immunol. Today,* 18, 140, 1997.
14. Catania, A., Garofalo, L., Cutuli, M., Gringeri, A., Santagostino, E., and Lipton, J. M., Melanocortin peptides inhibit production of proinflammatory cytokines in blood of HIV-infected patients, *Peptides,* in press.
15. Daynes, R. A., Robertson, B. A., Cho, B. H., Burnam, D. K., and Newton, R., α-Melanocyte stimulating hormone exhibits target cell selectivity in its capacity to affect interleukin-1-inducible responses *in vivo* and *in vitro, J. Immunol.,* 139, 103, 1987.
16. Martin, L. W., Deeter, L. B., and Lipton, J. M., Acute phase response to endogenous pyrogen in rabbit: effects of age and route of administration, *Am. J. Physiol.,* 257, R189, 1989.
17. Murphy, M. T., Richards, D. B., and Lipton, J. M., Antipyretic potency of centrally administered α-melanocyte-stimulating hormone, *Science,* 221, 192, 1983.
18. Shih, S. T. and Lipton, J. M., Intravenous α-MSH reduces fever in squirrel monkeys, *Peptides,* 6, 685, 1985.
19. Richards, D. B. and Lipton, J. M., Effect of α-MSH (11-13) (lysine-proline-valine) on fever in the rabbit, *Peptides,* 5, 815, 1984.
20. Deeter, L. B, Martin, L. W., and Lipton, J. M., Antipyretic properties of centrally administered α-MSH fragments in the rabbit, *Peptides,* 9, 1285, 1989.
21. Samson, W. K., Lipton, J. M., and Zimmer, J. A., The effect of fever on central α-MSH concentrations in the rabbit, *Peptides,* 2, 419, 1981.
22. Bell, R. C. and Lipton, J. M., Pulsatile release of antipyretic neuropeptide α-MSH from septum of rabbit during fever, *Am. J. Physiol.,* 252, R1152, 1987.
23. Shih, S. T., Khorram, O., Lipton, J. M., and McCann, S. M., Central administration of α-MSH antiserum augments fever in the rabbit, *Am. J. Physiol.,* 250, R803, 1986.
24. Lipton, J. M., Ceriani, G., Macaluso, A., McCoy, D., Carnes, K., Biltz, J., and Catania, A., Antiinflammatory effects of the neuropeptide α-MSH in acute, chronic, and systemic inflammation, *Ann. NY Acad. Sci.,* 741, 137, 1994.
25. Catania, A., Suffredini, A. F., and Lipton, J. M., Endotoxin causes α-MSH release in normal human subjects, *Neuroimmunomodulation,* 2, 258, 1995.
26. Catania, A., Airaghi, L., Garofalo, L., Cutuli, M., and Lipton, J. M., The neuropeptide α-MSH in AIDS and other conditions in humans, *Ann. NY Acad. Sci.,* 840, in press.
27. Tracey, K. J. and Cerami, A., Tumor necrosis factor: a pleiotropic cytokine and therapeutic target, *Annu. Rev. Med.,* 45, 491, 1994.
28. Murphy, M., Catania, A., Cutuli, M., Garofalo, L., Valenza, F., and Lipton, J. M., α-MSH reduces production of IL-1 and TNF in septic humans, *Intens. Care Med.,* 22, S336, 1996.
29. Ceriani, G., Diaz, J., Murphree, S., Catania, A., and Lipton, J. M., The neuropeptide α-MSH inhibits experimental arthritis in rats, *Neuroimmunomodulation,* 1, 28, 1994.

30. Catania, A., Gerloni, V., Procaccia, S., Airaghi, L., Manfredi, M. G., Lomater, C., Grossi, L., and Lipton, J. M., The anti-cytokine peptide α-MSH in synovial fluid of patients with rheumatic diseases: comparisons with other anti-cytokine molecules, *Neuroimmunomodulation,* 1, 321, 1994.
31. Smith, E. F., Egan, J. W., Bugelski, P. J., Hillegass, L. M., Hill, D. E., and Griswold, D. E., Temporal relation between neutrophil accumulation and myocardial reperfusion injury, *Am. J. Physiol.,* 255, H1060, 1988.
32. Jolly, S. R., Kane, W. J., Hook, B. G., Abrams, G. D., Kunkel, S. L., and Lucchesi, B. R., Reduction of myocardial infarct size by neutrophil depletion: effect of duration of occlusion, *Am. Heart J.,* 112, 682, 1986.
33. Romson, J., Hook, B., Rigot, V., Schork, A., Swanson, D., and Lucchesi, B., Reduction in the extent of ischemic myocardial injury by neutrophil depletion in the dog, *Circulation,* 67, 1016, 1987.
34. Lefer, A. M. and Crossley, K., Mechanisms of the optimal protective effects of ibuprofen in acute myocardial ischemia, *Adv. Shock Res.,* 3, 133, 1980.
35. Airaghi, L., Lettino, M., Manfredi, M. G., Lipton, J. M., and Catania, A., Endogenous cytokine antagonists during myocardial ischemia and thrombolytic therapy, *Am. Heart J.,* 130, 204, 1995.
36. Huh, S.-K., Lipton, J. M., and Batjer, H. H., The protective effects of α-melanocyte stimulating hormone on canine brainstem ischemia, *J. Neurosurgery,* 40, 132, 1997.
37. Mason, M. J. and Van Epps, D., Modulation of IL-1, tumor necrosis factor, and C5a-mediated murine neutrophil migration by α-melanocyte-stimulating hormone, *J. Immunol.,* 142, 1646, 1989.
38. Fox, J. A. E. T. and Kraicer, J., Immunoreactive α-melanocyte stimulating hormone, its distribution in the gastrointestinal tract of intact and hypophysectomized rats, *Life Sci.,* 28, 2127, 1981.
39. Rajora, N., Boccoli, G., Catania, A., and Lipton, J. M., α-MSH modulates experimental inflammatory bowel disease, *Peptides,* 18, 381, 1997.
40. Rosenberg, Z. F. and Fauci, A. S., Immunopathogenesis of HIV infection, *FASEB J.,* 5, 2382, 1991.
41. Clouse, K. A., Robbins, P. B., Fernie, B., Ostrove, J. M., and Fauci, A. S., Viral antigen stimulation of the production of human monokines capable of regulating HIV1 expression, *J. Immunol.,* 143, 470, 1989.
42. Poli, G., Kinter, A., Justement, J. S., Kehrl, J. H., Bressler, P., Stanley, S., and Fauci, A. S., Tumor necrosis factor α functions in an autocrine manner in the induction of human immunodeficiency virus expression, *Proc. Natl. Acad. Sci. U.S.A.,* 87, 782, 1990.
43. Catania, A., Airaghi, L., Manfredi, M. G., Vivirito, M. C., Milazzo, F., Lipton, J. M., and Zanussi, C., Proopiomelanocortin-derived peptides and cytokines: relations in patients with acquired immunodeficiency syndrome, *Clin. Immunol. Immunopathol.,* 66, 732, 1993.

7 Methionine Enkephalin in the Treatment of AIDS-Related Complex

Bernard Bihari and Nicholas P. Plotnikoff

CONTENTS

Abstract 77
7.1 Objectives 77
7.2 Methods 78
7.3 Statistical Methods 79
7.4 Results 80
7.5 Adverse Reactions 84
7.6 Discussion 84
Acknowledgment 87
References 90

Abstract

In light of the substantial evidence that the endorphinergic system plays a central role in homeostatic regulation of immune function, a 12-week placebo-controlled trial of 2 doses of i.v. methionine enkephalin (MEK) was carried out in 46 patients with AIDS-Related Complex (ARC). No significant toxicity was observed. The high-dose group (125 μ/kg/week) produced significant increases in IL2 receptors, CD56 NK and (LAK) cells, pokeweed mitogen-induced blastogenesis, lymphocyte percentage, and CD3 cell numbers (all as compared to baseline), a significant increase in CD4 cells and CD8 cells (as compared with placebo), and a significant reduction in total lymph node size. The results suggest that MEK is safe and may have a beneficial immune modulating effect in patients with ARC.

Key words: methionine enkephalin, immune modulators, HIV, AIDS-Related Complex, endorphins.

7.1 OBJECTIVES

In light of the growing evidence that the endorphinergic system plays a central role in the homeostatic regulation of immune function, we carried out a phase I-II double blind placebo-controlled trial of the safety, toxicity, and possible immune modulatory effects of methionine enkephalin (MEK), an endogenous opioid peptide produced primarily in the adrenal medulla and by mitogen-stimulated T lymphocytes.

Human T-cell rosette formation is enhanced by MEK and this effect is blocked by prior or simultaneous treatment with naloxone, a narcotic antagonist, thus demonstrating that T cells have functionally important opiate receptors.[1,2] Other studies have shown that the following immune functions involve opiate receptors and/or are facilitated by endorphins:

Lymphocyte blastogenesis[3,4]
T- and B-cell cooperation[5]
Lymphocyte mitogen responsiveness[6]
In vitro antibody response to sheep RBCs[7]
Natural killer (NK) cell activity[8]
Expression of cell surface markers involved in lymphocyte activation (such as OKT10, IL2, and Ia receptors)[9]
Monocyte chemotaxis[10]
Macrophage cytotoxicity[11]

Studies indicating that virus-infected lymphocytes produce β endorphin[12] and that both IL 1 and IL 2 stimulate pituitary synthesis of β endorphin suggest that the central nervous system and the immune system have complex endorphinergic feedback loops that may be important in homeostatic regulation of immune function.[14,15]

A 12-week double blind placebo-controlled trial of naltrexone, an opiate antagonist, in AIDS patients, given in small doses once a day (using a dose large enough to increase endogenous production of pituitary β endorphin and adrenal MEK but small enough to produce only a transient blockade of opiate receptors) appeared to show significant protection against opportunistic infections, stabilization of lymphocyte mitogen responses to CMV, and a significant decline in pathologically elevated levels of endogenously produced α-interferon as compared with a placebo.[16,17]

No significant toxicity was seen in two open label trials of MEK in patients with AIDS and ARC.[18-20] In the absence of a randomized, placebo-controlled research design, no conclusions could be drawn about immune modulation or clinical changes and the sample size was too small to be certain about safety and toxicity. This trial was, therefore, designed to evaluate the safety and toxicity of MEK in patients with 200 to 500 CD4s and an ARC diagnosis; a secondary objective was to look for evidence of immune modulation.

7.2 METHODS

The study, which was double blind and placebo controlled, involved randomized assignment to three arms, with patients on one arm receiving a weekly intravenous infusion of 60 μ/kg (low dose) for 12 weeks, those on a second arm receiving a weekly infusion of 60 μ/kg for 2 weeks, followed by 125 μ/kg for 10 weeks (high dose), and those on the third arm receiving a placebo infusion of normal saline weekly for 12 weeks. The study was designed to treat 20 patients in each arm for

12 weeks. The drug was packaged in 10 cc vials by the University of Iowa College of Pharmacy and shipped and stored frozen. Each sample was thawed overnight the night before the planned infusion. The trial was terminated after 17 weeks when it was noted for the first time that a precipitate was present in a thawed vial of MEK. At that point 46 patients had been randomized and had received at least one infusion; 20 had completed 12 weeks, 26 had completed 8 weeks, and 33 had completed 4 weeks of infusions. It was later determined that amongst the samples stored in Iowa, drug stored in vials at room temperature and under refrigeration showed no precipitate after 20 weeks, while drug stored in a freezer showed the same precipitate when thawed that was seen at the clinical site.

Eligibility included a positive HIV serology, a CD4 level between 200 and 500 at screening (within 7 days before the day of the baseline evaluation and first infusion), and a history of at least one ARC symptom. Exclusion criteria included treatment with AZT or α-interferon within the previous 30 days, a history of addiction to any drug including alcohol, and/or of abuse/recreational use of opioids even if the latter was not to the point of physiological addiction. PCP prophylaxis with aerosol pentamidine, but not Bactrim or dapsone, was permitted because of the potential of the latter two for bone marrow depression. Baseline evaluation included a complete physical exam, Karnovsky score, SMAC 24, CBC and differential, routine urinalysis, urine screen for toxicology, T cell subsets, CD56 (Leu 19) counts, lymphocyte blastogenic responses to CMV, PHA, and pokeweed mitogen, NK activity, T10 and DR and p24 antigen levels. All of these procedures were repeated every four weeks. A brief evaluation of signs and symptoms was done weekly before each infusion. The ACTG toxicity scale was used to evaluate laboratory signs of toxicity. Drug dosage was to be reduced if laboratory values reached Grade II toxicity. Random assignment was made, and was implemented by a mixing nurse who had no contact with the patients and was not otherwise involved in the trial. There were 46 subjects randomized who received at least one infusion: 41 were men and 5 were women, with an age range of 25 to 56 years. Twenty subjects completed 12 weeks of the trial (with 12 infusions). The other 26 were all active, receiving weekly infusions at the time the trial was terminated. There were no drop outs and no one discontinued treatment because of an adverse reaction.

7.3 STATISTICAL METHODS

Patients were stratified by baseline CD4 (200–299, 300–399, 400–499), and randomly assigned to treatment groups within their stratum.

Descriptive statistics for continuous variables are presented as medians and ranges, and means and standard deviations, where appropriate. The continuous variables represent parameters evaluated at screening, baseline, and at each of weeks 4, 8, and 12, at endpoint (patient's last visit or treatment), and at follow-up (occurring 1 month after the week 12 visit). Changes from baseline were evaluated in two ways: the first used an average of the patient's screening and baseline measures which was then subtracted from his corresponding value for a selected timepoint. The second used the difference between the patient's baseline and his corresponding value for a selected timepoint.

With regard to tests of hypotheses, changes from baseline were tested for significance using Wilcoxon Signed Rank Tests confirmed by paired t-tests. For selected primary efficacy parameters, a repeated measures analysis of variance was used to test differences in trends over time among the three treatment groups; a Bonferroni multiple comparison procedure was incorporated into the analysis to test the two contrasts of MEK (lower and higher doses) vs. placebo. Univariate analyses to assess the significance of differences among treatment groups were performed for changes at each timepoint and for measures obtained at each timepoint using Kruskal-Wallis Tests confirmed by one-way analyses of variance. All tests of significance were performed at the $\alpha = .05$ level.

A total of 20 patients completed week 12 — 6 on placebo, 6 on low-dose MEK, and 8 on high-dose MEK; 26 patients completed 8 weeks (8, 9, and 9, respectively) and 33 completed 4 weeks (10, 11, and 12). Thirteen received only one, two, or three infusions.

7.4 RESULTS

CD3 — There was a significant increase in CD3 cells and percent of lymphocytes at the 8-week treatment period in patients on the 125 μ/kg dose of MEK compared to the mean baseline. No significant differences from baseline were observed for the 60 μ/kg dose of MEK or the placebo group (Table 7.1). The differences among the groups were found for percentage of lymphocytes at weeks 8 and 12.

CD4 — Although not statistically significant, patients in the high dose MEK group experienced increases in CD4 cells at week 8, whereas patients in the placebo group experienced decreases for the same time period. The difference among groups with regards to changes in CD4 from baseline mean to 8 weeks was statistically significant ($p = .03$), reflecting differences between the high-dose MEK and placebo groups. No significant differences were observed between the low-dose group and either of the other groups (Table 7.2). However, for the low-dose group, CD4 percentage increased significantly at week 12.

CD8 — Significant increases in the number of CD8 cells was seen at the 8-week treatment period in the high-dose group. No changes were seen in the placebo or low-dose group compared to baseline (Table 7.3).

CD25 (IL2 receptors) — The high-dose group had a significant increase in CD25 and CD25% at 8 weeks compared to baseline. No significant changes were found for the low-dose or placebo groups compared to baseline (Table 7.4).

CD38 (OKT 10) — Significant increases in cells were found for CD38 and CD38% in the placebo group at 8 and 12 weeks, as well as for CD38 in the low dose group as 12 weeks and for CD38% at 8 weeks, compared to baseline. No significant changes were seen in the high dose group (Table 7.5). There was a significant difference in CD38 at week 12 among the groups ($p < .02$).

CD56 (Leu 19) (NK-K-LAK cells) — Significant increases in CD56 and CD56% were seen for the high-dose group at 8 and 12 weeks, and at 8 weeks for

TABLE 7.1
CD3

	Week	N	Mean	S.D.
	High Dose		CD3	
CD3	0	16	1151	342
	4	12	1198	408
	8	9	1273	298
	12	6	1204	324
DCD3	4D	12	106	356
	8D	9	121	120
	12D	6	94	281
	Placebo		CD3	
CD3	0	15	1254	585
	4	10	1507	584
	8	8	1095	331
	12	6	988	244
DCD3	4D	10	200	408
	8D	8	−29	227
	12D	6	−153	252

Note: Week 0 = Average of screening visit and baseline visit values. D = Difference from week 0 (i.e., 4D = difference between week 4 and week 0).

TABLE 7.2
CD4

	Week	N	Mean	S.D.
High Dose	0	16	347	80
	8	9	369	62
	8D	9	+40	61
Low Dose	0	14	342	68
	8	9	320	56
	8D	9	−44	95
Placebo	0	15	342	97
	8	8	299	75
	8D	8	−32	34

Note: See Table 7.1 for explanation of Week "0" and "D".

TABLE 7.3
CD8

	Week	N	Mean	S.D.
	High Dose		**CD8**	
CD8	0	16	782	304
	4	12	813	279
	8	9	887	320
	12	6	844	270
DCD8	4D	12	64	202
	8D	9	89	81
	12D	6	92	232
	Placebo		**CD8**	
CD8	0	15	910	515
	4	10	1043	467
	8	8	795	352
	12	6	670	213
DCD8	4D	10	96	274
	8D	8	–0. 4	205
	12D	6	–122	214

Note: See Table 7.1 for explanation of Week "O" and "D".

the low-dose group. No significant changes were observed in the placebo group compared to baseline (Table 7.6).

Natural killer cells NK (effector to target ratios of 60:1, 20:1, and 7:1) — Significant increases were recorded for all three ratios of effector to target cells in the low-dose group at 4 and 8 weeks as well as in the high-dose group at 8 weeks compared to baseline. No significant changes were seen in the placebo group (Table 7.7).

Phytohemagglutinin (PHA) blastogenesis — *High and medium concentrations*: significant increases were seen in the placebo group at week 8, in the low-dose group at weeks 4 and 8, and in the high-dose group at weeks 8 and 12. *Low concentration*: the high-dose group had significant increases at weeks 8 and 12 compared to baseline. No changes were seen in the low-dose group or the placebo group (Table 7.8).

Pokeweed mitogen (PWM) (medium concentration) — The high dose of MEK significantly increased PWM blastogenesis at 8 and 12 weeks of treatment. No significant changes were found in the low-dose group or the placebo group (Table 7.9).

Cytomegalovirus (CMV) blastogenesis — *High concentration*: there was a significant increase in the placebo group at week 8. No significant changes were seen in the other groups. *Medium concentration:* there was a significant increase in

TABLE 7.4
CD25 (IL2)

	Week	N	Mean	S.D.
	High Dose		**CD25**	
CD25	0	15	78	124
	4	11	70	49
	8	8	97	58
	12	6	72	35
DCD25	4D	10	−21	156
	8D	8	58	54
	12D	6	33	33
	Placebo		**CD25**	
CD25	0	15	78	85
	4	10	144	177
	8	8	109	149
	12	6	82	40
DCD25	4D	10	51	213
	8D	8	46	158
	12D	6	35	39

Note: See Table 7.1 for explanation of Week "O" and "D".

the placebo group at week 12. No significant changes were seen in the other groups. *Low concentration*: the high dose of MEK significantly increased CMV blastogenesis at weeks 8 and 12 compared to baseline. No significant changes were seen in the low-dose or placebo groups (Table 7.10).

Lymphocytes — Lymphocytes, expressed as a percent of white blood cells, were found to be increased by the high dose of MEK at weeks 8 and 12 (Table 7.11). No significant changes were seen in the other groups.

Lymph nodes — Very few patients had active ARC symptoms at baseline or during the trial; too few to evaluate any changes in symptoms. However, most patients had swollen lymph nodes. Total lymph node size (as measured by the mean longer diameter of the three largest nodes) decreased significantly on high-dose MEK at week 4 ($p = .01$) and week 12 ($p < .01$). Total lymph node size did not change significantly on either low-dose MEK or on placebo (Table 7.12). The change in lymph node size from baseline to week 8 differed significantly among groups ($p < .05$).

Body weight — Significant increases in body weight were recorded in the placebo group at weeks 4 and 8 compared to baseline. No significant changes were seen in the two MEK groups (Table 7.13). Differences among groups were significant at weeks 4 and 8.

TABLE 7.5
CD38

	Week	N	Mean	S.D.
	Low Dose		CD38	
CD38	0	14	509	333
	4	11	572	396
	8	8	687	290
	12	6	952	198
DCD38	4D	11	158	565
	8D	8	338	469
	12D	6	555	469
	Placebo		CD38	
CD38	0	15	559	499
	4	10	724	364
	8	8	642	184
	12	6	632	95
DCD38	4D	10	218	568
	8D	8	360	251
	12D	6	461	209

Note: See Table 7.1 for explanation of Week "0" and "D".

7.5 ADVERSE REACTIONS

There were five reported adverse reactions in placebo patients, two in the low-dose patients, and three in the high-dose patients.

The only reported adverse reaction possibly related to the drug occurred in a patient on high-dose MEK who developed dizziness and diaphoresis, accompanied by a rise in blood pressure and drop in pulse rate, during the twelfth infusion. The patient responded to supportive measures and discontinuance of the infusion.

There were no significant changes in hematological parameters or blood chemistries in the patients on drug except for the rise in total lymphocytes and lymphocyte percentage accompanied by a secondary decrease in polymorphonuclear percentage (with no absolute decrease in the latter).

7.6 DISCUSSION

MEK, at a weekly dose of 125 μ/kg compared to baseline appears to produce a rise in total lymphocytes, CD3, CD4, CD8, CD56/Leu 19 (lymphokine-activated killer cells plus NK-K cells), IL2 receptors and blastogenic response to PWM (medium), PHA (low), and CMV (low), accompanied by a significant decrease in total lymph node mass.

TABLE 7.6
CD56 (Leu 19)

	Week	N	Mean	S.D.
	High Dose		CD56	
CD56	0	15	65	67
	4	12	42	22
	8	8	61	40
	12	8	102	54
DCD56	4D	11	11	26
	8D	8	30	29
	12D	8	72	54
	Low Dose		CD56	
CDS6	0	14	44	45
	4	11	81	86
	8	8	96	58
	12	6	131	97
DCDS6	4D	11	39	81
	8D	8	45	43
	12D	6	99	128
	Placebo		CD56	
CD56	0	14	88	83
	4	10	89	108
	8	8	89	72
	12	6	135	86
DCD56	4D	9	21	76
	8D	7	S3	113
	12D	5	91	146

Note: See Table 7.1 for explanation of Week "0" and "D".

The low dose of MEK (60 μ/kg 1 × week) increased CD56 (Leu 19) numbers and percent of lymphocytes at week 8, as well as NK functional activity against the target cells (K562) at 4 and 8 weeks (effector/target ratios 60:1, 20:1, and 7:1). In addition the percent of CD4 cells was increased at week 12.

The increase in cytotoxic T cells as measured by CD8, CD56, and the NK functional assay against the target cells (K562), if it is present in follow-up Phase II studies, may represent an important effect of MEK. The phenotype CD56 includes Leu 11 NK cells as well as cross-antigenic phenotypes CD3 and CD8. These cytotoxic cells (NK-K-LAK cells) have been reported to have cytotoxic activities against HIV-infected cells.[21] Size reduction in swollen lymph nodes with MEK treatment may be a consequence of this effect of MEK.

TABLE 7.7
NK Activity 60:1

	Week	N	Mean	S.D.
	High Dose MEK		NK 60:1	
NK 60:1	0	15	6.9	1.6
	4	12	8.1	1.8
	8	9	10.5	1.9
	12	1	8.9	
DNK60:1	4D	11	1.0	2.0
	8D	8	3.7	3.3
	12D	1	4.8	
	Low Dose		NK 60:1	
NK60:1	0	14	6.4	1.1
	4	11	8.5	1.6
	8	9	9.7	2.3
	12	3	8.6	0.8
DNK60:1	4D	11	2.0	2.2
	8D	9	3.2	2.9
	12D	3	3.2	1.0
	Placebo		NK 60:1	
NK60:1	0	15	8.1	3.1
	4	10	9.3	1.8
	8	8	10.8	1.5
	12	1	8.5	
DNK60:1	4D	10	1.8	3.0
	8D	8	3.2	3.8
	12D	1	2.5	

Note: See Table 7.1 for explanation of Week "0" and "D".

There was a significant differences among treatment groups for numbers of CD38 (OKT10) cells, a phenotype associated with immature cells. The highest numbers were observed in the low-dose > placebo > high-dose groups (week 12). Compared to baseline, the placebo groups had significantly higher numbers at weeks 8 and 12. These findings suggest that the high dose of MEK stimulates the production of mature cells, as recorded by the increased cell numbers labeled by CD3, CD4, and CD8.

MEK appears to be safe at these doses and may have a beneficial immune modulating effect at the higher dose. Larger phase II trials of longer duration, with and without antiretroviral therapy, will be necessary to confirm the apparent immune modulating effects noted in this small trial and determine whether they have any clinical implications for the long-term management of HIV infection.

TABLE 7.8
PHA Blastogenesis — Low Concentration

	Week	N	Mean	S.D.
	High Dose			
PHA-L	0	15	13938	12701
	4	12	27554	24163
	8	9	51618	39852
	12	8	35311	16663
DPHA-L	4D	11	6025	19272
	8D	8	39572	37020
	12D	7	22302	22407
	Placebo			
PHA-L	0	14	43588	38264
	4	8	32788	28033
	8	7	78655	31102
	12	6	36526	15125
DPHA-L	4D	7	−14167	45739
	8D	6	44860	28780
	12D	6	−7417	20975

Note: See Table 7.1 for explanation of Week "0" and "D".

TABLE 7.9
Pokeweed Mitogen Blastogenesis—Medium Concentration

	High Dose		PWM: Medium	
	Week	N	Mean	S.D.
PWM-M	0	15	12251	11456
	4	10	16748	11375
	8	7	13718	9645
	12	6	14085	4352
DPWM-M	4D	10	6095	12865
	8D	7	5198	10747
	12D	6	4347	3984

Note: See Table 7.1 for explanation of Week "0" and "D".

ACKNOWLEDGMENT

Funding for this research provided by TNI Pharmaceuticals, Inc.

TABLE 7.10
CMV Blastogenesis — Low Concentration

	Week	N	Mean	S.D.
	High Dose		CMV: Low	
CMV-L	0	15	646	511
	4	12	654	452
	8	9	968	514
DCNV-L	4D	181	1366	635
	8D	8	506	576
	12D	7	746	461
	Placebo			
CMV-L	0	13	1530	2141
	4	8	977	392
	8	7	1464	837
	12	6	1648	851
DCMV-L	4D	6	−629	2685
	8D	5	1293	645
	12D	5	1114	1076

Note: See Table 7.1 for explanation of Week "0" and "D".

TABLE 7.11
Lymphocyte Percentage

	High Dose MEK			
Week	Percent Mean	S.D.	N	P
	31.4	6.1	10	
4	35.1	9.2	10	
8	34.5	4–7	9	
12	34.3	5.3	8	
4	3.8	10.8	10	0.30
8	3.9	4.3	9	0.03
12	4.4	4.8	8	0.04

Note: See Table 7.1 for explanation of Week "O" and "D".

TABLE 7.12
Lymph Node Size

Week	N	Mean	S.D.
High Dose			
0	16	2.3	0.9
4	12	1.8	0.5
8	9	1.3	0.8
12	8	1.5	0.8
4D	12	−0.3	0.4
8D	9	−0.8	0.5
12D	8	−0.8	0.8
Placebo			
0	15	2.1	1.5
4	10	2.2	1.1
8	8	2.1	1.3
12	6	1.4	0.5
4D	10	0	1.0
8D	8	−0.4	0.9
12D	6	−0.4	0.4

Note: See Table 7.1 for explanation of Week "O" and "D".

TABLE 7.13
Body Weight

Week	N	Mean	S.D.
High Dose			
16	75	9	
4	12	73	8
8	9	72	9
12	8	72	10
4D	12	0.49	
8D	9	0.82	2
12D	8	0.43	2
Low Dose			
14	74	12	
4	11	75	12
8	9	77	12
12	6	76	5
4D	11	−0.2	1
8D	9	0	2
12D	6	0.5	2
Placebo			
15	80	12	
4	10	84	10
8	8	86	11
12	6	86	12
4D	10	0.9	0.9
8D	8	1.5	1.4
12D	6	0.9	1.2

Note: See Table 7.1 for explanation of Week "O" and "D".

REFERENCES

1. Wybran, J., Appelboom, T., Famaey, J. P., and Govaerts, A., Suggestive evidence for receptors for morphine and methionine-enkephalin on normal human blood T lymphocytes, *J. Immunol.*, 123, 1068, 1979.
2. Hazum, E., Chang, K. J., and Cautrecasas, P., Specific nonopiate receptors for β-endorphin, *Science*, 205, 1033, 1979.
3. Plotnikoff, N. P., Miller, G. C., and Murgo, A. J., Enkephalins-endorphins: immunomodulators in mice, *Int. J. Immunopharmac.*, 4, 336, 1982.
4. Miller, G. C., Murgo, A. J., and Plotnikoff, N. P., The influence of leucine and methionine-enkephalin on immune mechanisms in humans, *Int. J. Immunopharmacol.*, 4, 367, 1982.
5. Wybran, J., Appelboom, T., and Famaey, J. P., Receptor for morphine and methionine-enkephalin in human T lymphocytes: the two hits opioid lymphocyte receptor

hypothesis, in *New Trends in Human Immunology and Cancer,* Serrou, B. and Resenfeld, Eds., Doin Publishers, Paris, 1980, 48.
6. Gilman, S. C., Schwartz, J. M., Milner, R. J., Bloom, F. E., and Feldman, J. D., β-endorphin enhances lymphocyte proliferative responses, *Proc. Natl. Acad. Sci. U.S.A.,* 79, 4226, 1982.
7. Johnson, A. M., Smith, E. M., Torres, B. A., and Blalock, J. E., Regulation of the vitro antibody response by neuro-endocrine hormones, *Proc. Natl. Acad. Sci.,* 79, 4171, 1983.
8. Mathews, P. M., Froelich, C. J., Sibbitt, W. L., Jr., and Bankhust, A. D., Enhancement of natural cytotoxicity by β-endorphin, *J. Immunol.,* 130, 1658, 1983.
9. Wybran, J., Enkephalins as molecules of lymphocyte activation and modifiers of the biological response, in *Enkephalins and Endorphins: Stress and the Immune System,* Plotnikoff, N. P., Faith, R. E., Murgo, A. J., and Good, R. A., Eds., Plenum Press, New York, 1986, 253.
10. Van Epps, E. E. and Saland, L., β-endorphin and met-enkephalin stimulate human peripheral blood mononuclear cell chemotaxis, *J. Immunol.,* 132, 3046, 1984.
11. Foris, G., Medgyes, G. A., Gyimesi, E., and Hauck, M., Met-enkephalin induced alterations of macrophage function, *Mol. Immunol.,* 21, 747, 1984.
12. Smith, E. M., Harbour-McMenamin, D. V., and Blalock, J. E., Lymphocyte production of endorphins and endorphin-mediated immunoregulatory activity, *J. Immunol.,* 135, 779, 1985.
13. Farrar, W. L., Relationships between lymphokine and opiatergic modulation of lymphocyte proliferation, in *Enkephalins and Endorphins: Stress and the Immune System,* Plotnikoff, N. P., Faith, R. E., Murgo, A. J., and Good, R. A., Eds., Plenum Press, New York, 1986, 241.
14. Moretti, C., Perricone, R., DeSanctis, G., Decarolis, C., Fabbri, A., Gnessi, L., Fraioli, F., and Fontana, L., Endorphins: evidence of multiple interactions with the immune system, in *Opioid Peptides in the Periphery,* Fraioli, A., Isidore, A., and Marretti, M., Eds., Elsevier, New York, 1984, 137.
15. Weber, R. J. and Pert, C. B., Opiatergic modulation of the immune system, in *Central and Peripheral Endorphins: Basic and Clinical Aspects,* Muller, E. E. and Genazzani, A. R., Eds., Raven Press, New York, 1984, 35.
16. Bihari, B., Drury, F., Ragone, V., et al., Low dose naltrexone in the treatment of AIDS, IV Int. Conf. AIDS: Stockholm, June, 1988 (Poster 3056).
17. Bihari, B., Drury, F., Ragone, V., et al., Low dose naltrexone in the treatment of AIDS: long-term followup results. V Int. Conf. on AIDS: Montreal, June, 1989 (Poster M.C.P. 62).
18. Zunich, K. and Kirkpatrick, C., Methionine enkephalin as immunomodulator therapy in human immunodeficiency virus infections, I., *Clin. Immunol.,* 8, 95, 1988.
19. Plotnikoff, N. P., Wybran, J., Nimeh, N. F., and Miller, G. C., Methionine enkephalin: Enhancement of T-cells in patients with Kaposi's Sarcoma, AIDS, and lung cancer, in *Enkephalins and Endorphins: Stress and the Immune System,* Plotnikoff, N. P. et al., Eds., Plenum Press, New York, 1986.
20. Wybran, J., Schandene, L., Van Vooren, J. P., Vandermoten, G., Latinne, D., Sonnet, J., DeBruyere, M., Taelman, H., and Plotnikoff, N. P., Immunologic properties of methionine-enkephalin, and therapeutic implications in AIDS, ARC, and cancer, *Ann. NY Acad. Sci.,* 496, 108, 1987.
21. Wybran, J. and Plotnikoff, N. P., Methionine enkephalin; A new lymphokine for the treatment of ARC patients, in *Stress and Immunity,* Plotnikoff, N. P., Murgo, A. J., Faith, R. E., and Wybran, J., Eds, CRC Press, Boca Raton, FL, 1991.

8 Central Nervous System Toxicities of Cytokine Therapy

Robert C. Turowski and Pierre L. Triozzi

CONTENTS

8.1 Introduction 94
8.2 Mechanisms of CNS Effects 94
8.3 Acute and Chronic CNS Toxicities 97
8.4 CNS Toxicities in Clinical Trials 97
    8.4.1 Interferons 97
        8.4.1.1 Interferon-α 97
        8.4.1.2 Interferon-β 99
        8.4.1.3 Interferon-γ 100
    8.4.2 T-Cell Growth Factors 101
        8.4.2.1 IL-2 101
        8.4.2.2 IL-4 102
        8.4.2.3 IL-12 102
    8.4.3 Proinflammatory Cytokines 102
        8.4.3.1 IL-1 102
        8.4.3.2 TNF 103
        8.4.3.3 IL-6 103
    8.4.4 Hematopoietins 103
        8.4.4.1 Erythropoietin 104
        8.4.4.2 Granulocyte Colony-Stimulating Factor (G-CSF) 104
        8.4.4.3 Granulocyte-Macrophage Colony-Stimulating Factor (GM-CSF) 105
        8.4.4.4 Macrophage Colony-Stimulating Factor (M-CSF) 105
        8.4.4.5 Stem Cell Factor (SCF) 105
        8.4.4.6 IL-3 106
        8.4.4.7 IL-11 106
        8.4.4.8 Thrombopoietin 106
8.5 Management of CNS Toxicities 106
8.6 Summary 107
References 108

8.1 INTRODUCTION

Cytokines are now commonly used pharmaceuticals. It has been over a decade since interferon α received approval in the U.S. for use in the treatment of hairy cell leukemia. While the therapeutic role of interferons continues to unfold, several additional cytokines have been brought to the clinic, and others are under investigation (Table 8.1). That systemic cytokine therapy affects the central nervous system (CNS) is clear. Psychomotor retardation, psychiatric disorders, and electroencephalographic (EEG) changes were among the initial effects noted in clinical studies.[1-3] In some instances, CNS toxicities are dose-limiting. The clinical experience provides direct evidence for both the influence of immune mediators in neuropsychiatric functions and for a link between the immune system and the CNS. Indeed, this link is underscored by the recent approval of interferon-β for the amelioration of the neurologic symptoms of multiple sclerosis. Herein we review the CNS effects of cytokines with special attention to those cytokines in clinical use.

8.2 MECHANISMS OF CNS EFFECTS

Several mechanisms have been proposed to account for the CNS effects of cytokine therapy. Diffusion through the blood-brain barrier (BBB) is unlikely due to the physiochemical properties of most cytokines. Clinical trials confirmed poor interferon α penetration of the CNS, with only trace amounts detectable in cerebrospinal fluids after intravenous doses greater than 50 million units.[4] The early, partially purified preparations studied were considered to harbor contaminants that penetrated the BBB. The presence of contaminants in current products has been ruled out for most preparations, and the nature and severity of CNS toxicities of partially purified formulations or highly purified recombinant products are similar.

Cytokines may communicate with the CNS through circumventricular organs (CVO), pockets of neuronal cell groups located at the edges of the brain's ventricular system. The BBB of CVO is "fenestrated" and allows peripheral cytokines to interact with receptors on astrocytes, microglial cells, and neurons located within the BBB. Interaction with these cells may lead to production of other substances, such as prostaglandin E_2, capable of penetrating the BBB. Alternatively, neurons within the CVO may extend centrally, providing direct connections to central brain structures.

There is evidence for carrier-mediated transport mechanisms capable of binding and moving cytokines, including interleukin-1 (IL-1), tumor necrosis factor (TNF) α, and IL-6, across the BBB.[5] Systemic cytokines may also interact with cerebral vascular endothelial cells, causing the release of neuroactive substances within the brain or altering the BBB to allow passage of factors normally excluded. Cytokines, such as IL-2, has been shown to alter endothelial integrity and cause a capillary leak syndrome with extravasation of intravascular fluid and contents into extravascular spaces, including those of the brain. The significance of increased brain water content is unknown. Patients with glioma on IL-2 do experience marked neurological deterioration associated with peritumoral edema.[6]

TABLE 8.1
Selected Cytokines in Clinical Use

Cytokine	Other Names	Indications
Interferon-α-2a (Roferon-A, Roche Labs.)	Interferon-α Leukocyte Interferon	Hairy cell leukemia AIDS-related Kaposi's sarcoma Chronic myelogenous leukemia
Interferon-α-2b (Intron-A, Schering Corp.)	Interferon-α Leukocyte Interferon	Hairy cell leukemia Malignant melanoma (adjuvant to surgery) Condylomata acuminata AIDS-related Kaposi's sarcoma Chronic hepatitis NANB/C Chronic hepatitis B
Interferon-α-n3 (Alferon N, Purdue Frederick Co.)	Interferon-α Human Leukocyte-Derived Interferon Leukocyte Interferon	Condylomata acuminata
Interferon β-1A (Avonex, Biogen)	Interferon-β Fibroblast Interferon	Multiple sclerosis
Interferon β 1B (Betaseron, Berlex Labs, Inc.)	Interferon-β Fibroblast Interferon	Multiple sclerosis
Interferon γ-1B (Actimmune, Genentech, Inc.)	Interferon-γ Immune Interferon	Chronic granulomatous disease
Aldesleukin (Proleukin, Chiron Therapeutics)	Interleukin-2 IL-2	Metastatic renal cell carcinoma
Epoetin α (Epogen, Amgen, Inc.; Procrit, Ortho Biotech, Inc.)	Erythropoietin Epo	Anemia associated with chronic renal failure, zidovudine therapy in HIV-infected patients, and cancer patients on chemotherapy
Filgrastim (Neupogen, Amgen, Inc.)	Granulocyte Colony-Stimulating Factor G-CSF	Neutropenia associated with myelosuppressive chemotherapy; bone marrow transplantations; severe congenital, cyclic, or idiopathic neutropenia Procurement of peripheral blood progenitor cells
Sargramostim (Leukine, Immunex, Corp.)	Granulocyte-Macrophage Colony-Stimulating Factor GM-CSF	Acceleration of myeloid recovery in patients undergoing autologous or allogeneic bone marrow transplantation and induction chemotherapy in acute myelogenous leukemia Procurement of autologous progenitor cells from peripheral blood

There also is evidence that systemic cytokines interact with peripheral neuronal receptors. Paraganglia, which serve as chemoreceptors, express IL-1 binding sites and form afferent synapses with vagal fibers. A proposed pathway of communication with the CNS involves cytokines interacting with peripheral receptors located on the subdiaphragmatic afferent nerves, which subsequently transmit via vagal fibers to the area postrema of the CNS.[7]

The CNS effects of cytokines appear to be global, although the hypothalamus does appear to be a key site of activity. Fever produced by cytokine therapy is proposed to occur when cytokines interact with CVO and induce prostaglandin E within the hypothalamus, resetting the thermoregulatory center.[8] Neuropsychiatric toxicities of cytokines may be a result of hypothalamus stimulation, which increases the formation of ACTH, cortisol, and β-endorphins from the hypothalamus-pituitary-adrenal axis.[9] A variety of EEG changes have been reported, particularly of the frontal lobe.[1] Cognitive impairment secondary to IL-2 and TNF has been associated with frontal lobe perfusion deficits observed on single photon emission tomography scan which was postulated to be caused by changes in hypothalamic and/or frontal subcortical function.[10]

The activities of cytokines overlap considerably, and a cytokine usually induces the secretion of other cytokines and mediators by immune effector cells, producing a cascade of biologic effects. The delay of symptoms following systemic or direct CNS injection and the attenuation of symptoms in patients taking immunosuppressive drugs suggest the importance of indirect mechanisms. The similarities in the CNS toxicities observed with the various agents also support the possibility of common final pathways. Most cytokines induce the production of proinflammatory cytokines, such as IL-1 and TNF, which play central roles in the regulation of fever, consciousness, appetite, and mental status.[11,12] The rarity of CNS toxicities with some cytokines, for example, granulocyte colony stimulating factor (G-CSF), may be related to their inability to activate immune effector cells to produce significant levels of IL-1 and TNF.[13] It is possible that a proinflammatory response to the cytokine protein itself may contribute; CNS effects, however, have been observed in the absence of evidence of immunogenicity.

The complexity of the cytokine system, alone, makes it difficult to study. Evaluation of toxicities is further complicated by the complexities of patients that have been studied. Concurrent medications with CNS effects, and the presence of organic brain abnormalities, such as brain metastases, may predispose to symptoms or may be directly responsible. It also must be emphasized that cytokines regulate not only immune responses and hematopoiesis but also many other biologic processes. Cytokine-induced cardiovascular, renal, or hepatic disorders also may contribute to CNS manifestations. Patients receiving interferons and IL-2 can develop hypothyroidism with symptoms of severe fatigue, speech changes, and somnolence.[14] Alterations in CNS function may be the consequence of a desired effect of the cytokine. For example, headaches may be due to an increase in the blood volume and resultant hypertension effected by erythropoietin. Finally, psychiatric disorders are commonly associated with underlying diseases treated with cytokines.[15,16] Certainly, endogenous cytokines may be playing a role in these observations, but there are obviously other confounding variables.

8.3 ACUTE AND CHRONIC CNS TOXICITIES

Although there is considerable variability, patients receiving cytokines usually experience two distinct CNS toxicity patterns: an acute phase followed by a chronic phase. In the acute phase, constitutional symptoms, or what has come to be called the flu-like syndrome, develop. Symptoms, including fever, chills, headache, and fatigue, typically begin 2 to 4 h and last for 6 to 12 h after the initial administration and are experienced for approximately the first 1 to 3 weeks of therapy. The severity of this acute phase depends on the dose, schedule, and route of administration, as well as on the agent. Regimens producing continuous serum levels generally are more toxic than those producing transient levels. For a particular dose, the likelihood of toxicities is greatest for continuous infusion and then decreases for subcutaneous, intramuscular, and intravenous bolus routes of administration.[17] Tachyphylaxis usually occurs within 1 to 3 weeks of continuous treatment, and acute toxicities seldom limit treatment. The mechanism of this tachyphylaxis is unknown. Although fever usually abates, low-grade fevers after administration may persist for months, and a lapse in treatment of 3 days or a dose increase may result in recurrence of acute symptoms.

A wide range of neurotoxicities may develop as cytokine treatment progresses beyond 1 to 2 weeks. Chronic-phase CNS toxicities occur with all dosages, schedules, and routes, but are more likely with higher doses and in patients treated for more than 2 months. Elderly patients and those with predisposing conditions, such as prior brain injury or hepatic dysfunction, are particularly prone. Prominent among chronic-phase toxicities is what is referred to as neurasthenia or fatigue syndrome. Symptoms include asthenia, malaise, lethargy, somnolence, headaches, low-grade fevers, and anorexia. The fatigue syndrome can be constant or intermittent in nature and can be dose-limiting. Psychomotor, cognitive, and psychiatric abnormalities also develop with chronic therapy as can delirium and coma. These more severe toxicities can also be observed very early in the course of therapy.

8.4 CNS TOXICITIES IN CLINICAL TRIALS

8.4.1 Interferons

Human interferons are differentiated by their source of production, chemistry, antiviral, antiproliferative, and immunomodulating properties. Interferons, interferon-α preparations in particular, have been the best-studied cytokines relative to CNS effects.

8.4.1.1 Interferon-α

Interferon-α has significant antiviral and antiproliferative effects in addition to immunologic activities. Several preparations have been evaluated clinically, including human recombinant α interferons and purified preparations. Interferon-α has been evaluated primarily in patients with neoplastic and viral diseases. The various preparations have demonstrated similar toxicities, including CNS toxicities.

Almost all patients receiving more than 1 million units of interferon-α will experience acute constitutional symptoms. The severity of these symptoms are related to dose and have been estimated to be severe in 16% of patients receiving up to 5 million units and in 26% of patients receiving up to 10 million units.[18] While unpleasant, constitutional toxicities elicited at these doses seldom limit the use of interferon. The potential severity of higher doses was confirmed in a recent trial employing interferon-α-2b at 20 million units/m^2/day in the adjuvant setting. In this study 45% of patients with melanoma experienced severe and 3% experienced life-threatening constitutional toxicities, resulting in a dose delay or reduction in 37% of patients during the induction phase.[19] Although differences were not observed in one retrospective study, most studies have indicated that patients older than 60 are less tolerant of the constitutional symptoms than younger patients.[20,21]

Chronic CNS toxicities can occur with all doses but are more likely to develop in patients receiving more than 18 to 20 million units per day.[22] Most patients, regardless of dose or age, experience fatigue which has been classified as severe in 15 to 23% of the cases.[23] Unlike acute constitutional toxicities, chronic fatigue often becomes intolerable, and is the most frequent reason for dose reduction or interruption of therapy.[24] Most studies indicate that patients older than 60 years of age tolerate the fatigue induced by chronic interferon therapy less well than do younger patients.[20]

Psychomotor retardation, described as diminished physical activity accompanied by a lack of desire or drive to participate in normal activities, has been reported to occur in 47 to 80% of patients treated with low doses; less than 5 million units.[25,26] Patients report a lower energy level and appear to be socially withdrawn, gesticulating, articulating, and even eating less than they are normally accustomed.

Cognitive changes occur in patients receiving as little as 9 million units per week. These begin as early as the first week of therapy but usually are most pronounced after 1 to 3 months and can include a decreased attention span, an inability to concentrate, impaired short-term memory, loss of decisiveness, or mental clouding. Patients can often be observed exhibiting periods of silence and vacant staring in mid-sentence.[24,27] Poor performance on tests of cognitive processing, such as verbal learning and recall, have been reported.[22] The reported incidence of cognitive changes is wide, ranging between 17 to 50% in patients receiving 21 to 56 million units interferon per week for chronic myelogenous leukemia or renal cell carcinoma.[26,28] While most cognitive changes are mild to moderate in severity and usually improve in 1 to 2 weeks with dosage reduction, persistent deficiencies in patients off interferon for up to 2 years have been reported.[29]

In some patients, fatigue, psychomotor retardation, and relatively minor cognitive dysfunction progress to states of profound somnolence, with patients sleeping up to 20 hours each day, accompanied by confusion and disorientation. Patients may experience expressive dysphasia and gait difficulties. Rarely, patients progress to a demented state.[24] These toxicities can rapidly develop, occurring in some cases after 2 to 10 doses of interferon at 5 to 10 million units/m^2, 3 times per week, with symptoms partially resolving over 1 to 2 weeks.[30] Hallucinations and losses of smell and taste have also been reported with high-dose intravenous interferon.[1-3]

Renault, et al. classified the psychiatric toxicities of interferon into an organic personality syndrome, organic affective syndrome, and delirium categories.[29] In their report, 10 of 58 patients (17%) with chronic viral hepatitis treated with interferon-α developed psychiatric disorders over 1 to 3 months of therapy. These disorders were more frequent in patients receiving 10 million units every other day than those receiving 5 million units daily. A prior history or family history of psychiatric disorders was not more frequent in patients developing psychiatric toxicities with interferon. The organic personality syndrome consisted of uncontrollable overreaction to a minor frustration, marked irritability, and short temper. Organic affective disorders were described as feelings of depression and hopelessness. Patients had difficulty interacting with others without tearfulness or crying. Delirium initially presented as clouding of consciousness, disorientation, inability to perform simple calculations, short-term memory problems, irritability, and mood changes. Patients became severely agitated, abusive, withdrawn, anxious, and exhibited suicidal ideation, persecutory delusions, and phobias. Patients manifesting delirium had coexisting liver dysfunction, a history of psychiatric illness, and/or previous brain injury, and did improve after stopping interferon.

8.4.1.2 Interferon-β

Human interferon-β is homologous to interferon-α; it interacts with a common receptor and has similar biologic activities. Three interferon-β preparations have been studied clinically, including naturally occurring β interferon and human recombinant interferons-β1a and β1b. Interferon-β1a is identical to natural β interferon. Interferon-β1b differs by the substitution of serine for cystine at position 17 and the lack of a methionine. These chemical differences in the recombinant preparations may account for differences in the incidence of neutralizing antibodies, 3% for β1a and 38% for β1b. The clinical significance of neutralizing antibodies is not known.[31,32]

Interferon-β lessens the frequency of neurologic attacks of multiple sclerosis. The mechanism of this therapeutic effect in what is considered to be an autoimmune disease is unknown. Activated T cells and macrophages can be found in the inflammatory lesions of multiple sclerosis. It has been suggested that endogenous interferon-γ, which is produced by antigen-stimulated T cells, upregulates the immune response by stimulating the synthesis and expression of major histocompatibility (MHC) molecules on antigen-presenting and glial cells; this mechanism may play a role in the active growth of multiple sclerosis lesions. Interferon-β has been shown to downregulate the MHC-inducing effects of interferon-γ on human glial cells.[33]

The CNS effects observed in patients with multiple sclerosis are similar for both recombinant interferon-β preparations. Overall, both formulations, which have been evaluated in placebo-controlled studies of over 300 patients, are well tolerated at the doses used to treat multiple sclerosis. Seldom do acute or chronic CNS toxicities result in withdrawing patients from therapy. On a per unit basis, interferon-β preparations appear to better tolerated than interferon-α preparations.

Mild acute constitutional symptoms were observed in 62 to 76% of patients with multiple sclerosis receiving 6 million units interferon-β1a intramuscularly once a week or 8 million units of interferon-β1b subcutaneously every other day. Constitutional symptoms were transient, usually abated after 1 week of therapy, and were more common in elderly patients. It should be noted that 40 to 56% of patients with multiple sclerosis treated with placebo experienced constitutional symptoms.[31,32]

Chronic neurasthenia occurred in 21% of patients treated with interferon-β1a compared to 13% of patients treated with placebo. Mean scores on the Beck Depression Inventory were not significantly different in patients receiving interferon-β1a and those receiving placebo. No suicide attempts occurred on the interferon-β1a arm.[32] Likewise, asthenia was reported in 49% of patients treated with interferon-β1b.[33] The incidence of depressive symptoms were higher for patients treated with interferon-β1b than placebo — 16.9% vs. 14.6% during the first year and 11.1% vs. 5.1% during the fifth year. Suicide attempts occurred in 1.6% of patients in the first year of treatment, all in the interferon arm. Although these data suggest that interferon-β1a is less likely to cause depression and suicide than β1b, comparative studies have not been performed, and this patient population has a high incidence of these problems. Confusion, somnolence, and emotional lability have been reported in patients treated with interferon-β, but are rare.[34] It should be noted that cognitive assessments might improve on interferon-β.[35]

CNS toxicities have been more prominent in patients with malignancies treated with interferon-β, as much higher doses, in the range of 90 to 600 million units/m^2, have been administered intravenously. Acute constitutional symptoms occurred in almost all patients but did become less severe with continued treatment. The exception was fatigue and malaise, which worsened with cumulative doses and frequently led to dose reductions. In one trial of interferon-β1b in 72 patients, 7% required dose reduction for mild to moderate neurotoxicities while 10% withdrew from the study due to severe neurotoxicities including headache, fatigue, apathy, dementia, agitation, disorientation, and personality change. While chronic neurotoxicities were usually reversible after cessation of therapy, death due to encephalopathy, seizures, and brain stem dysfunction, possibly related to radiation therapy, has been reported with large cumulative intravenous doses of interferon-β1b.[36,37] Of note, patients under 21 years of age were able to receive higher doses before reaching the maximum tolerated dose.[38]

8.4.1.3 Interferon-γ

Interferon-γ is distinct from interferons α and β in its chemistry, receptor, and biological activities. The antiviral and antiproliferative activities of interferon-γ are less potent; interferon-γ, however, plays a more central role in regulating immune responses, including induction of class II MHC molecules, activation of macrophages and neutrophils, and promotion of T-cell-mediated immunity. Interferon-γ has been evaluated in the setting of neoplastic, infectious, and immunologic diseases. Recombinant human interferon-γ, interferon-γ1b, has been shown to reduce the frequency of serious infections in patients with chronic granulomatous disease. Enhanced phagocyte function, including elevations of superoxide levels, and improved bactericidal activity have been observed in these patients.

As with the other interferons, acute constitutional symptoms occur in most patients at 50 μg/m^2 subcutaneously 3 times a week, the dose recommend for the treatment of chronic granulomatous disease. In this setting 60% of patients experienced occasional headaches and 30% experienced chills, each significantly more frequent than in placebo-treated patients.[39,40] Constitutional symptoms at this dose are usually mild and abate. Headaches and chills are more of a problem at doses greater than 250 µg/m^2 employed in several cancer trials, particularly when administered intravenously, and can be dose-limiting.[41] Infrequent chronic neurotoxicities reported with the higher doses include dizziness, slowing of thought processes, confusion, crying episodes, and Parkinson-like signs. These symptoms resolved over a period of 3 to 4 days with cessation of therapy and/or with medications.[41]

8.4.2 T-Cell Growth Factors

A number of cytokines have as their primary activity the modulation of the growth, differentiation, and function of T cells. These cytokines also have a variety of other effects, including effects on B-cell growth and function and on hematopoiesis.

8.4.2.1 IL-2

IL-2 stimulates lymphocyte cytolytic activities, including cytolytic T-cell (CTL), NK, and lymphokine-activated killer (LAK) activities, and stimulates lymphocyte T helper (Th) activities, including the production of interferon-γ, TNF, and IL-1. Human recombinant IL-2, aldesleukin, has been applied to a variety of malignancies. It is also being evaluated in the setting of infectious diseases, including human immunodeficiency virus (HIV-1) infection. Clinical trials have investigated a wide range of doses of IL-2 as well as IL-2 in combination with LAK and CTL infusions, other cytokines, and cytotoxic drugs.

Acute and chronic CNS toxicities have been prominent, particularly with high-dose intravenous bolus or continuous infusion programs.[42] In studies of aldesleukin at doses of 600,000 to 720,000 units/kg administered as 15-minute intravenous infusions every 8 h (approximately 150 million units per day) involving 255 patients, constitutional symptoms occurred in 97% of patients and were classified as severe or life-threatening in 19% and 5%, respectively. Mental status changes, primarily somnolence and disorientation, were observed in 82% of patients and classified as severe or life-threatening in 23% and 5%, respectively. Coma and seizures, although infrequent, were also problems.[43] Similar observations were made in other studies involving over 600 patients. The incidence of chills was greater in cancer patients receiving IL-2 combined with LAK cell infusions or interferons compared to IL-2 alone, but the incidence of somnolence, coma, or disorientation was similar.[44] Behavioral changes included agitation lasting less than 24 h, which required minimal supervision, to severe agitation with combativeness which required therapy with neuroleptics, while 50% of patients developed severe cognitive changes, meeting criteria for delirium. Delusions developed in 16%, and others reported hallucinatory confusional syndromes. In some instances, these persisted several days after the termination of therapy. Symptoms usually developed later in the course of therapy. The incidence of psychoneurotoxicities was associated with higher IL-2 doses but no predisposing patient conditions were identified.[45,46] IL-2 neurotoxicities may briefly worsen after treatment cessation before improvement is observed.

Low-dose IL-2, less than 10 million units/m^2/day, has been evaluated both in patients with cancer and with HIV-1 infection. Acute and chronic CNS toxicities are less of a problem with low-dose IL-2 regimens. Constitutional symptoms, particularly in HIV-1-infected patients, can still be dose-limiting.[47] Disorientation, hallucinations, concentration difficulties, and somnolence, however, are infrequently observed, and treatment cessation due to chronic CNS toxicities occurred in only 2% of cancer patients in one study.[48] Low-dose IL-2 in combination with standard doses of interferon-α2b was well tolerated in cancer patients. Acute and chronic CNS toxicities did occur but were usually mild. Although severe cognitive changes and disorientation were infrequently observed, coma and seizures were not, and symptoms usually resolved within 6 h of treatment cessation.[49,50]

8.4.2.2 IL-4

IL-4 acts on the proliferation and differentiation of T and B lymphocytes and enhances the function of NK cells, eosinophils, and mast cells. The spectrum of the CNS toxicities of IL-4, under investigation as an antitumor and hematopoietic stimulating agent, appears to be similar to those of IL-2, but less frequent and severe at the doses and schedules investigated. IL-4 therapy does cause moderate-grade fever, fatigue, anorexia, and headache in most patients. More serious CNS toxicities, including transient partial blindness, photophobia, and visual hallucinations, have also been reported, although the relationship to IL-4 therapy is uncertain.[51-53]

8.4.2.3 IL-12

IL-12 increases Th1 cytokine production, particularly that of interferon-γ, CTL, and NK cytotoxicity, and boosts the production of some subclasses of IgG antibodies. IL-12 also has antiangiogenic and antimetastatic effects. In animal models IL-12 appeared to be less toxic than IL-2.[54,55] IL-12 is currently being studied in the therapy of a variety of malignancies and also in the setting of HIV-1. Toxicities, including CNS toxicities, have not been completely characterized. In early-phase trials in cancer patients, IL-12 therapy resulted in generally acceptable toxicities, including fever, chills, fatigue, and headache.[56] In one study, however, an unexpected high incidence of serious adverse effects, including asthenia, as well as nausea, vomiting, gastrointestinal tract bleeding, and hepatotoxicity, was observed.[54]

8.4.3 PROINFLAMMATORY CYTOKINES

As noted, the induction of proinflammatory cytokines by cytokine therapy may function as final common pathways in the elicitation of CNS toxicities. Proinflammatory cytokines have been evaluated clinically, primarily for their antitumor activity but also for their hematopoietic activities. As could be predicted, CNS toxicities have been prominent. At present, no proinflammatory cytokine is approved for clinical use.

8.4.3.1 IL-1

Produced by activated macrophages, endothelial cells, and even cells of the CNS, IL-1 has far-reaching biological effects on the immune, hematopoietic, endocrine, and nervous systems.[57] IL-1 consists of two discrete polypeptides, IL-1α and IL-1β, produced by distinct genes. Although there is only 26% homology, IL-1α and IL-1β act through a common receptor and have similar biological activities. Recombinant IL-1α and IL-1β, under investigation as antitumor and hematopoietic stimulating agents, elicit similar CNS toxicities. Constitutional symptoms occurred in most patients given doses of 0.1 to 0.3 μg/kg within 30 min to 6 h of administration, depending on the infusion rate, and then usually resolved within 3 to 24 h. Although tachyphylaxis developed and symptoms were usually managed medically, these symptoms were often treatment-limiting, causing 12.5% of patients in one study to discontinue therapy.[58] Mild to moderate fatigue was reported by many patients, with headache increasing in intensity with dose.[59] Somnolence, confusion, agitation, delusional ideation, photophobia, blurred vision, and seizures have been reported at higher dose levels.[60-63]

8.4.3.2 TNF

Like IL-1, TNF has broad biological activities. Because of its capacity to induce necrosis of tumors, TNF has been primarily investigated as an antitumor agent. Constitutional toxicities, fever, rigors, chills, headache, myalgia, and fatigue occurred in most patients, no matter the route or schedule of administration. Fever and chills, which generally followed rapidly after the start a TNF infusion, did not appear to be dose-related, occurring with as little as 43 $\mu g/m^2$ administered by intravenous bolus. Fever was generally transient, resolving after a few hours. Headache, in particular, was a problem, even in patients treated with very low doses. Transient amnesia, aphasia, hallucinations, and diplopia have also been reported.[64,65]

8.4.3.3 IL-6

IL-6 has a wide range of biological activities much like IL-1, including effects on pituitary gland cells and astrocytes.[66] IL-6 plays a role in the acute-phase response being more proximally involved in the observed effects than other proinflammatory cytokines. In addition, IL-6 plays a role in hematopoiesis. IL-6 has been evaluated clinically as an antitumor and hematopoietic stimulating agent, primarily as stimulator of megakaryopoiesis. In early clinical trials, fever, headache, anorexia, and mild to moderate chilling and fatigue have been observed. The limited experience to date suggests that the CNS toxicities of IL-6, despite its broad biological effects, are not as prominent as those of IL-1 or TNF.[66-68]

8.4.4 HEMATOPOIETINS

Human hematopoietic agents are differentiated by their source of production, chemistry, and their ability to regulate the growth, differentiation, and function of specific hematopoietic cells. Although CNS toxicities have been observed, they have not been a major problem. As noted, the induction of proinflammatory cytokines is not a major effect of some agents. Another factor, however, is that hematopoietins are often not administered over long durations but rather over brief periods of 1 to 2 weeks to mediate the desired effect.

8.4.4.1 Erythropoietin

Erythropoietin induces maturation of red blood cells. The recombinant formulations approved in the U.S., identical in composition and source of production, are employed to treat anemias associated with chronic renal failure, HIV-1, and cancer and to provide blood for autologous transfusions. As noted, the mechanisms of CNS toxicities of erythropoietin may be unique and related to its primary biologic effect. Seizures may be related to a rapid correction of anemia causing hypertensive encephalopathy. Rapid increases in blood pressure, decreased cerebral blood flow, vasospasm, increased blood viscosity, and development of focal cerebral edema are proposed mechanisms.[69-71]

Acute CNS toxicities are infrequent and usually mild. Constitutional symptoms can develop 1 to 2 h after intravenous injection. These usually abate within 10 to 12 h, diminish with continued therapy, and rarely lead to cessation of therapy. The reported incidence of constitutional symptoms in erythropoietin-treated patients is 8 to 38%, which is comparable to incidence of these symptoms in placebo-treated patients.[72-74] The frequency of adverse CNS effects was also similar between erythropoietin and placebo in subjects in good general health treated to generate blood for autologous donations. CNS toxicities reported in this population included mild or moderate fatigue (20%), headache (20%), and dizziness (12%).[75] The incidence of constitutional symptoms can be reduced by administering intravenous erythropoietin over 1 to 2 min, or by giving it subcutaneously.

Seizures, the most significant CNS toxicity, were observed in the initial trials involving chronic renal failure patients at rates ranging from 0.8 to 10%.[69,76] The incidence of seizures is higher during the first 90 days of therapy. It is not certain whether seizures are related to erythropoietin or the underlying disease, as seizures are reported in 4 to 10% of the untreated dialysis population. In one controlled trial involving patients with end-stage renal disease, the incidence of seizures did not differ between erythropoietin- and placebo-treated patients.[77] Seizures in patients without renal failure are rare and usually observed in conjunction with predisposing disorders, such as brain metastases or meningitis.[78] Other CNS toxicities reported include transient ischemic attacks, depression, and hallucinations. The relation of these events to erythropoietin is uncertain.[79,80] It should be noted that erythropoietin therapy and correction of the anemia in dialysis patients improves fatigue, cognitive function, and depressive symptoms.[81,82]

8.4.4.2 Granulocyte Colony-Stimulating Factor (G-CSF)

G-CSF is a lineage-specific factor that stimulates neutrophil production and function. Human recombinant G-CSF, filgrastim, is principally employed to enhance neutrophil recovery in patients undergoing chemotherapy or bone marrow transplantation, treat idiopathic neutropenias, and to improve the yield of granulocytes and peripheral blood progenitor cells for use in transplantation.[83] As with erythropoietin, CNS toxicities associated with filgrastim are infrequent and mild. In patients with cancer undergoing cytotoxic chemotherapy, the reported incidence of constitutional symptoms with filgrastim trials was not different from the incidence in patients receiving placebo.[84] In one study of normal donors receiving filgrastim at 12 μg/kg per day subcutaneously for 4 to 6 days to enhance the collection of blood stem cell by apheresis, 28 of 40 patients (70%) experienced headache, which was considered severe in 3 patients, while 20% of the patients experienced fatigue which was severe in one patient. No patient experienced fever. Toxicities resolved usually within 2 to 4 days of filgrastim cessation.[85] Vertigo, dizziness, and sleep disturbances have also been reported.[83]

8.4.4.3 Granulocyte-Macrophage Colony-Stimulating Factor (GM-CSF)

GM-CSF is a multilineage factor with primary effects on the production and function of neutrophils and monocyte/macrophages. Human recombinant GM-CSF, sargramostim, is employed clinically to enhance myeloid recovery in patients undergoing chemotherapy and bone marrow transplantation. Through effects primarily on monocyte/macrophage, GM-CSF can induce proinflammatory cytokines, and while not as severe or frequent, constitutional toxicities associated with sargramostim more resemble those observed with interferon or IL-2 than those with other hematopoietins, such as erythropoietin or filgrastim. In comparative trials, however, toxicities were similar in sargramostim and filgrastim groups, with the exception of fever, which was usually more prominent in sargramostim-treated patients.[86,87]

Sargramostim toxicities are dose and route dependent. Doses greater than 10 to 15 μg/kg per day (500 to 750 μg/m^2 per day) are associated with dose-limiting CNS and non-CNS toxicities such as myalgia, bone pain, fever, headache, flu-like symptoms, rash, and pericardial/pleural effusions.[88,89] Doses less than 10 μg/kg per day, which appear to be as effective as higher doses, produce mild toxicities consisting primarily of fever, malaise, bone pain, and fatigue.[90] Intravenous bolus and continuous infusion programs are more toxic than subcutaneous administration.[91-93] Children and severely neutropenic patients may be able to tolerate higher doses before experiencing dose-limiting toxicities.[88]

8.4.4.4 Macrophage Colony-Stimulating Factor (M-CSF)

M-CSF modulates the growth and function of cells of the monocyte-macrophage (Mo/Mx) lineage. M-CSF has been shown to enhance direct and antibody-dependent cytolytic activity and the production of cytokines such as IL-1 and TNF by Mo/Mx. The toxicity of recombinant M-CSF preparations and also a purified preparation in clinical trials in patients with cancer has been mild. Acute constitutional symptoms including fever, fatigue, and insomnia have been noted in most patients but were not dose-limiting. Interestingly, ocular toxicity including bilateral eye pain, photosensitivity, blurred vision, scotomata, and erythema of the sclera has been observed and has been dose-limiting.[94,95] Progressive fatigue leading to treatment discontinuation has been observed, but other CNS toxicities were not prominent.

8.4.4.5 Stem Cell Factor (SCF)

SCF, also known as kit ligand, steel factor, and mast cell growth factor, acts on both primitive and mature hematopoietic progenitor cells. SCF is under investigation for mobilizing progenitor cells for transplantation and for treating various bone-marrow failure states. The clinical toxicity, in general, of recombinant SCF has not been problematic. Although allergic reactions have been observed, common acute CNS effects seen with other cytokines, such as fever, were not observed at the doses tested.[96,97]

8.4.4.6   IL-3

IL-3 affects the proliferation and maturation of erythrocyte, granulocyte, and megakaryocyte progenitor cells. Human recombinant IL-3 has been under study to ameliorate neutropenia and thrombocytopenia in patients undergoing myelosuppressive chemotherapy. CNS toxicities have been observed in early-phase clinical trials. At doses less than 10 μg/kg per day, mild to moderate toxicities of fever and headache occurred. Higher doses result in severe headache, constitutional symptoms, and fever.[98,99] Intravenous administration of IL-3, perhaps producing higher peak serum levels, was less well tolerated than subcutaneous administration, although the maximum tolerated dose was the same for either route.[100]

8.4.4.7   IL-11

IL-11 affects several different cell types in addition to hematopoietic cells such as hepatocytes, preadipocytes, synoviocytes, chondrocytes, osteoclasts, mucosal epithelial, and neuronal cells.[101] In early clinical trials, human recombinant IL-11 has been employed to ameliorate chemotherapy-induced thrombocytopenia. CNS toxicities have not been, as of yet, well characterized. Constitutional toxicities have been mild, transient, reversible, and more frequent and severe at doses greater than 50 μg/kg per day. Other CNS toxicities that have been observed include syncope of uncertain etiology.[102,103]

8.4.4.8   Thrombopoietin

Megakaryocyte growth and development factor (MGDF) is a recombinant polypeptide related to natural thrombopoietin. MGDF is under investigation as an agent to stimulate platelet production. As with IL-11, the CNS toxicities of MGDF have not been well characterized. In a small trial of patients with advanced cancer receiving MGDF without chemotherapy, reported toxicities were flushing, night sweats, and tremor but not lethargy, headache, myalgia, or fever.[104] When administered after myelosuppressive chemotherapy, the incidence of fatigue, rigors, flushing, myalgia, headache, dizziness, insomnia, and anxiety were not significantly different between MGDF and placebo-treated patients and were attributed to the effects of the underlying cancer and chemotherapy.[105]

8.5   MANAGEMENT OF CNS TOXICITIES

Management of the CNS toxicities of cytokines may require several approaches. Patients, family members, and healthcare professionals should be advised of possible CNS toxicities and directed to seek evaluation of severe symptoms. Reducing dosage or stopping cytokines once CNS toxicities occur usually reverses the toxicities in a few days, with more severe cases requiring up to 3 to 4 weeks. As noted, neurotoxicities may briefly worsen after treatment cessation before improvement is observed and may persist for years.[18,27] Reinstitution of cytokines, such as the interferons, at doses reduced by 25 to 50% can often be accomplished without recurrence of some toxicities. As noted, subcutaneous administration, may be as efficacious as intravenous administration with less toxicity in many situations.

Several medications have been employed to ameliorate the acute constitutional symptoms. Acetaminophen, aspirin, and nonsteroidal anti-inflammatory agents are often used prophylactically. Bedtime administration or gradual initiation of therapy at 50% of full dose for 2 weeks may ameliorate the acute constitutional symptoms. Meperidine may be used to relieve severe rigors associated with fevers.

Medical management of the chronic CNS toxicities is more difficult and is generally empiric, based on and supported by case reports or small trials. Metoclopramide, a dopamine antagonist, and methylphenidate, a dopamine agonist, have been reported to dramatically reverse some the neuropsychiatric toxicities of interferon.[24] Methylphenidate has been used to reduce the number and severity of irritability episodes in patients receiving interferon-α and to resolve sluggishness and fatigue associated with interferon-γ.[41] Naltrexone was reported to affect complete or moderate relief of side effects, including fatigue, in 77% of patients receiving interferon-α.[106] A variety of antidepressants have been employed to manage affective disorders developing during interferon therapy.[107,108] Patients with preexisting psychiatric illnesses can successfully complete interferon treatment with the assistance of a psychiatrist and maintenance of psychotropic drug therapy.[109] Temporary use of neuroleptics may be necessary to control severe agitation or paranoid delusions.

It is best to prevent the CNS effects of erythropoietin by closely monitoring the hematocrit. Although comparative data do not exist, subcutaneous erythropoietin administration may allow for a more gradual increase in hematocrit and lessen the aggravation of hypertension, which may be a contributing factor to the occurrence of seizures seen in this population.[110] Limiting the rate of hematocrit increase to four points in any 2-week period and monitoring blood pressure is recommend to minimize the likelihood of developing severe hypertension.

8.6 SUMMARY

Cytokines now play an important role in the therapy of neoplastic, viral, and immunologic diseases. The observations of significant CNS toxicities in clinical studies have emphasized the link between the immune system and CNS and have provided insight into this complex network. Cytokines induce CNS effects through a variety of direct and indirect mechanisms. Acute toxicities including fever and fatigue, and chronic toxicities including fatigue, psychomotor retardation, cognitive changes, and psychiatric manifestations, have been observed with most agents studied to date. In some instances, CNS toxicity is the dose-limiting toxicity. Whether CNS toxicities occur is dependent not only on the biologic activities of the cytokine but also is influenced by dose, schedule, and route of administration. CNS toxicities of some cytokines currently in clinical use, such as the interferons and IL-2, are commonly noted and may be a consequence of these agents' ability to induce proinflammatory mediators. CNS toxicities of other agents, such as G-CSF, have been less of a problem. This may reflect less capacity to induce proinflammatory mediators but may also be a consequence of the administration schedule. Management of the CNS toxicities may require several approaches, including dosage interruption, adjunctive medications, and psychiatric counseling. While our understanding of the etiologies of CNS toxicities and our ability to identify populations of patients at risk have expanded, much work still needs to be done. Further study should not only help improve the tolerance and thus the therapeutic benefit, but also should provide insight as to how these effects, themselves, can be exploited to mediate beneficial clinical effects.

REFERENCES

1. Rohatiner, A. Z., Prior, P. F., Burton, A. C., Smith, A. T., Balkwill, F. R., and Lister, T. A., Central nervous system toxicity of interferon, *Br. J. Cancer*, 47, 419, 1983.
2. Mattson, K. and Niiranen, R., Neurotoxicity of interferon, *Cancer Treat. Rep.*, 67, 958, 1983.
3. Smedly, H., Katarak, M., Sikora, K., and Wheeler, T., Neurological effects of recombinant human interferon, *Br. Med. J.*, 286, 262, 1983.
4. Smith, R. A., Norris, F., Palmer, D., Bernhardt, L., and Wills, R. J., Distribution of α interferon in serum and cerebrospinal fluid after systemic administration, *Clin. Pharmacol. Ther.*, 37, 85, 1985.
5. Watkins, L. R., Maier, S. K., and Goehler, L. E., Cytokine-to-brain communication: a review and analysis of alternative mechanisms, *Life Sci.*, 57, 1011, 1995.
6. Saris, S. C., Patronas, N. J., Rosenberg, S. A., Alexander, J. T., Frank, J., Schwartzentruber, D. J., Rubin, J. T., Barba, D., and Oldfield, E. H., The effect of intravenous interleukin-2 on brain water content, *J. Neurosurg.*, 71, 169, 1989.
7. Goehler, L. E., Busch, C. R., Tartaglia, N., Relton, J., Sisk, D., Maier, S. F., and Watkins, L. R., Blockade of cytokine induced conditioned taste aversion by subdiaphragmatic vagotomy: further evidence for vagal medication of immune-brain communication, *Neurosci. Lett.*, 185, 163, 1995.
8. Saper, C. B. and Breder, C. D., The neurologic basis of fever, *N. Engl. J. Med.*, 330, 1880, 1994.
9. Ho, B. T., Lu, J. G., Huo, Y. Y., Fan, S. H., Meyers, C. A., Tansey, L. W., Payne, R., and Levin, V. A., The opiod mechanism of interferon-α action, *Anticancer Drugs*, 5, 90, 1994.
10. Meyers, C. A., Valentine, A. D., Wong, F. C. L., and Leeds, N. E., Reversible neurotoxicity of interleukin-2 and tumor necrosis factor: correlation of SPECT with neuropsychological testing, *J. Neuropsychiatr. Clin. Neurosci.*, 6, 285, 1994.
11. Plata-Salaman, C. R., Immunomodulators and feeding regulation: a humoral link between the immune and nervous system, *Brain Behav. Immunity*, 3, 193, 1989.
12. Huddleston-Secor, V., The inflammatory/immune response in critical illness, *Crit. Care. Nurs. Clin. NA*, 6, 251, 1994.

13. Hartung, T., Docke, W. D., Ganter, F., Krieger, G., Sauer, A., Stevens, P., Völk, H. D., and Wendel, A., Effect of granulocyte colony-stimulating factor treatment on *ex vivo* blood cytokine response in human volunteers, *Blood*, 85, 2482, 1995.
14. Schwartzentruber, D. J., White, D. E., Zweig, M. H., Weintraub, B. D., and Rosenberg, S. A., Thyroid dysfunction associated with immunotherapy for patients with cancer, *Cancer*, 68, 2384, 1991.
15. Derogatis, L. R., Morrow, G. R., Fetting, J., Penman, D., Piasetsky, S., Schmale, A. M., Henrichs, M., and Carnicke, C. L. M., Jr., The prevalence of psychiatric disorders among cancer patients, *JAMA*, 249, 751, 1983.
16. Sadovnik, A. O., Eisen, K., Ebers, G. C., and Paty, D. W., Cause of death in patients attending multiple sclerosis clinics, *Neurology*, 41, 1192, 1991.
17. Bocci, V., Central nervous system toxicity of interferons and other cytokines, *J. Biol. Regul. Homeost. Agents*, 2, 107, 1988.
18. Spiegel, R. J., Intron® A (Interferon-α-2b): clinical overview and future directions, *Semin. Oncol.*, 13, (Suppl 2), 89, 1986.
19. Kirkwood, J. M., Strawderman, M. H., Ernstoff, M. S., Smith, T. J., Bordern, E. C., and Blum, R. H., Interferon-α-2b adjuvant therapy of high-risk resected cutaneous melanoma: the Eastern Cooperative Oncology Group trial EST 1684, *J. Clin. Oncol.*, 14, 7, 1996.
20. Cortes, J., Kantarjian, H., O'Brien, S., Robertson, L. E., Pierce, S., and Talpaz, M., Results of interferon-α therapy in patients with chronic myelogenous leukemia 60 years of age and older, *Am. J. Med.*, 100, 452, 1996.
21. Quesada, J. R., Talpaz, M., Rios, A., Kurzrock, R., and Gutterman, J. U., Clinical toxicity of interferons in cancer patients: a review, *J. Clin. Oncol.*, 4, 234, 1986.
22. Pavol, M. A., Meyers, C. A., Rexer, J. L., Valentine, A. D., Mattis, P. J., and Talpaz, M., Pattern of neurobehavioral deficits associated with interferon-α therapy for leukemia, *Neurology*, 54, 947, 1995.
23. Anon., Intron® A Package Insert, Schering Corporation, Kenilworth, NJ 07033, January 1996.
24. Adams, F., Quesada, J. R., and Gutterman, J. U., Neuropsychiatric manifestations of human leukocyte interferon therapy in patients with cancer, *JAMA*, 252, 938, 1984.
25. Iaffaiolo, R. V., Fiorenza, L., D'avino, M., Frasci, G., La Mura, G., and Grossi, D., Neurotoxic effects of long-term treatment with low-dose α 2b interferon, *Curr. Ther. Res.*, 48, 403, 1990.
26. Talpaz, M., Kantarjian, H. M., McCredie, K. G. B., Keating, M. J., Trujillo, J., and Gutterman, J., Clinical investigation of human α interferon in chronic myelogenous leukemia, *Blood*, 69, 1280, 1987.
27. Meyers, C. A., Scheiel, R. S., and Forman, A. D., Persistent neurotoxicity of systemically administered interferon-α, *Neurology*, 41, 672, 1991.
28. Merimsky, O., Reider-Groswasser, I., Inbar, M., and Chaitchik, S., Interferon-related mental deterioration and behavioral changes in patients with renal cell carcinoma, *Eur. J. Cancer*, 26, 596, 1990.
29. Renault, P. F., Hoofnagle, J. H., Park, Y., Mullen, K. D., Peters, M., Jones, B., Rustgi, V., and Jones, A., Psychiatric complications of long-term interferon-α therapy, *Arch. Intern. Med.*, 147, 1577, 1987.
30. Suter, G. C., Westmoreland, B. F., Sharbrough, F. W., and Hermann, R. C., Electroencephalographic abnormalities in interferon encephalopathy: a preliminary report, *Mayo Clin. Proc.*, 59, 847, 1984.

31. Anon., IFNB Multiple Sclerosis Study Group and the University of British Columbia MS/MRI Analysis Group, Interferon β-1b in the treatment of multiple sclerosis: final outcome of the randomized controlled trial, *Neurology*, 45, 1277, 1995.
32. Jacobs, L. S., Cookfair, D. L., Rudick, R. A., Herndon, R. M., Richert, J. R., Salazar, A. M., and Fischer, J. S., Intramuscular interferon β 1a for disease progression in relapsing multiple sclerosis, *Ann. Neurol.*, 39, 285, 1996.
33. Goodkin, D. E., Interferon β-1b, *Lancet*, 344, 1057, 1994.
34. Arnason, B. G. W., Short analytical review: Interferon β in multiple sclerosis, *Clin. Immunol. Immunopathol.*, 81, 1, 1996.
35. Pliskin, N. H., Towle, V. L., Hamer, D. P., Reder, A. T., Noronha, A., Pietre, S., and Arnason, B. G. W., The effects of interferon-β on cognitive function in multiple sclerosis, *Ann. Neurol.*, 36 (Abstr.), 326, 1994.
36. Yung, W. K. A., Prados, M., Levin, V. A., Fetel, M. R., Bennett, J., Mahaley, M. S., Salcman, M., and Etcubanas, E., Intravenous recombinant interferon β in patients with recurrent malignant gliomas: a phase I/II study, *J. Clin. Oncol.*, 9, 1945, 1991.
37. Packer, R. J., Prados, M., Phillips, P., Nicholson, H. S., Boyette, J. M., Goldwein, J., and Rorke, L. B., Treatment of children with newly diagnosed brain stem gliomas with intravenous recombinant β-interferon and hyperfractionated radiation therapy, *Cancer*, 77, 2150, 1996.
38. Allen, J., Packer, R., Bleyer, A., Zeltzer, P., Prados, M., and Nirenberg, A., Recombinant interferon β: a phase I-II trial in children with recurrent brain tumors, *J. Clin. Oncol.*, 9, 783, 1991.
39. Anon., The International Chronic Granulomatous Disease Cooperative Study Group., A controlled trial of interferon-γ to prevent infection in chronic granulomatous disease, *N. Engl. J. Med.*, 324, 509, 1991.
40. Bolinger, A. M. and Tafubelam, M. A., Recombinant interferon-γ for the treatment of chronic granulomatous disease and other disorders, *Clin. Pharmacol.*, 11, 834, 1992.
41. Kurzrock, R., Quesada, J. R., Rosenblum, M. G., Sherwin, S. A., and Gutterman, J. U., Phase I study of IV administered recombinant γ interferon in cancer patients, *Cancer Treat. Rep.*, 70, 1357, 1986.
42. Thompson, J. A., Lee, D. J., Lindgren, C. G., Benz, L. A., Collins, C., Levitt, D., and Fefer, A., Influence of dose and duration of infusion of interleukin-2 on toxicity and immunomodulation, *J. Clin. Oncol.*, 6, 669, 1988.
43. Fyfe, G., Fisher, R. I., Rosenberg, S. A., Sznol, M., Parkinson, D. R., and Louie, A. C., Results of treatment of 255 patients with metastatic renal cell carcinoma who received high-dose recombinant interleukin-2 therapy, *J. Clin. Oncol.*, 13, 688, 1995.
44. Rosenberg, S. A., Lotze, M. T., Yang, J. C., Aebersold, P. M., Linehan, W. M., Seipp, C. A., and White, D. E., Experience with the use of high-dose interleukin-2 in the treatment of 652 cancer patients, *Ann. Surg.*, 210, 474, 1989.
45. Denicoff, K. D., Rubinow, D. R., Papa, M. Z., Simpson, C., Seipp, C. A., Lotze, M. T., Chang, A. E., Rosenstein, D., and Rosenberg, S. A., The neuropsychiatric effects of treatment with interleukin-2 and lymphokine-activated killer cells, *Ann. Intern. Med.*, 107, 293, 1987.
46. Smith, M. J. and Khayat, D., Residual acute confusional and hallucinatory syndromes induced by interleukin-2/α interferon treatment, *Psycho-oncology*, 1, 115, 1992.
47. Kovacs, J. A., Vogel, S., and Jeffrem, M., Controlled trial of interleukin-2 infusion in patients infected with the human immunodeficiency virus, *N. Engl. J. Med.*, 335, 1350, 1996.

48. Buter, J., De Vries, E. G. E., Sleijer, D. T., Willemse, P. H. B., and Mulder, N. H., Neuropsychiatric symptoms during treatment with interleukin-2, *Lancet*, 341, 628, 1993.
49. Atzpodien, J. and Kirchner, H., The out-patient use of recombinant human interleukin-2 and interferon-α-2b in advanced malignancies, *Eur. J. Cancer*, 27 (Suppl. 4), S88, 1991.
50. Fenner, M. H., Hanninen, E. L., Kirchner, H. H., Poliwoda, H., and Atzpodien, J., Neuropsychiatric symptoms during treatment with interleukin-2 and interferon-α, *Lancet*, 341, 372, 1993.
51. Vial, T. and Descotes, J., Clinical toxicity of cytokines used as haemopoietic growth factors, *Drug Safety*, 13, 371, 1995.
52. Gilleece, M. H., Scarffe, J. H., Ghosh, A., Heyworth, C. M., Bonnem, E., Testa, N., Stern, P., and Dexter, T. M., Recombinant human interleukin 4 (IL-4) given as daily subcutaneous injections — a phase I dose toxicity trial, *Br. J. Cancer*, 66, 204, 1992.
53. Prendiville, J., Thatcher, N., Lind, M., McIntosh, R., Ghosh, A., Stern, P., and Crowther, D., Recombinant human interleukin-4 (rhu IL-4) administered by the intravenous and subcutaneous routes in patients with advanced cancer – a phase I toxicity study and pharmacokinetic analysis, *Eur. J. Cancer*, 12, 1700, 1993.
54. Lamont, A. G. and Adorini, L., IL-12: a key cytokine in immune regulation, *Immunol. Today*, 17, 214, 1996.
55. Banks, R. E., Patel, P. M., and Selby, P. J., Interleukin-12: a new clinical player in cytokine therapy, *Br. J. Cancer*, 71, 655, 1995.
56. Atkins, M. B., Robertson, M., Gordon, M. S., Lotze, M. T., Du Bois, J., Ritz, J., Sandler, A., Edington, H. D., and Sherman, M. L., Phase I evaluation of intravenous recombinant human interleukin-12 in patients with advanced malignancies, *Proc. Am. Soc. Clin. Oncol.*, 15 (Abstr.), 718, 1996.
57. Platanias, L. C. and Vogelzang, N. J., Interleukin-1: biology, pathophysiology, and clinical prospects, *Am. J. Med.*, 89, 621, 1990.
58. Weisdorf, D., Katsanis, E., Verfaillie, C., Ramsay, N. K. C., Haake, R., Garrison, L., and Blazar, B. R., Interleukin 1 α administered after autologous transplantation: a phase I/II clinical trial, *Blood*, 84, 2044, 1994.
59. Nemunaitis, J., Appelbaum, R., Lilleby, K., Buhles, W. C., Rosenfeld, C., Zeigler, Z. R., Shadduck, R. K., Singer, J. W., Meyer, W., and Buckner, C. D., Phase I study of recombinant human interleukin-1 β in patients undergoing autologous bone marrow transplant for acute myelogenous leukemia, *Blood*, 83, 3473, 1994.
60. Smith, J. W., II, Urba, W. J., Curti, B. D., Elwood, L. J., Steis, R. G., Janik, J. R., and Sharfman, W. H., The toxic and hematologic effects of interleukin-1 α administered in a phase I trial to patients with advanced malignancies, *J. Clin. Oncol.*, 10, 1141, 1992.
61. Redman, G., Abubakr, Y., Chou, T., Esper, P., and Flaherty, L. E., Phase II trial of recombinant interleukin-1 β in patients with metastatic renal cell carcinoma, *J. Immunother.*, 16, 211, 1994.
62. Rinehart, J. J., Hersh, E. M., Issell, B., Triozzi, P. L., Buhles, W., Nedihart, J. A., and Rothenberg, M. L., Phase I trial of recombinant human interleukin 1β (rhIL-1β), carboplatin, and etoposide in patients with solid cancer: evaluation of the rhIL-1β toxicity and hematologic effects in Southwest Oncology Group study 8940, *Cancer Invest.*, in press.
63. Curti, B. D. and Smith, J. W., II, Interleukin-1 in the treatment of cancer, *Pharmacol. Ther.*, 65, 291, 1995.

64. Hofman, F. M. and Hinton, D. R., Cytokine interactions in the central nervous system, *Reg. Immunol.*, 3, 268, 1990/1991.
65. Hieber, U. and Heim, M. E., Tumor necrosis factor for the treatment of malignancies, *Oncology*, 51, 142, 1994.
66. Kammuller, M. E., Recombinant human interleukin-6: safety issues of a pleiotropic growth factor, *Toxicology*, 105, 91, 1995.
67. Budd, G. T., Pelley, R., Samuels, B., Gockerman, J., Margolin, K., Zalupski, M., Manfreda, S., George, M., and Bukowski, R., Phase II randomized trial of simultaneous IL-6 and G-CSF following MAID chemotherapy, *Proc. Am. Soc. Clin. Oncol.*, 14 (Abstr.), 256, 1995.
68. Wos, E., Olencki, T., Budd, G. T., Peereboom, D., Finke, J., Wood, L., McLain, D., Tubbs, R., Redovan, C., and Bukowski, R. M., Phase IA/IB trial of recombinant human granulocyte macrophage colony-stimulating factor and interleukin-6 in patients with renal cell carcinoma: clinical and hematologic effects, *Proc. Am. Soc. Clinc. Oncol.*, 14 (Abstr.), 260, 1995.
69. Eschbach, J. W., Abdulhadi, M. H., Browne, J. K., Delano, B. G., Downing, M. R., and Egrie, J. C., Recombinant human erythropoietin in anemic patients with end-stage renal disease: results of a phase III multicenter trial, *Ann. Intern. Med.*, 111, 992, 1989.
70. Abraham, P. A., Practical approach to initiation of recombinant human erythropoietin therapy and prevention and management of adverse effects, *Am. J. Nephrol.*, 10 (Suppl. 2), 7 1990.
71. Raine, A. E. G., Seizures and hypertension events, *Semin. Nephrol.*, 10 (Suppl. 1), 40, 1990.
72. Abels, R. I., Use of recombinant human erythropoietin in the treatment of anemia in patients who have cancer, *Semin. Oncol.*, 19 (Suppl. 8), 29, 1992.
73. Henry, D. H., Beall, G. N., Benson, C. A., Carey, J., Cone, L. A., Eron, L. J., and Fiala, M., Recombinant human erythropoietin in the treatment of anemia associated with human immunodeficiency virus (HIV) infection and zidovudine therapy, *Ann. Intern. Med.*, 117, 739, 1992.
74. Eschbach, J. W., Egrie, J. C., Downing, M. R., Browne, J. K., and Adamson, J. W., The safety of epoetin-α: results of clinical trials in the United States, *Contrib. Nephrol.*, 88, 72, 1991.
75. Goodnough, L. T., Rudnick, S., Price, T. H., Price, T. H., Ballas, S. K., Collins, M. L., and Crowley, J. P., Increased preoperative collection of autologous blood with recombinant human erythropoietin therapy, *N. Engl. J. Med.*, 321, 1163, 1989.
76. Flaherty, K. K., Grimm, A. M., and Vlasses, P. H., Epoetin: human recombinant erythropoietin, *Clin. Pharmacol.*, 8, 769, 1989.
77. Bennett, W. M., A multicenter clinical trial of epoetin β for anemia of end-stage renal disease, *J. Am. Soc. Nephrol.*, 1, 990, 1991.
78. Case, D. C., Bukowski, R. M., Carey, R. W., Fishkin, E. H., Henry, D. H., Jacobson, R. J., and Jones, S. E., Recombinant human erythropoietin therapy for anemic cancer patients on combination chemotherapy, *J. Natl. Cancer Inst.*, 85, 801, 1993.
79. Paganini, E. P., Latham, D., and Abdulhadi, M., Practical considerations of recombinant human erythropoietin therapy, *Am. J. Kidney Dis.*, 14 (Suppl. 1), 19, 1989.
80. Steinberg, H., Erythropoietin and visual hallucinations, *N. Engl. J. Med.*, 325, 285, 1991.
81. Temple, R. M., Deary, I. J., and Winney, R. J., Recombinant erythropoietin improves cognitive function in patients maintained on chronic ambulatory peritoneal dialysis, *Nephrol. Dial.*, 10, 1733, 1995.

82. Laupacis, A., Canadian Erythropoietin Study Group. Changes in quality of life and functional capacity in hemodialysis patients treated with recombinant human erythropoietin, *Semin. Nephrol.*, 10 (Suppl. 1), 11, 1990.
83. Anderlini, P., Przepiorka, D., Champlin, R., and Korbling, M., Biologic and clinical effects of granulocyte colony-stimulating factor in normal individuals, *Blood*, 88, 2819, 1996.
84. Anon., Neupogen® Package Insert, Amgen Inc., Thousand Oaks, CA, December 1994.
85. Anderlini, P., Przepiorka, D., Seong, D., Miller, P., Sundberg, J., Lichtiger, B., Norfleet, F., Chan, K. W., Champlin, R., and Korbling, M., Clinical toxicity and laboratory effects of granulocyte-colony-stimulating factor (filgrastim) mobilization and blood stem cell apheresis from normal donors and analysis of charges for the procedures, *Transfusion*, 36, 590, 1996.
86. DePlacido, S., Perrone, F., Carlomagno, C., Lauria, R., Morabito, A., DeLaurentiis, M., Gallo, C., and Bianco, A. R., Toxicity and activity comparison of GM-CSF and G-CSF: ancillary results from two parallel phase I studies of high-dose chemotherapy for metastatic breast cancer, *Proc. Am. Soc. Clin. Oncol.*, 14 (Abstr.), 257, 1995.
87. Bregni, M., Siena, S., Di Nicola, M., Dodero, A., Peccatori, F., Ravagnani, F., Danesini, G., Laffranchi, A., Bonadonna, G., and Gianni, A. M., Comparative effects of granulocyte-macrophage colony-stimulating factor and granulocyte colony-stimulating factor after high-dose cyclophosphamide cancer therapy, *J. Clin. Oncol.*, 14, 628, 1996.
88. Neumunaitis, J., Granulocyte-macrophage colony-stimulating factor: a review from preclinical development to clinical application, *Transfusion*, 33, 70, 1993.
89. Neidhart, J. A., Hematopoietic colony-stimulating factors, *Cancer*, 70, 913, 1992.
90. Scarffe, J. H., Emerging clinical uses for GM-CSF, *Eur. J. Cancer*, 27, 1493, 1991.
91. Cebon, J., Lieschke, G. J., Bury, R. W., and Morstyn, G., The dissociation of GM-CSF efficacy from toxicity according to route of administration: a pharmacodynamic study, *Br. J. Haematol.*, 80, 144, 1992.
92. Honkoop, A. H., Hoekman, K., Wagstaff, J., van Groeningen, C. J., Vermokern, J. B., Boven, E., and Pinedo, H. M., Continuous infusion or subcutaneous injection of granulocyte-macrophage colony-stimulating factor: increased efficacy and reduced toxicity when given subcutaneously, *Br. J. Cancer*, 74, 1132, 1996.
93. O'Day, S. J., Rabinowe, S. N., Neuberg, D., Freedman, A. S., Soiffer, R. J., Spector, N. A., Robertson, M. J., Anderson, K., Whelan, M., Pesek, K., Ritz, J., and Nadler, L. M., A phase II study of continuous infusion recombinant human granulocyte-macrophage colony-stimulating factor as an adjunct to autologous bone marrow transplantation for patients with non-Hodgkin's lymphoma in first remission, *Blood*, 83, 2707, 1994.
94. Sanda, M. G., Yang, J. C., Topalian, S. L., Groves, E. S., Childs, A., Belfort, R., de Smet, M. D., Schwartzentruber, D. J., White, D. E., and Lotze, M. T., Intravenous administration of recombinant human macrophage colony-stimulating factor to patients with metastatic cancer: a phase I study, *J. Clin. Oncol.*, 10, 1643, 1992.
95. Bukowski, R. M., Budd, G. T., Gibbons, J. A., Bauer, R. J., Childs, A., Antal, J., Finke, J., Tuason, L., Lorenzi, V., McLain, D., Tubbs, R., Edinger, M., and Thomassen, M. J., Phase I trial of subcutaneous recombinant macrophage colony-stimulating factor: clinical and immunomodulatory effects, *J. Clin. Oncol.*, 12, 97, 1994.
96. Crawford, J., Lau, D., Erwin, R., Rich, W., McGuire, B., and Meyers, F., A phase I trial of recombinant methionyl human stem cell factor (SCF) in patients (PTS) with advanced non-small cell lung carcinoma (NSCLC), *Proc. Am. Soc. Clin. Oncol.*, 12 (Abstr.), 135, 1993.

97. Demetri, G., Costa, J., Hayes, D., Sledge, G., Galli, S., Hoffman, R., Merica, E., Rich, W., Harkins, B., McGuire, B., and Gordon, M., A phase I trial of recombinant methionyl human stem cell factor (SCF) in patients with advanced breast carcinoma pre- and post-chemotherapy with cyclophosphamide and doxorubicin, *Proc. Am. Soc. Clin. Oncol.*, 12 (Abstr.), 142, 1993.
98. Kudoh, S., Sawa, T., Kurihara, N., Furuse, K., Kurita, Y., Fukuoka, M., Takada, M., Takaku, F., Ogawa, M., and Ariyoshi, Y., The SDZ ILE 964 (IL-3) Study Group. Phase II study of recombinant human interleukin-3 administration following carboplatin and etoposide chemotherapy in small-cell lung cancer patients, *Cancer Chemother. Pharmacol.*, Suppl. 38, S89, 1996.
99. Rinehart, J., Margolin, K. A., Triozzi, P., Hersh, E., Campion, M., Resta, D., and Levitt, D., Phase I trial of recombinant interleukin-3 before and after carboplatin/etoposide chemotherapy in patients with solid tumors: a Southwest Oncology Group study, *Clin. Cancer Res.*, 1, 1139, 1995.
100. Newland, A. D., Is interleukin-3 active in anticancer drug-induced thrombocytopenia? *Cancer Chemother. Pharmacol.*, Suppl. 38, S83, 1996.
101. Du, X. X. and Williams, D. A., Interleukin-11: a multifunctional growth factor derived from the hematopoietic microenvironment, *Blood*, 83, 2023, 1994.
102. Tepler, I., Elias, L., Smith, J. W., II, Hussien, M., Rosen, G., Chang, A. Y.-C., Moore, J. O., Gordon, M. S., Kuca, B., Beach, K. J., Loewy, J. W., Garnick, M. B., and Kaye, J. A., A randomized placebo-controlled trial of recombinant human interleukin-11 in cancer patients with severe thrombocytopenia due to chemotherapy, *Blood*, 87, 3607, 1996.
103. Gordon, M. S., McCaskill-Stevens, W. J., Battiato, L. A., Loewy, J., Loesch, D., Breeden, E., Hoffman, R., Beach, K. J., Kuca, B., Kaye, J., and Sledge, G. W., Jr., A phase I trial of recombinant human interleukin-11 (neumega rhIL-11 growth factor) in women with breast cancer receiving chemotherapy, *Blood*, 87, 3615, 1996.
104. Basser, R. L., Rasko, J. E. J., Clarke, K., Cebon, J., Green, M. D., Hussein, S., Alt, C., Menchaca, D., Tomita, D., Marty, J., Fox, R. M., and Begley, C. G., Thrombopoietic effects of pegylated recombinant human megakaryocyte growth and development factor in patients with advanced cancer, *Lancet*, 348, 1279, 1996.
105. Fanucchi, M., Glaspy, J., Crawford, J., Garst, J., Figlin, R., Sheridan, W., Menchaca, D., Tomita, D., Ozer, H., and Harker, L., Effects of polyethylene glycol-conjugated recombinant human megakaryocyte growth and development factor on platelet counts after chemotherapy for lung cancer, *N. Engl. J. Med.*, 336, 404, 1997.
106. Valentine, A. D., Meyers, C. A., and Talpaz, M., Treatment of neurotoxic side effects of interferon-α with naltrexone, *Cancer Invest.*, 13, 561, 1995.
107. Goldman, L. S., Successful treatment of interferon-α-induced mood disorders with nortriptyline, *Psychosomatics*, 35, 412, 1994.
108. Levenson, J. and Fallon, H., Fluoxetine treatment of depression caused by interferon-α, *Am. J. Gastroenterol.*, 88, 760, 1993.
109. Van Thiel, D. H., Friedlander, L., Molloy, P. J., Fagiuoli, S., Kania, R. J., and Caraceni, P., Interferon-α can be used successfully in patients with hepatitis C virus-positive chronic hepatitis who have a psychiatric illness, *Eur. J. Gastroenterol. Hepatol.*, 7, 165, 1995.
110. Ashai, N. I., Paganini, E. P., and Wilson, J. M., Intravenous versus subcutaneous dosing of epoetin: A review of the literature, *Am. J. Kidney Dis*, 22 (Suppl. 1), 23, 1993.

9 Stress and Inflammation

Paul H. Black and Ari S. Berman

CONTENTS

9.1 Introduction 115
9.2 Inflammation 116
    9.2.1 Infection and Trauma..116
    9.2.2 Immunogenic and Neurogenic Inflammation..................................116
9.3 Cytokines and the Acute Phase Response 117
9.4 Mediators of the Acute Phase Response 119
    9.4.1 Nervous Innervation and Inflammation..119
    9.4.2 Substance P and the Inflammatory Response 120
        9.4.2.1 SP 120
        9.4.2.2 SP and Inflammation: Infection and Trauma 120
        9.4.2.3 SP and Inflammation: Immunogenic and
                Neurogenic Inflammation 121
        9.4.2.4 SP and Inflammatory Disease 121
9.5 Neurobiology of Stress...122
9.6 Stress Induces Cytokines 123
9.7 Stress Induces the Acute Phase Response 124
    9.7.1 Stress and Acute Phase Proteins 124
    9.7.2 Stress and Fever..125
9.8 Stress and SP 125
9.9 Stress and Inflammatory Disease 126
References...127

9.1 INTRODUCTION

There is now overwhelming evidence that the central nervous and immune systems interact. One consequence of this interaction is that mediators are released from immune cells subsequent to their activation by various neuropeptides, neurohormones, and/or neurotransmitter substances evoked by stress. Many of these mediators are the same as those released when an organism is infected, or sustains trauma and tissue injury; in these instances, the mediators produce an acute phase inflammatory response. We hypothesize that stress, by evoking these same mediators, may also cause a similar inflammatory response and may, therefore, be a factor participating in the onset, pathogenesis, and severity of certain inflammatory diseases of unknown etiology.

In this chapter we will review the evidence from both human and animal studies that many different stressors induce soluble mediators of inflammation including cytokines and acute phase reactants. We shall also consider the contribution made by a neuropeptide, substance P (SP), in cytokine induction, as well as the role of other mediators evoked by stress which participate in the inflammatory response.

9.2 INFLAMMATION

9.2.1 INFECTION AND TRAUMA

In response to an infectious agent or a nonspecific form of tissue injury, the host mounts an acute inflammatory reaction in the tissue affected. Chemotactic factors are produced by the host or are derived from infectious microorganisms which attract various cells to the perturbed site. Polymorphonuclear (PMN) leukocytes, mast cells, mononuclear phagocytes (macrophages), and platelets are important cellular components of the inflammatory reaction. These cells are stimulated by various factors derived from the host, such as SP, a neuropeptide derived from sensory neurons (see below), and/or derived from the microorganism, such as lipopolysaccharide (endotoxin), which is contained in the cell walls of Gram-negative bacteria. Once stimulated, these cells release inflammatory mediators such as prostaglandins from PMNs and/or macrophages or various cytokines from macrophages and/or lymphocytes. Generally, an early inflammatory response to a nonspecific injury is characterized by the production of thin exudates containing serous fluid, red blood cells, and fibrinogen. This reaction corresponds to the hemodynamic phase of inflammation in which blood flow and capillary permeability increase. It is followed by the cellular phase of inflammation during which phagocytic white blood cells (PMNs and macrophages) are elevated in the local area of disturbance.[1]

In contrast to the self-limiting process of acute inflammation, persistent or chronic inflammation is more prolonged and is usually caused by persistent irritants or microorganisms, most of which are relatively resistant to phagocytosis and other inflammatory mechanisms. Chronic inflammation is characterized by a relative increase in the number of mononuclear cells rather than granulocytes and proliferation of fibroblasts that cause scarring and deformity instead of the formation of an exudate.[1]

9.2.2 IMMUNOGENIC AND NEUROGENIC INFLAMMATION

The immunogenic inflammatory response occurs when foreign materials are encountered in a tissue. Immunogenic inflammation occurs when an antigen binds to an antibody or leukocyte receptor to trigger the inflammatory cascade.[1] Previous sensitization is required and the inflammatory response is of the immediate or cell-mediated hypersensitivity type. Neurogenic inflammation occurs when a chemical combines with the chemical receptors on sensory neurons, resulting in the release of SP and other neuropeptides.[2,3]

The neurogenic inflammatory response can also arise when a nerve impulse is transmitted down an axon, which results in a local release of SP and other peptides, which subsequently gives rise to edema and vasodilatation.[4] It has been shown in the skin,[3] eyes,[5] and joints[6] that simulation of sensory nerves induces neurogenic inflammation. Perhaps the most potent mediator of neurogenic inflammation is the neuropeptide, SP.[7] SP induces vasodilatation, increases vascular permeability, and contributes to the production of flare and plasma extravasation; these reactions correspond to the hemodynamic phase of inflammation. Further, it has been shown that capsaicin, a drug which is toxic to certain sensory nerves and is used to deplete SP in sensory nerve endings, can abolish urticarial reactions.[8]

There is an interrelationship between the mediators (i.e., SP and histamine) of immunogenic and neurogenic inflammatory reactions, SP mediates the neurogenic response by activating other sensory nerves and causing the degranulation of mast cells and subsequent release of histamine.[9] In addition to its role in mediating the ensuing inflammatory response, histamine, by binding to its receptor, can also activate sensory nerve endings which release SP, thus potentiating the inflammatory reaction.[10]

9.3 CYTOKINES AND THE ACUTE PHASE RESPONSE

Cytokines released from mononuclear cells (macrophages and lymphocytes) are important in deploying the immune system to contain the offending organism or irritant or to promote healing after an injury. The proinflammatory cytokines interleukins-1 and -6 (IL-1, IL-6), and tumor necrosis factor α (TNFα), have been characterized most thoroughly and are derived from macrophages. Many other cytokines (i.e., IL-2, -3, -4, -5, and -7 to -18) may also be involved in the inflammatory and/or immune response. In addition to the aforementioned roles, cytokines have many other systemic and/or metabolic effects. Of major importance is the impact of the proinflammatory cytokines on the liver and the induction of the acute phase response.

The body responds to tissue damage and infection by microorganisms by commencing a series of specific physiological reactions initiated by cells at the site of injury. This results in the recruitment of host defense mechanisms and the elaboration by the liver of a number of proteins — the acute phase reactants or acute phase proteins (APP). Together, this is called the acute phase response. Clinically, it is characterized by fever, leukocytosis, decreased appetite, altered sleep patterns, and malaise, also referred to as sickness behavior.[1]

The major cell involved in generating the acute phase response is the macrophage. A number of events including mast cell degranulation, platelet aggregation, and release of bacterial products will result both in the recruitment (chemotaxis) and activation of macrophages, such activated macrophages release cytokines. The early or alarm cytokines are IL-1 and TNFα which act locally on fibroblasts and endothelial cells to induce a secondary wave of cytokines (IL-6-type cytokines such as IL-11 and LIF), which act centrally through the induction of prostaglandins to induce fever and sickness behavior.[11] Both primary and secondary cytokines initiate other cellular and cytokine cascades that make up the acute phase response. Different cytokines may activate or inhibit cellular responses.

Cytokines are the primary stimulators of the liver to produce the acute phase proteins (APPs). These proteins have a wide range of activities, such as neutralizing inflammatory agents to minimize the extent of tissue damage or participating in the repair of host tissue that contribute to host defense.[12] For example, complement proteins are induced rapidly (see Table 9.1). Certain complement components promote the local accumulation of neutrophils, macrophages, and certain plasma proteins, which play a role in killing infectious agents, clearance of both foreign and cellular debris, and repair of damaged tissue.[11] Coagulation proteins, such as fibrinogen and von Willebrand factor likely play a role in wound healing. Proteinase inhibitors neutralize lysosomal hydrolases which are released during activation of phagocytic cells, thereby controlling the activity of certain proinflammatory enzyme cascades. They also participate in the regulation of the fibrinolytic pathway. Certain metal-binding proteins help prevent iron loss during infection and injury, diminishing the levels of heme iron available for bacterial uptake; metal-binding proteins also act as scavengers for potentially tissue-damaging oxygen free radicals. Elevation of the level of APPs vary, but they generally increase several fold.[12]

TABLE 9.1
Inflammatory Mediators Which Modulate Hepatic APR Synthesis in Humans

Complement Proteins	Coagulation Proteins
C2, C3, C4, C5, C9	Fibrinogen
Factor B	von Willebrand factor
C1 inhibitor	
C4 binding protein	
Metal-Binding Proteins	Other Proteins
Haptoglobin	Heme oxygenase
Hemopexin	Lipoprotein (a)
Managanese superoxide dismutase	Lipopolysaccharide binding-protein
Proteinase Inhibitors	Negative APRs
α2-Antiplasmin	Albumin
Heparin cofactor II	ApoAI
Plasminogen activator inhibitor I	ApoAII
Major APRs	Transferrin
Serum amyloid A	
C-reactive protein	
Serum amyloid P component	

Modified from Steel, D. M. and Whitehead, A. S., *Immunol. Today,* 15, 81, 1994. With permission.

In contrast, the major APPs [serum amyloid A (SAA) and C-reactive protein (CRP)] may increase approximately 1000-fold over normal levels. These proteins share homology and interact with other host defense systems. They have very short half lives, suggesting they may play an important role in host defense.[12] CRP can bind chromatin and histones to provide a clearing mechanism for necrotic material during inflammation, can act as an opsonin for bacteria, parasites, and immune complexes, and can activate the complement pathway.[12]

SAA also is massively induced. It is a small apolipoprotein that associates rapidly with the third fraction of HDL-3; it subsequently becomes the predominant lipoprotein on HDL-3, exceeding ApoA1 in quantity. Such displacement may interfere with cholesterol metabolism since SAA would enhance binding to macrophages relative to the diminished binding to hepatocytes at inflammatory sites; macrophages could then engulf cholesterol and lipid debris.[12] The excess cholesterol could thereby be redistributed for use in tissue repair or excretion. SAA can also induce collagenase synthesis in synovial fibroblasts, which suggests that it may have an autocrine role in inflammatory sites.[12] It should be stated that the precise role of SAA during inflammatory is not known for certain.

The acute phase response subsides in 24 to 48 h and the inflammatory response subsequently resolves. Corticosteroids provide an important negative feedback loop, which likely is coordinated with its role in evoking APPs (see below). Corticosteroids are produced by the hypothalamic-pituitary-adrenal axis. Cytokine feedback to the hypothalamus evokes corticotropin releasing factor (CRF) production which stimulates ACTH production from the anterior pituitary and thence corticosteroid synthesis in the adrenal cortex. Naturally occurring antagonists, such as the IL-6 receptor antagonist and soluble TNF receptor may also play a role in terminating the APR.[11] In addition, the "inhibitory cytokines" IL-4 and IL-10, both produced by T-helper lymphocytes (TH$_2$), may participate since these cytokines downregulate synthesis of several cytokines from macrophages and induce IL-1 receptor antagonist production.[11]

With persistence of the organism or tissue injury or by disruption of normal host control mechanisms, this normal pathway may convert to a chronic inflammation. There is some evidence that IL-4 may mediate the switch to a more monocytic phase of inflammation by increasing the synthesis of VCAM, an attachment protein on endothelial cells for monocytes, but the precise switch from acute to chronic inflammation remains unknown.[11]

9.4 MEDIATORS OF THE ACUTE PHASE RESPONSE

We have discussed the normal function of the proinflammatory cytokines IL-1, IL-6, and TNFα. It is evident that these cytokines are likely to participate in the induction of the APPs from the liver and are responsible for the febrile response and the sickness behavior that characterize the acute phase response. We shall now consider several other possible mediators that are responsible for both cytokine production from macrophages and the production of APPs from liver cells.

9.4.1 NERVOUS INNERVATION AND INFLAMMATION

We have considered neurogenic inflammation and its mediation by SP. We should now like to point out the relationship between nervous innervation and other inflammatory processes. Organs with a high density of neuropeptide receptors, such as the intestines and the lung, have been proposed to be more susceptible to perturbations from inflammation.[13] Indeed, nervous innervation of an organ is a requirement for establishing certain inflammatory reactions experimentally. For example, adjuvant arthritis can be elicited by the injection of Freund's adjuvant. Joints which are most densely innervated are the most susceptible to experimentally induced arthritis.[14] Moreover, if the nerve innervating a joint is cut, arthritis cannot be induced in the denervated joint.[14] This is reminiscent of clinical cases where rheumatoid inflammatory arthritis resolves on the side affected by a cerebral vascular accident. This relationship between nervous innervation and inflammation raises the question as to whether any inflammatory process has a requisite neurogenic component. We shall now consider certain neuropeptides and their role in inflammation.

9.4.2 Substance P and the Inflammatory Response

9.4.2.1 SP

SP is a neuropeptide containing 11 amino acids that is widely distributed in the central, peripheral, and enteric nervous systems of many species.[15-17] SP belongs to a family of bioactive peptides, the tachykinins, defined by their functionally significant conserved carboxyl-terminal sequences. This group also includes neurokinin A (substance K) and neurokinin B. SP functions in the central nervous system (CNS) as a neurotransmitter/neuromodulator and is localized in distinct areas of the brain. In the dorsal horn of the spinal cord, SP is involved in sensory, and most notably, nociceptive pathways.[18] In the periphery, SP has been identified in C-type sensory nerve endings and autonomic afferents throughout the body and mediates a neuroinflammatory response at the site of nerve trauma and injury.

SP has also been shown to affect nonneural tissue. SP plays a role in intestinal muscle contraction, arteriolar vasodilatation, and salivary gland secretion, and increases microvascular permeability which may facilitate traffic to sites of inflammation. In addition, the release of SP in response to electrical nerve stimulation, noxious heat, and chemical irritants has been demonstrated in the skin, eye, dental pulp, and lung.[7]

9.4.2.2 SP and Inflammation: Infection and Trauma

Long known as mediating neural inflammation, there is much evidence that SP is involved in nonneural inflammation as well as inflammatory diseases (see below). SP has pronounced effects on the immune system. SP-containing neurons have been identified in the spleen and in association with lymphocytes and macrophages.[19] Lymphocytes and macrophages, as well as leukocytes and mast cells, have receptors for SP and SP can affect the activation of these cells. Specifically, SP has been shown to induce the release of cytokines from lymphocytes,[20] from mononuclear leukocytes,[21] from bone marrow cells,[22] and from macrophages.[23-25] With respect to macrophages, SP can induce the release of the proinflammatory cytokines IL-1, IL-6, and TNFα.[26-28] SP can also stimulate macrophages to produce and secrete prostaglandin

E₂ (PGE$_2$) and thromboxane B2, important mediators of inflammation.[28] SP can also act as a chemoattractant for human and guinea pig macrophages.[28,29] These properties of SP lead to the production of known essential mediators of inflammation.

It is likely that SP is involved in the early or hemodynamic phase of inflammation (see above). SP can relax vascular tone and can increase local vascular permeability thereby promoting hyperemia and enhancing the diapedesis of inflammatory cells and fluid from the circulation.[30] Elevated levels of SP have been found at local sites of inflammation[31] and in granulomas of murine schistosomiasis mansoni.[13] In addition, it has been shown that eosinophils within the liver[13] and macrophages[32] can synthesize SP.

Finally, it has been demonstrated that SP can act as a priming substance for other mediators of inflammation or increased permeability.[33] It has been shown to be a priming factor for neutrophils,[34,35] eosinophils,[36,37] mast cells,[38] and macrophages.[25]

9.4.2.3 SP and Inflammation: Immunogenic and Neurogenic Inflammation

Inoculation of SP into the skin induces an inflammatory reaction composed of wheal, flare, pain, and pruritis.[7] SP has been implicated in the pathogenesis of contact urticaria,[39] has been shown to enhance irritant allergic contact dermatitis,[33] and has been shown to play a role in cutaneous and bronchial hypersensitivity.[40,41] Inoculation of capsaicin, a neurotoxin that depletes neurons of SP, abolishes an urticarial reaction. Somatostatin, a peptide that antagonizes SP and is anti-inflammatory, is found in the same nerve; its concentration is inversely proportional to SP.[42] Therefore, depending on the concentrations of SP and somatostatin present at a site of inflammation and the sensitivity of the immune cells to the effects of these peptides, the immune response could be inhibited or enhanced.

Nerves containing SP have been seen in association with mast cells in the gastrointestinal tract.[43] This structural association has led to the suggestion that mast cells and SP-containing nerves may form a functional unit involved in an axon reflex.[44] An axon reflex involves an afferent stimulus in one axon that is propagated to another efferent axon. This would explain how a reaction at one locus would cause symptoms at a locus distant from the allergen or irritant. In this instance, the axon reflex represents a combination of neurogenic and immunogenic inflammatory reactions involving SP and histamine, respectively.[45]

9.4.2.4 SP and Inflammatory Disease

A large body of evidence is accumulating which indicates that SP is involved in certain inflammatory diseases such as arthritis, asthma, and inflammatory bowel disease, i.e., Crohn's disease and ulcerative colitis.[30] SP has shown to increase the severity of adjuvant-induced arthritis in rats while SP antagonists reduce the severity of disease.[14] In addition, rats treated with capsaicin at birth fail to develop adjuvant-induced arthritis.[46] In the adjuvant-induced arthritis model, joints that had developed severe arthritis were more densely innervated by SP-containing neurons than were joints with less severe arthritis. Other studies indicate that synoviocytes from patients with rheumatoid arthritis release PGE_2 and collagenase in the presence of SP, where control synoviocytes do not.[47,48] In addition, macrophages from inflamed joints were shown to synthesize and secrete IL-1 and TNFα when incubated with SP.[49] These studies indicate that, in addition to its direct effect on cells, SP may also amplify other ongoing inflammatory reactions. A number of other studies have reported increases in both IL-1 and SP in the fluid from inflamed joints of patients with rheumatoid arthritis.[49] Indeed, much of the present therapy is directed at neutralizing or abrogating the increased cytokines present in this disease.[50]

In patients with both Crohn's disease and ulcerative colitis, mucosal-submucosal layer concentrations of SP were found to be significantly elevated.[51] It should be noted that other studies of surgical specimens from patients with these diseases have shown that, in addition to increases in levels of SP, an increased number of specific SP receptor binding sites are found on arterioles, venules, and regional lymph nodes of the inflamed areas.[52] The latter study was based on correlational data, but it suggests that SP may induce SP receptor expression in an autocrine or paracrine manner.

There is more direct evidence that SP ligand may induce SP receptor expression since the SP antagonist, CP96,345, but not its inactive enantiomer, prevents the induction of SP receptor mRNA in *Clostridium dificile* toxin A-induced enteritis in ileal loops in rats.[53] These studies indicate that SP participates in the inflammatory intestinal response to this toxin. More recent studies of this group confirm the involvement of SP in the *Clostridium difficile* toxin A-induced enteritis.[54] Injection of toxin A into rat ileum is followed by a rapid increase in SP content in lumbar dorsal root ganglia and mucosal scrapings 30 to 60 min later. Evidence of enteritis appears after 2 h. Macrophages from the lamina propria release greater amounts of TNFα spontaneously and after SP treatment than do control macrophages. Treatment of rats with an SP antagonist, CP96,345, but not the inactive enantiomer, prevents the TNFα release. These studies indicate that activated lamina propria macrophages, during toxin A enteritis, can lead to the release of cytokines by SP, suggesting an autocrine/paracrine regulation of cytokine secretion. Therefore, there is overwhelming evidence that SP is involved in nonneurogenic inflammatory processes. The question arises whether SP is involved in the response to stress.

9.5 NEUROBIOLOGY OF STRESS

We define stress as a state of disharmony or threatened homeostasis provoked by a psychological, environmental, or physiological stressor. Much evidence indicates that CRF, synthesized in the paraventricular nucleus in the hypothalamus, is the coordinator of the stress response. CRF stimulates the pituitary gland to produce corticotropin (ACTH), which causes the adrenal gland to produce corticosteroids. CRF also stimulates the locus coeruleus, a dense collection of autonomic nuclei in the brainstem, to secrete norepinephrine (NE). The adrenal medulla is also stimulated to secrete epinephrine (E). NE and E are catecholamines which, together with the corticosteroids, are the major stress hormones.

Stress and Inflammation

Many other molecules such as endorphins/enkephalins, α-melanocyte-stimulating hormone, arginine vasopressin, dopamine, and serotonin for which immune cells have receptors may also be released with stress.[55,56] However, the contributions of these mediators, insofar as they are known, will not be discussed in detail. Studies of the interactions of neuromediators evoked by stress and/or various emotions with the immune system, and the subsequent impact of immune mediators such as cytokines on the CNS which serve to downregulate the immune response are the subject matter of an emerging discipline called psychoneuroimmunology (see References 55 to 57 for review).

Thus far, we have considered inflammation and the acute phase response. We have also described, in brief, the neurobiology of stress. Our hypothesis states that stress can induce an inflammatory reaction or at least be a factor contributing to the pathogenesis of certain inflammatory diseases. What is the evidence that stress can induce an inflammatory reaction and the acute phase response?

9.6 STRESS INDUCES CYTOKINES

Many studies indicate that various stressors can influence proinflammatory cytokine secretion.[58] The majority of these studies show that stress increases plasma levels of cytokines; other studies show an augmented production of cytokines by cells (mostly macrophages) to *in vitro* stimulation by various substances (mostly endotoxin). It should be noted that epinephrine, the classic stress hormone, has been shown to increase cytokine secretion.[59,60]

There are several reports documenting the role of IL-1 in the stress response.[61] In animal studies, cold water stress has been shown to result in a spontaneous increase of both cell-associated and -secreted IL-1 from macrophages.[62] Inescapable foot shock has been shown to increase IL-1 from alveolar macrophages.[63] Finally, rotation stress has been found to increase plasma IL-1 levels.[64]

Levels of increased TNFα following stress have also been documented. In animal studies, the administration of many different stressors increased systemic production of TNFα. TNFα was also increased after *in vitro* stimulation of extirpated cells from stressed animals.[65] Cold-water stress has been shown to result in increased TNFα from stimulated macrophages.[23] Furthermore, increased amounts of TNFα have been obtained from alveolar macrophages of mice exposed to inescapable foot shock.[63] Finally, thermally injured rats have been shown to produce much greater amounts of TNFα than their noninjured counterparts.[66]

Many reports show enhanced IL-6 production in response to various stressors. In animal studies, exposure to physical and psychological stressors have been shown to elevate IL-6.[24,67-69] In human studies, increased levels of plasma IL-6 have been found in sepsis[70] and following surgery.[71-73] In addition, depression has been linked with increased IL-6.[74]

It is clear from these studies that various psychological, physical, and painful stressors, in addition to septic and surgical stressors, induce the production and secretion of the proinflammatory cytokines IL-1, TNFα, and IL-6. Both primary (IL-1 and TNFα), as well as the secondary cytokines (IL-6), are induced *in vivo* as well as *in vitro*. Although IL-6 can be induced in fibroblasts and endothelial cells by IL-1 and TNFα, much evidence indicates that macrophages are also an important source of IL-6.

9.7 STRESS INDUCES THE ACUTE PHASE RESPONSE

9.7.1 STRESS AND ACUTE PHASE PROTEINS

As stated above, the acute phase reaction is characterized by the dramatic increase of APP synthesized in the liver mainly in response to proinflammatory cytokines elaborated by activated macrophages. These cytokines also trigger the thermoregulatory center in the hypothalamus to produce fever, the most obvious sign of the acute phase response, and other hypothalamic nuclei to produce sickness behavior.

In animals, studies indicate that restraint stress causes an increase in plasma copper and fibrinogen and that their production is mediated by IL-1.[75] In humans, several investigators have examined the relationship between psychological stress and levels of certain acute phase proteins known to be involved in a disease state. For example, plasma fibrinogen has emerged as a major risk factor for coronary artery disease. Markowe et al.[76] found a positive correlation between occupational stress and fibrinogen levels, and Moller et al.[77] suggested that social class, cholesterol level, physical (in)activity, and psychological stress influence fibrinogen levels. In addition, Jern et al.[78] found mental stress (arithmetic and the Stroop color-word test) to increase both plasma coagulation and the fibrinolytic system. In the study by Jern et al., fibrinogen and the von Willebrand factor are acute phase reactants likely involved in coagulation; APPs, which are responsible for the simulation of fibrinolysis, are more difficult to discern since several APPs (α2 antiplasmin, plasminogen activator inhibitor, and heparin cofactor II) may have been involved (see Table 9.1).

Although cytokine levels in the majority of these studies were not measured, it is likely that they were elevated secondary to the stress and were the major inducers, presumably with the corticosteroids, of the APPs. Indeed, it is known that glucocorticoids stimulate the expression of most APPs directly. Moreover they, together with cytokines, evoke a strong synergistic enhancement for most APPs.[11]

Though normally tightly regulated,[79] the possible effects of increased levels of certain cytokines and APPs as consequences of stress should be considered. An extreme example of dysregulated cytokine production is found in mice made transgenic for overexpression of IL-6, which is produced constitutively under the control of a CNS promoter. Such mice sustain marked neurologic damage and eventually succumb to neurologic disease.[80] The effects of transient increases of cytokines associated with stress remain to be determined.

Certain increases in production of APP as well as some consequences of elevated APPs accompanying stress have been considered. For example, the consequences of binding SAA to HDL-3, which results in altered cholesterol metabolism, may be especially important in the pathogenesis of atherosclerosis (see Section 9.3). Continued documentation and the consequences of changes in cytokine and APP metabolism with stress will be of great interest as one considers the relationship between stress and the inflammatory response.

9.7.2 Stress and Fever

As fever is an integral component of the acute phase response, mediated by IL-1, TNFα, and IL-6, a number of studies have examined the relationship between psychological stress and fever. Exposure of rats to a novel environment has been shown to cause a rise in body temperature, with many similarities to LPS-induced fever.[81] Another study reports that this stress-induced fever can be blocked by pretreatment with antipyretic drugs.[82] Stress-induced hyperthermia has also been demonstrated in mice.[83]

Human studies have been more controversial.[84] It has been demonstrated, however, that medical residents sitting for their yearly exams show significant stress-related changes in body temperature, blood pressure, and heart rate when compared to a later date without an exam.[85]

9.8 STRESS AND SP

There is evidence that SP plays a role in the response to certain stressors, based on its known association with the sympathetic nervous system, in particular its colocalization with norepinephrine in certain sympathetic ganglia.[86] SP is also apparently essential to the maintenance of catecholamine secretion from the adrenal medulla in times of stress. In support of this, it has been shown in neonates that the neurotoxin capsaicin, which selectively acts on C-type afferent nerves containing SP, reduces the secretion of catecholamines in response to repeated stressors.[87]

Recently, it has been shown that SP may have additional though yet undefined roles in the response to stress. Centrally, alterations in SP concentrations have been reported in response to various stressors, including space flight,[88] the stress of being handled,[89] and short-term restraint stress.[90] In addition, it has been shown that the enhancement of synaptic transmission mediated by SP is involved in repeated cold-stress-induced hyperalgesia.[91,92] SP has also been implicated as a modulator of HPA axis function,[93] providing a pathway whereby centrally derived SP may be important in the immune response to stress.

There is other evidence that certain forms of stress due to somatosensory stimulation are contingent on the function of capsaicin-dependent afferent sensory nerves. Capsaicin-treated rats do not exhibit the increase in ACTH and corticoid production that accompany the stresses induced by cold, surgery, formalin, and opiate withdrawal.[94] Conversely, electrical nerve stimulation of somatic afferents results in increases in ACTH secretion from the pituitary gland. Thus, there is evidence that such sensory afferents may participate in the stress response.[46]

There have been several revealing animal studies which further support the role of SP in stress and its potential relationship to inflammation. One study shows that immobilization stress induces defecation and that defecation is inhibited by an SP antagonist, demonstrating the role of SP in stress-induced gut motility.[95] Another study shows that following a 4-day cold-water stress regimen in mice, elevated levels of SP are found in the peritoneal cavity. Furthermore, Zhu et al.[24] found SP to be responsible for the augmented proinflammatory cytokine response from elicited peritoneal macrophages in these stressed mice.[24] Their research shows that stress can increase levels of peritoneal SP and that SP is responsible for the augmentation of cytokine secretion from macrophages *in vitro*. They postulated that stress-induced SP in the peritoneal cavity induced SP receptor in elicited peritoneal macrophages,[24] and that SP acted as a priming agent for macrophages *in vitro* which produced high levels of IL-6 upon further stimulation by the secondary or triggering stimulus, endotoxin.[25] Capsaicin-treated stressed animals did not release IL-6 upon endotoxin treatment, and peritoneal macrophages from treated animals did not show the increase in SP binding. Such a sequence of events has been previously reported.[96]

9.9 STRESS AND INFLAMMATORY DISEASE

We have shown that SP is involved in the macrophage's response to stress by liberating cytokines and prostaglandins.[23,24,62] Catecholamines and corticosteroids, the two major stress hormones, augment the production of cytokines from macrophages and the production of APPs in the liver, respectively; epinephrine may also augment the production of certain APPs. SP is an important link between the nervous and immune systems in that it participates in the response to stress by inducing known inflammatory mediators; therefore, it is likely that the mediators induced by stress play a role in the pathogenesis of certain diseases of unknown etiology in which stress is suspected of playing some role. The question arises whether stress itself can produce an inflammatory response.

The literature reports isolated cases of stress-induced inflammation as well as controlled human and animal studies demonstrating the role of various stressors in the inflammatory response. In a notable case study, a dermatologist found that there exists a subset of patients who fail to respond to competent dermatologic treatment unless psychodynamically informed psychotherapy is added. The author reports four successful cases, including alopecia areata, psoriasis, atopic dermatitis, and acne vulgaris.[97] Another study notes that fear can mimic latex allergy in patients with dental phobia.[98]

In addition, there is copious literature indicating that stress is likely to play some role in the onset, pathogenesis, and severity of certain inflammatory diseases of unknown etiology. These diseases include:

- Inflammatory bowel disease, especially regional enteritis (Crohn's disease) and ulcerative colitis[99-101]
- Essential hypertension and coronary artery disease[102-105]
- Rheumatoid arthritis[106,107]
- Psoriasis[108,109]
- Asthma[110,111]

Although the precise mechanisms mediating these inflammatory states are unknown, the evidence is strong that stress, via the secretion of the classic stress hormones epinephrine and corticosteroids, as well as the neuropeptide SP and likely other neuromediators, can induce metabolic and endocrine changes that can lead to a neurogenic inflammatory response and the subsequent development of inflammatory disease.

REFERENCES

1. Fantone, J. C. and Ward, P. A., Inflammation, in *Pathology,* 2nd ed., Rubin, E. and Farber, J. L., Eds., J. B. Lippincott, Philadelphia, 1994, Chap. 2.
2. Meggs, W. J., Neurogenic inflammation and sensitivity to environmental chemicals, *Environ. Health Perspect.,* 101, 234, 1993.
3. Foreman, J. C., Peptides and neurogenic inflammation, *Br. Med. Bull.,* 43, 386, 1987.
4. Rang, H. P., Bevan, S., and Dray, A., Chemical activation of nociceptive peripheral neurons, *Br. Med. Bull.,* 47, 534, 1991.
5. Bill, A., Stjerschantz, J., Mandahl, A., Brodin, E., and Nilsson, G., Substance P: release on trigeminal nerve stimulation, effects in the eye, *Acta Physiol. Scand.,* 106, 371, 1979.
6. Arnalich, F., de Miguel, E., Perez-Ayala, C., Martinez, M., Vasquez, J. J., Gijon-Banos, J., and Hernanz, A., Neuropeptides and interleukin-6 in human joint inflammation: relationship between intraarticular substance P and interleukin-6 concentrations, *Neurosci. Lett.,* 170, 251, 1994.
7. Otsuka, M. and Yoshioka, K., Neurotransmitter functions of mammalian tachykinins, *Physiol. Rev.,* 73, 229, 1993.
8. Pincelli, C., Fantini, F., and Gianetti, A., Neuropeptides and skin inflammation, *Dermatology,* 187, 153, 1993.
9. Shanahan, F., Denburg, J. A., Fox, J., Bienenstock, J., and Befus, D., Mast cell heterogeneity: effects of neuroenteric peptides on histamine release, *J. Immunol.,* 135, 1331, 1985.
10. Foreman, J. C., Substance P and calcitonin gene-related peptide: effects on mast cells and in human skin, *Int. Arch. Allergy Appl. Immunol.,* 82, 366, 1987.
11. Baumann, H. and Gauldie, J., The acute phase response, *Immunol. Today,* 15, 74, 1994.
12. Steel, D. M. and Whitehead, A. S., The major acute phase reactants: C-reactive protein, serum amyloid P component and serum amyloid A protein, *Immunol. Today,* 15, 81, 1994.
13. Weinstock, J. V., Neuropeptides and the regulation of granulomatous inflammation, *Clin. Immunol. Immunopathol.,* 64, 17, 1992.
14. Levine, J. S., Clark, R., Devor, M., Helms, C., Moskowitz, M. A., and Basbaum, A. I., Intraneuronal substance P contributes to the severity of experimental arthritis, *Science,* 226, 547, 1984.
15. Leeman, S. E. and Mroz, E. A., Substance P, *Life Sci.,* 15, 2033, 1974.
16. Pernow, B., Substance P, *Pharmacol. Rev.,* 35, 85, 1983.
17. Payan, D. G., Neuropeptides and inflammation: the role of substance P., *Annu. Rev. Med.,* 40, 341, 1989.
18. Armstrong, M. J. and Leeman, S. E., Neurotensin and substance P, *Adv. Metab. Disord.,* 11, 469, 1988.
19. Lorton, D., Bellinger, D. L., Felten, S. Y., and Felten, D. L., Substance P innervation of spleen in rats: nerve fibers associated with lymphocytes and macrophages in specific compartments of the spleen, *Brain Behav. Immun.,* 5, 29, 1991.

20. Nio, D. A., Moylan, R. N., and Roche, J. K., Modulation of T lymphocyte function by neuropeptides: evidence for their role as local immunoregulatory elements, *J. Immunol.,* 150, 5281, 1993.
21. Kimball, E. S., Perisco, F. F., and Vaught, J. L., Substance P, neurokinin A, and neurokinin B induce generation of IL-1 like activity in P388D1 cells, possible relevance to arthritic disease, *J. Immunol.,* 141, 3564, 1988.
22. Rameshwar, P. and Gascon, P., Substance P mediates production of stem cell factor and interleukin-1 in bone marrow stroma: potential autoregulatory role for these cytokines in SP receptor expression and induction, *Blood,* 86, 482, 1995.
23. Chancellor-Freeland, C., Zhu, G. F., Kage, R., Beller, D. I., Leeman, S. E., and Black, P. H., Substance P and stress-induced immune-changes in macrophages, *Ann. NY Acad. Sci.,* 771, 472, 1995.
24. Zhu, G. F., Chancellor-Freeland, C., Berman, A. S., Kage, R., Leeman, S. E., Beller, D. I., and Black, P. H., Endogenous substance P mediates cold-water stress-induced increase in interleukin-6 secretion from peritoneal macrophages, *J. Neurosci.,* 16, 3745, 1996.
25. Berman, A. S., Chancellor-Freeland, C., Zhu, G. F., and Black, P. H., Substance P primes murine peritoneal macrophages for an augmented proinflammatory cytokine response to lipopolysaccharide, *Neuroimmunomodulation,* 3, 141, 1996.
26. Cozens, P. and Rowe, F., Substance P is a potent inducer of TNF and IL-1 secretion by macrophages: a potential role for TNF in the pathogenesis of asthma, *Immunobiology,* 175, 7, 1987.
27. Laurenzi, M. A., Persson, M. A., Dalsgaard, C. J., and Haegerstrand, A., The neuropeptide substance P stimulates production of interleukin-1 in human blood monocytes: activated cells are preferentially influenced by the neuropeptide, *Scand. J. Immunol.,* 31, 529, 1990.
28. Hartung, H. P., Activation of macrophages by neuropeptides, *Brain Behav. Immun.,* 2, 275, 1988.
29. Ruff, M. R., Wahl, S. M., and Pert, C. B., Substance P receptor-mediated chemotaxis of human monocytes, *Peptides,* 6 (Suppl. 2), 107, 1985.
30. Payan, D. G., Neuropeptides and inflammation: the role of substance P, *Annu. Rev. Med.,* 40, 341, 1989.
31. Payan, D. G. and Goetzl, E. J., Mediation of pulmonary immunity and hypersensitivity by neuropeptides, *Eur. J. Respir. Dis.,* Suppl. 144, 77, 1986.
32. Pascual, D. W. and Bost, K. L., Substance P production by P388D1 macrophages: a possible autocrine function for this neuropeptide, *Immunology,* 71, 52, 1990.
33. Gutwald, J., Goebler, M., and Sorg, C., Neuropeptides enhance irritant and allergic contact dermatitis, *J. Invest. Dermatol.,* 96, 695, 1991.
34. Perianin, A., Snyderman, R., and Malfroy, B., Substance P primes human neutrophil activation: a mechanism for neurological regulation of inflammation, *Biochem. Biophys. Res. Commun.,* 161, 520, 1989.
35. Lloyds, D., Brindle, N. P., and Hallett, M. B., Priming of human neutrophils by tumor necrosis factor-α and substance P is associated with tyrosin-phosphorylation, *Immunology,* 84, 220, 1995.
36. Kroegel, C., Giembycz, M. A., and Barnes, P. J., Characterization of eosinophil cell activation by peptides: differential effects of substance P, melittin, and FMET-Leu-Phe, *J. Immunol.,* 145, 2581, 1990.
37. Numao, T. and Agrawal, D. K., Neuropeptides modulate human eosinophil chemotaxis, *J. Immunol.,* 149, 3309, 1992.

38. Janiszewski, J., Bienenstock, J., and Blennerhassett, M. G., Picomolar doses of substance P trigger electrical responses in mast cells without degranulation, *Am. J. Physiol.,* 267, C138, 1994.
39. Wallengren, J. and Hakanson, R., Effects of capsaicin, bradykinin and prostaglandin E2 in the human skin, *Br. J. Dermatol.,* 126, 111, 1992.
40. Nakai, S., Iikura, Y., Akimoto, K., and Shiraki, K., Substance P-induced cutaneous and bronchial reactions in children with bronchial asthma, *Ann. Allergy,* 66, 155, 1991.
41. Van Oosterhout, A. J., van Ark, I., Hofman, G., Van der Linde, H. J., Fattah, D., and Nijkamp, F. P., Role of interleukin-5 and substance P in development of airway hyperreactivity to histamine in guinea-pigs, *Eur. Respir. J.,* 9, 493, 1996.
42. Karalis, K., Mastorakos, G., Sano, H., Wilder, R. L., and Chrousos, G. P., Somatostatin may participate in the anti-inflammatory actions of glucocorticoids, *Endocrinology,* 136, 4133, 1995.
43. McKay, D. M. and Bienenstock, J., Interaction between mast cells and nerves in the gastrointestinal tract, *Immunol. Today,* 15, 533, 1994.
44. Stead, R. H., Tomioka, M., Quinonez, G., Simon, G. T., Felten, S. Y., and Bienenstock, J., Intestinal mucosal mast cells in normal and nematode-infected rat intestines are in intimate contact with peptidergic nerves, *Proc. Natl. Acad. Sci. U.S.A.,* 84, 2975, 1987.
45. Meggs, W. J., Neurogenic switching: a hypothesis for a mechanism for shifting the site of inflammation in allergy and chemical sensitivity, *Environ. Health Perspect.,* 103, 54, 1995.
46. Donnerer, J., Eglezos, A., and Helme, R. D., Neuroendocrine and immune function in the capsaicin-treated rat: evidence for different neural modulation *in vivo,* in *The Neuroendocrine-Immune Network,* Freier, S., Ed., CRC Press, Boca Raton, FL, 1990, 70.
47. Lotz, M., Carson, D. A., and Vaughan, J. H., Substance P activation of rheumatoid synoviocytes: neural pathway in pathogenesis of arthritis, *Science,* 235, 893, 1987.
48. Payan, D. G., Brewster, D. R., and Goetzl, E. J., Specific stimulation of human T lymphocytes by substance P, *J. Immunol.,* 131, 1613, 1983.
49. Lotz, M., Vaughan, J., and Carson, D., Effect of neuropeptides on production of inflammatory cytokines by human monocytes, *Science,* 241, 1218, 1988.
50. Firestein, G. S. and Zvaifler, N. J., Anticytokine therapy in rheumatoid arthritis, *N. Engl. J. Med.,* 337, 195, 1997.
51. Koch, T. R., Carney, J. A., and Go, V. L. W., Distribution and quantification of gut neuropeptides in normal intestine and inflammatory bowel disease, *Dig. Dis. Sci.,* 32, 369, 1987.
52. Mantyh, C. R., Gates, T. S., Zimmerman, R. P., Welton, M. L., Passaro, E. P., Vigna, S. R., Maggio, J. E., Kruger, L., and Mantyh, P., Receptor binding sites for substance P, but not substance K or neuromedin K, are expressed in high concentrations by arterioles, venules, and lymph nodules in surgical specimens obtained from patients with ulcerative colitis and Crohn's disease, *Proc. Natl. Acad. Sci. U.S.A.,* 85, 3235, 1988.
53. Pothoulakis, C., Castagliuolo, I., LaMont, J. T., Jaffer, A., O'Keane, J., Snider, R. M., and Leeman, S. E., CP-96,345, a substance P antagonist, inhibits rat intestinal responses to Clostridium difficile toxin A but not cholera toxin, *Proc. Natl. Acad. Sci. U.S.A.,* 91, 947, 1994.
54. Castagliuolo, I., Keates, A. C., Qiu, B., Nikulasson, S., Leeman, S. E., and Pothoulakis, C., Increased substance P responses in dorsal root ganglia and intestinal macrophages during Clostridium difficile toxin A enteritis in rats, *Proc. Natl. Acad. Sci. U.S.A.,* 94, 4788, 1997.

55. Black, P. H., Central nervous system-immune system interactions: psychoneuroendocrinology of stress and its immune consequences, *Antimicrob. Agents Chemother.,* 38, 1, 1994.
56. Black, P. H., Immune system-central nervous system interactions: effect and immunoregulatory consequences of immune system mediators on the brain, *Antimicrob. Agents Chemother.* 38, 7, 1994.
57. Black, P. H., Psychoneuroimmunology: brain and immunity, *Sci. Med.,* 2, 16, 1995.
58. Abraham, E., Effects of stress on cytokine production, *Methods Achiev. Exp. Pathol.,* 14, 45, 1991.
59. Liao, J., Keiser, J. A., Scales, W. E., Kunkel, S. L., and Kluger, M. J., Role of epinephrine in TNF and IL-6 production from isolated perfused rat liver, *Am. J. Physiol.,* 268, R896, 1995.
60. DeRijk, R. H., Boelen, A., Tilders, F. J., and Berkenbosch, F., Induction of plasma interleukin-6 by circulating adrenaline in the rat, *Psychoneuroendocrinology,* 19, 155, 1994.
61. Shintani, F., Nakaki, T., Kanba, S., Kato, R., and Asai, M., Role of interleukin-1 in stress response: a putative neurotransmitter, *Mol. Neurobiol.,* 10, 47, 1995.
62. Jiang, G., Morrow-Tesch, J. L., Beller, D. I., Levy, E. M., and Black, P. H., Immunosuppression in mice induced by cold water stress, *Brain Behav. Immun.,* 4, 278, 1990.
63. Persoons, J. H., Schornagel, K., Breve, J., Berkenbosch, F., and Kraal, G., Acute stress affects cytokines and nitric oxide production by alveolar macrophages differently, *Am. J. Respir. Crit. Care Med.,* 152, 619, 1995.
64. Korneva, E. A., Rybakina, E. G., Formicheva, E. E., Kozinets, I. A., and Shkhinek, E. K., Altered interleukin-1 production in mice exposed to rotation stress, *Int. J. Tissue React.,* 14, 219, 1992.
65. Yamasu, K., Shimada, Y., Sakaizumi, M., Soma, G., and Mizuno, D., Activation of the systemic production of tumor necrosis factor after exposure to acute stress, *Eur. Cytokine Netw.,* 3, 391, 1992.
66. Minei, J. P., Williams, J. G., Hill, S. J., McIntyre, K., and Bankey, P. E., Augmented tumor necrosis factor response to lipopolysaccharide after thermal injury is regulated posttranscriptionally, *Arch. Surg.,* 129, 1198, 1994.
67. van Gool, J., van Vugt, H., Helle, M., and Aarden, L. A., The relation among stress, adrenalin, interleukin-6 and acute phase reactants in the rat, *Clin. Immunol. Immunopathol.,* 57, 200, 1990.
68. Zhou, D., Kusnecov, A. W., Shurin, M. R., DePaoli, M., and Rabin, B. S., Exposure to physical and psychological stressors elevates plasma interleukin-6: relationship to the activation of hypothalamic-pituitary-adrenal axis, *Endocrinology,* 133, 2523, 1993.
69. Takaki, A., Huang, Q. H., Somogyvari-Vigh, A., and Arimura, A., Immobilization stress may increase plasma interleukin-6 via central and peripheral catecholamines, *Neuroimmunomodulation,* 1, 335, 1994.
70. Hack, C. E., DeGroot, E. R., Felt-Bersma, R. J., Nuijens, J., Van Schijndel, R. J., Eerenberg-Belmer, A. J., Thijs, L. G., and Aarden, L. A., Increased plasma levels of interleukin-6 in sepsis, *Blood,* 74, 1704, 1989.
71. Sweed, Y., Puri, P., and Reen, D. J., Early induction of IL-6 in infants undergoing major abdominal surgery, *J. Pediatr. Surg.,* 27, 1033, 1992.
72. Tsukada, K., Katoh, H., Shiojima, M., Suzuki, T., Takenoshita, S., and Nagamachi, Y., Concentrations of cytokines in peritoneal fluid after abdominal surgery, *Eur. J. Surg.,* 159, 475, 1993.

73. Jones, M. O., Pierro, A., Hasim, I. A., and Lloyd, D. A., Postoperative changes in resting energy expenditure and interleukin-6 levels in infants, *Br. J. Surg.*, 81, 536, 1994.
74. Maes, M., Scarpé, S., Meltzer, H. Y., Bosmans, E., Suy, E., Calabrese, J., and Cosyns, P., Relationship between interleukin-6 activity, acute phase proteins, and function of the hypothalamic-pituitary-adrenal axis in severe depression, *Psychiatry Res.*, 49, 11, 1993.
75. Morimoto, A., Watanabe, T., Myogin, T., and Murakami, N., Restraint induced stress elicits acute phase response in rabbits, *Pfluegers Arch. Eur. J. Physiol.*, 410, 554, 1987.
76. Markowe, H. L., Marmot, M. G., Shipley, M. J., Bulpitt, C. J., Meade, T. W., Striling, Y., Vickers, M. V., and Semmence, A., Fibrinogen: a possible link between social class and coronary heart disease, *Br. Med. J.*, 291, 1312, 1985.
77. Moller, L. and Kristensen, T. S., Plasma fibrinogen and ischemic heart disease risk factors, *Arterioscler. Thromb.*, 11, 344, 1991.
78. Jern, C., Eriksson, E., Tengborn, L., Risberg, B., Wadenvik, H., and Jern, S., Changes of plasma coagulation and fibrinolysis in response to mental stress, *Thromb. Haemostasis*, 62, 767, 1989.
79. Tracey, K. J. and Cerami, A., Tumor necrosis factor, other cytokines and disease, *Annu. Rev. Cell Biol.*, 9, 317, 1993.
80. Campbell, I. L., Abraham, C. R., Masliah, E., Kemper, P., Inglis, J. D., Oldstone, M. B., and Mucke, L., Neurologic disease induced in transgenic mice by cerebral overexpression of interleukin-6, *Proc. Natl. Acad. Sci. U.S.A.*, 90, 10061, 1993.
81. Kluger, M. J., O'Reilley, B., Shope, T. R., and Vander, A. J., Further evidence that stress hyperthermia is a fever, *Physiol. Behav.*, 39, 763, 1987.
82. Briese, E. and Cabanac, M., Stress hyperthermia: physiological arguments that it is a fever, *Physiol. Behav.*, 49, 1153, 1991.
83. Zethof, T. J., van der Heyden, J. A., Tolboom, J. T., and Olivier, B., Stress-induced hyperthermia in mice: a methodological study, *Physiol. Behav.*, 55, 109, 1994.
84. Hasan, M. K. and White, A. C., Psychogenic fever: entity or non-entity? *Postgrad. Med.*, 66, 152, 1979.
85. Marazziti, D., Di Muro, A., and Castrogiovanni, P., Psychological stress and body temperature changes in humans, *Physiol. Behav.*, 52, 393, 1992.
86. Kessler, J. A., Differential regulation of peptide and catecholamine characters in cultured sympathetic neurons, *Neuroscience*, 15, 827, 1985.
87. Livett, B., Zhou, X., Khalil, Z., Wan, D., Bunn, S., and Marley, P., Endogenous neuropeptides maintain adrenal catecholamine output during stress, in *Molecular Biology of Stress*, Alan R. Liss, New York, 1989, 179.
88. Zhu, X. and Desiderio, D. M., Effects of spaceflight stress on proopiomelanocortin, proenkephalin A, and tachykinin neuropeptidergic systems in the rat posterior pituitary, *Life Sci.*, 55, 347, 1994.
89. Brodin, E., Rosen, A., Schott, E., and Brodin, K., Effects of sequential removal of rats from a group cage, and of individual housing of rats, on substance P, cholecystokinin and somatostatin levels in the periaqueductal gray and limbic regions, *Neuropeptides*, 26, 253, 1994.
90. Rosen, A., Brodin, K., Eneroth, P., and Brodin, E., Short-term restraint stress and s.c. saline injection alter the tissue levels of substance P and cholecystokinin in the periaqueductal grey and limbic regions of rat brain, *Acta Physiol. Scand.*, 146, 341, 1992.
91. Satoh, M., Kuraishi, Y., and Kawamura, M., Effects of intrathecal antibodies to substance P, calcitonin gene-related peptide and galanin on repeated cold-stress-induced hyperalgesia: comparison with carrageenan-induced hyperalgesia, *Pain*, 49, 273, 1992.

92. Okano, K., Kuraishi, Y., and Satoh, M., Effects of intrathecally injected glutamate and substance P antagonists on repeated cold stress-induced hyperalgesia in rats, *Biol. Pharm. Bull.,* 18, 42, 1995.
93. Jessop, D. S., Chowdrey, H. S., Larsen, P. J., and Lightman, S. L., Substance P: multifunctional peptide in the hypothalamo-pituitary system, *J. Endocrinol.,* 132, 331, 1992.
94. Donnerer, J., Amann, R., Skofitsch, G., and Lembeck, F., Substance P afferents regulate ACTH-coticosterone release, *Ann. NY Acad. Sci.,* 632, 296, 1991.
95. Castagliuolo, I., Lamont, J. T., Qiu, B., Fleming, S. M., Bhaskar, K. R., Nikulasson, S. T., Kornetzky, C., and Pouthalakis, C., Acute stress causes mucin release from rat colon: role of corticotropin releasing factor and mast cells, *Am. J. Physiol.,* 271, G884, 1996.
96. Mayer, N. R., Gamse, R., and Lembeck, F., Effect of capsaicin pretreatment on substance P binding to synaptic vesicles, *J. Neurochem.,* 35, 1238, 1980.
97. Koblenzer, C. S., Psychotherapy for intractable inflammatory dermatoses, *J. Am. Acad. Dermatol.,* 32, 609, 1995.
98. Longley, A. J., Fiset, L., Getz, T., Van Arsdel, P. P., and Weinstein, P., Fear can mimic latex allergy in patients with dental phobia, *Gen. Dentistry,* 42, 236, 1994.
99. Porcelli, P., Leoci, C., and Guerra, V., A prospective study of the relationship between disease activity and psychological distress in patients with inflammatory bowel disease, *Scand. J. Gastroenterol.,* 31, 792, 1996.
100. Talal, A. H. and Drossman, D. A., Psychosocial factors in inflammatory bowel disease, *Gastroenterol. Clin. N. Am.,* 24, 699, 1995.
101. Ramchandani, D., Schindler, B., and Katz, J., Evolving concepts of psychopathology in inflammatory bowel disease: implications for treatment, *Med. Clin. N. Am.,* 78, 1321, 1994.
102. Pickering, T. G., Schwartz, J. E., and James, G. D., Ambulatory blood pressure monitoring for evaluating the relationship between lifestyle, hypertension and cardiovascular risk, *Clin. Exp. Pharmacol. Physiol.,* 22, 226, 1995.
103. Rabkin, S. W., Non-pharmacologic therapy in the management of hypertension: an update, *Can. J. Pub. Health,* 85 (Suppl. 2), S44, 1994.
104. Gullette, E. C., Blumenthal, J. A., Babyak, M., Jiang, W., Waugh, R. A., O'Connor, C. M., Morris, J. J., and Krantz, D. S., Effects of mental stress on myocardial ischemia during daily life, *J. Am. Med. Assoc.,* 277, 1521, 1997.
105. Williams, R. B. and Littman, A. B., Psychosocial factors: role in cardiac risk and treatment strategies, *Cardiol. Clin.,* 14, 97, 1996.
106. O'Leary, A., Stress, emotion and human immune function, *Psychol. Bull.,* 108, 363, 1990.
107. Harrington, L., Affleck, G., Urrows, S., Tennen, H., Higgins, P., Zautra, A., and Hoffman, S., Temporal covariation of soluble interleukin-2 receptor levels, daily stress, and disease activity in rheumatoid arthritis, *Arthritis Rheum.,* 36, 199, 1993.
108. Farber, E. M., Rein, G., and Lanigan, S. W., Stress and psoriasis, psychoneuroimmunologic mechanisms, *Int. J. Dermatol.,* 30, 8, 1991.
109. Harvima, I. T., Vinamaki, H., Naukkarinen, A., Paukkonen, K., Neittaanmaki, H., Harvima, R., and Horsmanheimo, M., Association of cutaneous mast cells and sensory nerves with psychic stress in psoriasis, *Psychother. Psychosom.,* 60, 168, 1993.
110. Leigh, D., Allergy and the psychiatrist, *Int. Arch. Allergy,* 4, 22, 1953.
111. Crocco, J. A., Stress and the lung: hypersensitivity of airways, neurogenic pulmonary edema, and hyperventilation syndrome, *Mt. Sinai J. Med.,* 54, 63, 1987.

10 Neuropeptides, Cytokines, and Cancer — Interrelationships*

Anthony J. Murgo, Robert E. Faith, and Nicholas P. Plotnikoff

CONTENTS

10.1 Introduction 133
10.2 Cytokines and Neuropeptides as Autocrine/Paracrine Tumor Growth Factors 134
10.3 Cytokines — Source and General Properties 134
10.4 Cytokines as Tumor Growth Factors 141
10.5 Clinical Manifestations of Cytokine–Cancer Relationships 141
10.6 Cancer and Endogenous Opioids 143
10.7 Cancer and Other Tumor-Associated Neuropeptides 145
10.8 Conclusions 147
References 147

> While there are several chronic diseases more destructive to life than cancer, none is more feared — Charles H. Mayo, *Annals of Surgery,* 83, 357, 1926.

> ...is it conceivable that emotional factors could possibly play a part in the precancerous or cancerous process? — Max Cutler, in, *The Psychological Variables in Human Cancer,* Joseph A. Gangerelli and Frank J. Kirkner, Eds., University of California Press, Berkeley, 1954.

10.1 INTRODUCTION

The centuries-old notion that emotional stress can affect the growth and development of human tumors remains a subject of continued interest and controversy. Although the influence of stress on tumor growth is strongly supported by the results of

---

* This chapter represents the views solely of the authors and no government agency. No official endorsement by the Uniformed Services University of the Health Sciences or the National Cancer Institute is intended or should be inferred.

extensive animal experimentation,[1-4] incontrovertible human data are lacking. Results of clinical studies are often inconclusive or conflicting which is, in part, explained by the difficulty in designing and conducting well-controlled clinical trials that adequately test this principle.[5-14] Nevertheless, improved knowledge of tumor biology and tumor-host relationships has led to a better understanding of how neuroendocrine and immune factors can mediate the effects of stress on tumor growth and development.[4,15-21]

Conversely, nothing is more a stimulus of physical and psychological stress in the human host than a malignant tumor. The direct consequences of tumor growth, such as invasion and replacement of normal tissues, are clearly apparent. Less appreciated are the indirect or remote effects of tumors. Cancer can result in a variety of systemic symptoms and signs caused by the release of known and unknown tumor-derived humoral factors or by the perturbation of host-derived factors. Pain, anorexia, weight loss, fever, and depression are frequent consequences of cancer. These symptoms can also result from the administration of cytokines and neurohormones to cancer patients or in response to other types of stress-provoking interventions such as chemotherapy, surgery, or radiotherapy. Our understanding of the indirect effects of cancer, or so-called paraneoplastic phenomena, continues to improve as we learn more about the potential interactions between tumor and host. The purpose of this chapter is to provide a framework for interactions between cancer and the neuroendocrine and immune systems, and describe some potential clinical sequelae and manifestations of these relationships.

10.2 CYTOKINES AND NEUROPEPTIDES AS AUTOCRINE/PARACRINE TUMOR GROWTH FACTORS

Many different types of human tumors have been shown to express a variety of neuropeptide hormones and cytokines (Table 10.1), and their receptors (Table 10.2). This provides a network for interactions between tumor cells and the host's neuroendocrine and immune systems. Since some tumor cells produce cytokine or neuropeptide growth factors for which they also possess specific functional receptors, these factors can also function in an autocrine fashion. Paradigms of autocrine tumor growth regulation are shown in Table 10.3. Elucidation of autocrine and paracrine interactions is considered an important area of cancer research because the information can be used to identify specific therapeutic targets for clinical development. Furthermore, the host response to stress stimuli includes the release of a variety of humoral factors including neuroendocrine hormones and cytokines that can influence immune response and tumor growth.

10.3 CYTOKINES — SOURCE AND GENERAL PROPERTIES

Cytokines represent a large group of hormone-like, low molecular weight proteins that regulate many different types of immune responses, and also act as hematopoi-

TABLE 10.1
Production of Cytokines and Neuropeptides by Various Types of Human Tumor Cells

Tumor Type	Cytokine/Neuropeptide	Reference(s)
Angiosarcoma	VEGF	136
Bladder (urinary) carcinoma	IL-6	30, 137
	β-hCG	125
	VEGF	138
Brain glioma	GM-CSF	139
	bFGF	140, 141
	HGF/SF	44, 142
	Met-Enk	143
Brain meningioma	bFGF	140
	Met-Enk	143
Breast carcinoma	BN/GRP	144, 145
	CCK-B/Gastrin	146
	β-Endorphin	26
	Leu-Enk	26
	Met-Enk	26
	VEGF	46, 47
Cervical carcinoma	IL-6	32
Colorectal carcinoma	CCK-B/Gastrin	147
	HGF/SF	148
	IL-1β	149, 150
	IL-6	151
	IL-8	149
	IL-10	149
	MIP-1	149
	TGF-β1	149
	TNFα	149
	VEGF	42, 45, 152
Endometrial cancer	GnRH	128
Gastric carcinoma	bFGF	153
	IL-1α	154
	VEGF	42
Head/neck squamous cell carcinoma	Met-Enk	93
Hepatocellular carcinoma	bFGF	48
	VEGF	48,155
Ileal — small intestine carcinoid	Substance K	156
	Substance P	156
Kaposi's sarcoma	bFGF	157
	IL-1α	41
	IL-1β	40, 41
	IL-6	41
	Oncostatin M	41
	PDGF	41

TABLE 10.1 (CONTINUED)
Production of Cytokines and Neuropeptides by Various Types of Human Tumor Cells

Tumor Type	Cytokine/ Neuropeptide	Reference(s)
	VEGF	158
Leukemia — lymphoid	CCK-B/Gastrin	159
	GM-CSF	139
	IL-1β	160
	TGF-β1	161
Leukemia —myeloid	GM-CSF	162, 163
	Gastrin	159
	IL-1β	160, 162–164
	VEGF	165
Lymphoma — Sézary	IL-4	161
Lung — small-cell carcinoma	ACTH	102, 166
	Adrenomedullin	167
	BN/GRP	114, 115, 168–170
	CCK-B/Gastrin	146
	Dynorphin	89
	β-Endorphin	89, 100, 102, 171
	Leu-Enk	89
	Neurokinin A	170
	Neurokinin B	170
	Neurotensin	170, 172–174
	Somatostatin	170
	Substance P	115
	Vasopressin	175–177
Lung cancer — non-small-cell type	Adrenomedullin	167
	BN/GRP	178
	Dynorphin	89
	β-endorphin	89
	G-CSF	179, 180
	GM-CSF	139, 179
	Leu-Enk	89
	M-CSF	179
	IL-6	179, 180
	IL-8	181
	VIP	182
Lung — bronchial carcinoid	ACTH	183, 184
	CRH	183
	Endothelin-1	183
	Neuropeptide Y	183
Ovarian adenocarcinoma	IL-6	151
	VEGF	185
Melanoma	VEGF	186
	EGF	186

Tumor Type	Cytokine/ Neuropeptide	Reference(s)
	bFGF	186
	GM-CSF	23
	IL-1α	23
	IL-1β	23, 186
	IL-6	23, 186, 187
	IL-7	186
	IL-8	186
	IL-10	188, 189
	IL-12	186
	LIF	186
	NGF	186
	TGFα	186
	TGFβ	186
Myeloma	IL-6	36, 190
	TNFβ	191
Neuroblastoma	Met-Enk	192, 193
	VIP	194
Pancreatic carcinoma	VEGF	42, 43
Paraganglioma — extra-adrenal	Met-enk	195
	Neuropeptide Y	195
	Somatostatin	195
Prostate carcinoma	IL-6	25, 29
Renal cell carcinoma	IL-6	31, 196, 197
Sarcoma — leiomyosarcoma	IL-6	151
Thymic carcinoid	ACTH	198
	CRH	198
Thyroid carcinoma	CRH	199

Abbreviations

ACTH = adrenocorticotropin hormone
BN/GRP = bombesin-like/gastrin-releasing peptide
CCK-B/Gastrin = cholecystokinin-B/Gastrin
CRH = corticotropin-releasing hormone
EGF = epidermal growth factor
bFGF = basic-fibroblast growth factor
GM-CSF = granulocyte-macrophage colony-stimulating factor
GnRH = gonadotropin-releasing hormone
HGF/SF = hepatocyte growth factor/scatter factor
β-hCG = β-human choreogonadotropin
Leu-Enk = leucine-enkephalin
LIF = leukemia inhibitory factor
Met-Enk = methionine-enkephalin
MIP-1 = macrophage inhibitory protein 1
NGF = Nerve growth factor
PDGF = platelet-derived growth factor
TGF-α = transforming growth factor α
TGF-β = transforming growth factor β
VEGF = vascular endothelial growth factor
VIP = vasoactive intestinal peptide

---

etic and tumor growth factors. They are released by lymphocytes, monocytes, and a variety of cells within the reticuloendothelial system (e.g., mast cells, fibroblasts, epithelial cells, and endothelial cells) and mediate their effects by interaction with specific receptors on target cells. They are involved in the pathogenesis of a variety

TABLE 10.2
Cytokine/Neuropeptide Receptors Expressed by Various Human Tumors

Tumor	Receptor	Reference(s)
Angiosarcoma	VEGF	136
Bladder carcinoma	β-hCG	125
	IL-6	30
Brain — astrocytoma	CCK-B/Gastrin	146
	Opioid	87
	Somatostatin	121
	SP	121, 200
Brain — glioblastoma	HGF/SF	44, 142
	Opioid	87
	SP	121
Brain — glioma	bFGF	140
Brain — meningioma	bFGF	140
	CCK-A	146
Breast cancer	GRP	144
	LH/hCG	126
	Opioid	85, 88, 201
	Somatostatin	201
	SP	121
Carcinoid — intestinal	Somatostatin	124
	VIP	124
Cervical carcinoma	IL-6	32
Colorectal adenocarcinoma	Gastrin	120
	HGF/SF	148
	IL-1β	149, 150
	Opioid	202
	Somatostatin	124
	VIP	124
Ganglioneuroblastoma	Somatostatin	121
	SP	121
Gastric adenocarcinoma	bFGF	153
	Met-Enk	203
	VIP	124
Endometrial cancer	GnRH	128
	Opioid	201
Head and neck — squamous cell carcinoma	Opioid	93
Insulinoma — primary	Somatostatin	124
	VIP	124
Kaposi's sarcoma	IL-1β	40
	IL-6	35, 40, 204
	bFGF	157

Tumor	Receptor	Reference(s)
	PDGF	41
	VEGF	158
Leukemia — lymphoid	CCK-B/Gastrin	159
Leukemia — myeloid	CCK-B/Gastrin	159
	GM-CSF	162, 163
	IL-1β	160, 162–164
	Opioid	205
Lung cancer — small-cell carcinoma	Adrenomedullin	167
	BN/GRP	114, 206, 207
	CCK-B/Gastrin	118, 119, 146
	EGF	170
	Galanin	177
	Neurotensin	177, 208
	Opioid	86, 89, 171, 209
	Tachykinin	207
Lung cancer — non-small-cell carcinoma	BN/GRP	206
	CCK-B/Gastrin	119
	IL-8	181
	Opioid	89
Myeloma	IL-6	36, 190
	VIP	210
Neuroblastoma	Opioid	84, 211
Ovary — stromal	CCK-B/Gastrin	212
	VEGF	185
Pancreatic adenocarcinoma	VIP	124, 213
	Secretin	214
Prostate carcinoma	IL-6	25
Renal cell carcinoma	IL-6	31, 197
Thyroid — medullary carcinoma	CCK-B/Gastrin	146, 212
	SP	121

of immunological and inflammatory diseases, as well as in neoplasia. In addition to their ability to modulate tumor immunity, cytokines can influence the growth behavior of tumors in a variety of different ways. These include effects on tumor cell differentiation, proliferation, motility and adherence, and modulation of angiogenesis. It is becoming increasingly apparent that cytokines are involved in stress responses and that they have major effects on the hypothalamic-pituitary-adrenal axis and mediate interactions between the immune and neuroendocrine systems.[4,15,16,19-22]

A variety of cytokines are expressed by many different of types of neoplastic cells of hematologic and nonhematologic origin. Some types of tumor cells express multiple cytokines, and there is considerable heterogeneity in the cytokines produced among tumors of the same cell type.[23] Many tumor-associated cytokines are produced by tumor cells themselves (Table 10.1) or by nonneoplastic cells in the surrounding stromal tissue,[24-27] and sometimes it is difficult to make this distinction.[28]

TABLE 10.3
Autocrine Regulation by Cytokines and Neuropeptides for Various Human Tumors

Tumor Type	Cytokine/ Neuropeptide	Autocrine Role	Reference(s)
Angiosarcoma	VEGF	Stimulation	136
Bladder carcinoma	β-hCG	Stimulation	125
	IL-6	Stimulation	30
Brain — malignant glioma	bFGF	Stimulation	140
	HGF/SF	Stimulation	44, 142
Brain — meningioma	bFGF	Stimulation	140
Breast carcinoma	BN/GRP	Stimulation	144, 145
Cervical carcinoma	IL-6	Stimulation	32
Colorectal carcinoma	Gastrin	Stimulation	147, 215
	HGF/SF	Stimulation	148
	IL-1β	Stimulation	149, 150
Gastric carcinoma	bFGF	Stimulation	153
Kaposi's sarcoma	IL-1β	Stimulation	40, 41
	IL-6	Stimulation	34, 35
	bFGF	Stimulation	157
	PDGF	Stimulation	41
	VEGF	Stimulation	158
Leukemia — lymphoid	CCK-B/Gastrin	Stimulation	159
Leukemia — myeloid	CCK-B/Gastrin	Stimulation	159
	GM-CSF	Stimulation	162, 163
	IL-1β	Stimulation	160, 162-164
Lung cancer — small-cell type	Adrenomedullin	Stimulation	167
	BN/GRP	Stimulation	113, 169, 170, 172, 216
	Bradykinin	Stimulation	177
	CCK-B/Gastrin	Stimulation	146, 177
	β-Endorphin	Stimulation	100, 171, 216, 217
	Neurokinin A	Inhibition	170
	Neurokinin B	Inhibition	170
	Neurotensin	Stimulation	173, 177, 217
	Vasopressin	Stimulation	177
Lung cancer — non-small-cell	Adrenomedullin	Stimulation	167
	BN/GRP	Stimulation	178
	IL-8	Inhibition	181
	VIP	Stimulation	182
Myeloma	IL-6	Stimulation	36, 190
Neuroblastoma	Met-Enk	Inhibition	192
	VIP	Stimulation	194
Ovarian carcinoma	VEGF	Stimulation	185
Prostate carcinoma	IL-6	Stimulation	25
Renal cell carcinoma	IL-6	Stimulation	31, 197

10.4 CYTOKINES AS TUMOR GROWTH FACTORS

Since tumor cells can express functional cytokine receptors, cytokines can act as autocrine or paracrine growth factors as mentioned above. Interleukin-6 (IL-6) was first recognized as a T-cell-derived cytokine that induces B cells to mature into antibody-forming cells. Subsequently, IL-6 and its receptor (IL-6r) have been found to be normally expressed in a variety of cells and tissues. The diverse nature of functional IL-6r could explain the myriad effects of IL-6. There is substantial evidence that IL-6r is expressed in various types of tumor cells and that IL-6 can influence cell proliferation. Findings that some tumors express both IL-6 and IL-6r have led investigators to conclude that IL-6 can function as an autocrine growth factor. *In vitro* evidence to support a functional autocrine regulatory role of IL-6 has been found in prostate cancer cells,[25,29] bladder carcinoma cells,[30] renal cancer cells,[31] cervix cancer,[32] lymphoma,[33] Kaposi's sarcoma,[34,35] and multiple myeloma.[36,37]

In regards to myeloma, there has been some disagreement as to the origin and specific role of IL-6. The expression of IL-6 by some myeloma cell lines[36,37] but not others[38,39] suggested that IL-6 plays a paracrine rather than an autocrine role. Additionally, recent evidence indicates that dendritic cells in the bone marrow of myeloma patients, but not in the marrow of normal patients, are infected Kaposi's sarcoma-associated herpes virus (KSHV).[27] KSHV is reputed to be an etiologic agent for Kaposi's sarcoma[34] and its genome encodes for the human equivalent of IL-6.[28] Taken together, these results suggest that, in myeloma, IL-6 may function both as an autocrine and as a paracrine growth factor. In Kaposi's sarcoma (KS), KSHV-derived IL-6 has been found to be primarily expressed in KSHV-infected hematopoietic cells rather than in the KS lesions.[28] The exact role of IL-6 in the development of KS and whether this factor functions in a autocrine or paracrine manner requires further elucidation.[28,35,40,41]

The primary role of some cytokines in tumor development is to influence cells in the tissue surrounding the tumor rather than to modulate tumor immunity or tumor cell proliferation directly. Cytokines, such as vascular endothelial growth factor (VEGF), basic fibroblast growth factor (bFGF), and hepatocyte growth factor or scatter factor (HGF/SF), enhance tumor growth and invasion by their influence on the surrounding supportive tissues.[42-46] One such effect is the promotion of new blood vessel formation or angiogenesis, which is required by tumors to supply oxygen and nutrients for continued growth. The degree of expression of factors that stimulate angiogenesis in individual tumor specimens has been somewhat predictive of clinical behavior (e.g., vascularity, invasiveness, and metastasis).[24,47,48]

10.5 CLINICAL MANIFESTATIONS OF CYTOKINE–CANCER RELATIONSHIPS

Given their diverse biological activity, perturbations in the production and release of cytokines could contribute to some of the systemic manifestations of malignancy. Frequent constitutional findings in patients with advanced malignancy include fever, anorexia, weight loss, metabolic disturbances, and depression. These symptoms are also frequent side effects of cytokines when administered to patients for therapeutic purposes.

Cachexia is one of the most troubling symptoms that a patient with cancer can experience. The cancer anorexia-cachexia syndrome, as it is usually called, consists of marked loss of appetite, progressive weight loss and tissue wasting, and often portends serious consequences such as infection and death. The biochemical and metabolic abnormalities, which have been extensively studied both in the laboratory and in patients, are very complex.[49] Similarly complex is the etiology of this syndrome. Although cancer cachexia attributed to the release of humoral factors derived from tumors or stimulated in the host has been based on animal models, the various factors involved in cancer cachexia in humans are still being elucidated.[50] Considerable data suggest that the inflammatory cytokines, including tumor necrosis factor-α (TNFα), IL-1, IL-6, and interferon-γ (IFN-γ), play a role in cancer cachexia and that the levels of these factors are variably increased in cancer patients. The relationship between the inflammatory cytokines, neurohormones, and other factors associated in weight control, such as leptin, is an area of continuing research.[51]

Fever is a common symptom experienced by cancer patients and often occurs in the absence of infection. A variety of humoral factors are involved in the febrile response.[52-54] Included among the endogenous pyrogens are the proinflammatory cytokines, IL-1β, IL-6, and TNFβ. Fever may be mediated directly by these cytokines released within the hypothalamus or indirectly by cytokines released in the circulation and which feed back on the CNS.[52,54] Within the CNS, IL-1β and IL-6 may induce fever by acting directly on the thermoregulatory center or by stimulating specialized endothelial cells in the periventricular organs to release arachidonic acid metabolites and prostaglandin E_2 (PGE$_2$).[54]

Development of destructive bone lesions is a frequent complication of different types of cancer and can result as a consequence of increased osteoclastic resorption with an inhibition of new bone formation. Although the pathogenesis of bone lesions in cancer is a complicated process, cytokines are clearly involved. These humoral factors are often referred to as osteoclast activating factors (OAFs) because they directly or indirectly (e.g., via interaction with osteoblasts) stimulate bone resorption by osteoclasts. Several different cytokines have been shown to play a role in bone destruction, including IL-1, IL-3, IL-6, IL-11, TNFα, TNFβ, LIF, M-CSF, and GM-CSF.[55-57] The cytokines involved in the production of bone lesions are derived locally from diverse sources including the tumor cells themselves, tumor-associated lymphocytes, monocytes/macrophages, stromal cells, or by osteoblasts.

Psychological problems are common in patients with chronic illness including those with cancer.[12,58-60] Clinical depression in cancer patients is often considered an appropriate reaction to a devastating illness. Given the fact that some cytokines and neuropeptides can affect the CNS, the production of these factors by tumor cells may contribute to emotional manifestations in patients with cancer. CNS and emotional effects, including depression and psychological abnormalities, are common complications of treatment with biological response modifiers including interferon-α.[61,62] Interactions between cytokines and the CNS are addressed in several other chapters of this book.

10.6 CANCER AND ENDOGENOUS OPIOIDS

The endogenous opioid system consists of three families of opioid peptides and their receptors: proopiomelanocortin (POMC) from which β-endorphin is derived, proenkephalin from which methionine-enkephalin and leucine-enkephalin are derived, and prodynorphin from which dynorphin is derived. That the endogenous opioid system is involved in many different physiological responses, including stress, is well established.[63] Opioid peptides mediate their effects by interaction with specific subtypes of membrane receptors.[64,65] β-Endorphin has potent analgesic properties and exhibits preferential binding to epsilon (ε) opiate receptors and to mu (μ) receptors, and binds with less affinity to delta (δ) receptors. Methionine-enkephalin and leucine-enkephalin bind with high affinity to δ receptors, bind less well to μ receptors, and have little or no binding to kappa (κ) or ε receptors. Methionine-enkephalin and leucine-enkephalin possess minimal or no analgesic activity, perhaps because of their relatively short half-life. Dynorphin has high affinity for κ receptors, and much less affinity for μ and δ. The endogenous opioid peptides and their receptors are widely distributed within the peripheral and CNS, the pituitary and other neuroendocrine tissues.[63] In addition, a variety of cells of the immune system and hematopoietic system express endogenous opioids and possess opioid-binding sites similar to those expressed in the nervous system.[66,67] Mitogen-stimulated lymphocytes, both of rat and human origin, have been shown to produce and release proenkephalin-derived opioid peptides, including the pentapeptide methionine-enkephalin.[68,69]

There is mounting evidence that the endogenous opioid system may influence tumor growth and development, and there are basically two distinct mechanisms by which this can occur: (1) opioids can directly interact with tumor cells to cause a cytotoxic or antiproliferative effect, and (2) opioids can modulate host antitumor immune mechanisms.[66,70]

The antitumor effects of endogenous opioids have been demonstrated *in vitro* and *in vivo* in a number of animal models.[66,70-76] The endogenous opioid peptides, including β-endorphin, methionine-enkephalin, and leucine-enkephalin, enhance a variety of immunological mechanisms involved in antitumor defense including the generation of cytotoxic T cells and natural killer (NK) cell activity.[66,74,77,78] Methionine-enkephalin and leucine-enkephalin have been shown to enhance NK cell activity of human peripheral blood mononuclear cells (PBMC) *in vitro*,[79-82] and the administration of methionine-enkephalin increases NK cell activity in healthy volunteers and cancer patients *in vivo*.[74] Treatment of mice with methionine-enkephalin or leucine-enkephalin stimulates NK cell activity of splenic lymphocytes and inhibits tumor growth and metastasis.[71,73,76] These results suggest that stimulation of NK cell activity represents at least one mechanism which accounts for the antitumor effects of these two endogenous opioid peptides. Because methionine-enkephalin has potent immunomodulatory properties and is produced by cells of the immune system, Plotnikoff et al.[74] have proposed that this neuropeptide be designated an immunomodulatory cytokine.

Opioid receptors or binding sites have been described in a wide variety of animal and human tumor cells.[70,83-89] Tumors in which opioid receptor subtypes have been described are shown in Table 10.4. The tumor-associated receptors described thus far include classical opioid receptors of the δ, κ, μ subtypes, and the σ receptor, and a more recently described receptor designated zeta (ζ) by Zagon et al.[90] The ζ receptor is purported to represent a receptor, distinct from δ and other previously classified opioid receptor subtypes, that mediates the growth inhibitory effects of methionine-enkephalin.[91] It has been proposed that through its specific interaction with this receptor, methionine-enkephalin may function to tonically inhibit tumor growth in an autocrine and/or paracrine manner.[92-94] Methionine-enkephalin has been shown to inhibit the growth of human pancreatic and colorectal carcinoma cells *in vitro*.[92,95] Similarly, a variety of opioids have been shown to have inhibitory effects against human lung and breast cancer cells,[88,96] and these effects may involve nonconventional opioid binding sites.[96] There are data from several laboratories that suggest that at least one mechanism for the direct antitumor effects of opioids is induction of cell death by apoptosis.[97,98] However, there also are data to suggest mechanisms other than apoptosis may be involved in opioid-mediated direct tumor cell killing.[99]

TABLE 10.4
Opioid Receptor Subtypes in Human Tumors

Tumor cell type or origin	Receptor subtype(s)	Reference(s)
Astrocytoma	κ	87
Glioblastoma	κ, σ	87
Breast	δ, κ, μ	85, 88, 218
Colon	δ (ζ)[a]	202
Endometrial	δ, κ	219
Gastric adenocarcinoma	δ (ζ)	203
Lung, non-small-cell	δ, κ, μ	89
Lung, small-cell	δ, μ	86, 89
Neuroblastoma	δ, μ, ORL$_1$[b]	86, 89, 209
Squamous cell of head and neck	δ (ζ)	93

[a] ORL$_1$ = opioid receptor-like$_1$
[b] ζ has been the designation used by Zagon et al. as the receptor for met-enkephalin on tumor cells (*J. Natl. Cancer Inst.*, 82, 325, 1990.).

β-Endorphin has been reported to have variable effects on tumor growth. β-Endorphin has been shown to stimulate the proliferation of a human small-cell carcinoma cell line *in vitro* by binding to nonopioid sites.[100] Since these cells also produce β-endorphin, an autocrine regulatory role for this opioid peptide in lung cancer has been proposed.[100] It has been suggested that this stimulating effect of β-endorphin is due to its nonopioid binding to the integrin vibrinectin and induction of a scattering-like effect on tumor cells which may be necessary for cell proliferation.[100] Both β-endorphin and morphine have been shown to inhibit blood vessel formation in a chicken chorioallantoic membrane model.[101] Based on these findings, it would be important to study whether these and other opioids can inhibit angiogenesis in an *in vivo* tumor model.

Small-cell lung carcinoma cell lines have been shown to express proopiomelanocortin (POMC) mRNA and secrete β-lipoprotein, ACTH, and β-endorphin.[100,102-104] In some patients with lung cancer, tumor cells produce large amounts of pituitary-like ACTH, resulting in symptoms of Cushing's syndrome due to ACTH excess. However, POMC mRNA has been found in a variety of lung tumor types and in normal lung, without the expression of large amounts of pituitary-like ACTH.[105] With the possible exception of lung tumors with features of neuroendocrine differentiation, there are differences in the way POMC is processed in lung cancer cells compared to pituitary cells.[104-107] The ectopic production of the POMC family of peptides and of CRF has been associated with different types of cancer in addition to those arising in lung.[106] Elevated blood levels of β-endorphin immunoreactivity have been found in patients with small-cell lung cancer but not in control subjects or in patients with other types of carcinoma or non-Hodgkin's lymphoma.[108] It is conceivable that the increased plasma level of β-endorphin in patients with small-cell lung cancer is derived from the tumor cells, but this requires further study. The role of opioid peptide expression in lung carcinoma cells is unclear, but there is suggestive evidence some opioid peptides inhibit the effects of nicotine which stimulates tumor proliferation via specific nicotine receptors.[89,97]

Interestingly, the proto-oncogene c-*fos* in pituitary cells is induced by corticotropin-releasing factor (CRF) and stimulates the transcription of the POMC gene.[109] Morphine has been shown to increase the expression of c-*fos* in the rat hypothalamus.[110] Also of interest is the finding that the products of oncogenes c-*fos* and c-*jun*, Fos and Jun, stimulate the transcription of proenkephalin mRNA.[111] It remains to be shown if functional relationships exist in human cancer between neuropeptides and oncogene expression.

10.7 CANCER AND OTHER TUMOR-ASSOCIATED NEUROPEPTIDES

Tumor-associated neuropeptides can result in a variety of systemic manifestations that are remote from the primary tumor and sites of metastases; these aggregates of symptoms and signs are called *paraneoplastic syndromes*. The clinical situations that are probably best understood are those in which the neuropeptide is produced by a tumor originating from cells or tissues that constituently produce that particular hormone, but in excessive and in a manner that is independent of normal feedback regulatory mechanisms. A variety of tumors of endocrine origin fall into the latter category. Examples include pheochromocytoma; medullary thyroid carcinoma; carcinoid glucagonomas; gastrinomas; vasoactive intestinal peptide tumors (VIPomas); and tumors of pituitary, parathyroid, and adrenal origin.[112] The clinical manifestations of the endocrine tumors vary depending upon the neuropeptide or hormone produced. A discussion of neuroendocrine tumors and paraneoplastic syndromes that result from these tumors is beyond the scope of this treatise. This chapter will provide some examples of human "nonendocrine" tumors that express neuropeptides (i.e., ectopic expression).

Gastrin-releasing peptide (GRP) is a member of the bombesin-like family of bioactive peptides, and is the mammalian counterpart to bombesin (BN) which is of amphibian origin. GRP is a 27-amino-acid peptide with a carboxyl-terminal heptapeptide sequence identical to that of bombesin. Since GRP and BN have similar pharmacological effects and interact with the same receptors, the designation bombesin-like peptides or BN/GRP are often used for the family of neuropeptides.[113,114] The expression of bombesin-like peptide in human lung cancer cells was first reported by Moody et al.[114,115] These investigators found high levels of bombesin-like peptide in a panel of cell lines derived from patients with small-cell carcinoma of the lung, but not in a variety of cell lines of non-small-cell lung origin. Of note, small-cell carcinoma of the lung makes up about 25% of all cases of lung cancer and is usually distinguished from non-small-cell lung cancer by the presence of neurosecretory granules. It is important to note that lung cancer, and in particular small-cell lung cancer, is one of the human tumors more frequently associated with ectopic hormone production and the paraneoplastic syndromes that result. However, some patients have lung carcinoma of the non-small cell type but with features of neuroendocrine differentiation, and the clinical course of these patients is similar to those of patients with small-cell carcinoma.[116] Because tumor cells derived from patients with lung carcinoma express a variety of neuropeptides which can act to stimulate tumor growth, means of inhibiting these growth factors, with broad neuropeptide antagonists, are being considered for their therapeutic potential.[117]

Gastrin and CCK possess the same five amino acids at their COOH terminus. The COOH terminus is the biologically active site, but the actions of these two peptides are mediated by two different types of receptors. CCK-A has a low affinity for gastrin whereas CCK-B has a high affinity for gastrin. CCK-B/gastrin receptors are expressed by small-cell lung cancer cells but not by non-small-cell lung cancer cells.[118,119] Gastrin receptors have been found in tumor cells obtained from patients with colon carcinoma and, for reasons that are unclear, their expression correlates with a relatively better prognosis.[120]

Substance P (SP), substance K, and neuromedin K are members of the tachykinin family of neuropeptides of mammalian origin. These neuropeptides interact with varying affinities with tachykinin receptor subtypes NK-1, NK-2, and NK-3. SP preferentially binds to NK-1 subtype receptors which are widely expressed in tissues of the central and peripheral nervous system. Receptors for SP and other members of this group are also expressed in a variety of human tumors.[121]

Somatostatin is a peptide hormone that is produced by a wide variety of normal cells within the neuroendocrine system and within the immune system,[21] and by some tumor cells. Within the neuroendocrine system, somatostatin potently inhibits the secretion of a variety of other hormones including growth hormone (GH). GH up-regulates insulin-like growth factor 1 (IGF-1) which has been shown to stimulate the growth of breast cancer cells *in vitro*. Octreotide is a somatostatin analog with greater potency as an inhibitor of GH secretion and antitumor activity *in vitro* and *in vivo* against a variety of tumors. It was developed for clinical use because of its relatively longer plasma half-life compared to somatostatin. Although octreotide has been effective in controlling the symptoms associated with some neuroendocrine tumors, clinical trials of octreotide as an antitumor agents for other tumors have met with disappointing results.[122,123] A variety of other somatostatin analogs with potent inhibitory activity against a broad range of tumors are being developed. Some of the analogs tested thus far are more selective for tumors in that they lack the ability to inhibit GH secretion, and appear to inhibit tyrosine kinase and result in tumor cell death by apoptosis.[123]

Radiolabeled octreotide and VIP have been administered to patients to image tumors that express receptors for those neuropeptides. This technique has been successful in determining the extent of primary tumor and metastatic involvement in a variety of gastrointestinal tumors, including adenocarcinomas of the colorectum, stomach, and pancreas, and in intestinal carcinoid and primary insulinoma.[124]

Human chorionic gonadotropin hormone (β-hCG) has been implicated as a autocrine/paracrine growth factor in various tumors.[125-128] Although the mechanism is not clearly understood, β-hCG inhibits the growth of Kaposi's sarcoma cell lines *in vitro* and results in tumor regressions when administered to patients with AIDS-related Kaposi's sarcoma.[129,130]

A variety of normal and neoplastic cells express a glycoprotein antigen on the cell surface called CD/10 neutral endopeptidase 24.11 (NEP).[131,132] Previous designations for this factor based on its antigenic properties on hematopoietic cells include the common acute lymphocytic leukemia antigen (CALLA) and common designation 10 (CD10).[132] Among the substrates hydrolyzed by NEP include opioid peptides, fMLP, substance P, bombesin-like peptides, atrial natriuretic factor, endothelin, oxytocin, bradykinin, and angiotensins I and II.[133] Although the functional properties of this metalloproteinase is becoming somewhat better understood,[131,133-135] the role of this enzyme in modulating the effects of neuropeptides on tumor cells requires further research.

10.8 CONCLUSIONS

Neuropeptides and cytokines, and receptors for these humoral factors, are expressed by many different types of human tumors. This provides a complex network for bidirectional interactions between the neuroendocrine system, the immune system, and cancer. It is becoming increasingly apparent that both neuropeptides and cytokines act as paracrine and/or autocrine growth factors in a variety of tumors. The effects of stress on tumor development and growth may be attributed to the influence of neuropeptides and cytokines released by the host's neuroendocrine and immune systems. Similarly, many of the remote effects of cancer on the host are due to neuropeptides and cytokines released by tumor cells. A better understanding of these relationships may help to develop strategies for improving cancer treatment and control.

REFERENCES

1. Riley, V., Psychoneuroendocrine influences on immunocompetence and neoplasia, *Science*, 212, 1100, 1981.
2. Riley, V., Mouse mammary tumors: alteration of incidence as apparent function of stress, *Science*, 189, 465, 1975.

3. Giraldi, T., Perissin, L., Zorzet, S., Rapozzi, V., and Rodani, M. G., Metastasis and neuroendocrine system in stressed mice, *Int. J. Neurosci.*, 74, 265, 1994.
4. Besedovsky, H. O., Herberman, R. B., Temoshok, L. R., and Sendo, F., Psychoneuroimmunology and cancer: 15th Sapporo Cancer Seminar, *Cancer Res.*, 56, 4278, 1996.
5. Andersen, B. L., Stress and quality of life following cervical cancer, *J. Natl. Cancer Inst. Monogr.*, 65, 1996.
6. Andersen, B. L., Kiecolt-Glaser, J. K., and Glaser, R., A biobehavioral model of cancer stress and disease course, *Am. Psychol.*, 49, 389, 1994.
7. Andersen, B. L., Surviving cancer, *Cancer.*, 74, 1484, 1994.
8. Cassileth, B. R., Walsh, W. P., and Lusk, E. J., Psychosocial correlates of cancer survival: a subsequent report 3 to 8 years after cancer diagnosis [see comments], *J. Clin. Oncol.*, 6, 1753, 1988.
9. Spiegel, D., Psychological distress and disease course for women with breast cancer: one answer, many questions [editorial; comment], *J. Natl. Cancer Inst.*, 88, 629, 1996.
10. Bleiker, E. M., van der Ploeg, H. M., Hendriks, J. H., and Ader, H. J., Personality factors and breast cancer development: a prospective longitudinal study, *J. Natl. Cancer Inst.*, 88, 1478, 1996.
11. Creagan, E. T., Attitude and disposition: do they make a difference in cancer survival?, *Mayo Clin. Proc.*, 72, 160, 1997.
12. Fife, A., Beasley, P. J., and Fertig, D. L., Psychoneuroimmunology and cancer: historical perspectives and current research, *Adv. Neuroimmunol.*, 6, 179, 1996.
13. Jasmin, C., Le, M. G., Marty, P., and Herzberg, R., Evidence for a link between certain psychological factors and the risk of breast cancer in a case-control study. Psycho-Oncologic Group (P.O.G.), *Ann. Oncol.*, 1, 22, 1990.
14. Tross, S., Herndon, J., II, Korzun, A., Kornblith, A. B., Cella, D. F., Holland, J. F., Raich, P., Johnson, A., Kiang, D. T., Perloff, M., Norton, L., Wood, W., and Holland, J. C., Psychological symptoms and disease-free and overall survival in women with stage II breast cancer. Cancer and Leukemia Group B [see comments], *J. Natl. Cancer Inst.*, 88, 661, 1996.
15. Sternberg, E. M., Chrousos, G. P., Wilder, R. L., and Gold, P. W., The stress response and the regulation of inflammatory disease, *Ann. Intern. Med.*, 117, 854, 1992.
16. Besedovsky, H. O. and del Rey, A., Immune-neuro-endocrine interactions: facts and hypotheses, *Endocr. Rev.*, 17, 64, 1996.
17. Pert, C. B., Ruff, M. R., Weber, R. J., and Herkenham, M., Neuropeptides and their receptors: a psychosomatic network, *J. Immunol.*, 135, 820s, 1985.
18. Faith, R. E., Murgo, A. J., and Plotnikoff, N. P., Interactions between the immune system and the nervous system, in *Stress and Immunity,* Plotnikoff, N., Murgo, A., Faith, R., and Wybran, J., Eds., CRC Press, Boca Raton, Fl., 1991, 287.
19. Blalock, J. E., A molecular basis for bidirectional communication between the immune and neuroendocrine systems, *Physiol. Rev.*, 69, 1, 1989.
20. Chrousos, G. P., The hypothalamic-pituitary-adrenal axis and immune-mediated inflammation [see comments], *N. Engl. J. Med.*, 332, 1351, 1995.
21. Reichlin, S., Neuroendocrine-immune interactions, *N. Eng. J. Med.*, 329, 1246, 1993.
22. Dinarello, C. A., The biological properties of interleukin-1, *Eur. Cytokine Network*, 5, 517, 1994.
23. Armstrong, C. A., Tara, D. C., Hart, C. E., Kock, A., Luger, T. A., and Ansel, J. C., Heterogeneity of cytokine production by human malignant melanoma cells, *Exp. Dermatol.*, 1, 37, 1992.

24. Jin, L., Yuan, R. Q., Fuchs, A., Yao, Y., Joseph, A., Schwall, R., Schnitt, S. J., Guida, A., Hastings, H. M., Andres, J., Turkel, G., Polverini, P. J., Goldberg, I. D., and Rosen, E. M. Expression of interleukin-1β in human breast carcinoma, *Cancer*, 80, 421, 1997.
25. Okamoto, M., Lee, C., and Oyasu, R., Interleukin-6 as a paracrine and autocrine growth factor in human prostatic carcinoma cells *in vitro*, *Cancer Res.*, 57, 141, 1997.
26. Chatikhine, V. A., Chevrier, A., Chauzy, C., Duval, C., d'Anjou, J., Girard, N., and Delpech, B., Expression of opioid peptides in cells and stroma of human breast cancer and adenofibromas, *Cancer Lett.*, 77, 51, 1994.
27. Rettig, M. B., Ma, H. J., Vescio, R. A., Pold, M., Schiller, G., Belson, D., Savage, A., Nishikubo, C., Wu, C., Fraser, J., Said, J. W., and Berenson, J. R., Kaposi's sarcoma-associated herpes virus infection of bone marrow dendritic cells from multiple myeloma patients [see comments], *Science*, 276, 1851, 1997.
28. Moore, P. S., Boshoff, C., Weiss, R. A., and Chang, Y., Molecular mimicry of human cytokine and cytokine response pathway genes by KSHV, *Science*, 274, 1739, 1996.
29. Siegall, C. B., Schwab, G., Nordan, R. P., Fitzgerald, D. J., and Pastan, I., Expression of the interleukin-6 receptor and interleukin-6 in prostate carcinoma cells, *Cancer Res.*, 50, 7786, 1990.
30. Okamoto, M., Hattori, K., and Oyasu, R., Interleukin-6 functions as an autocrine growth factor in human bladder carcinoma cell lines *in vitro*, *Int. J. Cancer*, 72, 149, 1997.
31. Miki, S., Iwano, M., Miki, Y., Yamamoto, M., Tang, B., Yokokawa, K., Sonoda, T., Hirano, T., and Kishimoto, T. Interleukin-6 (IL-6) functions as an *in vitro* autocrine growth factor in renal cell carcinomas, *FEBS Lett.*, 250, 607, 1989.
32. Eustace, D., Han, X., Gooding, R., Rowbottom, A., Riches, P., and Heyderman, E., Interleukin-6 (IL-6) functions as an autocrine growth factor in cervical carcinomas *in vitro*, *Gyneco.l Oncol.*, 50, 15, 1993.
33. Yee, C., Biondi, A., Wang, X. H., Iscove, N. N., de Sousa, J., Aarden, L. A., Wong, G. G., Clark, S. C., Messner, H. A., and Minden, M. D., A possible autocrine role for interleukin-6 in two lymphoma cell lines, *Blood*, 74, 798, 1989.
34. Staskus, K. A., Zhong, W., Gebhard, K., Herndier, B., Wang, H., Renne, R., Beneke, J., Pudney, J., Anderson, D. J., Ganem, D., and Haase, A. T., Kaposi's sarcoma-associated herpesvirus gene expression in endothelial (spindle) tumor cells, *J. Virol.*, 71, 715, 1997.
35. Murakami-Mori, K., Taga, T., Kishimoto, T., and Nakamura, S., The soluble form of the IL-6 receptor (sIL-6R α) is a potent growth factor for AIDS-associated Kaposi's sarcoma (KS) cells; the soluble form of gp130 is antagonistic for sIL-6R α-induced AIDS-KS cell growth, *Int. Immunol.*, 8, 595, 1996.
36. Schwab, G., Siegall, C. B., Aarden, L. A., Neckers, L. M., and Nordan, R. P., Characterization of an interleukin-6-mediated autocrine growth loop in the human multiple myeloma cell line, U266, *Blood*, 77, 587, 1991.
37. Bataille, R., Jourdan, M., Zhang, X. G., and Klein, B., Serum levels of interleukin-6, a potent myeloma cell growth factor, as a reflect of disease severity in plasma cell dyscrasias, *J. Clin. Invest.*, 84, 2008, 1989.
38. Klein, B., Zhang, X. G., Jourdan, M., Content, J., Houssiau, F., Aarden, L., Piechaczyk, M., and Bataille, R., Paracrine rather than autocrine regulation of myeloma-cell growth and differentiation by interleukin-6, *Blood*, 73, 517, 1989.
39. Zhang, X. G., Klein, B., and Bataille, R., Interleukin-6 is a potent myeloma-cell growth factor in patients with aggressive multiple myeloma, *Blood*, 74, 11, 1989.

40. Louie, S., Cai, J., Law, R., Lin, G., Lunardi-Iskandar, Y., Jung, B., Masood, R., and Gill, P., Effects of interleukin-1 and interleukin-1 receptor antagonist in AIDS-Kaposi's sarcoma, *J. Acquir. Immune Defic. Syndr. Hum. Retrovirol.*, 8, 455, 1995.
41. Sturzl, M., Brandstetter, H., Zietz, C., Eisenburg, B., Raivich, G., Gearing, D. P., Brockmeyer, N. H., and Hofschneider, P. H., Identification of interleukin-1 and platelet-derived growth factor-B as major mitogens for the spindle cells of Kaposi's sarcoma: a combined *in vitro* and *in vivo* analysis, *Oncogene,* 10, 2007, 1995.
42. Brown, L. F., Berse, B., Jackman, R. W., Tognazzi, K., Manseau, E. J., Senger, D. R., and Dvorak, H. F., Expression of vascular permeability factor (vascular endothelial growth factor) and its receptors in adenocarcinomas of the gastrointestinal tract, *Cancer Res.*, 53, 4727, 1993.
43. Itakura, J., Ishiwata, T., Friess, H., Fujii, H., Matsumoto, Y., Buchler, M. W., and Korc, M., Enhanced expression of vasular endothelial growth factor in human pacreatic cancer correlates with local progression, *Clin. Cancer Res.,* 3, 1309, 1997.
44. Koochekpour, S., Jeffers, M., Rulong, S., Taylor, G., Klineberg, E., Hudson, E. A., Resau, J. H., and Vande Woude, G. F., Met and hepatocyte growth factor/scatter factor expression in human gliomas, *Cancer Res.,* 57, 5391, 1997.
45. Takahashi, Y., Kitadai, Y., Bucana, C. D., Cleary, K. R., and Ellis, L. M., Expression of vascular endothelial growth factor and its receptor, KDR, correlates with vascularity, metastasis, and proliferation of human colon cancer, *Cancer Res.*, 55, 3964, 1995.
46. Yoshiji, H., Gomez, D. E., Shibuya, M., and Thorgeirsson, U. P., Expression of vascular endothelial growth factor, its receptor, and other angiogenic factors in human breast cancer, *Cancer Res.*, 56, 2013, 1996.
47. Gasparini, G., Toi, M., Gion, M., Verderio, P., Dittadi, R., Hanatani, M., Matsubara, I., Vinante, O., Bonoldi, E., Boracchi, P., Gatti, C., Suzuki, H., and Tominaga, T., Prognostic significance of vascular endothelial growth factor protein in node-negative breast carcinoma, *J. Natl. Cancer Inst.*, 89, 139, 1997.
48. Mise, M., Arii, S., Higashituji, H., Furutani, M., Niwano, M., Harada, T., Ishigami, S., Toda, Y., Nakayama, H., Fukumoto, M., Fujita, J., and Imamura, M., Clinical significance of vascular endothelial growth factor and basic fibroblast growth factor gene expression in liver tumor, *Hepatology*, 23, 455, 1996.
49. Nelson, K. A., Walsh, D., and Sheehan, F. A., The cancer anorexia-cachexia syndrome [see comments], *J. Clin. Oncol.*, 12, 213, 1994.
50. Todorov, P., Cariuk, P., McDevitt, T., Coles, B., Fearon, K., and Tisdale, M., Characterization of a cancer cachectic factor, *Nature,* 379, 739, 1996.
51. Schwartz, M. W. and Seeley, R. J., Seminars in medicine of the Beth Israel Deaconess Medical Center. Neuroendocrine responses to starvation and weight loss, *N. Engl. J. Med.*, 336, 1802, 1997.
52. Kluger, M. J., Kozak, W., Leon, L. R., Soszynski, D., and Conn, C. A., Cytokines and fever, *Neuroimmunomodulation,* 2, 216, 1995.
53. Sakata, T., Yoshimatsu, H., and Kurokawa, M., Hypothalamic neuronal histamine, implications of its homeostatic control of energy metabolism, *Nutrition,* 13, 403, 1997.
54. Lesnikov, V. A., Efremov, O. M., Korneva, E. A., Van Damme, J., and Billiau, A., Fever produced by intrahypothalamic injection of interleukin-1 and interleukin-6, *Cytokine*, 3, 195, 1991.
55. Martin, T. J. and Ng, K. W., Mechanisms by which cells of the osteoblast lineage control osteoclast formation and activity, *J. Cell. Biochem.*, 56, 357, 1994.
56. Bataille, R., The mechanisms of bone lesions in human plasmacytomas, *Stem Cells (Dayt).*, 13 (Suppl. 2), 40, 1995.

57. Dinarello, C. A., Biologic basis for interleukin-1 in disease, *Blood*, 87, 2095, 1996.
58. Spiegel, D., Cancer and depression, *Br. J. Psychiatry,* Suppl: 109, 1996.
59. McDaniel, J. S., Musselman, D. L., Porter, M. R., Reed, D. A., and Nemeroff, C. B., Depression in patients with cancer: diagnosis, biology, and treatment, *Arch. Gen. Psych.*, 52, 89, 1995.
60. Breitbart, W., Psycho-oncology, depression, anxiety, delirium, *Semin. Oncol.*, 21, 754, 1994.
61. Dusheiko, G., Side effects of α interferon in chronic hepatitis C. *Hepatol.*, 26, 112S, 1997.
62. Triozzi, P. L., Kinney, P., and Rinehart, J. J., Central nervous system toxicity of biological response modifiers, *Ann. NY Acad. Sci.*, 594, 347, 1990.
63. Olson, G. A., Olson, R. D., and Kastin, A. J., Endogenous opiates, 1995, *Peptides,* 17, 1421, 1996.
64. Yamada, K. and Nabeshima, T., Stress-induced behavioral responses and multiple opioid systems in the brain, *Behav. Brain Res.*, 67, 133, 1995.
65. Borsodi, A. and Toth, G., Characterization of opioid receptor types and subtypes with new ligands, *Ann. NY Acad. Sci.*, 757, 339, 1995.
66. Murgo, A. J., Faith, R. E., and Plotnikoff, N. P., Enkephalins, Mediators of stress-induced immunomodulation, in *Stress and the Immune System,* Plotnikoff, N. P., Faith, R. E., Murgo, A. J., and Good R. A., Eds., Plenum Press, New York, 1986, 221.
67. Carr, D. J., The role of endogenous opioids and their receptors in the immune system, *Proc. Soc. Exp. Biol. Med.*, 198, 710, 1991.
68. Padros, M. R., Vindrola, O., Zunszain, P., Fainboin, L., Finkielman, S., and Nahmod, V. E., Mitogenic activation of the human lymphocytes induce the release of proen-kephalin derived peptides, *Life Sci.*, 45, 1805, 1989.
69. Zurawski, G., Benedik, M., Kamb, B. J., Abrams, J. S., Zurawski, S. M., and Lee, F. D., Activation of mouse T-helper cells induces abundant preproenkephalin mRNA synthesis, *Science,* 232, 772, 1986.
70. Zagon, I. S. and McLaughlin, P. J., The role of endogenous opioids and opioid receptors in human and animal cancer, in *Stress and Immunity*, Plotnikoff, N. P., Murgo, A., Faith, R., and Wyban, J., Eds., CRC Press, Boca Raton, FL, 1991, 343.
71. Murgo, A. J., Inhibition of B16-BL6 melanoma growth in mice by methionine-enkephalin, *J. Natl. Cancer Inst.*, 75, 341, 1985.
72. Plotnikoff, N. P. and Miller, G. C., Enkephalins as immunomodulators, *Int. J. Immunopharmacol.*, 5, 437, 1983.
73. Faith, R. E. and Murgo, A. J., Inhibition of pulmonary metastases and enhancement of natural killer cell activity by methionine-enkephalin, *Brain Behav. Immun.*, 2, 114, 1988.
74. Plotnikoff, N. P., Faith, R. E., Murgo, A. J., Herberman, R. B., and Good, R. A., Methionine enkephalin, a new cytokine--human studies, *Clin. Immunol. Immunopathol,* 82, 93, 1997.
75. Murgo, A. J., Faith, R. E., and Plotnikoff, N. P., Enhancement of tumor resistance in mice by enkephalin, in *Stress and Immunity,* Plotnikoff, N., Murgo, A., Faith, R., and Wyban, J., Eds., CRC Press, Boca Raton, FL, 1991, 357.
76. Scholar, E. M., Violi, L., and Hexum, T. D., The antimetastatic activity of enkephalin-like peptides, *Cancer Lett.*, 35, 133, 1987.
77. Herberman, R. B. and Holden, H. T., Natural killer cells as antitumor effector cells, *J. Natl. Cancer Inst.*, 62, 441, 1979.
78. Carr, D. J. and Klimpel, G. R., Enhancement of the generation of cytotoxic T cells by endogenous opiates, *J. Neuroimmunol.*, 12, 75, 1986.

79. Mathews, P. M., Froelich, C. J., Sibbitt, W. L., Jr., and Bankhurst, A. D., Enhancement of natural cytotoxicity by β-endorphin, *J. Immunol.*, 130, 1658, 1983.
80. Faith, R. E., Plotnikoff, N. P., and Murgo, A. J., Effects of opiates and neuropeptides on immune functions, *NIDA Res. Monogr.*, 54, 300, 1984.
81. Faith, R. E., Liang, H. J., Plotnikoff, N. P., Murgo, A. J., and Nimeh, N. F., Neuroimmunomodulation with enkephalins, *in vitro* enhancement of natural killer cell activity in peripheral blood lymphocytes from cancer patients, *Nat. Immun. Cell Growth Regul.*, 6, 88, 1987.
82. Puente, J., Maturana, P., Miranda, D., Navarro, C., Wolf, M. E., and Mosnaim, A. D., Enhancement of human natural killer cell activity by opioid peptides, similar response to methionine-enkephalin and β-endorphin, *Brain Behav. Immun.*, 6, 32, 1992.
83. Zagon, I. S., McLaughlin, P. J., Goodman, S. R., and Rhodes, R. E., Opioid receptors and endogenous opioids in diverse human and animal cancers, *J. Natl. Cancer Inst.*, 79, 1059, 1987.
84. Toll, L., Polgar, W. E., and Auh, J. S., Characterization of the δ-opioid receptor found in SH-SY5Y neuroblastoma cells, *Eur. J. Pharmacol.*, 323, 261, 1997.
85. Kampa, M., Loukas, S., Hatzoglou, A., Martin, P., Martin, P. M., and Castanas, E., Identification of a novel opioid peptide (Tyr-Val-Pro-Phe-Pro) derived from human α S1 casein (α S1-casomorphin, and α S1-casomorphin amide), *Biochem. J.*, 319, 903, 1996.
86. Campa, M. J., Schreiber, G., Bepler, G., Bishop, M. J., McNutt, R. W., Chang, K. J., and Patz, E. F., Jr., Characterization of δ opioid receptors in lung cancer using a novel nonpeptidic ligand, *Cancer Res.*, 56, 1695, 1996.
87. Thomas, G. E., Szucs, M., Mamone, J. Y., Bem, W. T., Rush, M. D., Johnson, F. E., and Coscia, C. J., Sigma and opioid receptors in human brain tumors, *Life Sci.*, 46, 1279, 1990.
88. Maneckjee, R., Biswas, R., and Vonderhaar, B. K., Binding of opioids to human MCF-7 breast cancer cells and their effects on growth, *Cancer Res.*, 50, 2234, 1990.
89. Maneckjee, R. and Minna, J. D., Opioid and nicotine receptors affect growth regulation of human lung cancer cell lines, *Proc. Natl. Acad. Sci. U.S.A.*, 87, 3294, 1990.
90. Zagon, I. S., Goodman, S. R., and McLaughlin, P. J., Characterization of zeta (ζ), a new opioid receptor involved in growth, *Brain Res.*, 482, 297, 1989.
91. Zagon, I. S., Goodman, S. R., and McLaughlin, P. J., Demonstration and characterization of zeta (ζ), a growth-related opioid receptor, in a neuroblastoma cell line, *Brain Res.*, 511, 181, 1990.
92. Zagon, I. S., Hytrek, S. D., Smith, J. P., and McLaughlin, P. J., Opioid growth factor (OGF) inhibits human pancreatic cancer transplanted into nude mice, *Cancer Lett.*, 112, 167, 1997.
93. Levin, R. J., Wu, Y., McLaughlin, P. J., and Zagon, I. S., Expression of the opioid growth factor, [Met5]-enkephalin, and the ζ opioid receptor in head and neck squamous cell carcinoma, *Laryngoscope*, 107, 335, 1997.
94. Zagon, I. S., Wu, Y., and McLaughlin, P. J., The opioid growth factor, [Met5]-enkephalin, and the ζ opioid receptor are present in human and mouse skin and tonically act to inhibit DNA synthesis in the epidermis, *J. Invest. Dermatol.*, 106, 490, 1996.
95. Zagon, I. S., Hytrek, S. D., and McLaughlin, P. J., Opioid growth factor tonically inhibits human colon cancer cell proliferation in tissue culture, *Am. J. Physiol.*, 271,R511, 1996.

96. Maneckjee, R. and Minna, J. D., Nonconventional opioid binding sites mediate growth inhibitory effects of methadone on human lung cancer cells, *Proc. Natl. Acad. Sci. U.S.A.*, 89, 1169, 1992.
97. Maneckjee, R. and Minna, J. D., Opioids induce while nicotine suppresses apoptosis in human lung cancer cells, *Cell Growth Differ.*, 5, 1033, 1994.
98. Mernenko, O. A., Blishchenko, E. Y., Mirkina, I. I., and Karelin, A. A., Met-enkephalin induces cytolytic processes of apoptotic type in K562 human erythroid leukemia cells, *FEBS Lett.*, 383, 230, 1996.
99. Blishchenko, E. Y., Merenko, O. A., Mirkina, I. I., Satpaev, D. K., Ivanov, V. S., Tchikin, L. D., Ostrovsky, A. G., Karelin, A. A., and Ivanov, V. T., Tumor cell cytolysis mediated by valorphin, an opioid-like fragment of hemoglobin β-chain, *Peptides*, 18, 79, 1997.
100. Melzig, M. F., Nylander, I., Vlaskovska, M., and Terenius, L., β-endorphin stimulates proliferation of small cell lung carcinoma cells *in vitro* via nonopioid binding sites, *Exp. Cell. Res.*, 219, 471, 1995.
101. Pasi, A., Qu, B. X., Steiner, R., Senn, H. J., Bar, W., and Messiha, F. S., Angiogenesis, modulation with opioids, *Gen. Pharmacol.*, 22, 1077, 1991.
102. Bertagna, X. Y., Nicholson, W. E., Sorenson, G. D., Pettengill, O. S., Mount, C. D., and Orth, D. N., Corticotropin, lipotropin, and β-endorphin production by a human nonpituitary tumor in culture, evidence for a common precursor, *Proc. Natl. Acad. Sci. U.S.A.*, 75, 5160, 1978.
103. Bergh, J., Nilsson, K., Ekman, R., and Giovanella, B., Establishment and characterization of cell lines from human small cell and large cell carcinomas of the lung, *Acta. Pathol. Microbiol. Immunol. Scand. [A].*, 93, 133, 1985.
104. White, A., Stewart, M. F., Farrell, W. E., Crosby, S. R., Lavender, P. M., Twentyman, P. R., Rees, L. H., and Clark, A. J., Pro-opiomelanocortin gene expression and peptide secretion in human small-cell lung cancer cell lines, *J. Mol. Endocrinol.*, 3, 65, 1989.
105. Texier, P. L., de Keyzer, Y., Lacave, R., Vieau, D., Lenne, F., Rojas-Miranda, A., Verley, J. M., Luton, J. P., Kahn, A., and Bertagna, X., Proopiomelanocortin gene expression in normal and tumoral human lung, *J. Clin. Endocrinol. Metab.*, 73, 414, 1991.
106. Schteingart, D. E., Ectopic secretion of peptides of the proopiomelanocortin family, *Endocrinol. Metab. Clin. N. Am.*, 20, 453, 1991.
107. Chang, A. C., Israel, A., Gazdar, A., and Cohen, S. N., Initiation of pro-opiomelanocortin mRNA from a normally quiescent promoter in a human small cell lung cancer cell line, *Gene*, 84, 115, 1989.
108. Pasi, A., Amsler, U., and Messiha, F. S., β-endorphin-like immunoreactivity, assessment of blood levels in patients with tumors of different origin, *J. Med.*, 22, 327, 1991.
109. Boutillier, A. L., Sassone-Corsi, P., and Loeffler, J. P., The protooncogene *c-fos* is induced by corticotropin-releasing factor and stimulates proopiomelanocortin gene transcription in pituitary cells, *Mol. Endocrinol.*, 5, 1301, 1991.
110. Chang, S. L. and Harlan, R. E., The FOS Proto-oncogene protein, regulation by morphine in the rat hypothalamus, *Life Sci.*, 46, 1825, 1990.
111. Sonnenberg, J. L., Rauscher, F. J. D., Morgan, J. I., and Curran, T., Regulation of proenkephalin by Fos and Jun, *Science*, 246, 1622, 1989.
112. Schally, A. V., Oncological applications of somatostatin analogues [published erratum, appears in *Cancer Res.* 1989 Mar 15, 49(6), 1618], *Cancer Res.*, 48, 6977, 1988.
113. Carney, D. N., Cuttitta, F., Moody, T. W., and Minna, J. D., Selective stimulation of small cell lung cancer clonal growth by bombesin and gastrin-releasing peptide, *Cancer Res.*, 47, 821, 1987.

114. Moody, T. W., Bertness, V., and Carney, D. N., Bombesin-like peptides and receptors in human tumor cell lines, *Peptides*, 4, 683, 1983.
115. Moody, T. W., Pert, C. B., Gazdar, A. F., Carney, D. N., and Minna, J. D., High levels of intracellular bombesin characterize human small-cell lung carcinoma, *Science*, 214, 1246, 1981.
116. Schleusener, J. T., Tazelaar, H. D., Jung, S. H., Cha, S. S., Cera, P. J., Myers, J. L., Creagan, E. T., Goldberg, R. M., and Marschke, R. F., Jr., Neuroendocrine differentiation is an independent prognostic factor in chemotherapy-treated nonsmall cell lung carcinoma, *Cancer*, 77, 1284, 1996.
117. Langdon, S., Sethi, T., Ritchie, A., Muir, M., Smyth, J., and Rozengurt, E., Broad spectrum neuropeptide antagonists inhibit the growth of small cell lung cancer *in vivo*, *Cancer Res.*, 52, 4554, 1992.
118. Sethi, T., Herget, T., Wu, S. V., Walsh, J. H., and Rozengurt, E., CCKA and CCKB receptors are expressed in small cell lung cancer lines and mediate Ca2+ mobilization and clonal growth, *Cancer Res.*, 53, 5208, 1993.
119. Matsumori, Y., Katakami, N., Ito, M., Taniguchi, T., Iwata, N., Takaishi, T., Chihara, K., and Matsui, T., Cholecystokinin-B/gastrin receptor, a novel molecular probe for human small cell lung cancer, *Cancer Res.*, 55, 276, 1995.
120. Upp, J. R., Jr., Singh, P., Townsend, C. M., Jr., and Thompson, J. C., Clinical significance of gastrin receptors in human colon cancers, *Cancer Res.*, 49, 488, 1989.
121. Hennig, I. M., Laissue, J. A., Horisberger, U., and Reubi, J. C., Substance-P receptors in human primary neoplasms, tumoral and vascular localization, *Int. J. Cancer*, 61, 786, 1995.
122. Goldberg, R. M., Moertel, C. G., Wieand, H. S., Krook, J. E., Schutt, A. J., Veeder, M. H., Mailliard, J. A., and Dalton, R. J., A phase III evaluation of a somatostatin analogue (octreotide) in the treatment of patients with asymptomatic advanced colon carcinoma. North Central Cancer Treatment Group and the Mayo Clinic [see comments], *Cancer*, 76, 961, 1995.
123. Keri, G., Erchegyi, J., Horvath, A., Mezo, I., Idei, M., Vantus, T., Balogh, A., Vadasz, Z., Bokonyi, G., Seprodi, J., Teplan, I., Csuka, O., Tejeda, M., Gaal, D., Szegedi, Z., Szende, B., Roze, C., Kalthoff, H., and Ullrich, A., A tumor-selective somatostatin analog (TT-232) with strong *in vitro* and *in vivo* antitumor activity, *Proc. Natl. Acad. Sci. U.S.A.* 93, 12513, 1996.
124. Virgolini, I., Raderer, M., Kurtaran, A., Angelberger, P., Banyai, S., Yang, Q., Li, S., Banyai, M., Pidlich, J., Niederle, B., et al., Vasoactive intestinal peptide-receptor imaging for the localization of intestinal adenocarcinomas and endocrine tumors, *N. Engl. J. Med.*, 331, 1116, 1994.
125. Gillott, D. J., Iles, R. K., and Chard, T., The effects of β-human chorionic gonadotrophin on the *in vitro* growth of bladder cancer cell lines, *Br. J. Cancer*, 73, 323, 1996.
126. Meduri, G., Charnaux, N., Loosfelt, H., Jolivet, A., Spyratos, F., Brailly, S., and Milgrom, E., Luteinizing hormone/human chorionic gonadotropin receptors in breast cancer, *Cancer Res.*, 57, 857, 1997.
127. Acevedo, H. F. and Hartsock, R. J., Metastatic phenotype correlates with high expression of membrane-associated complete β-human chorionic gonadotropin *in vivo*, *Cancer*, 78, 2388, 1996.
128. Chatzaki, E., Bax, C. M., Eidne, K. A., Anderson, L., Grudzinskas, J. G., and Gallagher, C. J., The expression of gonadotropin-releasing hormone and its receptor in endometrial cancer, and its relevance as an autocrine growth factor, *Cancer Res.*, 56, 2059, 1996.

129. Gill, P. S., Lunardi-Ishkandar, Y., Louie, S., Tulpule, A., Zheng, T., Espina, B. M., Besnier, J. M., Hermans, P., Levine, A. M., Bryant, J. L., and Gallo, R. C., The effects of preparations of human chorionic gonadotropin on AIDS-related Kaposi's sarcoma [see comments], *N. Engl. J. Med.*, 335, 1261, 1996.
130. Harris, P. J., Intralesional human chorionic gonadotropin for Kaposi's sarcoma [letter; comment], *N. Engl. J. Med.*, 336, 1187, 1997.
131. Shipp, M. A. and Look, A. T., Hematopoietic differentiation antigens that are membrane-associated enzymes, cutting is the key!, *Blood*, 82, 1052, 1993.
132. LeBien, T. W. and McCormack, R. T., The common acute lymphoblastic leukemia antigen (CD10)-emancipation from a functional enigma, *Blood*, 73, 625, 1989.
133. Ganju, R. K., Sunday, M., Tsarwhas, D. G., Card, A., and Shipp, M. A., CD10/NEP in non-small cell lung carcinomas. Relationship to cellular proliferation, *J. Clin. Invest.*, 94, 1784, 1994.
134. Cohen, A. J., Bunn, P. A., Franklin, W., Magill-Solc, C., Hartmann, C., Helfrich, B., Gilman, L., Folkvord, J., Helm, K., and Miller, Y. E., Neutral endopeptidase, variable expression in human lung, inactivation in lung cancer, and modulation of peptide-induced calcium flux, *Cancer Res.*, 56, 831, 1996.
135. Cohen, A. J., Gilman, L. B., Moore, M., Franklin, W. A., and Miller, Y. E., Inactivation of neutral endopeptidase in lung cancer, *Chest*, 109, 12S, 1996.

FURTHER READING

Allen, A. E., Carney, D. N., and Moody, T. W., Neurotensin binds with high affinity to small cell lung cancer cells, *Peptides*, 9 (Suppl. 1), 57, 1988.

Bailer, R. T., Ng-Bautista, C. L., Ness, G. M., and Mallery, S. R., Expression of interleukin-6 receptors and NF-kappa B in AIDS-related Kaposi sarcoma cell strains, *Lymphology*, 30, 63, 1997.

Bataille, R., Klein, B., Jourdan, M., Rossi, J. F., and Durie, B. G., Spontaneous secretion of tumor necrosis factor-β by human myeloma cell lines, *Cancer*, 63, 877, 1989.

Bepler, G., Rotsch, M., Jaques, G., Haeder, M., Heymanns, J., Hartogh, G., Kiefer, P., and Havemann, K., Peptides and growth factors in small cell lung cancer, production, binding sites, and growth effects, *J. Cancer Res. Clin. Oncol.*, 114, 235, 1988.

Bertagna, X. Y., Nicholson, W. E., Pettengill, O. S., Sorenson, G. D., Mount, C. D., and Orth, D. N., Ectopic production of high molecular weight calcitonin and corticotropin by human small cell carcinoma cells in tissue culture, evidence for separate precursors, *J. Clin. Endocrinol. Metab.*, 47, 1390, 1978.

Boocock, C. A., Charnock-Jones, D. S., Sharkey, A. M., McLaren, J., Barker, P. J., Wright, K. A., Twentyman, P. R., and Smith, S. K., Expression of vascular endothelial growth factor and its receptors flt and KDR in ovarian carcinoma, *J. Natl. Cancer Inst.*, 87, 506, 1995.

Bradbury, D., Rogers, S., Kozlowski, R., Bowen, G., Reilly, I. A., and Russell, N. H., Interleukin-1 is one factor which regulates autocrine production of GM-CSF by the blast cells of acute myeloblastic leukaemia, *Br. J. Haematol.*, 76, 488, 1990.

Bradbury, D., Bowen, G., Kozlowski, R., Reilly, I., and Russell, N., Endogenous interleukin-1 can regulate the autonomous growth of the blast cells of acute myeloblastic leukaemia by inducing autocrine secretion of GM-CSF, *Leukemia*, 4, 44, 1990.

Chen, Q., Daniel, V., Maher, D. W., and Hersey, P., Production of IL-10 by melanoma cells, examination of its role in immunosuppression mediated by melanoma, *Int. J. Cancer*, 56, 755, 1994.

Cozzolino, F., Rubartelli, A., Aldinucci, D., Sitia, R., Torcia, M., Shaw, A., and Di Guglielmo, R. Interleukin 1 as an autocrine growth factor for acute myeloid leukemia cells, *Proc. Natl. Acad. Sci. U.S.A.*, 86, 2369, 1989.

Crowell, S. L., Burgess, H. S., and Davis, T. P., The effect of mycoplasma on the autocrine stimulation of human small cell lung cancer *in vitro* by bombesin and β-endorphin, *Life Sci.*, 45, 2471, 1989.

Cuttitta, F., Carney, D. N., Mulshine, J., Moody, T. W., Fedorko, J., Fischler, A., and Minna, J. D., Bombesin-like peptides can function as autocrine growth factors in human small-cell lung cancer, *Nature*, 316, 823, 1985.

Davidson, A., Moody, T. W., and Gozes, I., Regulation of VIP gene expression in general. Human lung cancer cells in particular, *J. Mol. Neurosci.*, 7, 99, 1996.

Davis, T. P., Burgess, H. S., Crowell, S., Moody, T. W., Culling-Berglund, A., and Liu, R. H., B-endorphin and neurotensin stimulate *in vitro* clonal growth of human SCLC cells, *Eur. J. Pharmacol.*, 161, 283, 1989.

Davis, T. P., Crowell, S., McInturff, B., Louis, R., and Gillespie, T., Neurotensin may function as a regulatory peptide in small cell lung cancer, *Peptides*, 12, 17, 1991.

Doran, T., Stuhlmiller, H., Kim, J. A., Martin, E. W., and Triozzi, P. L., Oncogene and cytokine expression of human colorectal tumors responding to immunotherapy, *J. Immunother.*, 20, 372, 1997.

Ezaki, K., Tsuzuki, M., Katsuta, I., Maruyama, F., Kojima, H., Okamoto, M., Nomura, T., Wakita, M., Miyazaki, H., Sobue, R., et al., Interleukin-1 β (IL-1 β) and acute leukemia, *in vitro* proliferative response to IL-1 β, IL-1 β content of leukemic cells and treatment outcome, *Leuk. Res.*, 19, 35, 1995.

Fathi, Z., Corjay, M. H., Shapira, H., Wada, E., Benya, R., Jensen, R., Viallet, J., Sausville, E. A., and Battey, J. F., BRS-3, a novel bombesin receptor subtype selectively expressed in testis and lung carcinoma cells, *J. Biol. Chem.*, 268, 5979, 1993.

Fielder, W., Graeven, U., Ergun, S., Verago, S., Kilic, N., Stockschlader, M., and Hossfeld, D. K., Expression of FLT4 and its ligand VEGF-C in acute myeloid leukemia, *Leukemia*, 11, 1234, 1997.

Finch, R. J., Sreedharan, S. P., and Goetzl, E. J., High-affinity receptors for vasoactive intestinal peptide on human myeloma cells, *J. Immunol.*, 142, 1977, 1989.

Folkesson, R., Monstein, H. J., Geijer, T., Påhlman, S., Nilsson, K., and Terenius, L., Expression of the proenkephalin gene in human neuroblastoma cell lines, *Brain Res.*, 427, 147, 1988.

Fried, G., Wikstrom, L. M., Hoog, A., Arver, S., Cedermark, B., Hamberger, B., Grimelius, L., and Meister, B., Multiple neuropeptide immunoreactivities in a renin-producing human paraganglioma, *Cancer*, 74, 142, 1994.

Gross, A. J., Steinberg, S. M., Reilly, J. G., Bliss, D. P., Jr., Brennan, J., Le, P. T., Simmons, A., Phelps, R., Mulshine, J. L., Ihde, D. C., et al., Atrial natriuretic factor and arginine vasopressin production in tumor cell lines from patients with lung cancer and their relationship to serum sodium, *Cancer Res.*, 53, 67, 1993.

Gustin, T., Bachelot, T., Verna, J. M., Molin, L. F., Brunet, J. F., Berger, F. R., and Benabid, A. L., Immunodetection of endogenous opioid peptides in human brain tumors and associated cyst fluids, *Cancer Res.*, 53, 4715, 1993.

Halmos, G., Wittliff, J. L., and Schally, A. V., Characterization of bombesin/gastrin-releasing peptide receptors in human breast cancer and their relationship to steroid receptor expression, *Cancer Res.*, 55, 280, 1995.

Hashimoto, M., Ohsawa, M., Ohnishi, A., Naka, N., Hirota, S., Kitamura, Y., and Aozasa, K., Expression of vascular endothelial growth factor and its receptor mRNA in angiosarcoma, *Lab. Invest.*, 73, 859, 1995.

Hatzoglou, A., Bakogeorgou, E., Hatzoglou, C., Martin, P. M., and Castanas, E., Antiproliferative and receptor binding properties of α- and β-casomorphins in the T47D human breast cancer cell line, *Eur. J. Pharmacol.*, 310, 217, 1996.

Hatzoglou, A., Gravanis, A., Margioris, A. N., Zoumakis, E., and Castanas, E., Identification and characterization of opioid-binding sites present in the Ishikawa human endometrial adenocarcinoma cell line, *J. Clin. Endocrinol. Metab.*, 80, 418, 1995.

Hatzoglou, A., Ouafik, L., Bakogeorgou, E., Thermos, K., and Castanas, E., Morphine cross-reacts with somatostatin receptor SSTR2 in the T47D human breast cancer cell line and decreases cell growth, *Cancer Res.*, 55, 5632, 1995.

Hiscox, S. E., Hallett, M. B., Puntis, M. C. A., Nakamura, T., and Jiang, W. G., Expression of the HGF/SF receptor, *c-met*, and its ligand in human colorectal cancers, *Clin. Invest.*, 15, 513, 1997.

Hoosein, N. M., Kiener, P. A., Curry, R. C., Rovati, L. C., McGilbra, D. K., and Brattain, M. G., Antiproliferative effects of gastrin receptor antagonists and antibodies to gastrin on human colon carcinoma cell lines, *Cancer Res.*, 48, 7179, 1988.

Hytrek, S. D., Smith, J. P., McGarrity, T. J., McLaughlin, P. J., Lang, C. M., and Zagon, I. S., Identification and characterization of ζ-opioid receptor in human colon cancer, *Am. J. Physiol.*, 271, R115, 1996.

Iser, G., Pfohl, M., Dorr, U., Weiss, E. M., and Seif, F. J., Ectopic ACTH secretion due to a bronchopulmonary carcinoid localized by somatostatin receptor scintigraphy, *Clin. Investig.*, 72, 887, 1994.

Ito, R., Kitadai, Y., Kyo, E., Yokozaki, H., Yasui, W., Yamashita, U., Nikai, H., and Tahara, E., Interleukin 1 α acts as an autocrine growth stimulator for human gastric carcinoma cells, *Cancer Res.*, 53, 4102, 1993.

Iwata, N., Murayama, T., Matsumori, Y., Ito, M., Nagata, A., Taniguchi, T., Chihara, K., Matsuo, Y., Minowada, J., and Matsui, T., Autocrine loop through cholecystokinin-B/gastrin receptors involved in growth of human leukemia cells, *Blood*, 88, 2683, 1996.

Jiang, S., Kopras, E., McMichael, M., Bell, R. H., Jr., and Ulrich, C. D., II, Vasoactive intestinal peptide (VIP) stimulates *in vitro* growth of VIP-1 receptor-bearing human pancreatic adenocarcinoma-derived cells, *Cancer Res.*, 57, 1475, 1997.

Jiang, S. and Ulrich, C., Molecular cloning and functional expression of a human pancreatic secretin receptor, *Biochem. Biophys. Res. Commun.*, 207, 883, 1995.

Katsumata, N., Eguchi, K., Fukuda, M., Yamamoto, N., Ohe, Y., Oshita, F., Tamura, T., Shinkai, T., and Saijo, N., Serum levels of cytokines in patients with untreated primary lung cancer, *Clin. Cancer Res.*, 2, 553, 1996.

Kawano, M., Hirano, T., Matsuda, T., Taga, T., Horii, Y., Iwato, K., Asaoku, H., Tang, B., Tanabe, O., Tanaka, H., et al., Autocrine generation and requirement of BSF-2/IL-6 for human multiple myelomas, *Nature*, 332, 83, 1988.

Kazmi, S. M. and Mishra, R. K., Comparative pharmacological properties and functional coupling of μ and δ opioid receptor sites in human neuroblastoma SH-SY5Y cells, *Mol. Pharmacol.*, 32, 109, 1987.

Lahm, H., Petral-Malec, D., Yilmaz-Ceyhan, A., Fischer, J. R., Lorenzoni, M., Givel, J. C., and Odartchenko, N., Growth stimulation of a human colorectal carcinoma cell line by interleukin-1 and -6 and antagonistic effects of transforming growth factor β 1, *Eur. J. Cancer*, 28A, 1894, 1992.

Lilling, G., Wollman, Y., Goldstein, M. N., Rubinraut, S., Fridkin, M., Brenneman, D. E., and Gozes, I., Inhibition of human neuroblastoma growth by a specific VIP antagonist, *J. Mol. Neurosci.*, 5, 231, 1994.

Lou, L. G. and Pei, G., Modulation of protein kinase C and cAMP-dependent protein kinase by δ-opioid, *Biochem. Biophys. Res. Commun.*, 236, 626, 1997.

Mano, H., Nishida, J., Usuki, K., Maru, Y., Kobayashi, Y., Hirai, H., Okabe, T., Urabe, A., and Takaku, F., Constitutive expression of the granulocyte-macrophage colony-stimulating factor gene in human solid tumors, *Jpn. J. Cancer Res.*, 78, 1041, 1987.

Martinez, A., Miller, M. J., Catt, K. J., and Cuttitta, F., Adrenomedullin receptor expression in human lung and in pulmonary tumors, *J. Histochem. Cytochem.*, 45, 159, 1997.

Masood, R., Cai, J., Zheng, T., Smith, D. L., Naidu, Y., and Gill, P. S., Vascular endothelial growth factor/vascular permeability factor is an autocrine growth factor for AIDS-Kaposi sarcoma, *Proc. Natl. Acad. Sci. U.S.A.*, 94, 979, 1997.

Matsuguchi, T., Okamura, S., Kawasaki, C., Shimoda, K., Omori, F., Hayashi, S., Kimura, N., and Niho, Y., Constitutive production of granulocyte colony-stimulating factor and interleukin-6 by a human lung cancer cell line, KSNY, gene amplification and increased mRNA stability, *Eur. J. Haematol.*, 47, 128, 1991.

Mattei, S., Colombo, M. P., Melani, C., Silvani, A., Parmiani, G., and Herlyn, M., Expression of cytokine/growth factors and their receptors in human melanoma and melanocytes, *Int. J. Cancer,* 56, 853, 1994.

McLaughlin, P. J. and Zagon, I. S., Modulation of human neuroblastoma transplanted into nude mice by endogenous opioid systems, *Life Sci.*, 41, 1465, 1987.

Meyers, F. J., Gumerlock, P. H., Kawasaki, E. S., Wang, A. M., deVere White, R. W., and Erlich, H. A., Bladder cancer. Human leukocyte antigen II, interleukin-6, and interleukin-6 receptor expression determined by the polymerase chain reaction, *Cancer*, 67, 2087, 1991.

Moody, T. W., Carney, D. N., Cuttitta, F., Quattrocchi, K., and Minna, J. D., High affinity receptors for bombesin/GRP-like peptides on human small cell lung cancer, *Life Sci.*, 37, 105, 1985.

Moody, T. W., Carney, D. N., Korman, L. Y., Gazdar, A. F., and Minna, J. D., Neurotensin is produced by and secreted from classic small cell lung cancer cells, *Life Sci.*, 36, 1727, 1985.

Morrison, R. S., Yamaguchi, F., Bruner, J. M., Tang, M., McKeehan, W., and Berger, M. S., Fibroblast growth factor receptor gene expression and immunoreactivity are elevated in human glioblastoma multiforme, *Cancer Res.*, 54, 2794, 1994.

Mouawad, R., Benhammouda, A., Rixe, O., Antoine, E. C., Borel, C., Weil, M., Khayat, D., and Soubrane, C., Endogenous interleukin-6 levels in patients with metastatic malignant melanoma, correlation with tumor burden, *Clin. Cancer Res.,* 2, 1404, 1996.

Murakami, O., Takahashi, K., Sone, M., Totsune, K., Ohneda, M., Itoi, K., Yoshinaga, K., and Mouri, T., An ACTH-secreting bronchial carcinoid, presence of corticotropin-releasing hormone, neuropeptide Y and endothelin-1 in the tumor tissue, *Acta Endocrinol (Copenhagen),* 128, 192, 1993.

O'Brien, T., Cranston, D., Fuggle, S., Bicknell, R., and Harris, A. L., Different angiogenic pathways characterize superficial and invasive bladder cancer, *Cancer Res.*, 55, 510, 1995.

Ozawa, Y., Tomoyasu, H., Takeshita, A., Shishiba, Y., Yamada, S., Kovacs, K., and Matsushita, H., Shift from CRH to ACTH production in a thymic carcinoid with Cushing's syndrome, *Horm Res.*, 45, 264, 1996.

Reubi, J. C., Mazzucchelli, L., Hennig, I., and Laissue, J. A., Local up-regulation of neuropeptide receptors in host blood vessels around human colorectal cancers, *Gastroenterology*, 110, 1719, 1996.

Reubi, J. C., Schaer, J. C., and Waser, B., Cholecystokinin (CCK)-A and CCK-B/gastrin receptors in human tumors, *Cancer Res.*, 57, 1377, 1997.

Rosen, E. M., Laterra, J., Joseph, A., Jin, L., Fuchs, A., Way, D., Witte, M., Weinand, M., and Goldberg, I. D., Scatter factor expression and regulation in human glial tumors, *Int. J. Cancer*, 67, 248, 1996.

Roth, K. A. and Barchas, J. D., Small cell carcinoma cell lines contain opioid peptides and receptors, *Cancer*, 57, 769, 1986.

Roth, K. A., Makk, G., Beck, O., Faull, K., Tatemoto, K., Evans, C. J., and Barchas, J. D., Isolation and characterization of substance P, substance P 5-11, and substance K from two metastatic ileal carcinoids, *Regul. Pept.*, 12, 185, 1985.

Samaniego, F., Markham, P. D., Gallo, R. C., and Ensoli, B., Inflammatory cytokines induce AIDS-Kaposi's sarcoma-derived spindle cells to produce and release basic fibroblast growth factor and enhance Kaposi's sarcoma-like lesion formation in nude mice, *J. Immunol.*, 154, 3582, 1995.

Sato, T., McCue, P., Masuoka, K., Salwen, S., Lattime, E. C., Mastrangelo, M. J., and Berd, D., Interleukin 10 production by human melanoma, *Clin. Cancer Res.*, 2, 1383, 1996.

Sausville, E., Carney, D., and Battey, J., The human vasopressin gene is linked to the oxytocin gene and is selectively expressed in a cultured lung cancer cell line, *J. Biol. Chem.*, 260, 10236, 1985.

Scopa, C. D., Mastorakos, G., Friedman, T. C., Melachrinou, M., Merino, M. J., and Chrousos, G. P., Presence of immunoreactive corticotropin releasing hormone in thyroid lesions, *Am. J. Pathol.*, 145, 1159, 1994.

Sethi, T., Langdon, S., Smyth, J., and Rozengurt, E., Growth of small cell lung cancer cells, stimulation by multiple neuropeptides and inhibition by broad spectrum antagonists *in vitro* and *in vivo*, *Cancer Res.*, 52, 2737s, 1992.

Sethi, T. and Rozengurt, E., Multiple neuropeptides stimulate clonal growth of small cell lung cancer, effects of bradykinin, vasopressin, cholecystokinin, galanin, and neurotensin, *Cancer Res.*, 51, 3621, 1991.

Shahabi, N. A., Peterson, P. K., and Sharp, B., B-endorphin binding to naloxone-insensitive sites on a human mononuclear cell line (U937), effects of cations and guanosine triphosphate, *Endocrinology*, 126, 3006, 1990.

Siegfried, J. M., Han, Y. H., DeMichele, M. A., Hunt, J. D., Gaither, A. L., and Cuttitta, F., Production of gastrin-releasing peptide by a non-small cell lung carcinoma cell line adapted to serum-free and growth factor-free conditions, *J. Biol. Chem.*, 269, 8596, 1994.

Singh, P., Owlia, A., Varro, A., Dai, B., Rajaraman, S., and Wood, T., Gastrin gene expression is required for the proliferation and tumorigenicity of human colon cancer cells, *Cancer Res.*, 56, 4111, 1996.

Suzuki, K., Hayashi, N., Miyamoto, Y., Yamamoto, M., Ohkawa, K., Ito, Y., Sasaki, Y., Yamaguchi, Y., Nakase, H., Noda, K., Enomoto, N., Arai, K., Yamada, Y., Yoshihara, H., Tujimura, T., Kawano, K., Yoshikawa, K., and Kamada, T., Expression of vascular permeability factor/vascular endothelial growth factor in human hepatocellular carcinoma, *Cancer Res.*, 56, 3004, 1996.

Tabibzadeh, S. S., Poubouridis, D., May, L. T., and Sehgal, P. B., Interleukin-6 immunoreactivity in human tumors, *Am. J. Pathol.*, 135, 427, 1989.

Takenawa, J., Kaneko, Y., Okumura, K., Yoshida, O., Nakayama, H., and Fujita, J., Inhibitory effect of dexamethasone and progesterone *in vitro* on proliferation of human renal cell carcinomas and effects on expression of interleukin-6 or interleukin-6 receptor, *J. Urol.*, 153, 858, 1995.

Takenawa, J., Kaneko, Y., Fukumoto, M., Fukatsu, A., Hirano, T., Fukuyama, H., Nakayama, H., Fujita, J., and Yoshida, O., Enhanced expression of interleukin-6 in primary human renal cell carcinomas, *J. Natl. Cancer Inst.*, 83, 1668, 1991.

Takuwa, N., Takuwa, Y., Ohue, Y., Mukai, H., Endoh, K., Yamashita, K., Kumada, M., and Munekata, E., Stimulation of calcium mobilization but not proliferation by bombesin and tachykinin neuropeptides in human small cell lung cancer cells, *Cancer Res.*, 50, 240, 1990.

Taylor, J. E., Human small cell lung cancer cells express high affinity naloxone-insensitive [125I]-endorphin binding sites, *Life Sci.*, 56,PL97, 1995.

Tendler, C. L., Burton, J. D., Jaffe, J., Danielpour, D., Charley, M., McCoy, J. P., Pittelkow, M. R., and Waldmann, T. A., Abnormal cytokine expression in Sezary and adult T-cell leukemia cells correlates with the functional diversity between these T-cell malignancies, *Cancer Res.*, 54, 4430, 1994.

Ueba, T., Takahashi, J. A., Fukumoto, M., Ohta, M., Ito, N., Oda, Y., Kikuchi, H., and Hatanaka, M., Expression of fibroblast growth factor receptor-1 in human glioma and meningioma tissues, *Neurosurgery*, 34, 221, 1994.

Ueki, T., Koji, T., Tamiya, S., Nakane, P. K., and Tsuneyoshi, M., Expression of basic fibroblast growth factor and fibroblast growth factor receptor in advanced gastric carcinoma, *J. Pathol.*, 177, 353, 1995.

Wang, J., Huang, M., Lee, P., Komanduri, K., Sharma, S., Chen, G., and Dubinett, S. M., Interleukin-8 inhibits non-small cell lung cancer proliferation, a possible role for regulation of tumor growth by autocrine and paracrine pathways, *J. Interferon Cytokine Res.*, 16, 53, 1996.

Warren, R. S., Yuan, H., Matli, M. R., Gillett, N. A., and Ferrara, N., Regulation by vascular endothelial growth factor of human colon cancer tumorigenesis in a mouse model of experimental liver metastasis, *J. Clin. Invest.*, 95, 1789, 1995.

Weber, C. J., O'Dorsio, T. M., McDonald, T. J., Howe, B., Koschitzky, T., and Merriam, L., Gastrin-releasing peptide-, calcitonin gene-related peptide-, and calcitonin-like immunoreactivity in human breast cyst fluid and gastrin-releasing peptide-like immunoreactivity in human breast carcinoma cell lines, *Surgery*, 106, 1134, 1989.

Zagon, I. S., Gibo, D., and McLaughlin, P. J., Expression of zeta (ζ), a growth-related opioid receptor, in metastatic adenocarcinoma of the human cerebellum, *J. Natl. Cancer Inst.*, 82, 325, 1990.

11 Cytokines, Stress Hormones, and Immune Function

Robert E. Faith, Nicholas P. Plotnikoff, and Anthony J. Murgo

CONTENTS

11.1 Introduction 161
11.2 Stress and Host Defense 165
11.3 Conclusions 168
References 169

11.1 INTRODUCTION

Immunity, the ability of an individual to resist infectious and neoplastic disease, is provided by a complex of innate and acquired mechanisms. Innate immunity is comprised of those nonspecific defense mechanisms such as physical and chemical barriers, and phagocytic cells that an individual is born with. Acquired immunity is comprised of specific defense mechanisms that develop following contact with a "foreign" material. Acquired immunity results from the functions of various subpopulations of lymphocytes and monocytes. The specific immune system is divided into humoral and cellular arms with both arms being further subdivided. A number of exogenous factors, including stress, have the ability to affect immune function.

Stress is a complex concept that has both mental and physiological components. There has long been an understanding that stress may affect the immune system or host defense mechanisms, sometimes profoundly. The dictionary definition of stress is "a physical, chemical, or emotional factor that causes bodily or mental tension and may be a factor in disease causation."[1] For the purposes of this review it may be helpful to consider stress as a stimulus or succession of stimuli of such magnitude as to tend to disrupt the homeostasis of the organism. Immunity as used here refers to all those physiologic mechanisms used by the host to recognize materials as foreign to itself and to neutralize, eliminate, or metabolize them, with or without injury to its own tissues. Stress is not always negative and in fact may be beneficial, depending on how it is perceived and handled by the host. For this reason Selye coined the term "eustress" to indicate that some level of stress is required for well-being.[2] There is an individual component to stress, with different individuals reacting differently to the same stressful event. The same stressor may result in an extreme reaction from some individuals and a mild-to-no reaction in others.

A number of stressful situations have been investigated in both humans and animals for their effects on immune function. Various stressors or significant life events have been shown to impact one or more aspects of immune function. Stressors, including test taking, unemployment, bereavement, separation/divorce, and depression, have been observed to result in decreased aspects of immune function in human populations. There is a close interrelationship between the immune system, central nervous system (CNS), and the endocrine system. All three systems have the ability to affect the functioning of the other two through a series of feedback regulatory loops. Events that affect one system can secondarily affect the other two through these pathways, and it appears that stress effects on immune function are mediated through these same pathways. While it is clear that stress may impact immune function, it still remains to be shown what the real significance of this is to most individuals. This review will focus primarily on studies involving animals rather than humans, especially those in which the stressor appears to be rather benign while having significant effects on host defense mechanisms.

Clinicians have been making observations for more than 2000 years indicating that mood states may affect susceptibility to physical illness.[3] In 1919, Ishigami performed the first scientific study of the effects of emotional stress on host defense mechanisms.[4] He studied the opsonization of tubercule bacilli among chronic tuberculous patients during active and inactive phases of the disease. His findings indicated that episodes of emotional distress resulted in decreased phagocytic activity. From this he postulated that stressful events led to immunodepression and consequently increased susceptibility to tuberculosis. Since this initial investigation there have been a number of studies performed in humans and animals which demonstrate that stressful events, sometimes seemingly quite minor in nature, can have significant effects on immune function. A limited number of these studies are reviewed here.

The stress system consists of both brain elements and their peripheral effectors.[5] The brain elements include corticotropin-releasing hormone (CRH) and the locus ceruleus-norepinephrine autonomic systems, and the peripheral effectors include the pituitary-adrenal axis and the autonomic nervous system. These elements function to coordinate the stress response. The stress system is closely integrated with other central nervous system elements involved in the regulation of behavior and emotion, and with the axes responsible for reproduction, growth, and immunity.[5]

During stress a large number of changes occur in the periphery and the CNS. There may be increases or decreases in many neurotransmitters in the brain, especially the hypothalamus, where norepinephrine levels drop markedly. Peripherally, catecholamines are released from the sympathetic nerve endings and the adrenal medulla within seconds and reach peak values after 10 to 20 min. These increases can reach levels of 10 to 20 times the resting values. Corticosteroids are released somewhat later and increase two to fourfold above resting levels. The levels of many other chemicals, including endorphins, enkephalins, and amino acids, also change during stress.[6] The activities of cellular elements of the immune system, lymphocytes and macrophages, are altered by stressors in a complex way that depends on the type of immune response, the physical and psychological characteristics of the stressor, and the timing of stress relative to the immune response.[7] Individuals who are most likely to show negative health effects in response to these changes resulting from stress effects are those whose immune function is already compromised to some extent by an immunosuppressive disease such as AIDS or by a natural process, like aging, that is associated with immunological impairments.

Experimental and nonexperimental stress-inducing stimuli may have profound physiological consequences. Stress may activate the adrenal cortex via the pituitary, resulting in biochemical, cellular, and tissue alterations. Increased serum corticosterone levels may lead to many alterations in experimental data. A direct consequence of an increase in corticosteroids is injury to elements of the immune system, which may increase susceptibility to latent infectious and oncogenic agents, newly transformed cancer cells, or other potentially pathogenic processes normally held in balance by a competent immune system.[8] Other effects of increased corticosteroid levels include changes in protein metabolism, reduced growth rates in young animals, and loss of weight in adults.

Stress has a number of effects on various elements of the immune system. Stress may result in a decrease in the number and percentage of peripheral blood lymphocytes, and an increase in the number and percentage of peripheral blood neutrophils. Neutrophil function may be impaired. Stress may affect the number and function of monocytes, T- and B-cell ratios, CD4/CD8 ratios, and E-rosette formation. That stress may adversely affect immune function is demonstrated by reduced responsiveness of lymphocytes to mitogen stimulation, reduced antibody responses, and reduced natural killer (NK) cell activity.[9-11] Many of these effects of stress on immune function have been attributed to the action of the glucocorticoids; therefore, a brief discussion of the effects of glucocorticoids follows.

There are indications that the glucocorticoids function in the control of the numbers of peripheral blood lymphocytes (pbls). There is a diurnal cycle in the number of pbls, with the numbers being highest around midnight and lowest around 8:00 a.m. The circulating levels of glucocorticoids show a diurnal cycle that is inverse to that of pbls.[8,12] Additionally, there is an inverse relationship between endogenous levels of glucocorticoids and splenic cellularity and mass. In studies performed in mice, lowered blood levels of corticosterone resulted in increased numbers of immunoglobulin-secreting cells in the spleen, while high corticosterone levels resulted in reduced numbers of immunoglobulin-secreting cells.[13] From these studies it was concluded that endogenous levels of glucocorticoids contribute to the control of B-cell activity and possibly to the interaction of these cells with other immunologic cells. Additionally, there is considerable evidence that a physiologic amount of cortisol is necessary for the development and maintenance of normal immunity in humans.[14]

Selye was the first to demonstrate thymus atrophy in rats following administration of adrenocortical extracts.[15] This finding was soon verified by Carriere et al.[16] and Ingle.[17] Since these early studies a number of investigations have shown glucocorticoids to have several effects on the immune system. In reviewing the effects of glucocorticoids on immune functions, it is important to remember that there are steroid-sensitive and steroid-resistant species. The mouse, rat, hamster, and rabbit are steroid sensitive, and the guinea pig, monkey, and humans are steroid resistant.[8] Glucocorticoids easily lyse the lymphoid cells and inhibit antibody production in steroid-sensitive species. In contrast, the lymphoid cells of resistant species are not easily lysed by glucocorticoids, and it is difficult to demonstrate inhibition of circulating antibody by steroids in these species. Responses to glucocorticoids are mediated through cellular receptors, as are responses to other hormones. Lymphoid cells in bone marrow, thymus, spleen, lymph nodes, and circulation all have glucocorticoid receptors.[8] Activation of lymphocytes by mitogens or antigen leads to an increase in the number of steroid receptors on these cells. Injection of corticosteroids prior to inoculation with a normally avirulent dose of a virus can induce 100% mortality in adult mice.[18,19] In addition, glucocorticoids have the ability to affect a number of functions of immune cells. They inhibit mitogen-induced lymphocyte proliferation, tumoricidal activity of activated macrophages, the production of lymphokines, including lymphocyte activating factor (LAF), macrophage activating factor (MAF), T cell growth factor (TCGF), macrophage activation by MAF, and they depress or enhance NK cell activity.[8,20] The TCGF effects provide a possible explanation for the fact that glucocorticoids are much more effective in suppressing an immune response (*in vivo* or *in vitro*) early rather than late in the response. Early, they prevent TCGF-dependent clonal expansion; when presented later, they are ineffective because clonal expansion has already occurred.[20]

In addition to the glucocorticoids, the adrenals also produce the catecholamines, epinephrine, and norepinephrine which function as stress hormones. In a study of first-time parachutists, the sympathetic-adrenal hormones were monitored before, during, and shortly after jumping.[21] There was a significant increase in plasma epinephrine and norepinephrine during the jump and a significant increase in plasma cortisol shortly after jumping. Examination of the functional capacity of NK cells revealed an increase immediately after jumping, followed by a decrease significantly below the starting values one hour later. The changes in NK activity correlated to plasma concentrations of norepinephrine. The authors suggest that the quick mobilization of NK cells is one major mechanism for the effective adaptation of the immune system to stress situations. Injection of epinephrine into normal volunteers resulted in a significant decrease in $CD3^+$ and $CD4^+$ T cells 5 to 60 min after injection.[22] Additionally, these studies demonstrated a significant increase in NK cell numbers and activity following injection of epinephrine or norepinephrine, showing that both sympathetic-adrenal hormones are modulators of natural immunity.

Many of the effects of stress on immune function have been attributed to the actions of glucocorticoids. However, stress-induced immunosuppression occurs in adrenalectomized as well as normal rats. Therefore, other factors are involved in mediating the effects of stress on immune function. One of the other main factors involved in the mediation of stress-induced modulation of immune function is corticotropin-releasing hormone (CRH). CRH plays a central role in mediating stress-induced immunosuppression. The major effects of CRH are to regulate the activities of the hypothalamic-pituitary-adrenal axis (HPA), activate central noradrenergic neurons, and generate many of the physiological and behavioral changes observed in stress.[23] CRH functions centrally to elicit changes in neuroendocrine, autonomic, and behavioral activity similar to those observed after stress.[24] The administration of CRH centrally results in a number of changes in host defense mechanisms. These include reduced phagocytic responses, reduced lymphocyte responsiveness to mitogen stimulation, reduced splenic and blood NK activity, and increased corticosteroid levels.[23,24] However, central CRH treatment appears not to alter numbers of splenic lymphocytes or T-cell subpopulations. Irwin[25] has shown that central administration of CRH significantly slows the induction of the specific antibody response to the T-cell-dependent antigen keyhole limpet hemocyanin following either primary or secondary immunization with a low threshold dose of the antigen. This suppression could be overcome by a 100-fold increase in the dose of antigen. CRH inhibited the antibody response when given 20 min before immunization, but not when given 24 h following immunization, suggesting that CRH may alter initial antigen processing. The central actions of CRF are said to be independent of the pituitary.[26] However, receptors for CRF and interleukin-1 (IL-1) have been demonstrated in the brain, endocrine, and immune tissues, specifically in the anterior and intermediate lobes of the pituitary, in brain areas involved in mediating stress responses, and in the macrophage-enriched marginal zones of the spleen.[27] The immune effects of CRH appear to be mediated by the sympathetic branch of the autonomic nervous system. Both chemical sympathectomy and pharmacological blockade of β-adrenergic receptors reverse the immune effects of CRH.[28]

It also was demonstrated in the studies mentioned above that central administration of IL-1β can inhibit the antibody response similarly to CRH.[25] Weiss et al.[29] furthered these findings by demonstrating that central IL-1 rapidly suppresses a variety of immune responses measured in peripheral lymphocytes. Extremely small amounts of IL-1 acting centrally can cause suppression of cellular immunity rapidly and for a prolonged period of time.[30] Additionally, it was shown that central IL-1 stimulates CRF in the CNS, and that, in turn, results in activation of both the pituitary-adrenal axis and the autonomic nervous system, leading to suppression of cellular immune responses.[29] It is now felt that activation of the HPA axis to release CRH, ACTH, and glucocorticoids during stress is a key function of IL-1.[31]

11.2 STRESS AND HOST DEFENSE

Focus will now be directed to reviewing several different examples of model stressors and their effects on elements of host defenses. All of the stressors discussed here are seemingly rather mild stimuli, but they have significant effects. In 1981 Riley reported a series of studies in mice in which rotation was used as a stressor.[32] The stressor was very simple; he placed animals on a record turntable and rotated them at 16, 33, 45, or 78 rpm. This produced an anxiety stress which resulted in increased levels of corticosterone directly correlated with rotational speed. Rotation at 45 rpm resulted in plasma corticosterone concentrations of almost 500 ng/ml, and this speed was chosen as the stressor for further studies. Rotation at 45 rpm for 10 min/h over a 5-h period resulted in a 50% leukocytopenia by the end of the second hour. This leukocytopenia was maintained throughout the 5 h of the intermittent stress. This same type of intermittent stress applied over a 24-h period resulted in thymus atrophy which was most severe at day 2 and then recovered slowly over several days. This stressor was then used to study the effect of stress on tumor growth. C3H/He mice, implanted with a transplantable tumor which they reject when unstressed, are unable to reject the tumor when subjected to intermittent rotational stress on days 4, 5, and 6 following tumor implantation. Finally, intermittent rotational stress was shown to enhance the growth of Maloney sarcoma virus-induced tumors in BALB/c female mice.

Glasser et al.[33] studied the effects of examination stress in medical students. They found a significant reduction in the percentages of IL-2 receptor-positive cells in peripheral blood leukocytes during examinations. Additionally IL-2 receptor messenger RNA in peripheral blood leukocytes was significantly decreased in a subset of the subjects during examinations.[33] These studies provide evidence that stress effects may be observed at the level of gene expression. Examination stress also has been shown to depress the percentages of total T lymphocytes, helper T cells, and suppressor T cells in peripheral blood, and to significantly lower the response of T lymphocytes to stimulation with the mitogens phytohemagglutinin and concanavilin-A.[34] Other studies of examination stress have shown that this stressor causes a reduction in the numbers and function of NK cells and a suppression of interferon production by mitogen-stimulated peripheral bloods leukocytes.[35]

Restraint is another factor that has been used as a stressor in the study of the effects of stress on host defense mechanisms. Stress induced by restraint has been observed in all species studied, including humans.[36] The most commonly observed changes following restraint stress were behavioral, including a period of anxiety and hyperactivity accompanied by reduced food intake and loss of body mass and followed by lethargy. Consistent findings include lymphopenia and neutropenia.[36,37] Restraint stress also results in organ and body mass changes, an increase in blood acidity, plasma protein, calcium, and magnesium, and increased plasma corticosterone.[38-40] Restraint stress has been shown to increase the susceptibility of mice to viral infections[41-44] and to delay the production of virus-specific antibody production in mice infected with influenza virus.[45] A restraint of 12 h/day for 2 days following immunization with sheep red blood cells (SRBC) resulted in reduction of antibody titers and the number of antibody-forming spleen cells. Adrenalectomy and chemical sympathectomy blocked the immunosuppression induced by this stressor.[46-48] Restraint stress results in decreased body weights, increased adrenal gland weight, and thymic atrophy in experimental mice when compared to unrestrained controls.[41] Bonneau et al.[41] demonstrated that repeated, prolonged periods of restraint (16 h/day for a varying number of days around the time of inoculation) resulted in suppression of NK cell activity and the primary development of HSV-specific cytotoxic T lymphocytes. They also recovered higher titers of infectious HSV at the site of injection in restrained mice when compared to unrestrained mice. In further studies it was shown that restraint inhibits the *in vitro* activation and/or migration of HSV-specific cytotoxic T lymphocytes from previously primed mice, indicating that activation of memory cells may be inhibited by stress.[42] Further studies by these authors indicate that both adrenal-dependent and -independent mechanisms contribute to stress-induced modulation of HSV immunity.[49] Dobbs, et al.[50] confirmed these findings by demonstrating that both corticosterone and catecholamine-mediated mechanisms operate in the stress-induced suppression of antiviral cellular immunity.

Laboratory animals may encounter a number of nonexperimental stress events including isolation, overcrowding, social rank interactions, handling, and transportation. Being handled is something that happens to most research animals, and it may act as a significant stressor. When groups of mice are removed from their home shelf and randomly captured and bled over a period of time, tenfold increases in plasma corticosterone levels can occur in 30 min. The physiological response to handling may occur very rapidly. Riley[32] found that reliable baseline measurements of plasma corticosterone can only be obtained within 3 to 5 min following initial disturbance of animals by handling them. Furthermore, what he called "contagious anxiety" affects the levels of corticosterone in other mice that remained in the cage during the sequential capture of each mouse. Henning[51] demonstrated that the simple act of individually removing rat pups from their home cage in sequence results in elevated corticosterone levels in remaining pups.

Another seemingly minor thing that may occur in the environment of laboratory animals is a shift in the light-dark cycle. Kort and Weijma[52] reported that shifting the light-dark cycle by inversion (i.e., light 7:00 a.m. to 7:00 p.m. one week and 7:00 p.m. to 7:00 a.m. the next week) for 35 weeks resulted in a significant decrease of cellular immune responses and body weight. Further studies by these authors showed that this type of stressor did not significantly influence spontaneous tumor rates in rats.[53] Transportation is another factor that is commonplace in the life of laboratory animals. In mice the mere transference from cage to transit box may result in loss of body weight.[54] Other studies have shown that shipment of research animals results in loss of body weight;[55] it generally takes at least a week to recover from this stress. The simple act of moving mice from one room to another resulted in significant elevation of corticosterone levels.[56]

Exposure of animals to cold water has been used as a source of stress in several studies. Cold-water stress has been shown to cause reduced numbers of thymocytes and splenocytes, decreased T-cell blastogenesis in response to mitogen stimulation, reduced NK activity, and altered macrophage function.[57,58] The effects of cold-water stress on macrophage function are quite interesting. Exposure of mice to cold water (4°C) for 1 min twice a day on 4 consecutive days resulted in activation of macrophages in an apparently dysregulated manner.[58] Some functions such as IL-1 and prostaglandin E_2 (PGE_2) production were enhanced, while other functions, such as the ability to induce Ia antigen expression by interferon-γ (INF-γ) were suppressed. This is interesting since stress can cause elevated levels of glucocorticoids, and glucocorticoids can block the production of prostaglandins and IL-1 by macrophages. The expression of less Ia antigen by macrophages from stressed mice may explain at least one of the mechanisms of stress-induced immunosuppression. This diminished Ia antigen expression may compromise macrophage functions such as antigen presentation and other accessory cell functions.

Attention will now be turned to studies involving shock as a stressor. Both foot shock and tail shock have been used as stressors in the study of the immune effects of stress. Shock stress has been shown to inhibit immune responses, including lymphocyte responsiveness to mitogen stimulation NK cell activity, and antibody production.[59-63] Foot shock in rats has been shown to result in markedly elevated plasma corticosterone levels.[62] Laudenslager et al.[64] reported the results of studies that indicated inescapable, but not escapable, foot shock resulted in depressed mitogen responsiveness of lymphocytes. The exposure of male Lewis rats to one session of mild electric foot shock resulted in a decrease in production of INF-γ by splenocytes in response to concanavalin-A (Con A) when compared to splenocytes from control animals.[65] This effect could be inhibited by a β-adrenergic receptor antagonist, suggesting that catecholamines mediate shock-induced suppression of IFN-γ production. Foot shock also significantly decreased the number of splenic mononuclear cells expressing class II histocompatibility (Ia) antigens on their surfaces. Inescapable intermittent foot shock results in significantly increased concentrations of β-endorphin in splenocytes, peripheral blood mononuclear cells, and lymph node cells.[66] Pretreatment with a CRH receptor antagonist prevented the increase in immunocyte β-endorphin and shock-induced immunosuppression, indicating that CRH plays a central role in mediating the immune effects of intermittent foot shock stress. One very interesting finding that has resulted from studies on foot shock stress is that simple exposure to the odors of stressed mice provoked an immunosuppression that appeared to be as marked as that provoked by foot shock.[63] This confirms that anxiety can be communicated between individuals, as indicated by Riley in studies discussed above. Pezzone et al.[62] have shown that the paraventricular nucleus plays a role in the immune changes induced by foot shock. They reported that lesioning of the paraventricular nucleus actuates the changes in immune function and elevation of corticosterone levels resulting from foot shock in intact animals.

11.3 CONCLUSIONS

Studies wherein tail shock was the stressor have provided results similar to those which used foot shock. Inescapable tail shock causes a reduction in the ability of lymphocytes to proliferate in response to mitogen stimulation and decreased the antibody responses.[61,67] Escapable electric tail shock of identical intensities, durations, and distributions, did not reduce these responses. The difference in response to escapable and inescapable shock suggests that the immune changes observed in the inescapably shocked animals were not produced by the physical stressor per se, but rather by a psychological stressor, the inescapability of the shock. Thus, if one can cope with a stressful situation, as in being able to escape the shock, the physiological consequences of the stress may be lessened or absent. The salience of coping is illustrated by the study of Dorian et al.[68] They demonstrated that students who reacted with overt distress to examinations manifested greater immunosuppression than those who coped with the same stress in an asymptomatic fashion.

We have now returned to the fact that there is a significant individual component to stress reactions, as mentioned early in this review. This was eloquently stated by

Vogel and Bower:[69] "Stress is in the mind of the beholder. We create our own stress, make our own stressful events, and cause our own diseases. These processes are based on specific genetic vulnerabilities, individual experiences, and environmental circumstances. This explains the well-known observations that some individuals are quite stress-resistant while others are stress-susceptible, that stress-susceptible individuals will experience different stress reactions, and that the nature and extent of stress-related diseases vary greatly among individuals."[70,71]

REFERENCES

1. Webster's Medical Desk Dictionary, Merriam-Webster, Springfield, MA, 1986.
2. Selye, H., *Stress in Health and Disease,* Butterworth, London, 1975.
3. Kronfol, Z., Silva, J., Jr., Greden, J., Denbinski, S., Gordner, R., and Carroll, B., Impaired lymphocyte function in depressive illness, *Life Sci.,* 33, 241, 1983.
4. Ishigami, T., The influence of psychic acts on the progress of pulmonary tuberculosis, *Am. Rev. Tuberculosis,* 2, 470, 1919.
5. Johnson, E. O., Kamilaris, T. C., Chrousos, G. P., and Gold, P. W., Mechanisms of stress: a dynamic overview of hormonal and behavioral homeostasis, *Neurosci. Biobehav. Rev.,* 16, 115, 1992.
6. Khansari, D. N., Murgo, A. J., and Faith, R. E., Effects of stress on the immune system, *Immunol. Today,* 11, 170, 1990.
7. Dantzer, R. and Kelley, K. W., Stress and immunity: an integrated view of relationships between the brain and the immune system, *Life Sci.* 44, 1995, 1989.
8. Berczi, I., The influence of pituitary-adrenal axis on the immune system. in *Pituitary Function and Immunity,* Berczi, I., Ed., CRC Press, Boca Raton, FL, 1986.
9. Ader, R. and Cohen, N., Psychoneuroimmunology: conditioning and stress, *Annu. Rev. Psychol.,* 44, 53, 1993.
10. Monjan, A. A., Stress and immunologic competence: studies in animals, in *Psychoneuroimmunology,* Ader, R., Ed., Academic Press, New York, 1981, 185.
11. Palmblad, J. E. W., Stress-related modulation of immunity: a review of human studies, *Cancer Detec. Prev.,* Suppl. 1, 57, 1987.
12. Thomson, S. P., McMahon, L. J., and Nugent, C. A., Endogenous cortisol: a regulator of the number of lymphocytes in peripheral blood, *Clin. Immunol. Immunopathol.,* 17, 506, 1980.
13. Del Ray, A., Besedovsky, H., and Sorkin, E., Endogenous blood levels of corticosterone control the immunologic cell mass and B cell activity in mice, *J. Immunol.,* 133, 572, 1984.
14. Jefferies, W. M., Cortisol and immunity, *Med. Hypotheses.,* 34, 198, 1991.
15. Selye, H., A syndrome produced by diverse nocuous agents, *Nature (London),* 138, 32, 1936.
16. Carriere, G., Morel, J., and Gineste, P.-J., Modifications histo-physiologiques du thymus du rat albinos sous l'influence de la progestine et de l'hormone gonadotrope ou antelobine, *C.R. Soc. Biol.,* 126, 44, 1937.
17. Ingle, D. J., Atrophy of the thymus in normal and hypophysectomized rats following administration of cortin, *Proc. Soc. Exp. Biol., Med.,* 38, 443, 1938.
18. Kilbourne, E. D. and Horsfall, F. L., Jr., Lethal injection with Coxsackie virus of adult mice given cortisone, *Proc. Soc. Exp. Biol. Med.,* 77, 135, 1951.

19. Boring, W. D., Angevine, D. M., and Walker, D. L., Factors influencing host-virus interactions, *J. Exp. Med.,* 102, 753, 1955.
20. Munck, A., Guyre, P. M., and Holbrook, N. J., Physiological functions of glucocorticoids in stress and their relation to pharmacological actions, *Endo. Rev.,* 5, 25, 1984.
21. Schedlowski, M., Jacobs, R., Stratmann, G., Richter, S., Hadicke, A., Tewes, U., Wagner, T. O., and Schmidt, R. E., Changes of natural killer cells during acute psychological stress, *J. Clin. Immunol.,* 13, 119, 1993.
22. Schedlowski, M., Falk, A., Rohne, A., Wagner, T. O., Jacobs, R., Tewes, U., and Schmidt, R. E., Catecholamines induce alterations of distribution and activity of human natural killer (NK) cells, *J. Clin. Immunol.,* 13, 344, 1993.
23. Song, C., Earley, B., and Leonard, B. E., Behavioral, neurochemical, and immunological responses to CRF administration, *Ann. NY Acad. Sci.,* 771, 55, 1995.
24. Strausbaugh, H. and Irwin, M., Central corticotropin-releasing hormone reduces cellular immunity, *Brain Behav. Immun.,* 6, 11, 1992.
25. Irwin, M., Brain corticotropin-releasing hormone and interleukin-1 β-induced suppression of specific antibody production, *Endocrinology,* 133, 1352, 1993.
26. Rothwell, N. J., Central effects of CRF on metabolism and energy balance, *Neurosci. Biobehav. Rev.,* 14, 263, 1990.
27. DeSouza, E. B., Corticotropin-releasing factor and interlukin-1 receptors in the brain-endocrine-immune axis. Role in stress response and infection, *Ann. NY Acad. Sci.,* 697, 9, 1993.
28. Friedman, E. M., and Irwin, M. R., A role for CRH and the sympathetic nervous system in stress-induced immunosuppression, *Ann. NY Acad. Sci.,* 771, 396, 1995.
29. Weiss, J. M., Quan, N., and Sundar, S. K., Widespread activation and consequences of interleukin-1 in the brain, *Ann. NY Acad. Sci.,* 741, 338, 1994.
30. Weiss, J. M., Sundar, S. K., Becker, K. J., and Cierpial, M. A., Behavioral and neural influences on cellular immune responses: effects of stress and interleukin-1, *J. Clin. Psychiatry,* 50, 43, 1989.
31. Payne, L. C., Weigent, D. A., and Blalock, J. E., Induction of pituitary sensitivity to interleukin-1: a new function for corticotropin-releasing hormone, *Biochem. Biophys. Res. Commun.,* 198, 480, 1994.
32. Riley, V., Psychoneuroendocrine influences on immunocompetence and neoplasia, *Science,* 212, 1100, 1981.
33. Glasser, R., Kennedy, S., Lafuse, W. P., Bonneau, R. H., Speicher, C., Hillhouse, J., and Kiecolt-Glasser, J. K., Psychological stress-induced modulation of interleukin-2 receptor gene expression and interleukin-2 production in peripheral blood leukocytes, *Arch. Gen. Psychiatry,* 47, 707, 1990.
34. Glasser, R., Kiecolt-Glasser, J. K., Stout, J. C., Tarr, K. L., Speicher, C. E., and Holliday, J. E., Stress-related impairments in cellular immunity, *Psychiatry Res.,* 16, 233, 1985.
35. Glasser, R., Rice, J., Speicher, C. E., Stout, J. C., and Kiecolt-Glasser, J. K., Stress depresses interferon production by leukocytes concomitant with a decrease in natural killer cell activity, *Behav. Neurosci.,* 100, 675, 1986.
36. Smith, A. H., Response of animals to reduced acceleration fields, in *Principles of Gravitational Physiology,* NASA Contract NSR-09-010-027, National Aeronautics and Space Administration, Washington, D.C., 1970.
37. Besch, E. L., Smith, A. H., Burton, R. R., and Sluka, S. J., Physiological limitations of animal restraint, *Aerosp. Med.,* 38, 1130, 1967.
38. Besch, E. L., Burton, R. R., and Smith, A. H., Organ and body mass changes in restrained and fasted domestic fowl, *Proc. Soc. Biol. Med.,* 141, 456, 1972.

39. Upton, P. K. and Morgan, D. J., The effect of sampling technique on some blood parameters in the rat, *Lab. Anim.*, 9, 85, 1975.
40. Tache, Y., Ducharme, J. R., Charpenet, G., Haour, F., Saez, J., and Collu, R., Effect of chronic intermittent immobilization stress on hypophyso-gonadal function of rats, *Acta Endrocrinol.*, 93, 168 1980.
41. Bonneau, R. H., Sheridan, J. F., Feng, N., and Glasser, R., Stress-induced suppression of herpes simplex virus (HSV)-specific cytotoxic T lymphocytes and natural killer cell activity and enhancement of acute pathogenesis following local HSV infection., *Brain Behav. Immun.*, 5, 170, 1991.
42. Bonneau, R. H., Sheridan, J. F., Feng, N., and Glasser, R., Stress-induced effects on cell mediated innate and adaptive memory components of the murine immune response to herpes simplex virus infection, *Brain Behav. Immun.*, 5, 274, 1991.
43. Rasmussen, A. F., Marsh, J. T., and Brill, N. O., Increased susceptibility to herpes simplex in mice subjected to avoidance-learning stress or restraint, *Proc. Soc. Exp. Biol. Med.*, 96, 183, 1957.
44. Seifter, E., Rettura, G., Zisblatt, M., Levinson, S., Levine, N., et al., Enhancement of tumor development in physically stressed mice incubated with an oncogenic virus, *Experientia*, 29, 1379, 1973.
45. Feng, N., Pagniano, R., Tovar, A., Bonneau, R., Glasser, R., and Sheridan, J. F., The effect of restraint stress on the kinetics, magnitude, and isotype of the humoral immune response to influenza virus infection, *Brain Behav. Immun.*, 5, 370, 1991.
46. Okimura, T. and Nigo, Y., Stress and immune responses. I. Suppression of T cell functions in restraint-stressed mice, *Jpn. J. Pharmacol.*, 40, 505, 1986.
47. Okimura, T., Ogawa, M., Yamauchi, T., and Satomi-Sasaki, Y., Stress and immune responses. IV. Adrenal involvement in the alteration of antibody responses in restraint-stressed mice, *Jpn. J. Pharmacol.*, 41, 237, 1986.
48. Okimura, T., Satomi-Sasaki, Y., and Okimura, S., Stress and immune responses. II. Identification of stress-sensitive cells in murine spleen cells, *Jpn. J. Pharmacol.*, 40, 513, 1986.
49. Bonneau, R. H., Sheridan, J. F., Feng, N., and Glasser, R., Stress-induced modulation of the primary cellular immune response to herpes simplex virus infection is mediated by both adrenal-dependent and independent mechanisms, *J. Neuroimmunol.*, 42, 167, 1993.
50. Dobbs, C. M., Vasquez, M., Glasser, R., and Sheridan, J. F., Mechanisms of stress-induced modulation of viral pathogenesis and immunity, *J. Neuroimmunol.*, 48, 151, 1993.
51. Henning, S. J., Plasma concentrations of total and free corticosterone during development in the rat, *Am. J. Physiol.* 4, E451, 1978.
52. Kort, W. J. and Weijma. J. M., Effect of chronic light-dark shift stress on the immune response of the rat, *Physiol. Behav.* 29, 1083, 1982.
53. Kort, W. J., Zondervan, P. E., Hulsman, L. O. M., Weijma, I. M., and Westbroek, D. L., Light-dark shift stress, with special reference to spontaneous tumor incidence in female BN rats, *J. Natl Canc. Inst.*, 76, 439, 1986.
54. Wallace, M. E., Effects of stress due to deprivation and transport in different genotypes of house mouse, *Lab. Anim.*, 10, 335, 1976.
55. Salvia, M., Weisbroth, S. H., and Paganelli, R. G., Technical notes: lab animals show less "shipment stress" given drinking water, *Lab Anim.*, May–June, 38, 1979.
56. Tuli, J. S., Smith, J. A., and Morton, D. B., Stress measurements in mice after transportation, *Lab. Anim.*, 29, 132, 1994.

57. Aarstad, H. J., Gardernack, G., and Seljelid, R., Stress causes reduced natural killer activity in mice, *Scand. J. Immunol.*, 18, 461, 1983.
58. Jiang, C. G., Morrow-Tesch, J. L., Beller, D. I., Levy, E. M., and Black, P. H., Immunosuppression in mice induced by cold water stress, *Brain Behav. Immun.*, 4, 278, 1990.
59. Cunnick, J. E., Lysle, D. T., Armfield, A., Fowler, H., and Rabin, B. S., Shock-induced immune suppression of lymphocyte responsiveness and natural killer activity: differential mechanisms of induction, *Brain Behav. Immun.*, 2, 102, 1988.
60. Lysle, D. T., Cunnick, J. E., Fowler, H., and Rabin, B. S., Pavlovian conditioning of shock-induced suppression of lymphocyte reactivity: acquisition, extinction and pre-exposure effects, *Life Sci.*, 42, 2185, 1988.
61. Maier, S. F. and Laudenslager, L. M., Inescapable shock, shock controllability, and mitogen-stimulated lymphocyte proliferation, *Brain Behav. Immun.*, 2, 87, 1988.
62. Pezzone, M. A., Dohanics, J., and Rabin, B. S., Effects of foot shock stress upon spleen and peripheral blood lymphocyte mitogenic responses in rats with lesions of the paraventricular nuclei, *J. Neuroimmunol.*, 53, 39, 1984.
63. Zalcman, S., Kerr, L., and Anisman, H., Immunosuppression elicited by stressors and stress-related odors, *Brain Behav. Immun.*, 5, 262, 1991.
64. Laudenslager, M. L., Ryan, S. M., Drugan, R. C., Hyson, R. L., and Maier, S. F., Coping and immunosuppression: inescapable but not escapable shock suppresses lymphocyte proliferation, *Science*, 221, 568, 1983.
65. Sonnenfeld, G., Cunnick, J. E., Armfield, A. V., Wood, P. G., and Rabin, B. S., Stress-induced alterations in interferon production and class II histocompatibility antigen expression, *Brain Behav. Immun.*, 6, 170, 1992.
66. Sacerdote, P., Manfredi, B., Bianchi, M., and Panerai, A. E., Intermittent but not continuous inescapable foot shock stress affects immune responses and immunocyte β-endorphin concentrations in the rat, *Brain Behav. Immun.*, 8, 251, 1994.
67. Laudenslager, M. L., Fleshner, M., Hofstadter, P., Held, P. E., Simons, L., and Maier, S. F., Suppression of specific antibody production by inescapable shock: stability under varying conditions, *Brain Behav. Immun.*, 2, 92, 1988.
68. Dorian, B., Garfinkel, G., Brown, G., Shore, A., Glladman, D., and Keystone, E., Aberrations in lymphocyte subpopulations and function during psychological stress, *Clin. Exp. Immunol.*, 50, 132, 1982.
69. Vogel, W. H. and Bower, D. B., Stress, immunity, and cancer, in *Stress and Immunity*, Plotnikoff, N., Murgo, A., Faith, R., and Wybran, J., Eds., CRC Press, Boca Raton, FL, 1991, 493.
70. Plotnikoff, N. P., Faith, R. E., Murgo, A. J., and Good, R. A., Eds., *Enkephalins and Endorphins — Stress and the Immune System*, Plenum Press, New York, 1986.
71. Plotnikoff, N., Murgo, A., Faith, R., and Wybran, J., Eds., *Stress and Immunity*, CRC Press, Boca Raton, FL, 1991.

12 Bidirectional Communication Between the Immune and Neuroendocrine Systems

Douglas A. Weigent and J. Edwin Blalock

CONTENTS

12.1 Introduction 173
12.2 Influence of the Neuroendocrine System on Cells of
    the Immune System 174
12.3 Influence of the Immune System on the Brain 179
12.4 Conclusion 182
Acknowledgments 182
References 183

12.1 INTRODUCTION

The past 15 years have seen the emergence of data defining some of the mechanisms for bidirectional communication between the immune and neuroendocrine systems. Most of the influence of the brain on the immune system seems to be exerted via hormones released from the neuroendocrine system as well as by sympathetic innervation of immune organs. Most of the influence of the immune system on the brain seems to be exerted via hormones, including cytokines, released by lymphocytes and monocytes. Increasingly, data obtained from both *in vitro* and *in vivo* studies suggest that shared ligands and receptors are used as the chemical language between the immune and neuroendocrine systems.[1,2] On the one hand, the immune system recognizes noncognitive stimuli such as bacteria and viruses and converts this into an array of cytokines which act on receptors in the neuroendocrine system to alter its function. On the other hand, psychological or physical stimuli appear to induce secretion of a pattern of neurotransmitters and hormones which act on receptors of the immune system to alter its function. Taken together, the evidence suggests an immunoregulatory role for the brain and a sensory function for the immune system. In this review, we will briefly discuss the evidence for brain-immune system interactions.

TABLE 12.1
Modulation of Immune Responses by Pituitary Hormones

Hormone	Modulating Effect
Corticotropin	Antibody synthesis
	IFN-γ production
	B lymphocyte growth
Endorphins	Antibody synthesis
	Mitogenesis
	Natural killer cell activity
Thyrotropin	Increased antibody synthesis
	Comitogenic with Con A
GH	Cytotoxic T cells
	Mitogenesis
LH and FSH	Proliferation
	Cytokine production
PRL	Comitogenic with Con A
	Induces IL-2 receptors

Note: See text for description and Weigent and Blalock (1995).[2]

12.2 INFLUENCE OF THE NEUROENDOCRINE SYSTEM ON CELLS OF THE IMMUNE SYSTEM

Much of the influence of the brain on immune events is mediated through the secretions of the hypothalamus and pituitary.[1] The hypothalamus is the efferent link of the brain receiving information from the periphery and integrating it with the internal environment. The hypothalamus stimulates the pituitary gland through a variety of polypeptide hormones including corticotropin releasing hormone (CRH), thyrotropin releasing hormone (TRH), growth hormone releasing hormone (GHRH), luteinizing hormone releasing hormone (LHRH), somatostatin, and dopamine. These control the release of corticotropin (ACTH), thyrotropin (TSH), growth hormone (GH), luteinizing hormone (LH), and prolactin (PRL), respectively. A large number of reports support the idea that neuroendocrine hormones modulate the immune response (Table 12.1). There is also a reasonable amount of evidence to support the existence of neuroendocrine hormone-specific binding sites on cells of the immune system (for review see References 2 and 3). The effects of these hormones on cells of the immune system are briefly reviewed below.

ACTH and endorphins have been found to influence many aspects of the immune response. The ability of ACTH to suppress the *in vitro* antibody response to T-dependent antigens is more effective than it is to T-independent antigens, suggesting that ACTH may interfere with the production or action of helper T-cell signals.

ACTH has also been shown to suppress major histocompatibility complex class II expression by murine peritoneal macrophage, stimulate natural killer (NK) cell activity, modulate the rise in intracellular-free Ca^{2+} concentration after T-cell activation, suppress the production of interferon-γ (IFN-γ), function as a late-acting B-cell growth factor that can synergize with IL-5, and stimulate the growth and differentiation of human tonsillar B cells.[4] The endogenous opiates β-, γ-, and α-endorphins contained in the polyprotein POMC have also been shown to modulate the activity of cells in the immune system.[3] The α-endorphin is a potent inhibitor of the anti-sheep red blood cell plaque-forming cell response, whereas β- and γ-endorphin are mild inhibitors.[5] It has been suggested that an increase in cytokine production seen in inflammation induces β-endorphin production in both the pituitary and lymphocytes. The increase in β-endorphin may act directly on B cells to inhibit antibody synthesis or block antibody synthesis by inhibiting T helper cell function or through enhancing NK cell activity.[6] Thus, β-endorphins may play a role in limiting the progression of autoimmune reactions. Many other aspects of immunity are modulated by the opiate peptides including: (1) enhancement of the natural cytotoxicity of lymphocytes and macrophages toward tumor cells; (2) enhancement or inhibition of T-cell mitogenesis; (3) enhancement of T-cell rosetting; (4) stimulation of human peripheral blood mononuclear cells; and (5) inhibition of major histocompatibility class II antigen expression.[3]

The effects of TSH on immunity were first recognized when it was shown that TSH could augment both T-dependent and T-independent antibody production.[4] TSH had to be present early for enhancement of the antibody response to occur. TRH enhanced the antibody plaque-forming cell response and also induced splenocyte production of TSH.[7] This enhancement by TRH was specifically blocked by antibodies to the TSH β-subunit, which demonstrated that the action of TRH was through its ability to induce TSH production by lymphocytes. This was the first demonstration that a pituitary hormone could function as an autocrine or paracrine regulator of the immune system.[7] Subsequently, studies have suggested that TSH may elevate cAMP levels and influence differentiation in B-cell lines.[8] In another report it was shown that although TSH increased DNA synthesis and intracellular cAMP levels of FRTL-5 rat thyroid cells, it did not have much stimulatory effect on lymphocytes.[9] In this report, however, it was confirmed that TSH caused a moderate increase in Ig production by activated B cells. Another study reported that TSH at various concentrations increased the proliferative response of mouse lymphocytes to T-cell mitogens and stimulated IL-2-induced NK cell activity without modifying the basal levels of cytotoxicity.[10] Recently it was demonstrated that TRH and TSH upregulate the development of certain subsets of intestinal intraepithelial lymphocytes in athymic mice while thyroxine in normal euthymic mice was immunosuppressive.[11] These same investigators now have shown that intestinal epithelial cells express receptors for TRH and can be a source of TSH as well as lymphocytes. The locally produced TSH can act on intraepithelial lymphocytes which bear the TSH receptor and modulate their development.[12] The data support the idea that the composition and distribution of gut T cells are regulated by neuroendocrine hormones. Taken together, these results support the immunoregulatory role of TSH on both T and B cells.

LH has been shown to modulate cytokine and γ globulin secretion in mice.[13] In addition, LH at various concentrations increased the proliferative response to mitogens. Overall, it appears that LH modulates both humoral and cellular immunity. It has been shown that inhibin and activin, whose major function is the control of FSH release from the pituitary, can also modulate lymphocyte function.[14] A significant dose-related increase in monocyte chemotaxis can be induced by inhibin. Although activin increased the migrational activity of monocytes, inhibin significantly decreased IFN-γ production and the effect was reversed by activin. Inhibin and/or activin had no significant effect on either PHA-induced lymphocyte proliferation or lymphocyte cytotoxic capability. These findings suggest that inhibin and activin may affect selected immune parameters and suggest a possible involvement of these hormones along with LH and possibly FSH in regulating cell-mediated immune function.

It has also been shown that GH and PRL have potent effects on cells of the immune system.[15] The potential role of GH in immunoregulation has been demonstrated for numerous immune functions, including stimulation of DNA and RNA synthesis in the spleen and thymus of normal and hypophysectomized rats. GH also affects neutrophil differentiation, erythropoiesis, and thymic development. GH affects the functional activity of T cells and NK cells[16] and stimulates the production of superoxide anion from macrophages.[17] Several observations suggest that GH may also stimulate the local production of IGF-1 which acts to promote tissue growth and function in a paracrine/autocrine fashion.[18] Many studies have been conducted *in vivo* in normal and GH-deficient animals that demonstrate the ability of GH to modulate immune function. Injections of GH have been found to increase thymic size, stimulate thymocyte proliferation, induce c-myc, and augment antibody synthesis and skin graft rejection. GH can stimulate the production of IL-1, IL-2, TNFα and thymulin, enhance NK activity, and restore the architecture of the thymus in aged animals.[19] GH also promotes lymphocyte engraftment in immunodeficient mice.[20] The function of lymphocyte-derived GH is less clear, but it has been shown to increase IGF-1 production by lymphocytes and stimulate proliferation.[21]

The studies with PRL show that suppression of PRL secretion in mice with bromocriptine: (1) increases the lethality of a *Listeria* challenge, (2) abrogates T-lymphocyte-dependent activation of macrophages as well as the production of lymphocyte IFN after inoculation with *Listeria* or mycobacteria, and (3) suppresses T-lymphocyte proliferation without affecting the production of IL-2. All of these changes could be prevented *in vivo* by the administration of PRL.[22] Nb2 cells are of T cell origin and can be stimulated to grow by both PRL and IL-2.[23] The effects of the two hormones are additive and appear to be mediated by different receptors. IL-2 stimulated [^{3}H]thymidine incorporation into splenocytes but the IL-2-stimulated increase in [^{3}H]thymidine incorporation was blocked in the presence of antibodies to PRL. Thus, Nb2 cell proliferation was shown to be a PRL receptor-mediated event and antibodies to the PRL receptor were able to abolish PRL-induced proliferation of Nb2 cells. There are other data from dwarf mice showing that PRL and GH can correct the defective induction of IL-2 receptor expression after Con A stimulation.[24] Stimulation of a cloned T cell with PRL was shown to induce the expression of IFN regulatory factor-1, suggesting that PRL may regulate cell proliferation by enhancing expression of some genes required for entry into the S phase.[25] Finally, it has been shown that PRL appears to exert counterregulatory actions which may modify glucocorticoid actions on immune tissues.[26] It is clear from all the studies to date that PRL and GH have important and sometimes similar effects on the immune system. Both animal studies and the available clinical data in humans suggest that absence of either PRL or GH can lead to deficiencies in both cell-mediated and humoral immunological functions. The deficiencies can be corrected by replacement therapy with PRL or GH.

The receptors and effects of hypothalamic-releasing hormones on cells of the immune system has been studied for CRF, GHRH, LHRH, TRH, and SOM. The data support the view that these peptides may play important roles in immunity (Table 12.2). The most important function of CRF may be to stimulate production of lymphocyte POMC peptides. Thus, the effects of ACTH and endorphins discussed above may be initiated in the immune system via hypothalamic releasing hormones. Some studies suggest that CRF modulates the immune response to stress in the rat by inhibiting lymphocyte proliferation and NK cell activity.[27] Another report on cultured human peripheral blood mononuclear cells treated with CRF showed that this hormone increased IL-6 release, decreased IFN-γ levels, and had no significant effect on IL-1β production or lymphocyte proliferation.[28] The neuropeptide α-melanocyte-stimulating hormone (α-MSH) has powerful antipyretic and anti-inflammatory properties.[29] It appears CRF mimics the immunosuppressive effects of MSH but has a longer duration of action. The activation of granulocytes can be suppressed by ACTH and MSH.[30] The GHRH receptor also has been identified on cells of the immune system which is associated with the stimulation of lymphocyte proliferation.[31] In addition, LHRH receptors have been functionally identified on thymocytes and LHRH agonists were found to diminish NK activity, stimulate T-cell proliferation, and rejuvenate the involuted thymic tissue in aged rats.[32]

Leukocytes have been shown to respond to TRH treatment by producing TSH mRNA and protein. The TSH produced by lymphocytes may influence antibody production similarly to what has been described earlier for pituitary or exogenous TSH.[7] SOM has been shown to inhibit Molt-4 lymphoblast proliferation and PHA stimulation of human T lymphocytes, and nanomolar concentrations are able to inhibit the proliferation of both spleen-derived and Peyer's patch-derived lymphocytes.[33] Other immune responses, such as SEA-stimulated IFN-γ secretion, endotoxin-induced leukocytosin, and colony-stimulating activity release are also inhibited by SOM.[34]

In addition to neuroendocrine hormones from the brain influencing immune function, it is also highly likely that the local production of these same peptides by cells of the immune system will have dramatic effects on immunity. The evidence that cells of the immune system can produce these peptides is overwhelming (see Table 12.3; for review see Reference 2). Although in all cases only small amounts of hormone are produced and certainly function by autocrine/paracrine mechanisms, there is also good evidence, at least in the case of ACTH, that endocrine effects are also possible.[2] The impact that the lymphocyte production

TABLE 12.2
Modulation of Immune Responses by Hypothalamic Hormones

Hormone	Modulating Effect
CRF	IL-1 production
	Enhanced NK cell activity
	Immunosuppressive
TRH	Increased antibody synthesis
GHRH	Stimulates proliferation
	Inhibits NK cell activity
	Inhibits chemotactic response
SOM	Inhibits proliferation
	Reduces IFN-γ production

Note: See text for description and Weigent and Blalock (1995).[2]

TABLE 12.3
Neuroendocrine Hormones Produced in the Immune System

Corticotropin (ACTH) and endorphins	PRL
	GH
[Met]-enkephalin	SOM
TSH	CRF
LH	GHRH
FSH	LHRH

of these hormones has on the immune system and the mechanism employed is currently under investigation.

Additional evidence for brain-immune system interactions has also been derived from lesioning experiments. Electrolytic lesions of the hypothalamus have been shown to produce either inhibition or enhancement of various immune functions.[35] Lesioning of the pituitary gland can also alter immune function, and the changes produced by hypothalamic lesioning can be prevented by hypophysectomy, indicating that the pituitary gland mediates the hypothalamic effects. It has also been observed that neurotransmitters and neuropeptides affect immune function both *in vivo* and *in vitro* and that receptors for these same molecules are present on cells of the immune system.[36] The sympathetic nervous system innervates both primary and secondary lymphoid tissues.[37] This latter finding suggests that the nervous system can transmit signals neurogenically to lymphoid tissue. Finally, a number of studies show that the immune system can be conditioned. Conditioned immunosuppression can be observed when a sensory stimulus such as camphor is paired with an immunosuppressive drug such as cyclosporine.[38] The observations suggest that the thermoregulatory pathway in the brain can be conditioned and that IFN-β, prostaglandin E2, CRH, and ACTH function in a common pathway to modulate core body temperature and NK cell activity.[39] Taken together, the body of evidence clearly suggests that the nervous system influences the immune system.

12.3 INFLUENCE OF THE IMMUNE SYSTEM ON THE BRAIN

The invasion of the body by microorganisms activates the immune system, which results in the release of a complex variety of soluble mediators from cells of the immune system. The mediators include immunoglobulins, complement proteins, and cytokines. Cytokines include interleukins, interferons, colony stimulating factors, tumor necrosis factors, and transforming growth factors. Cytokines not only modulate the activity in the immune system but also influence the brain. These substances, at the present time, constitute the main players in the second arm of bidirectional communication; that is, the influence of the immune system on the nervous system.[40] Although the blood-brain barrier (BBB) excludes proteins from the brain, this barrier is absent in the area of the preoptic nucleus of the hypothalamus. Thus, cytokines may bind to neurons in this region that have cytokine receptors and pass into the brain. Activated lymphocytes may synthesize small amounts of neuroendocrine hormones, including CRF, ACTH, SOM, TSH, GH, GHRH, PRL, and β-endorphin, which also may influence the brain or the peripheral nervous system. Much more work is needed to establish this latter possibility.

The potential action of lymphocyte products on the brain was realized when altered hormone levels were observed in animals in response to antigenic challenge, which peaked with the immune response to the particular antigen.[41] Although supernatant fluids from activated leukocytes could mimic this phenomenon, the actual effector molecules were unknown. The effects of cytokines and the presence of their receptors in the neuroendocrine system is currently the topic of much research. Of particular interest have been IL-1, IL-2, IL-6, TNF, and IFN-γ, with particular emphasis on IL-1. In general, immune stimulation has numerous effects on neuroendocrine hormone secretion (Table 12.4). A common pattern is an increase in ACTH secretion and suppression of TSH release while the pattern for other hormones is less consistent.[42]

The earliest report with IL-1 showed that it was able to stimulate the release of ACTH.[43] Later it was found that IL-1 could stimulate the synthesis of CRH mRNA and release of CRH from hypothalamic fragments and that antibodies to CRH could block the IL-1-induced *in vivo* release of ACTH.[44] Several pathways appear to mediate the influence of IL-1 on neuroendocrine neurons and depend on the route of administration and on whether the cytokine acts at the level of the release and/or the biosynthesis of CRF.[45] The increase in CRF mRNA appears to be dependent upon IL-1 in the CNS because central administration of the IL-1 receptor antagonist completely blocked the expression of CRF transcripts in the periventricular

TABLE 12.4
Effect of Cytokines on the Neuroendocrine System

	Neuroendocrine Hormone				
Cytokine	ACTH	TSH	PRL	GH	LH
IL-1	S	I	S	S	I
IL-2	S	S	S	I	I
IL-6	S	I	0	I	0
IFN-γ	S	I	0	I	?
TNF	S	I	S	S	?

Note: S, stimulation; I, inhibition; 0, no effect; ?, not determined.

nucleus.[46] These studies have been extended, and it has been reported that IL-1 receptors can be demonstrated on the pituitary gland[47] and that CRH can upregulate IL-1 receptors on AtT-20.[48] A more recent finding by us has revealed that low levels of exogenous or endogenous CRH can sensitize the pituitary gland to the direct releasing activity of IL-1.[49] Therefore, one can conclude that IL-1 functions as a neuromodulator in the hypothalamus to enhance CRH release into the hypophyseal portal blood and that both IL-1 and CRH can sensitize the corticotroph, thus facilitating the release of ACTH.

More recent evidence suggests that both IL-1 and CRH activate corticotrophs, but they elicit different patterns in the regulation of POMC.[50] Thus, IL-1 evoked an early release of β-lipotropin and an intermediate release of β-endorphin while CRH caused an early β-endorphin secretory response. Such a distinct pattern allows the pituitary to be specifically activated and therefore determine the interaction with the immune system. In addition to its effects on the hypothalamic-pituitary-adrenal axis (HPA), IL-1 has been shown to inhibit the hypothalamic-thyroid axis and the hypothalamic-gonadal axis. Thus, IL-1 inhibits the ovarian steroid-induced LH surge and release of hypothalamic LHRH in rats.[51] It also decreases plasma thyroid hormone and TSH levels in rats, probably by suppressing hypothalamic TRH secretion.[52]

IL-2 is the most potent regulator of pituitary ACTH secretion and is more active than the classical hypothalamic regulator, CRH.[53] *In vitro,* it has been shown to enhance POMC gene expression and induce ACTH release in the AtT-20 pituitary cell line and rat anterior pituitary cell cultures.[54] In rat pituitary cell cultures at low concentrations, IL-2 elevated ACTH, PRL, and TSH release and inhibited the release of FSH, LH, and GH from hemipituitaries *in vitro*.[53] It appears that both IL-2 and IL-6, in addition to their effects on hormone secretion, may participate in anterior pituitary cell growth regulation.[55] Both cytokines were found to stimulate the growth of the GH_3 cell line and inhibit the proliferation of normal rat anterior pituitary cells.[55] It seems clear that some of the effects of IL-2 may occur directly at the level of the hypothalamus. Thus, IL-2 stimulates the release of GHRH from medial hypothalamic fragments and can stimulate the release of SOM.[56] IL-2 given centrally induces a somewhat different pattern of response than the other cytokines because it stimulated instead of inhibited TSH release and also inhibited GH release, which was stimulated by low dosages of IL-1 and cachectin. As in the case of IL-1, IL-2 inhibited LH release but it also inhibited FSH release. Thus, this cytokine has powerful actions at the hypothalamic level to alter pituitary hormone release as well as direct actions on the pituitary.

Another cytokine, initially identified by its antiviral activity, known to influence hormone secretion in the neuroendocrine system is IL-6. The intracerebral ventricular injection of IL-6 results in an increase in plasma ACTH along with elevated temperature and a decrease in TSH.[57] Another group reported that IL-6 could stimulate the *in vitro* release of PRL, GH, and LH by dispersed pituitary cells.[58] The stimulation of PRL release by TRH could be enhanced by IL-6, indicating these hormones may have different intracellular mechanisms to stimulate hormone release.[59] In the case of IL-6 stimulating plasma ACTH levels, it has been observed that IL-6 may increase CRH release from medial hypothalamic tissue.[60] The *in vivo* effect of IL-6 could be blocked by the prior administration of antibody against CRH, thus demonstrating that IL-6 stimulates ACTH secretion through the production of CRH.[61]

IFN-γ also has been shown to influence the secretory activity of anterior pituitary cells in culture. Initial studies within IFN-γ on hemipituitaries revealed no effect on the release of PRL, TSH, and GH, but a stimulation of ACTH release was observed.[62] *In vivo* injection of IFN-γ stimulated ACTH, with no effect on PRL and a delayed inhibition of GH and TSH release.[63] Intracerebroventricular administration of IFN-γ results in the suppression of TSH secretion, which may be mediated by increased SOM.[62] IFN-γ at physiological concentrations has been shown to inhibit stimulated secretion of ACTH, PRL, and GH of pituitary cells cultured *in vitro* and stimulated with hypothalamic factors.[64] The inhibition was evident with dosages of IFN-γ as small as those needed for the antiviral or macrophage-activating effects. These results indicate that IFN-γ may modulate GH secretion from the pituitary gland by both a direct suppressive effect at the level of the pituitary and indirect hypothalamic suppression involving stimulation of SOM release.[62]

TNF is a cytokine produced by activated macrophages or monocytes that is important in the hormonal response to shock.[65] It has been shown that after only one hour of incubation, cachectin was capable of stimulating the release of ACTH, GH, TSH, and PRL from either overnight-cultured dispersed pituitary cells or hemipituitaries; however, the dose for these actions of cachectin was 100-fold greater with dispersed cells than with hemipituitaries.[66] In another report by a different group, TNF-α did not influence basal secretion but did inhibit the hormonal response of the pituitary gland to hypothalamic releasing factors.[67] *In vivo,* TNF-α has been reported to stimulate the release of ACTH, PRL, and GH similar to what has been observed with IL-1.[68] The stimulatory effect on ACTH release was completely inhibited by previous injection of CRH antiserum, suggesting that endogenous CRH serves as a mediator of the response.

Alternatively, TNF-α could activate the HPA axis by stimulating local secretion of IL-1, IL-2, or IL-6. It appears, however, that TNF-α is much less potent than IL-1 with respect to effects on hormone secretion.[42] In addition to these effects, TNF-α has been suggested to inhibit the hypothalamic-pituitary-thyroid axis at multiple levels[52] and the hypothalamic-pituitary-gonadal axis.[67] Overall, these findings suggest that TNF-α inhibits the secretion of pituitary hormones and particularly suppresses the corticotroph cells.

Several recent studies also indicate that cytokines may synergize in the CNS. A form of motivation known as social investigation was used to demonstrate synergy between centrally injected (i.c.v.) IL-1 and TNF-α.[69] More recently, IL-6 and its soluble receptor, when injected i.c.v., have been shown to interact in a way that potentiates fever and anorexia.[70] Thus, the biological activity of cytokines may be dependent on the presence (or absence) of soluble receptors, which may exhibit either agonistic or antagonistic activity.

12.4 CONCLUSION

The immune system is constantly interacting with the neuroendocrine system. This interaction assures that immune and inflammatory responses are in harmony with other bodily functions. A rapidly expanding body of evidence supports the concept that crosstalk or bidirectional communication between the neuroendocrine and immune systems is achieved by shared ligands and receptors. Cytokines have actions at both the hypothalamic and pituitary level. Several cytokines, especially IL-1, IL-2, IL-6, and TNF, activate the adrenal axis, whereas IL-1 and TNF inhibit the gonadal axis and TNF and IFN-γ have inhibitory actions on the thyroid axis. Overall, the organism's response may be influenced both by the circulating cytokines derived from the immune system and/or those endogenously produced cytokines within the neuroendocrine system. The pathway of immune activation of the neuroendocrine system may serve to limit the severity of the immune response via the glucocorticoids produced, while enhancing other hormones that promote immunity at the same time. The presence of neuroendocrine hormone receptors on immune cells and their immunomodulatory effects has been known for some time. It is now clear that cells of the immune system produce neuroendocrine hormones and contain specific receptors for these hormones. Consequently, neuroendocrine hormones have profound effects on almost all aspects of the immune response. Overall, the recognition of the immune system as our sixth sense may ultimately provide the new understanding of physiology required for successful diagnostic and therapeutic programs against disease and stress involving immune-neuroendocrine communication.[40]

ACKNOWLEDGMENTS

The research was supported in large part by NIH grants NS24636 (JEB), MH52527 (JEB), and AI41651 (DAW). We also thank Diane Weigent for excellent editorial assistance and typing the manuscript.

REFERENCES

1. Blalock, J. E., The syntax of immune-neuroendocrine communication, *Immunol. Today*, 15, 504, 1994.
2. Weigent, D. A., and Blalock, J. E., Associations between the neuroendocrine and immune systems, *J. Leukocyte Biol.*, 58, 137, 1995.
3. Carr, D. J. J., Neuroendocrine peptide receptors on cells of the immune system, *Neuroimmunoendocrinology*, Blalock, J. E., Ed., Karger, Basel, Switzerland, 1992, 49.
4. Johnson, H. M., Downs, M. O., and Pontzer, C. H., Neuroendocrine peptide hormone regulation of immunity, *Chem. Immunol.*, 52, 49, 1992.
5. Johnson, H. M., Smith, E. M., Torres, B. A., and Blalock, J. E., Neuroendocrine hormone regulation of *in vitro* antibody formation, *Proc. Natl. Acad. Sci. U.S.A.*, 79, 4171, 1982.
6. Morch, H. and Pedersen, B. K., β-endorphin and the immune system — possible role in autoimmune diseases, *Autoimmunity*, 21, 161, 1995.
7. Kruger, T. E., Smith, L. R., Harbour, D. V., and Blalock, J. E., Thyrotropin: an endogenous regulator of the *in vitro* immune response, *J. Immunol.*, 142, 744, 1989.
8. Harbour, D. V., Leon, S., Keating, C., and Hughes, T. K., Thyrotropin modulates B-cell function through specific bioactive receptors, *Prog. Neuro. Endocrin. Immunol.*, 3, 266, 1990.
9. Coutelier, J. P., Kehrl, J. H., Bellur, S. S., Kohn, L. D., Notkins, A. L., and Prabhakar, B. S., Binding and functional effects of thyroid stimulating hormone on human immune cells, *J. Clin. Immunol.*, 10, 204, 1990.
10. Provinciali, M., Di Stefano, G., and Fabris, N., Improvement in the proliferative capacity and natural killer cell activity of murine spleen lymphocytes by thyrotropin, *Int. J. Immunopharmacol.*, 14, 865, 1992.
11. Wang, J. and Klein, J. R., Hormone regulation of murine T cells: potent tissue-specific immunosuppressive effects of thyroxine targeted to gut T cells, *Int. Immunol.*, 8, 231, 1996.
12. Wang, J., Whetsell, M., and Klein, J. R., Local hormone networks and intestinal T cell homeostasis, *Science*, 275, 1937, 1997.
13. Rouabhia, M., Chakir, J., and Deschaux, P., Interaction between the immune and endocrine systems: immunomodulatory effects of luteinizing hormone, *Prog. Neuro. Endocrin. Immunol.*, 4, 86, 1991.
14. Petraglia, F., Sacerdote, P., Cossarizza, A., Angioni, S., Genazzani, A. D., Franceschi, C., Muscettola, M., and Grasso, G., Inhibin and activin modulate human monocyte chemotaxis and human lymphocyte interferon-γ production, *J. Clin. Endocrin. Metab.*, 72, 496, 1991.
15. Johnson, R. W., Arkins, S., Dantzer, R., and Kelley, K. W., Hormones, lymphohemopoietic cytokines and the neuroimmune axis, *Comp. Biochem. Physiol.*, 116A, 183, 1997.
16. Gala, R. R., Prolactin and growth hormone in the regulation of the immune system, *Proc. Soc. Exp. Biol. Med.*, 198, 513, 1991.
17. Fu, Y. K., Arkins, S., Fuh, G., Cunningham, B. C., Wells, J. A., Fong, S., Cronin, M. J., Dantzer, R., and Kelley, K. W., Growth hormone augments superoxide anion secretion of human neutrophils by binding to the prolactin receptor, *J. Clin. Invest.*, 89, 451, 1992.
18. Johnson, E. W., Jones, L. A., and Kozak, R. W., Expression and function of insulin-like growth factor receptors on anti-CD3-activated human T lymphocytes, *J. Immunol.*, 148, 63, 1992.

19. Kelley, K. W., Growth hormone, lymphocytes and macrophages. [Review], *Biochem. Pharmacol.*, 38, 705, 1989.
20. Murphy, W. J., Durum, S. K., and Longo, D. L., Human growth hormone promotes engraftment of murine or human T cells in severe combined immunodeficient mice, *Proc. Natl. Acad. Sci. U.S.A.*, 89, 4481, 1992.
21. Baxter, J. B., Blalock, J. E., and Weigent, D. A., Expression of immunoreactive growth hormone in leukocytes *in vivo*, *J. Neuroimmunol.*, 33, 43, 1991.
22. Bernton, E. W., Bryant, H. U., and Holaday, J. W., Prolactin and immune function, *Psychoneuroimmunology*, Ader, R., Felten, D. L., and Cohen, N., Eds., Academic Press, San Diego, 1991, 403.
23. Croze, F., Walker, A., and Friesen, H. G., Stimulation of growth of Nb2 lymphoma cells by interleukin-2 in serum-free and serum-containing media, *Mol. Cell. Endocrinol.*, 55, 253, 1988.
24. Gala, R. R. and Shevach, E. M., Influence of prolactin and growth hormone on the activation of dwarf mouse lymphocytes *in vivo*, *Proc. Soc. Exp. Biol. Med.*, 204, 224, 1993.
25. Clevenger, C. V., Sillman, A. L., Hanley-Hyde, J., and Prystowsky, M. B., Requirement for prolactin during cell cycle regulated gene expression in cloned T-lymphocytes, *Endocrinology*, 130, 3216, 1992.
26. Bernton, E., Bryant, H., Holaday, J., and Dave, J., Prolactin and prolactin secretagogues reverse immunosuppression in mice treated with cysteamine, glucocorticoids, or cyclosporin-A, *Brain. Behav. Immun.*, 6, 394, 1992.
27. Jain, R., Zwickler, D., Hollander, C. S., Brand, H., Saperstein, A., Hutchinson, B., Brown, C., and Audhya, T., Corticotropin-releasing factor modulates the immune response to stress in the rat, *Endocrinology*, 128, 1329, 1991.
28. Angioni, S., Petraglia, F., Gallinelli, A., Cossarizza, A., Franceschi, C., Muscettola, M., Genazzani, A. D., Surico, N., and Genazzani, A. R., Corticotropin-releasing hormone modulates cytokines release in cultured human peripheral blood mononuclear cells, *Life. Sci.*, 53, 1735, 1993.
29. Lipton, J. M., Neuropeptide α-melanocyte-stimulating hormone in control of fever, the acute phase response, and inflammation, *Neuroimmune Networks: Physiology and Diseases*, John Wiley & Sons, New York, 1989, 243.
30. Smith, E. M., Hughes, T. K., Hashemi, F., and Stefano, G. B., Immunosuppressive effects of corticotropin and melanotropin and their possible significance in human immunodeficiency virus infection, *Proc. Natl. Acad. Sci. U.S.A.*, 89, 782, 1992.
31. Guarcello, V., Weigent, D. A., and Blalock, J. E., Growth hormone releasing hormone receptors on thymocytes and splenocytes from rats, *Cell. Immunol.*, 136, 291, 1991.
32. Blalock, J. E. and Costa, O., Immune neuroendocrine interactions: implications for reproductive physiology, *Ann. NY Acad. Sci*, 564, 261, 1989.
33. Stanisz, A. M., Befus, D., and Bienenstock, J., Differential effects of vasoactive intestinal peptide, substance P and somatostatin on immunoglobulin synthesis and proliferation by lymphocytes from Peyer's patches, mesenteric lymph nodes and spleen, *J. Immunol.*, 136, 152, 1986.
34. Muscettola, M. and Grasso, G., Somatostatin and vasoactive intestinal peptide reduce interferon-γ production by human peripheral blood mononuclear cells, *Immunobiology*, 180, 419, 1990.
35. Ader, R., Historical perspectives on psychoneuroimmunology, *Psychoneuroimmunology, Stress, and Infection*, Friedman, H., Klein, T. W., and Friedman, A. L., Eds., CRC Press, Boca Raton, FL, 1996, 1.

36. Goetzl, E. J. and Sreedharan, S. P., Mediators of communication and adaptation in the neuroendocrine and immune systems, *FASEB J.*, 6, 2646, 1992.
37. Felten, D. L., Felten, S. Y., Carlson, S. L., Olschowka, J. A., and Livnat, S., Noradrenergic and peptidergic innervation of lymphoid tissue, *J. Immunol.*, 135, 755s, 1985.
38. Ader, R., Felten, D. L., and Cohen, N., *Psychoneuroimmunology,* Academic Press, New York, 1991,
39. Hiramoto, R., Rogers, C., Demissie, S., Hsueh, C.-M., Hiramoto, N., Lorden, J., and Ghanta, V., The use of conditioning to probe for CNS pathways that regulate fever and NK cell activity, *Int. J. Neurosci.*, 84, 229, 1996.
40. Blalock, J. E., The immune system: Our sixth sense, *Immunologist*, 2, 8, 1994.
41. Besedovsky, H., Sorkin, E., Keller, M., and Muller, J., Changes in blood hormone levels during the immune response, *Proc. Soc. Exp. Biol. Med*, 150, 466, 1975.
42. McCann, S. M., Karanth, S., Kamat, A., Dees, W. L., Lyson, K., Gimeno, M., and Rettori, V., Induction of cytokines of the pattern of pituitary hormone secretion in infection, *Neuroimmunomodulation*, 1, 2, 1994.
43. Woloski, B. M., Smith, E. M., Meyer, W. J., III, Fuller, G. M., and Blalock, J. E., Corticotropin-releasing activity of monokines, *Science*, 230, 1035, 1985.
44. Sapolsky, R., Rivier, C., Yamamoto, G., Plotsky, P., and Vale, W., Interleukin-1 stimulates the secretion of hypothalamic corticotropin-releasing factor, *Science*, 238, 522, 1987.
45. Rivest, S., Molecular mechanisms and neural pathways mediating the influence of interleukin-1 on the activity of neuroendocrine CRF motoneurons in the rat, *Int. J. Devl. Neurosci.*, 13, 135, 1995.
46. Kakucska, I., Qi, Y., Clark, B. D., and Lechan, R. M., Endotoxin-induced corticotropin-releasing hormone gene expression is the hypothalamic paraventricular nucleus is mediated centrally by interleukin-1, *Endocrinology*, 133, 815, 1993.
47. DeSouza, E. B., Corticotropin-releasing factor and interleukin-1 receptors in the brain-endocrine-immune axis, *Ann. NY Acad. Sci.*, 697, 9, 1993.
48. Webster, E. L., Tracey, D. E., and De Souza, E. B., Upregulation of interleukin-1 receptors in mouse AtT-20 pituitary tumor cells following treatment with corticotropin-releasing factor, *Endocrinology*, 129, 2796, 1991.
49. Payne, L. C., Weigent, D. A., and Blalock, J. E., Induction of pituitary sensitivity to interleukin-1: a new function for corticotropin-releasing hormone, *Biochem. Biophy. Res. Commun.*, 198, 480, 1994.
50. Ruzicka, B. and Huda, A. K. L., Differential cellular regulation of POMC by IL-1 and corticotrophin-releasing hormone, *Neuroendocrinology*, 61, 136, 1995.
51. Kalra, P. S., Sahu, A., and Kalra, S. P., Interleukin-1 inhibits the ovarian steroid-induced luteinizing hormone surge and release of hypothalamic luteinizing hormone-releasing hormone in rats, *Endocrinology*, 126, 2145, 1990.
52. Dubuis, J. M., Dayer, J. M., Siegrist-Kaiser, C. A., and Burger, A. G., Human recombinant interleukin-1 β decreases plasma thyroid hormone and thyroid stimulating hormone levels in rats, *Endocrinology*, 123, 2175, 1988.
53. Karanth, S. and McCann, S. M., Anterior pituitary hormone control by interleukin 2, *Proc. Natl. Acad. Sci. U.S.A.*, 88, 2961, 1991.
54. Blalock, J. E., Production of peptide hormones and neurotransmitters by the immune system, *Chem. Immunol.*, 52, 1, 1992.
55. Arzt, E., Stelzer, G., Renner, U., Lange, M., Muller, A., and Stalla, G. K., Interleukin-2 and interleukin-2 receptor in human corticotrophic adenoma and murine pituitary cell cultures, *J. Clin. Invest.*, 90, 1944, 1992.

56. Karanth, S., Aguila, M. C., and McCann, S. M., The influence of interleukin-2 on the release of somatostatin and growth hormone-releasing hormone by mediobasal hypothalamus, *Neuroendocrinology*, 58, 185, 1993.
57. Spangelo, B. L. and MacLeod, R. M., Regulation of acute phase response and neuroendocrine function by interleukin 6, *Prog. Neuro. Endocrin. Immunol.*, 3, 167, 1990.
58. Bernton, E. W., Beach, J. E., Holaday, J. W., Smallridge, R. C., and Fein, H. G., Release of multiple hormones by a direct action of interleukin-1 on pituitary cells, *Science*, 238, 519, 1987.
59. Spangelo, B. L., Judd, A. M., Isakson, P. C., and MacLeod, R. M., Interleukin-6 stimulates anterior pituitary hormone release *in vitro*, *Endocrinology*, 125, 575, 1989.
60. Lyson, K. and McCann, S. M., Involvement of arachidonic acid cascade pathways in interleukin-6-stimulated corticotropin-releasing factor release *in vitro*, *Neuroendocrinology*, 55, 708, 1992.
61. Naitoh, Y., Fukata, J., Tominaga, T., Nakai, Y., Tamai, S., Mori, K., and Imura, H., Interleukin-6 stimulates the secretion of adrenocorticotropic hormone in conscious, freely-moving rats, *Biochem. Biophys. Res. Commun.*, 155, 1459, 1988.
62. Gonzalez, M. C., Aguila, M. C., and McCann, S. M., *In vitro* effects of recombinant human γ-interferon on growth hormone release, *Prog. Neuro. Endocrin. Immunol.*, 4, 222, 1991.
63. Gonzalez, M. C., Riedel, M., Rettori, V., Yu, W. H., and McCann, S. M., Effect of recombinant human γ-interferon on the release of anterior pituitary hormones, *Prog. Neuro. Endocrin. Immunol.*, 3, 49, 1990.
64. Vankelecom, H., Carmeliet, P., Heremans, H., Van Damme, J., Dijkmans, R., Billiau, A., and Denef, C., Interferon-γ inhibits stimulated adrenocorticotropin, prolactin, and growth hormone secretion in normal rat anterior pituitary cell cultures, *Endocrinology*, 126, 2919, 1990.
65. Kelley, K. W., Arkins, S., and Li, Y. M., Growth hormone, prolactin, and insulin-like growth factors: new jobs for old players. *Brain. Behav. Immun.*, 6, 317, 1992.
66. Milenkovic, L., Rettori, V., Snyder, G. E., Reutler, B., and McCann, S. M., Cachectin alters anterior pituitary hormone release by a direct action *in vitro*, *Proc. Natl. Acad. Sci. U.S.A.*, 86, 2418, 1989.
67. Gaillard, R. C., Turnill, D., Sappino, P., and Muller, A. F., Tumor necrosis factor α inhibits the hormonal response of the pituitary gland to hypothalamic releasing factors, *Endocrinology*, 127, 101, 1990.
68. Bernardini, R., Kammilaris, T. C., Calogero, A. E., Johnson, E. O., Gomez, M. T., Gold, P. W., and Chrousos, G. P., Interactions between tumor necrosis factor-α, hypothalamic corticotropin-releasing hormone, and adrenocorticotropin secretion in the rat, *Endocrinology*, 126, 2876, 1990.
69. Bluthe, R. M., Pawloski, M., Suarez, S., Parnet, P., Pittman, Q., Kelley, K. W., and Dantzer, R., Synergy between tumor necrosis factor α and interleukin-1 in the induction of sickness behavior in mice, *Psychoneuroendocrinology*, 19, 197, 1994.
70. Schobitz, B., Pezeshki, G., Pohl, T., Hemmann, U., Heinrich, P. C., Holsboer, F., and Reul, J. M. H. M., Soluble interleukin-6 (IL-6) receptor augments central effects of IL-6 *in vivo*, *FASEB J.*, 9, 659, 1995.

13 The Thymus-Pituitary Axis: A Paradigm to Study Immunoneuroendocrine Connectivity in Normal and Stress Conditions

Wilson Savino and Eduardo Arzt

CONTENTS

13.1 The Interdependence of the Neuroendocrine and
      Immune Systems 187
13.2 Cytokines and Thymic Hormone Modulate Hypothalamic-Pituitary
      Functions 188
13.3 Hypothalamic-Pituitary Control of Thymus Physiology 191
13.4 Thymus-Pituitary Similarities for Cytokine and Hormone Production ... 193
13.5 How Two Systems Can Be Endocrinally and Paracrinally Controlled
      by Similar Molecules 194
13.6 Thymus-Pituitary Connectivity Under Distinct Stress Situations 196
13.7 Conclusions 197
Acknowledgments 197
References 197

13.1 THE INTERDEPENDENCE OF THE NEUROENDOCRINE AND IMMUNE SYSTEMS

The crosstalk between the neuroendocrine and immune systems is now largely demonstrated. These systems use similar ligands and receptors to establish a physiological intra and intersystem communication circuitry which plays a relevant role in homeostasis.[1-3] Accordingly, not only classical hormones can be produced by cells of the immune system, but also a variety of cytokines (originally described as being produced by monocytes and lymphocytes) are synthesized and released by a variety of endocrine glands and nervous tissues. Moreover, specific receptors for such distinct molecular families can be detected in both immune and neuroendocrine systems. In the present review, we shall focus the discussion on the thymus-pituitary axis, which can be regarded as a paradigm to analyze the connectivity between these two systems in both normal and stress situations.

13.2 CYTOKINES AND THYMIC HORMONES MODULATE HYPOTHALAMIC-PITUITARY FUNCTIONS

Cytokines are now recognized as playing important roles in modulating the neuroendocrine system, particularly the hypothalamic-pituitary-adrenal (HPA) axis, and in this respect interleukin-1 (IL-1) is a potent regulatory molecule. In spite of some discrepancies in the literature, it seems to act at hypothalamic, pituitary, and adrenal levels, inducing corticotropin-releasing factor (CRH), adrenocorticotropin (ACTH), and cortisol production, respectively.[4-6] When the influence on hormone secretion was compared, IL-1α was less potent than IL-1β.[7-8] Specifically at the pituitary level, however, the effects of IL-1 are controversial. Growth hormone (GH), prolactin (PRL), luteinizing hormone (LH), and follicle stimulating hormone (FSH) secretion of rat pituitary cells have been reported not to be influenced by IL-1β,[9] whereas others have shown a stimulation of GH, LH, and thyroid stimulating hormone (TSH) secretion,[10] together with an inhibition of PRL release.[11,12] In rat pituitary cell cultures, ACTH secretion was stimulated[7,9,10,12-14] or not influenced[15-16] by IL-1β, whereas secretion of this pituitary hormone is enhanced by IL-1β in AtT-20 cells[17-20] and in human corticotroph pituitary adenoma cell cultures.[17] Lastly, IL-1 further increases ACTH secretion by corticotropes previously exposed to CRH.[21]

Importantly, other inflammatory cytokines such as IL-6 and tumor necrosis factor-α (TNF-α) share to a certain extent the effects of IL-1, acting at the three levels of the HPA axis, stimulating CRH production by the hypothalamus, but displaying particular stimulatory patterns.[5-7,22] Concerning the pituitary, IL-6 stimulates the release of ACTH, PRL, GH, LH, and FSH from normal rat anterior pituitary cells *in vitro*[22-24] and stimulates ACTH secretion from AtT20 cells.[18] In the same model, IL-6 inhibits stimulation of adenylate cyclase by vasoactive intestinal peptide (VIP) and suppresses the enhancement in inositol phosphate and intracellular calcium induced by thyrotropin releasing hormone (TRH).[25] In addition, IL-6 modulates proliferation of pituitary cells; it inhibits normal rat pituitary cell growth[26,27] and stimulates growth of GH_3 and MtT/E rat pituitary tumor cell lines.[26,28] TNF-α stimulates ACTH, GH, TSH and PRL secretion of rat pituitary cells *in vitro*.[29-30] In contrast, TNF-α also was found to inhibit the release of ACTH and other hormones in response to hypothalamic factors, by acting directly on pituitary cells.[31]

Classical T-cell-derived cytokines, such as IL-2 and interferon-γ also act on the HPA axis. When administered to human cancer patients, IL-2 increases β-endorphin, ACTH and cortisol levels.[32,33] Rat IL-2 enhances ACTH secretion *in vivo*,[34] and induces corticosterone production by a direct action on rat adrenocortical cells[35] as well as CRH release from hypothalamic neurons.[36]

In normal rat pituitaries, IL-2 has been shown to stimulate proopiomelanocortin mRNA expression[37,38] as well as ACTH, PRL, and TSH secretion, but to inhibit LH,

TABLE 13.1
Cytokine Effects Upon Pituitary Hormone Production[a]

Cytokine	ACTH	PRL	GH	TSH	FSH
IL-1	NO or →	NO or →	NO or ↑	↑	NO
IL-2	↑	↑	→	↑	→
IL-6	↑	↑	↑	ND	↑
IFN-γ	→	↑ or →	→	ND	ND
TNF-α	↑ or →	↑	↑	↑	ND

[a] Increase (↑), decrease (→), or no effect (NO) on hormone production. ND; not determined.

FSH, and GH release.[39-42] The release pattern of pituitary hormones after IL-2 stimulation mimics almost completely the alterations in pituitary hormone secretion in response to stress.[39] The IL-2-induced PRL secretion is blocked by dopamine.[41] IL-2 also enhances ACTH secretion in AtT20 cells[40] and stimulates PRL release from GH_3 cells;[43] in the latter case antiestrogens inhibit IL-2-induced PRL secretion.

Interferon-γ also stimulates PRL release in rat anterior pituitary cell cultures, probably acting via the release of IL-6 from folliculostellate cells.[44] On the other hand, it was shown that these cells mediate an inhibitory action of interferon-γ on PRL secretion, as well as on ACTH and GH release.[45] The various cytokine effects upon hypothalamic/pituitary functions are summarized in Table 13.1.

In addition to cytokines, it is noteworthy that thymic hormones modulate the production of classic hypothalamic-pituitary hormones and neuropeptides. Initial experiments on this subject revealed that neonatal thymectomy promotes a decrease in the numbers of secretory granules in acidophilic cells of the adenopituitary.[46] In the same vein, athymic nude mice exhibit significantly low levels of various pituitary hormones including PRL, GH, as well as the gonadotropics LH and FSH.[47] In respect to thymic peptides, it was shown that thymosin-β4, when perfused intraventricularly, stimulates LH and its hypothalamic releasing hormone LH-RH.[46] Additionally, another thymosin component, the MB-35 peptide, enhances PRL and GH production.[48] Interestingly, *in vivo* studies in children, showed that administration of thymopoietin (a further chemically defined thymic hormone) increases GH and cortisol serum levels. Moreover, thymopentin (the synthetic biologically active peptide of thymopoietin) enhances *in vitro* the production of proopiomelanocortin derivatives such as ACTH, β-endorphin, and β-lipotropin.[49] Interestingly, thymosin-α1 apparently downregulates TSH, ACTH, and PRL secretion *in vivo,* with no effects on GH levels.[50] Lastly, thymulin exhibits an *in vitro* stimulatory effect upon perfused rat pituitaries, enhancing GH, PRL, and to a lesser extent TSH and LH release.[51]

Together, these findings point to a complex circuitry involving the role of distinct thymic peptides upon the hypothalamus-pituitary axis, as schematically depicted in Figure 13.1.

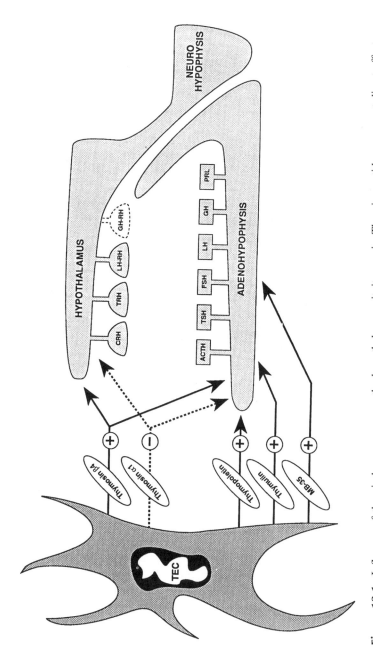

Figure 13.1 Influence of thymic hormones upon the hypothalamus-pituitary axis. Thymic peptides can exert direct effects on both hypophyseal and hypothalamic cells, with hormonal release being positively or negatively modulated depending on the thymic peptide.

13.3 HYPOTHALAMIC-PITUITARY CONTROL OF THYMUS PHYSIOLOGY

Studies on the neuroendocrine control of the immune function are comprised in the distinct levels of the immune system organization, including primary and secondary lymphoid organs, as well as sites of effector immunological activities. The thymus gland is a primary lymphoid organ that has been studied extensively in the last few years. Within this compartment, bone-marrow-derived T-cell precursors undergo a complex process of maturation that includes selection of the T-cell repertoire, with positively selected cells eventually migrating to the T-dependent areas of peripheral lymphoid organs where they will further expand.[52,53] Intrathymic T-cell differentiation is driven by interactions with the *thymic microenvironment,* a tridimensional network composed of various cell types including epithelial cells, dendritic cells, and macrophages, as well as extracellular matrix elements.[54-55] The thymic microenvironment controls thymocyte migration and differentiation through distinct ways — secretion of a variety of polypeptides, such as thymic hormones and cytokines;[51] cell-cell contacts and interactions occurring through classical adhesion molecules;[56] and expression of peptide-bound major histocompatibility complex (MHC) class I and class II proteins by cells that interact with the T-cell receptor within the context of CD8 or CD4 molecules. Lastly, microenvironmental cells bind to and interact with maturing thymocytes via extracellular matrix (ECM) ligands and receptors.[55] In fact, the intrathymic ECM network may function as a substrate onto which thymocytes migrate in a ordered fashion.[57]

One biological activity of thymic epithelial cells (TEC) which is under neuroendocrine control is the secretion of thymic hormones. Data from different laboratories have definitely demonstrated that secretion of thymulin, a zinc-containing nonapeptide strictly produced by TEC,[58,59] is modulated by various pituitary hormones. We showed that both *in vivo* and *in vitro* treatment with PRL upregulates thymulin secretion,[60] an effect that can be obtained even in aging mice that normally present low levels of the circulating thymic hormone.[61] In the same vein, patients with prolactinomas exhibited high thymulin circulating levels, as compared to normal age-matched control individuals.[62] Conversely, administration of bromocriptine (an agonist of the dopamine receptor, largely used as an inhibitor of PRL synthesis by adenopituitary cells) decreases thymulin serum levels in mice, an effect which is reserved after exogenous administration of PRL.[60]

Thymulin secretion also is enhanced by GH in various mammalian species including mice, rats, dogs, and humans.[63] In GH-related pituitary hyperfunction such as acromegalism, abnormally high levels of circulating thymulin are detected which decrease upon appropriate therapy of the patients.[64] Conversely, GH deficiency in children was accompanied by low thymulin levels, whereas GH treatment consistently restored this thymic endocrine function.[65] In keeping with these observations, treatment of murine and human TEC cultures with GH enhances thymulin contents in the culture supernatants.[66] Lastly, exogenous GH enhances thymulin production in old animals.[68]

It should be pointed out that the control of thymulin secretion by GH appears to be mediated by insulin-like growth factor 1 (IGF-1), since GH-induced enhancing of thymulin production *in vitro* can be prevented when TEC cultures are subjected to antibodies specific for IGF-1 or IGF-1 receptor.[64] Additionally, IGF-1 alone stimulates thymulin production by cultured TEC. Moreover, there is a clear-cut positive correlation between serum levels of thymulin and IGF-1 in acromegalic patients.[64]

An important concept regarding the pituitary control TEC physiology is the pleiotropic nature of the effects. For example, PRL upregulates the expression of high molecular weight cytokeratins by medullary TEC.[60] Epithelial growth is also increased *in vitro* following PRL and GH treatments.[60,64] Furthermore, extracellular matrix ligands and receptors were shown to be enhanced by these pituitary hormones.[65]

One further relevant aspect of intrathymic T-cell differentiation concerns direct cell-cell interactions between thymocytes and thymic microenvironmental cells. We recently demonstrated that adhesion of thymocytes to cultured TEC can be enhanced by treating the latter cell type with PRL, GH, or IGF-1.[65-66] Again, the effects of GH in this system could be abrogated by anti-IGF-1 or anti-IGF-1 receptor antibodies. Additionally, we observed that these pituitary hormones enhance thymocyte release by cultured thymic nurse cells, a lymphoepithelial complex that partially supports thymocyte differentiation.[67]

Since pituitary hormones affect functions of microenvironmental cells related to thymocyte differentiation, it is apparent that the latter process also is under neuroendocrine control. However, besides the indirect influences mediated by the thymic microenvironment, direct effects have been reported. For example, synthetic TRH enhances bromodeoxyuridine uptake by thymic cell suspensions,[68] an effect apparently shared by PRL likely being mediated by an enhancement of IL-2 production and IL-2 receptor expression.[69] In a second vein, GH was shown to be comitogenic for thymocyte proliferation.[70,71]

A series of *in vivo* experiments also evidenced that important changes in thymocyte differentiation occur under neuroendocrine influence. It was shown that GH injections in aging mice increased total thymocyte numbers and the percentage of CD3-bearing cells,[72] in keeping with our data showing an enhanced concanavalin-A mitogenic response as well as Il-6 production by thymocytes from GH-treated animals.[73] Interestingly, similar findings were observed in animals treated with IGF-1.[74] Additionally, IGF-1 was able to induce a repopulation of the atrophic thymus from diabetic rats.[75] Moreover, mouse substrains, selected for bearing high or low IGF-1 circulating levels, exhibit differential thymus developmental patterns that positively correlated with IGF-1 levels.[76]

The role of GH on thymus development also is stressed by the findings obtained with GH-deficient dwarf mice. In these animals, besides the precocious decline in thymulin serum values,[77] there is a progressive thymic hypoplasia with decreased numbers of CD4/8 double-positive thymocytes. Such defects were largely restored by long-term treatment with GH.[78,79]

13.4 THYMUS-PITUITARY SIMILARITIES FOR CYTOKINE AND HORMONE PRODUCTION

The thymus-pituitary crosstalk can be observed by the common production of several soluble mediators. Accordingly, the pituitary gland is a site of origin of cytokines also produced intrathymically. IL-1β immunoreactive material and the respective mRNA were found in rat pituitaries, increasing after *in vivo* treatment with bacterial lipopolysaccharide.[80] In keeping with these findings, IL-1β mRNA expression was detected by RT-PCR in a series of pituitary adenomas cultured *in vitro*.[81]

Production of IL-6 by cells from rat anterior pituitary glands and expression of corresponding mRNA was also demonstrated,[82-84] being stimulated, among other cytokines, by IL-1.[85-86] Concerning the human anterior pituitary, IL-6 mRNA expression has been detected in corticotrophic adenoma cell cultures[87] and IL-6 was found to be secreted from 7 out of 10 pituitary tumors cultured *in vitro*.[88] By immunocytochemistry, the presence of IL-6 was demonstrated in almost all pituitary adenomas tested.[89]

The expression of IL-2 and its receptor (IL-2R) by pituitary cells of different species also was described.[90] In the mouse AtT-20 pituitary tumor cell line, we detected IL-2 mRNA expression after stimulation with CRH or phorbol myristate acetate. In human corticotrophic adenoma cells, basal IL-2 mRNA expression as well as IL-2 secretion were further stimulated by PMA.[91]

Other cytokines also are expressed in the pituitary gland. TNF-α gene expression has been demonstrated by RT-PCR in pituitary adenoma tissue and culture.[92] Leukemia inhibitory factor (LIF) has been shown to be secreted by bovine pituitary follicular cells in culture.[93] LIF protein and the respective mRNA as well as LIF-binding sites have been demonstrated in the developing human fetal pituitary and in normal and adenomatous adult human tissue.[94] Macrophage-migration inhibitory factor (MIF), which plays a central role in the response to endotoxemia, is also expressed in the pituitary and this expression increases after LPS treatment.[95]

Intrathymically, the production of typical pituitary hormones has been demonstrated by several research groups. In this respect, it is noteworthy that specific immunoreactivity for typical pituitary hormones such as PRL, GH, TSH, ACTH, FSH, and LH, as well as oxytocin and vasopressin (together with respective neurophysins), were detected in thymic cells (reviewed in Reference 96). Moreover, at least regarding GH, PRL, LH, oxytocin, vasopressin and somatostatin, specific messenger RNAs were also detected in the thymus.[97-102] Additionally, typical hypothalamic releasing hormones (Table 13.2), namely CRH and LH-RH, as well as neuropeptides such as β-endorphin, VIP (vasoactive intestinal peptide), and substance y were evidenced in the organ.[103-106]

It should be pointed out that, similar to what is found in the pituitary gland, there is a certain degree of cell-type specificity regarding the production of different hormones. For example, oxytocin and vasopressin appear to be exclusively produced by TEC,[107] whereas PRL expression is apparently restricted to thymocytes. Differently, however, GH can be produced by both thymocytes and epithelial cells (Mello-Coelho, et al., manuscript in preparation).

TABLE 13.2
Intrathymic Production of Pituitary and Hypothalamic Hormones, and Expression of Respective Receptors[a]

Hormone[c]	Hormone Production[b]		Receptor Expression[b]	
	Thymocyte	TEC	Thymocyte	TEC
GH	+	+	+	+
PRL	+	−	+	+
ACTH	+	+	+	+
Oxytocin	−	+	+	+
Vasopressin	−	+	+	+
CRH	+	+	+	+
LH-RH	+	+	+	ND
Corticosterone	ND	+	+	+

[a] Expression determined by techniques including ligand binding, immunocytochemistry, peptide sequencing, immunoblotting, Northern blotting, and/or polymerase chain reaction.
[b] Positive: +; negative: −; not determined: ND.
[c] GH: growth hormone; PRL: prolactin; ACTH: corticotropin; CRH: corticotropin releasing hormone; LH-RH: gonadotropin releasing hormone.

Taken together, the data discussed above clearly shows that the thymus and pituitary share common secretory products, as depicted in Figure 13.2.

13.5 HOW TWO SYSTEMS CAN BE ENDOCRINALLY AND PARACRINALLY CONTROLLED BY SIMILAR MOLECULES

Cytokines have been shown not only to act as lymphocyte-derived messengers but also constitute autocrine and paracrine factors in the regulation of pituitary hormone secretion and pituitary cell growth. Similarly, hormones such as PRL and GH not only act as endocrine messengers from the pituitary gland, but also appear to represent autocrine and paracrine factors involved in the general regulation of the thymus. The molecular basis for these actions is provided by two complementary lines of evidence: (1) as discussed above, various cytokines and pituitary hormones are produced in both organs, and (2) receptors for these molecules are also expressed in distinct cell types in the pituitary and in the thymus. For example, IL-1 receptors and respective mRNA were characterized in mouse pituitary cells and AtT-20 corticotrophs.[108-109] Furthermore, IL-6 receptors are expressed in rat anterior pituitary cells.[110]

The Thymus-Pituitary Axis

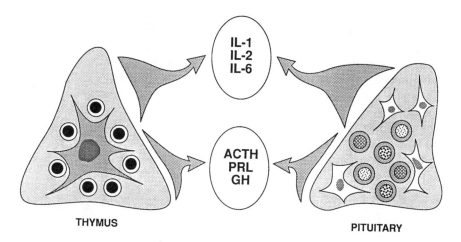

Figure 13.2 Thymus-pituitary similarities for cytokine and hormone production. To illustrate this concept, three classical cytokines (IL-1, IL-2, and IL-6) and three classical pituitary hormones (ACTH, PRL, and GH) were chosen. Both organs are able to produce all these peptides, which may act via endocrine or paracrine/autocrine pathways.

We also found detectable amounts of IL-2 receptor mRNA and expression of the receptor on cell membranes in both adenoma and AtT-20 cells.[111] This receptor was detected in human pituitary adenomas in culture using RT-PCR,[112] and a related protein has been detected in AtT-20 cells.[40] The IL-2 receptor α chain is expressed in cultures of normal rat pituitary cells including ACTH, GH, and PRL-producing cells.[27] The presence of the receptor in these cells is in agreement with previous studies showing the action of IL-2 on the secretion of these hormones.[39,41]

Intrathymic expression of pituitary hormone receptors also have been demonstrated by distinct approaches. We showed PRL receptors to be present in human and murine thymic epithelial cells by means of immunocytochemistry, immunoblotting, and Northern blot.[113] The PRL receptor gene was also seen by PCR in thymocyte-derived extracts. In this regard, we further demonstrated by tricolor cytofluorometry that PRL receptors are expressed in the various CD4-CD8 defined thymocyte subsets in both mice and humans.[114-115] Interestingly, mitogenic stimulation by concanavalin-A resulted in an enhancement of PRL receptor expression in thymocytes, as ascertained by flow cytometry.[116]

In addition to the PRL receptor, GH binding sites were initially detected in murine and human TEC,[117] and then confirmed by immunocytochemistry, *in situ* hybridization, and PCR techniques.[118-119] Very recently, we noticed that only CD4⁻CD8⁻ immature human thymocytes express GH receptor, suggesting that this differentiation stage is the main target for GH action (Mello-Coelho et al., manuscript in preparation).

The data discussed above should now be placed in the following context. The three criteria used to establish an autocrine or paracrine role for a substance are followed by cytokines at the level of the anterior pituitary and by hormones at the thymus level: (1) the cytokine (or the hormone) that has an effect that modulates some aspect of pituitary physiology also modulates some aspect of thymus physiology; (2) the cytokine is produced in the anterior pituitary or the hormone is produced at the thymus; and (3) receptors for the cytokine or the hormone are expressed in the pituitary or the thymus, respectively. For example, IL-1, IL-2, and IL-6 as well as their respective receptors are expressed in the pituitary, and are able to influence the growth and function of pituitary cells. Reciprocally, GH and PRL and their receptors are expressed in the thymus and these hormones regulate the physiology of this gland. By consequence, all these molecules fulfill the criteria for autocrine or paracrine regulators of pituitary and thymus functions.

13.6 THYMUS-PITUITARY CONNECTIVITY UNDER DISTINCT STRESS SITUATIONS

It is well recognized and described in other chapters that the HPA axis is an essential component of the stress response. CRH acts as a key mediator of the stress response and stress has been shown to be almost always associated with elevated levels of circulating glucocorticoids.[120] Moreover, the HPA axis is essential in coupling the stress response to the immune system. CRH, ACTH, and glucocorticoids affect lymphocyte activation and other immune functions and also regulate the expression and synthesis of cytokines (extensively reviewed in Reference 5). PRL, which is also elevated during stress, is essential for the T-cell response, acting as a comitogen of IL-2.[121]

Having proved that the anterior pituitary and the thymus share common molecules, several questions arise concerning how their connectivity can be influenced by stress.

During the stress response some pituitary hormones like ACTH and PRL are elevated. In what way does this elevation in the anterior pituitary influence the action and expression of the hormones in the thymus? Reciprocally, during stress and as a consequence of glucocorticoid elevation, cytokines produced by lymphocytes are inhibited. How does this inhibition influence the action and expression of cytokines in the pituitary? In view of the multiple levels of connectivity between these two glands we anticipate that a coordinated response should occur during the adaptation of the immune-neuroendocrine physiology to stress.

Acute infectious diseases can be relevant examples to study this issue. In particular, we have previously shown that in acute experimental Chagas' disease (a parasitic infection caused by the flagellate protozoan *Trypanosoma cruzi*) mice exhibit a progressive increase in corticosterone levels.[122] Moreover, a cortical thymocyte depletion occurs, in the context of an intrathymic infection by the parasite.[123] Interestingly, however, thymocyte depletion was still seen in adrenalectomized infected animals.[122] In this respect, one important point is to evaluate the recently reported intrathymic production of corticosterone in mice[124,125] in order to determine if a putative ACTH-thymic corticosterone axis could be engaged.

Another question open to investigation concerns the hypothetical *in vivo T. cruzi* infection of the hypothalamic-pituitary axis, which may generate abnormal patterns of hormone secretion with consequences in the thymus. Also, it will be relevant to define whether or not the hypothalamus and hypophysis are targets for an autoimmune process in Chagas' disease, similar to what has been demonstrated for thymocytes and thymic epithelial cells.[123,126]

13.7 CONCLUSIONS

In conclusion, a better understanding of the thymus-pituitary axis in distinct stress situations, particularly during infectious diseases, will be necessary for designing neuroendocrine-based therapy in such affections. However, much effort has to be done before such procedures become clinical routines.

ACKNOWLEDGMENTS

This work was partially funded with grants from CNPq (Brazil) and the University of Buenos Aires (Argentina). The authors are indebted to Martine Netter for the computer drawings.

REFERENCES

1. Blalock, J. E., The syntax of immune-neuroendocrine communication, *Immunol. Today,* 15, 504, 1994.
2. Savino, W. and Dardenne, M., Immunoneuroendocrine interactions, *Immunol. Today,* 7, 318, 1995.
3. Dardenne, M. and Savino, W., Interdependence of the endocrine and immune systems, *Adv. Neuroimmunol.,* 6, 297, 1996.
4. Besedovsky, H. and Del Rey, A., Immune-Neuroendocrine circuits: integrative role of cytokines, *Front. Neuroendocrinol.,* 13, 61, 1992.
5. Bateman, A., Singh, A., Kral, T., and Solomon, S., The immune-hypothalamic-pituitary-adrenal axis, *Endocrine Rev.,* 10, 92, 1989.
6. Hermus, A. R. M. M. and Sweep, C. G. J., Cytokines and the hypothalamic-pituitary-adrenal axis, *J. Steroid Biochem. Mol. Biol.,* 37, 867, 1990.
7. Rivier, C., Vale, W., and Brown, M., In the rat, interleukin-1α and -β stimulate adrenocorticotropin and catecholamin release, *Endocrinology,* 125, 3102, 1989.
8. Matta, S. G., Linne, K. M., and Sharp, B. M., Interleukin-1α and interleukin-1β stimulate adrenocorticotropin secretion in the rat through a similar hypothalamic receptor(s): effects of interleukin1 receptor antagonist protein, *Neuroendocrinology,* 57, 14, 1993.
9. Uehara, A., Gillis, S., and Arimura, A., Effects of interleukin-1 on hormone release from normal rat pituitary cells in primary culture, *Neuroendocrinology,* 45, 43, 1987.
10. Bernton, E. W., Beach, J. E., Holaday, J. W., Smallridge, R. C., and Fein, H. G., Release of multiple hormones by a direct action of interleukin-1 on pituitary cells, *Science,* 238, 519, 1987.
11. Florio, T., Meucci, O., Landolfi, E., Grimaldi, M., Ventra, C., Scorziello, A., et al., Interleukin 1 modulation of anterior pituitary function: effect on hormone release and second messenger systems, *Pharmacol. Res.,* 21 (Suppl. 1), 35, 1989.

12. Kehrer, P., Turnill, D., Dayer, J. M., Muller, A. F., and Gaillard, R. C., Human recombinant interleukin-β and α, but not recombinant tumor necrosis factor α stimulate ACTH release from rat anterior pituitary cells *in vitro* in a prostaglandin E2 cAMP independent manner, *Neuroendocrinology,* 48, 160, 1988.
13. Cambronero, J. C., Rivas, F. J., Borrel, J., and Guaza, C., Interleukin-1β induces pituitary adrenocorticotropin secretion: evidence for glucocorticoid modulation, *Neuroendocrinology,* 55, 648, 1992.
14. Beach, J. E., Smallridge, R. C., Kinzer, C. A., Bernton, E. W., Holaday, J. W., and Fein, H. G., Rapid release of multiple hormones from rat pituitaries perfused with recombinant interleukin-1, *Life Sci.,* 144, 1, 1989.
15. Parsadaniantz, S. M., Lenoir, V., Terlain, B., and Kerdelhue, B., Lack of effect of interleukin-1α and β, during in vitro perfusion, on anterior pituitary release of adrenocorticotropin hormone and β-endorphin in the male rat, *J. Neurosci. Res.,* 34, 315, 1993.
16. Renner, U., Newton, C. J., Pagotto, U., Sauer, J., Artz, E., and Stalla, G. K., Involvement of interleukin-1 and interleukin-1 receptor antagonist in rat pituitary cell growth regulation, *Endocrinology,* 136, 3186, 1995.
17. Malarkey, W. B. and Zvara, B. J., Interleukin-1β and other cytokines stimulate adrenocorticotropin release from cultured pituitary cells of patients with Cushing's disease. *J. Clin. Endocrinol. Metab.,* 69, 196, 1989.
18. Fukata, J., Usui, T., Naitoh, Y., Nakai, Y., and Imura, H., Effects of recombinant human interleukin-1α, -1β, 2 and 6 on ACTH synthesis and release in the mouse pituitary tumor cell line AtT-20, *J. Endocrinol.,* 122, 33, 1989.
19. Woloski, B. M. R. N. J., Smith, E. M., Meyer, W. J., Fuller, G. M., and Blalock, J. E., Corticotropin-releasing activity of monokines, *Science,* 230, 1035, 1985.
20. Gwosdow, A. R., Spencer, J. A., O'Connell, N. A., and Abou-Samra, A. B., Interleukin-1 activates protein kinase A and stimulates adrenocorticotropin hormone release from AtT20 cells, *Endocrinology,* 132, 710, 1993.
21. Payne, L. C., Weigent, D. A., and Blalock, J. E., Induction of pituitary sensitivity to interleukin-1: a new function for corticotropin-releasing hormone, *Biochem. Biophys. Res. Commun.,* 198, 480, 1994.
22. Spangelo, B. L., Judd, A. M., Isakson, P. C., and MacLeod, R. M., Interleukin-6 stimulates anterior pituitary hormone release *in vitro, Endocrinology,* 125, 575, 1989.
23. Lyson, K. and McCann, S. M., The effect of interleukin-6 on pituitary hormone release in vivo and *in vitro, Neuroendocrinology,* 54, 262, 1991.
24. Yamaguchi, M., Matsuzaki, N., Hirota, K., Miyake, A., and Tanizawa, O., Interleukin 6 possibly induced by interleukin-1β in the pituitary gland stimulates the release of gonadotropins and prolactin, *Acta Endocrinol.,* 122, 201, 1990.
25. Grimaldi, M., Meucci, O., Scorziello, A., Florio, T., Ventra, C., and De Mercato, R., Interleukin-6 modulation of second messenger systems in anterior pituitary cells, *Life Sci.,* 51, 1243, 1992.
26. Arzt, E., Buric, R., Stelzer, G., Stalla, J., Sauer, J., Renner, U., and Stalla, G. K., Interleukin involvement in anterior pituitary cell growth regulation: effects of interleukin-2 (IL-2) and IL-6, *Endocrinology,* 132, 459, 1993.
27. Arzt, E., Sauer, J., Buric, R., Stalla, J., Renner, U., and Stalla, G. K., Characterization of Interleukin-2 (IL-2) receptor expression and action of IL-2 and IL-6 on normal anterior pituitary cell growth, *Endocrine,* 3, 113, 1995.
28. Sawada, T., Koike, K., Kanda, Y., Ikegami, H., Jikihara, T., and Maeda, T., Interleukin-6 stimulates cell proliferation of rat anterior pituitary clonal cell lines *in vitro, J. Endocrinol. Invest.,* 18, 83, 1995.

29. Milenkovic, L., Rettori, V., Snyder, G. D., Beutler, B., and McCann, S. M., Cachectin alters anterior pituitary hormone release by a direct action *in vitro*, *Proc. Natl. Acad. Sci. U.S.A.*, 86, 2418, 1989.
30. Koike, K., Hirota, K., Ohmichi, M., Kadowaki, K., Ikegami, H., and Yamaguchi, M., Tumor necrosis factor-α increases release of arachidonate and prolactin from rat anterior pituitary cells, *Endocrinology*, 128, 2791, 1991.
31. Gaillard, R. C., Turnill, D., Sappino, P., and Muller, A. F., Tumor necrosis factor a inhibits the hormonal response of the pituitary gland to hypothalamic releasing factors, *Endocrinology*, 127, 101, 1990.
32. Lotze, M. T., Frana, L. W., Sharrow, S. O., Robb, R. J., and Rosenberg, S. A., *In vivo* administration of purified human interleukin 2. I. Half-life and immunologic effects of the Jurkat cell line-derived IL-2, *J. Immunol.*, 134, 166, 1985.
33. Denicoff, K. D., Durkin, T. M., Lotze, M. T., Quinlan, P. E., Davis, C. L., Listwak, S. J., Rosenberg, S. A., and Rubinow, D. R., The neuroendocrine effects of interleukin-2 treatment, *J. Clin. Endocrinol. Metab.*, 69, 402, 1989.
34. Naito, Y., Fukata, J., Tominaga, T., Masui, Y., Hirai, Y., Murakami, N., Tamai, S., Mori, K., and Imura, H., Adrenocorticotropic hormone-releasing activities of interleukins in a homologous *in vivo* system, *Biochem. Biophys. Res. Commun.*, 164, 1262, 1989.
35. Tominaga, T., Fukata, J., Naito, Y., Usui, T., Murakami, N., Fukushima, M., Nakai, Y., Hirai, Y., and Imura, H., Prostaglandin-dependent *in vitro* stimulation of adrenocortical steroidogenesis by interleukins, *Endocrinology*, 128, 526, 1991.
36. Cambronero, J. C., Rivas, F. J., Borrell, J., and Guaza, C., Interleukin-2 induces corticotropin-releasing hormone release from superfused rat hypothalami: influence of glucocorticoids, *Endocrinology*, 131, 677, 1992.
37. Brown, S. L., Smith, L. R., and Blalock, J. E., Interleukin-1 and Interleukin-2 enhance proopiomelanocortin gene expression in pituitary cells, *J. Immunol.*, 139, 3181, 1987.
38. Harbuz, M. S., Stephanou, A., Knight, R. A., Chover-Gonzalez, A. J., and Lightman, S. L., Action of interleukin-2 and interleukin-4 on CRF mRNA in the hypothalamus and POMC mRNA in the anterior pituitary, *Brain Behav. Immun.*, 6, 214, 1992.
39. Karanth, S. and McCann, S. M., Anterior pituitary hormone control by interleukin 2, *Proc. Natl. Acad. Sci. U.S.A.*, 88, 2961, 1991.
40. Smith, L. R., Brown, S. L., and Blalock, J. E., Interleukin-2 induction of ACTH secretion: presence of an interleukin-2 receptor a-chain-like molecule on pituitary cells, *J. Neuroimmunol.*, 21, 249, 1989.
41. Karanth, S., The influence of dopamine (DA) on interleukin-2 induced release of prolactin (PRL), luteinizing hormone (LH) and follicle stimulating hormone (FSH) by the anterior pituitary, 73rd Annu. Meet. Endocrine Society, Washington, D.C., 1991, 210 (Abstr.).
42. Karanth, S. and McCann, S. M., Influence of dopamine on the altered release of prolactin, luteinizing hormone and follicle stimulating hormone induced by interleukin-2 *in vitro*, *Neuroendocrinology*, 56, 871, 1992.
43. Newton, C. J., Arzt, E., and Stalla, G. K., Involvement of the estrogen receptor in the growth response of pituitary cells to interleukin-2, *Biochem. Biophys. Res. Commun.*, 205, 1930, 1994.
44. Yamaguchi, M., Koike, K., Matsuzaki, N., Yoshimoto, Y., Taniguchi, T., and Miyake, A., The interferon family stimulates the secretions of prolactin and interleukin-6 by the pituitary gland *in vitro*, *J. Endocrinol. Invest.*, 14, 457, 1991.
45. Vankelecom, H., Andries, M., Billiau, A., and Denef, C., Evidence that folliculostellate cells mediate the inhibitory effect of interferon-γ on hormone secretion in rat anterior pituitary cell cultures, *Endocrinology*, 130, 3537, 1992.

46. Goya, R.G., Sosa, Y. E., Console, G. M., and Dardenne, M., Altered thyrotropic and somatotropic responses to environmental challenges in congenitally athymic mice, *Brain Behav. Immun.,* 9, 79, 1995.
47. Daneva, T., Spinedi, E., Hadid, R., and Gaillard, R., Impaired hypothalamo-pituitary-adrenal axis function in Swiss nude athymic mice, *Neuroendocrinology,* 62, 79, 1995.
48. Badamchian, M., Spangelo, B. L., Damavandy, T., MacLeod, R. M., and Goldstein, A. L., Complete amino acid sequence of a peptide isolated from the thymus that enhances release of growth hormone and prolactin, *Endocrinology,* 128, 1580, 1991.
49. Malaise, M. G., Hazee-Hagelstein, M. T., Reuter, A. M., Vrinds-Gevaert, Y., Goldstein, G., and Franchimont, P., Thymopoietin and thymopentin enhance the levels of ACTH, beta-endorphin and beta-lipotropin from rat pituitary cells *in vitro, Acta Endocrinol.,* 115, 455, 1987.
50. Milenkovic, L. and McCann, S. M., Effects of thymosin alpha-1 on pituitary hormone release, *Neuroendocrinology,* 55, 14, 1992.
51. Goya, R. G., Sosa, Y. E., Brown, O. A., and Dardenne, M., *In vitro* studies on the thymus-pituitary axis in young and old rats, *Ann. NY Acad. Sci.,* 741, 108, 1994.
52. van Ewijk, W., T-cell differentiation is influenced by thymic microenvironments, *Annu. Rev. Immunol.,* 9, 591, 1991.
53. Anderson, G., Moore, N. C., Owen, J. J. T., and Jenkinson, E. J., Cellular interactions in thymocyte development, *Annu. Rev. Immunol.,* 14, 73, 1996.
54. Boyd, R. L., Tucek, C. L., Godfrey, D. I., Izon, D. J., Wilson, T. J., Davidson, N. J., Bean, A. G. D., Ladyman, H. M., Ritter, M. A., and Hugo, P., The thymic microenvironment, *Immunol. Today,* 14, 445, 1993.
55. Savino, W., Villa-Verde, D. M. S., and Lannes-Vieira, J., Extracellular matrix proteins in intrathymic T cell migration and differentiation, *Immunol. Today,* 14, 158, 1993.
56. Patel, D. D. and Haynes, B. F., Cell adhesion molecules involved in intrathymic T cell development, *Semin. Immunol.,* 5, 283, 1993.
57. Savino, W., Dardenne, M., and Carnaud, C., Conveyor belt model for intrathymic cell migration, *Immunol. Today,* 27, 97, 1996.
58. Savino, W., Dardenne, M., Papiernik, M., and Bach, J. F., Thymic hormone containing cells. Characterization and localization of serum thymic factor in young mouse thymus studied by monoclonal antibodies, *J. Exp. Med.,* 156, 628, 1982.
59. Dardenne, M., Savino, W., Berrih, S., and Bach, J. F., Evidence for a zinc-dependent epitope on the molecule of thymulin, a thymic hormone, *Proc. Natl. Acad. Sci. U.S.A.,* 82, 7035, 1985.
60. Dardenne, M., Savino, W., Gagnerault, M. C., Itoh, T., and Bach, J. F., Neuroendocrine control of thymic hormonal production. I. Prolactin stimulates *in vivo* and *in vitro* the production of thymulin by human and murine thymic epithelial cells, *Endocrinology,* 125, 3, 1989.
61. Savino, W., Dardenne, M., and Bach, J. F., Thymic hormone containing cells. II. Evolution of cells containing the serum thymic factor (FTS or thymulin) in normal and autoimmune mice, as revealed by anti-FTS monoclonal antibodies. Relationship with Ia-bearing cells, *Clin. Exp. Immunol.,* 52, 1, 1983.
62. Timsit, J., Safieh, B., Gagnerault, M. C., Savino, W., Lobetzki, J., Bach, J. F., and Dardenne, M., Augmentation des taux circulants de thymuline au cours de l'hyperprolactinemie et de l'acromegalie, *C. R. Acad. Sci. Paris,* 310 (Serie III), 7, 1989.
63. Mello-Coelho, V., Savino, W., Postel-Vinay, M. C., and Dardenne, M., Role of prolactin and growth hormone on thymus physiology, *Dev. Immunol.,* in press, 1997.

64. Timsit, J., Savino, W., Safieh, B., Chanson, P., Gagnerault, M. C., Bach, J. F., and Dardenne, M., Effects of growth hormone and insulin-like growth factor 1 in thymic hormonal function in man, *J. Clin. Endocrinol. Metab.,* 75, 183, 1992.
65. Mello-Coelho, V., Villa-Verde, D. M. S., Dardenne, M., and Savino, W., Pituitary hormones modulate cell-cell interactions between thymocytes and thymic epithelial cells, *J. Neuroimmunol.,* 76, 39, 1997.
66. Savino, W., Mello-Coelho, V., and Dardenne, M., Control of the thymic microenvironment by growth hormone/IGF-1-mediated circuits, *NeuroImmuno-Modulation,* 2, 313, 1995.
67. Villa Verde, D. M. S., Mello-Coelho, V., Lagrota-Cândido, J. M., and Savino, W., The thymic nurse cell complex: an *in vitro* model for extracellular matrix-mediated intrathymic T cell migration, *Braz. J. Med. Biol. Res.,* 28, 907, 1995.
68. Pawlikowski, M., Zerek-Melen, G., and Winczyk, K., Thyroliberin (TRH) increases thymus cell proliferation in rats, *Neuropeptides,* 23, 199, 1992.
69. Viselli, S. M., Stanek, E. M., Mukherjee, P., Hymer, W. C., and Mastro, A. M., Prolactin-induced mitogenesis of lymphocytes from ovariectomized rats, *Endocrinology,* 129, 983, 1991.
70. Sabharwal, P. and Varma, S., Growth hormone synthesized and secreted by human thymocytes acts via insulin-like growth factor I as an autocrine and paracrine growth factor, *J. Clin. Endocrinol. Metab.,* 81, 2663, 1996.
71. Postel-Vinay, M. C., Mello-Coelho, V., Gagnerault, M. C., and Dardenne, M., Growth hormone stimulates the proliferation of activated mouse T lymphocytes, *Endocrinology,* 138, 1816, 1997.
72. Li, Y. M., Brunke, D. L., Dantzer, R., and Kelley, K. W., Pituitary epithelial cell implants reverse the accumulation of $CD4^-CD8^-$ lymphocytes in thymus glands of aged rats, *Endocrinology,* 130, 2703, 1992.
73. Goya, R. G., Gagnerault, M. C., Leite-de-Moraes, M. C., Savino, W., and Dardenne, M., *In vivo* effects of growth hormone on thymus function in aging mice, *Brain Behav. Immun.,* 6, 341, 1992.
74. Clarck, R., Strasser, J., McCabe, S., Robbins, K., and Jardieu, P., Insulin-like growth factor-I stimulation of lymphopoiesis, *J. Clin. Invest.,* 92, 540, 1993.
75. Binz, K., Joller, P., Froesch, P., Binz, H., Zapf, J., and Froesch, E. R., Repopulation of atrophied thymus in diabetic rats by insulin-like growth factor-I, *Proc. Natl. Acad. Sci. U.S.A.,* 87, 3690, 1990.
76. Siddiqui, R. A., McCutcheon, S. N., Blair, H. T., Mackenzie, D. D., Morel, P. C., Breier, B. H., and Gluckman, P. D., Growth allometry of organs, muscles and bones in mice from lines divergently selected on the basis of plasma insulin-like growth factor-I, *Growth Dev. Aging,* 56, 531, 1992.
77. Pelletier, M., Montplaisir, S., Dardenne, M., and Bach, J. F., Thymic hormone activity and spontaneous autoimmunity in dwarf mice and their littermates, *Immunology,* 30, 783, 1976.
78. Murphy, W. J., Durum, S. K., and Longo, D. L., Role of neuroendocrine hormones in murine T cell development. Growth hormone exerts thymopoietic effects *in vivo, J. Immunol.,* 149, 3851, 1992.
79. Knyszynski, A., Adler-Kunin, S., and Globerson, A., Effects of growth hormone on thymocyte development from progenitor cells in the bone marrow, *Brain Behav. Immun.,* 6, 327, 1992.

80. Koenig, J. I., Snow, K., Clark, B. D., Toni, R., Cannon, J. G., Shaw, A. R., Dinarello, C. A., Reichlin, S., Lee, S. L., and Lechan, R. M., Intrinsic pituitary IL-1β is induced by bacterial LPS, *Endocrinology,* 126, 3053, 1990.
81. Schneider, J. H., Hofman, F. M., Weiss, M. H., and Hinton, D. R., Cytokine expression in pituitary adenomas, *Int. Congr. Pituitary Adenomas,* Marina del Rey, CA, 1993, MP-10 (Abstr.).
82. Vankelecom, H., Carmeliet, P., Van Damme, J., Billiau, A., and Denef, C., Production of IL-6 by folliculo-stellate cells of the anterior pituitary gland in a histiotypic cell aggregate culture system, *Neuroendocrinology,* 49, 102, 1989.
83. Spangelo, B. L., MacLeod, R. M., and Isakson, P. C., Production of interleukin-6 by anterior pituitary cells *in vitro, Endocrinology,* 126, 582, 1990.
84. Spangelo, B. L., Judd, A. M., MacLeod, R. M., Goodman, D. W., and Isakson, P. C., Endotoxin-induced release of interleukin-6 from rat medial basal hypothalami, *Endocrinology,* 127, 1779, 1990.
85. Spangelo, B. L., Judd, A. M., Isakson, P. C., and MacLeod, R. M., Interleukin-1 stimulates interleukin-6 release from rat anterior pituitary cells *in vitro, Endocrinology,* 128, 2685, 1991.
86. Yamaguchi, M., Matsuzaki, N., Hirota, K., Miyake, A., and Tanizawa, O., Interleukin 6 possibly induced by interleukin 1β in the pituitary gland stimulates the release of gonadotropins and prolactin, *Acta Endocrinol. (Copenhdgen),* 122, 201, 1990.
87. Velkeniers, B., D'Haens, G., Smets, G., Vergani, P., Vanhaelst, L., and Hooghe-Peters, E. L., Expression of IL-6 mRNA in corticotroph cell adenomas, *J. Endocrinol. Invest.,* 14 (Suppl. 1), 31, 1991.
88. Jones, T. H., Justice, S., Price, A., and Chapman, K., Interleukin-6 secreting human pituitary adenomas *in vitro, J. Clin. Endocrinol. Metab.,* 73, 207, 1991.
89. Tsagarakis, S., Kontogeorgeos, G., and Giannou, P., Interleukin-6, a growth promoting cytokine, is present in human pituitary adenomas: an immunocytochemical study, *Clin. Endocrinol.,* 37, 163, 1992.
90. Arzt, E., Stelzer, G., Renner, U., Lange, M., Müller, O. A., and Stalla, G. K., Interleukin-2 and IL-2 receptor expression in human corticotrophic adenoma and murine pituitary cell cultures, *J. Clin. Invest.,* 90, 1944, 1992.
91. Arzt, E., Buric, R., Stelzer, G., Stalla, J., Sauer, J., Renner, U., and Stalla, G. K., Interleukin involvement in anterior pituitary cell growth regulation: effects of IL-2 and IL-6, *Endocrinology,* 132, 459, 1993.
92. Todd, V. L., Atkin, S. L., Speirs, V., and White, M. C., PCR expression of cytokines in anterior pituitary adenomas, *J. Endocrinol.,* (Suppl. 144), P285, 1995.
93. Ferrara, N., Winer, J., and Henzel, W. J., Pituitary follicular cells secrete an inhibitor of aortic endothelial cell growth: identification as leukemia inhibitory factor, *Proc. Natl. Acad. Sci. U.S.A.,* 89, 698, 1992.
94. Akita, S., Webster, J., Ren, S.-G., Takino, H., Said, J., Zand, O., and Melmed, S., Human and pituitary expression of leukemia inhibitory factor. Novel intrapituitary regulation of adrenocorticotropin hormone synthesis and secretion, *J. Clin. Invest.,* 95, 1288, 1995.
95. Bernhagen, J., Calandra, T., Mitchell, R. A., Martin, S. B., Tracey, K. L., Voelter, W., Manogue, K. R., Cerami, A., and Bucala, R., MIF is a pituitary-derived cytokine that potentiates lethal endotoxaemia, *Nature,* 365, 756, 1993.
96. Dardenne, M. and Savino, W., Neuroendocrine control of thymus physiology by peptidic hormones and neuropeptides, *Immunol. Today,* 15, 518, 1994.

97. de Leeuw, F. E., Jansen, G. H., Batanero, E., van Wichen, D. F., Huber, J., and Schuurman, H. J., The neural and neuro-endocrine component of the human thymus. I. Nerve-like structures, *Brain Behav. Immun.*, 6, 234, 1992.
98. Montgomery, D. W., Shen, G. K., Ulrich, E. D., Steiner, L. L., Parrish, P. R., and Zukoski, C. F., Human thymocytes express a prolactin-like messenger ribonucleic acid and synthesize bioactive prolactin-like proteins, *Endocrinology*, 131, 3019, 1992.
99. Wu, H., Devi, R., and Malarkey, W. B., Expression and localization of prolactin messenger ribonucleic acid in the human immune system, *Endocrinology*, 137, 349, 1996.
100. Maggiano, N., Piantelli, M., Ricci, R., Larocca, L. M., Capelli, A., and Ranelletti, F. O., Detection of growth hormone-producing cells in human thymus by immunohistochemistry and non-radioactive *in situ* hybridization, *J. Histochem. Cytochem.*, 42, 1349, 1994.
101. Sabharwal, P. and Varma, S., Growth hormone synthesized and secreted by human thymocytes acts via insulin-like growth factor I as an autocrine and paracrine growth factor, *J. Clin. Endocrinol. Metab.*, 81, 2663, 1996.
102. Robert, F., Geenen, V., Schoenen, J., Burgeon, E., De Groote, D., Defresne, M. P., Legros, J. J., and Franchimont, P., Colocalization of immunoreactive oxytocin, vasopressin and interleukin-1 in human thymic epithelial neuroendocrine cells, *Brain Behav. Immun.*, 5, 102, 1991.
103. Jessop, D. S., Renshaw, D., Lightman, S. L., and Harbuz, M. S., Changes in ACTH and β-endorphin immunoreactivity in immune tissues during a chronic inflammatory stress are not correlated with changes in corticotropin-releasing hormone and arginine vasopressin, *J. Neuroimmunol.*, 60, 29, 1995.
104. Gomariz, R. P., Lorenzo, M. J., Cacicedo, L., Vicente, A., and Zapata, A. G., Demonstration of immunoreactive vasoactive intestinal peptide (IR-VIP) and somatostatin (IR-SOM) in rat thymus, *Brain Behav. Immun.*, 4, 151, 1990.
105. al-Shawaf, A. A., Kendall, M. D., and Cowen, T., Identification of neural profiles containing vasoactive intestinal polypeptide, acetylcholinesterase and catecholamines in the rat thymus, *J. Anat.*, 174, 131, 1991.
106. Jessop, D., Biswas, S., D'Souza, L., Chowdrey, H., and Lightman, S., Neuropeptide Y immunoreactivity in the spleen and thymus of normal rats and following adjuvant-induced arthritis, *Neuropeptides*, 23, 203, 1992.
107. Moll, U. M., Lane, B. L., Robert, F., Geenen, V., and Legros, J. J., The neuroendocrine thymus. Abundant occurrence of oxytocin-, vasopressin-, and neurophysin-like peptides in epithelial cells, *Histochemistry*, 89, 385, 1988.
108. De Souza, E. B., Webster, E. L., Grigoriadis, D. E., and Tracey, D. E., Corticotropin-releasing factor (CRF) and interleukin-1 (IL-1) receptors in the brain-pituitary-immune axis, *Psychopharmacol. Bull.*, 25, 299, 1989.
109. Bristulf, J., Simoncsits, A., and Bartfai, T., Characterization of a neuronal interleukin-1 receptor and the corresponding mRNA in the mouse anterior pituitary cell line AtT-20. *Neurosci. Lett.*, 128, 176, 1991.
110. Ohmichi, M., Hirota, K., Koike, K., Kurachi, H., Ohtsuka, S., Matsuzaki, N. M., Miyake, A., and Tanizawa, O., Binding sites for interleukin-6 in the anterior pituitary gland, *Neuroendocrinology*, 55, 199, 1992.
111. Arzt, E., Steizer, G., Renner, U., Lange, M., Muller, O. A., and Stalla, G. K., Interleukin-2 and interleukin-2 receptor expression in human corticotrophic adenoma and murine pituitary cell cultures, *J. Clin. Invest.*, 90, 1944, 1992.

112. Schneider, J. H., Hofman, F. M., Weiss, M. H., and Hinton, D. R., Cytokine expression in pituitary adenomas, 3rd Int. Pituitary Congr., 1993, MP10.
113. Arzt, E., Sauer, J., Buric, R., Stalla, J., Renner, U., and Stalla, G. K., Characterization of interleukin-2 (IL-2) receptor expression and action of IL-2 and IL-6 on normal anterior pituitary cell growth, *Endocrine,* 3, 113, 1995.
114. Dardenne, M., Kelly, P. A., Bach, J. F., and Savino, W., Identification and functional activity of prolactin receptors in thymic epithelial cells, *Proc. Natl. Acad. Sci. U.S.A.,* 88, 9700, 1991.
115. Gagnerault, M. C., Touraine, P., Savino, W., Kelly, P. A., and Dardenne, M., Expression of prolactin receptors on murine lymphoid cells in normal and autoimmune conditions, *J. Immunol.,* 151, 1, 1993.
116. Dardenne, M., Leite-de-Moraes, M. C., Kelly, P. A., and Gagnerault, M. C., Prolactin receptors expression in human hematopoietic tissues analysed by flow cytofluorometry, *Endocrinology,* 134, 2108, 1993.
117. Ban, E., Gagnerault, M. C., Jammes, H., Postel-Vinay, M. C., Haour, F., and Dardenne, M., Specific binding sites for growth hormone in cultured mouse thymic epithelial cells, *Life Sci.,* 48, 2141, 1991.
118. Gagnerault, M. C., Postel-Vinay, M. C., and Dardenne, M., Expression of growth hormone receptors in murine lymphoid cells analyzed by flow cytofluorometry, *Endocrinology,* 137, 1719, 1996.
119. Mertani, H. C., Delehaye-Zervas, M. C., Martini, J. F., Postel-Vinay, M. C., and Morel, G., Localization of growth hormone receptor messenger RNA in human tissues, *Endocrine,* 3, 135, 1995.
120. Reisine, T., Affolter, H.-U., Rougon, G., and Barbet, J., New insights into molecular mechanisms of stress, *Trends Neurosci.,* 9, 574, 1986.
121. Clevenger, C. V., Sillman, A. L., Hanley-Hyde, J., and Prystowsky, M. B., Requirement for prolactin during cell cycle regulated gene expression in cloned T-lymphocytes, *Endocrinology,* 130, 3216, 1992.
122. Leite-de-Moraes, M. C., Hontebeyrie-Joskowicz, M., Leboulanger, F., Savino, W., Dardenne, M., and Lepault, F., Studies on the thymus in Chagas' disease. II. Thymocyte subset fluctuations in *Trypanosoma cruzi*-infected mice: relationship to stress, *Scand. J. Immunol.,* 33, 267, 1991.
123. Savino, W., Leite-de-Moraes, M. C., Hontebeyrie-Joskowicz, M., and Dardenne, M., Studies on the thymus in Chagas' disease. I. Changes in the thymic microenvironment in mice acutely infected with *Trypanosoma cruzi, Eur. J. Immunol.,* 19, 1727, 1989.
124. Vacchio, M. S., Papadopoulos, V., and Ashwell, J. D., Steroid production in the thymus: implications for thymocyte selection, *J. Exp. Med.,* 179, 1835, 1994.
125. Vacchio, M. S. and Ashwell, J. D., Thymus-derived glucocorticoids regulate antigen-specific positive selection, *J. Exp. Med.,* 185, 2033, 1997.
126. Savino, W., Silva, J. S., Silva-Barbosa, S. D., Dardenne, M., and Ribeiro-dos-Santos, R., Anti-thymic cell autoantibodies in human and murine chronic Chagas' disease. *EOS J. Immunol. Immunopharmacol.,* 10, 204, 1990.

14 New Insights into the Hypothalamic Control of FSH and LH by Cytokines and Nitric Oxide

S. M. McCann, M. Kimura, A. Walczewska,
S. Karanth, V. Rettori, and W. H. Yu

CONTENTS

Abstract 206
14.1 Hypothalamic Control of Gonadotropin Secretion 206
    14.1.1 Role of Nitric Oxide (NO) in Control of LHRH Release 209
    14.1.2 Effect of Cytokines (IL-1 and GMCSF) on NOergic
        Control of LHRH Release 212
    14.1.3 Role of NO in Mating Behavior 212
    14.1.4 Effect of NO in the Release of Other Hypothalamic Peptides 212
    14.1.5 Action of NO to Control Release of Anterior Pituitary
        Hormones 213
14.2 Potential Role of Leptin in Reproduction 213
    14.2.1 Effect of Leptin on LH Release 214
    14.2.2 Effect of Leptin on FSH Release 214
    14.2.3 Effect of Leptin on Prolactin Release 215
    14.2.4 Effect of Leptin on LHRH Release 215
    14.2.5 Effect of Intraventricularly Injected Leptin on Plasma
        Gonadotropin Concentrations in Ovariectomized,
        Estrogen-Primed Rats 215
    14.2.6 Mechanism of Action of Leptin on the Hypothalamic-Pituitary
        Axis 215
Acknowledgments 217
References 217

Key Words: nitric oxide synthase, cyclooxygenase, guanylate cyclase, norepinephrine, glutamic acid, oxytocin

ABSTRACT

Gonadotropin secretion by the pituitary gland is under the control of luteinizing hormone-releasing hormone (LHRH) and the putative follicle-stimulating hormone-releasing factor (FSHRF). Lamprey III LHRH is a potent FSHRF in the rat and appears to be resident in the FSH controlling area of the rat hypothalamus. It is an analog of mammalian LHRH and may be the long-sought FSHRF. Gonadal steroids feed back at the hypothalamic and pituitary levels to either inhibit or stimulate the release of LH and FSH, which is also affected by inhibin and activin secreted by the gonads. Important control is exercised by acetylcholine, norepinephrine (NE), dopamine, serotonin, melatonin, and glutamic acid (GA). Furthermore, LH and FSH also act at the hypothalamic level to alter secretion of gonadotropins. More recently, growth factors have been shown to have an important role. Many peptides act to inhibit or increase release of LH and the sign of their action is often reversed by estrogen. A number of cytokines act at the hypothalamic level to suppress acutely the release of LH but not FSH. NE, GA, and oxytocin stimulate LHRH release by activation of neural nitric oxide synthase (nNOS). The pathway is as follows: oxytocin and/or GA activate NE neurons in the medial basal hypothalamus (MBH) that activate NOergic neurons by α_1 (α_1) receptors. The NO released diffuses into LHRH terminals and induces LHRH release by activation of guanylate cyclase (GC) and cyclooxygenase. NO not only controls release of LHRH bound for the pituitary, but also that which induces mating by actions in the brain stem. An exciting recent development has been the discovery of the adipocyte hormone, leptin, a cytokine related to tumor necrosis factor α (TNFα) In the male rat, leptin exhibits a high potency to stimulate FSH and LH release from hemipituitaries incubated *in vitro*, and increases the release of LHRH from MBH explants. LHRH and leptin release LH by activation of NOS in the gonadotropes. The NO released activates GC that releases cyclic GMP which induces LH release. Leptin induces LH release in conscious, ovariectomized estrogen-primed female rats, presumably by stimulating LHRH release. At the effective dose of estrogen to activate LH release, FSH release is inhibited. Leptin may play an important role in induction of puberty and control of LHRH release in the adult as well.

14.1 HYPOTHALAMIC CONTROL OF GONADOTROPIN SECRETION

The control of gonadotropin secretion is extremely complex, as is revealed by the research of the past 35 years since the discovery of luteinizing hormone-releasing hormone (LHRH)[1], now commonly called gonadotropin-releasing hormone (GnRH).[2] This was the second of the hypothalamic-releasing hormones to be characterized, and it was then shown to have an effect on follicle-stimulating hormone (FSH) release, albeit smaller than that on LH release. For this reason, it was renamed GnRH.[2,3] Overwhelming evidence indicates that there must be a separate FSH-releasing factor (FSHRF) since pulsatile release of LH and FSH can be dissociated. In the castrated male rat, roughly half of the FSH pulses occur in the absence of LH pulses and only a small fraction of the pulses of both gonadotropins are coincident. LHRH antisera or antagonists can suppress release of LH without altering FSH release.[4] LH but not FSH pulses can be suppressed by alcohol,[5] δ-9-tetrahydrocannabinol, and cytokines such as interleukin-1 α (IL-1α).[6] In addition, a number of peptides inhibit LH but not FSH release and a few stimulate FSH without affecting LH.[4,7]

The hypothalamic areas controlling LH and FSH are separable. Stimulation in the dorsal anterior hypothalamic area can cause selective FSH release, whereas lesions in this area selectively suppress the pulses of FSH and not LH.[8] Contrariwise, stimulations or lesions in the medial preoptic region can augment or suppress LH release without affecting FSH release. Electrical stimulation in the preoptic region releases only LH, whereas lesions in this area inhibit LH release without inhibiting FSH release. The medial preoptic area contains the perikarya of the LHRH neurons. The axons of these neurons project from the preoptic region to the anterior and mid-portions of the median eminence. Extracts of the anterior mid-median eminence contain LH-releasing activity commensurate with the content of immunoassayable LHRH, whereas extracts of the caudal median eminence and organum vasculosum lamina terminalis contain more FSH-releasing activity than can be accounted for by the content of LHRH.[4]

Lesions confined to the rostral and mid-median eminence can selectively inhibit pulsatile LH release without altering FSH pulsations, whereas lesions which destroy the caudal and mid-median eminence can selectively block FSH pulses in castrated male rats.[4,9,10] Therefore, it appears that the putative FSHRF is produced in neurons with perikarya in the dorsal anterior hypothalamic area, with axons which project to the mid and caudal median eminence to control FSH release selectively.

We, followed by several other groups, reported FSH-releasing activity in the stalk-median eminence. The activity was purified and separated from the LH-releasing activity in 1965 as measured by *in vivo* bioassays.[11] Later, using radioimmunoassay for identification of FSH and LH release, it was reported that these activities could not be separated; however, it has now been clearly shown that the FSH-releasing activity can be separated from bioactive and radioimmunoassayable LHRH by gel filtration through Sephadex G-25 on the same column used in the earlier research. FSH- and LH-releasing activity were assayed by the increase in plasma FSH and LH, respectively, in ovariectomized, estrogen-progesterone-blocked rats. The separation was confirmed by radioimmunoassay of the LHRH in the fractions.[11] The separation of the two activities was also demonstrable by assay of FSH and LH released from hemipituitaries incubated *in vitro*.[12] In both assay systems, FSHRF emerged from the column just before elution of LHRH.

The final isolation of FSHRF has not yet been achieved, but in the meantime an analog of LHRH was shown to release FSH selectively over a 50-fold dose range and the gonadotropin-releasing hormone-associated peptide (GAP) was also shown to have slightly selective FSH-releasing activity *in vivo* in ovariectomized, estrogen-progesterone-blocked rats. However, GAP_{1-13} had selective FSH-releasing activity and this was augmented by blocking peptidic digestion by inserting D-trp in position 9 of the molecule. In this connection, rat (r) D-trp-^{9} GAP_{1-13} was more potent than human (h) D-trp-^{9} GAP_{1-13} on assay in ovariectomized estrogen-progesterone-blocked rats. In this preparation, r D-trp-^{9} GAP_{1-13} produced a sustained increase in plasma FSH without altering LH.[13] Furthermore, r GAP_{1-13} was also active on assay *in vitro* using hemipituitaries. However, the selectivity was less than *in vivo*.[14]

In the search for FSHRF, we first believed that it might be an analogue of LHRH and we had many such analogues synthesized and tested for the forms of LHRH that were known to exist in lower species. We had not tested lamprey (l)-III LHRH, but when we realized that antiserum which cross-reacted with l-III and l-I LHRH immunostained neural fibers in the arcuate nucleus proceeding to the median eminence of human brain, it occurred to us that l-III LHRH could be the FSHRF since l-I LHRH had little activity to release either LH of FSH. Indeed, (l)-III LHRH was a potent FSH-releasing factor with little or no LH-releasing activity both *in vitro* when incubated with hemipituitaries and *in vivo* when injected into ovariectomized, estrogen-progesterone-blocked rats. The lowest dose tested in that preparation (10 pmol) produced a highly significant increase in plasma FSH with no rise in LH. Preliminary immunocytochemistry indicates that this peptide is present in the rat hypothalamus in the region which had previously been shown to control FSH release. Therefore, it is either FSHRF or a very closely related peptide.[12]

In addition to control by LH and the putative FSHRH, FSH secretion from the pituitary gland is also under inhibitory control by inhibin and stimulatory control by activin secreted from the gonads.[2,3] The generally accepted form of inhibin is a 32-Kda α,β dimer which acts on the pituitary gland with a fairly long latency (over 1 h *in vivo*) to suppress FSH and not LH release.[2,3] Another form of inhibin, α inhibin-92, was isolated and its structure determined by Li and co-workers.[16] It has a clear action *in vivo* to suppress FSH release in a dose-related manner within 10 min without affecting LH. This inhibin was isolated from human seminal fluid but has been found in the gonads and may well be very important in the moment-to-moment control of FSH release.[15] It is equally active on a molar basis with the 32-KDa inhibin and its fragments are also active, but much less so, with the major activity in the mid-portion of the molecule.[17] α Inhibin-92 may be of practical significance since it can be made by standard peptide synthetic methods, whereas synthesis of the 32-KDa inhibin requires molecular genetics techniques. Because of a complex carbohydrate in the molecule, it has been very difficult to obtain in useful quantities.

FSH and LH themselves have intrahypothalamic actions to alter gonadotropin secretion which are probably of physiologic significance.[18] Growth factors are also involved.[19] Gonadal steroids play an important role in controlling LHRH release and pituitary responsiveness to the peptide.[2-4] In the male, the influence is inhibitory by androgens at both the hypothalamic and pituitary level, whereas in females there is a biphasic effect of estrogen to first suppress the release of LHRH and the pituitary responsiveness to it, and then, after a delay, to stimulate the release of LHRH and to augment the responsiveness of the pituitary to the peptide. Furthermore, there is a self-priming action of LHRH to further augment the responsiveness of the gonadotropes to the peptide when the gland is under the influence of estrogen.[2,3,7]

The pulse frequency and amplitude of pulses of LH are also altered by gonadal steroids and this plays an important role in the control of the menstrual cycle and in the induction of puberty.[2,3,4] In turn, pulsatile release of LH and FSH is under the control of a host of classical transmitters and peptides as further discussed below.

14.1.1 ROLE OF NITRIC OXIDE (NO) IN CONTROL OF LHRH RELEASE

NO is formed in the body by NO synthase (NOS), an enzyme which converts arginine in the presence of oxygen and several cofactors into equimolar quantities of citrulline and NO. There are three isoforms of the enzyme. One of these, neural (n) NOS, is found in the cerebellum and various regions of the cerebral cortex and also in various ganglion cells of the autonomic nervous system. Large numbers of nNOS-containing neurons, termed NOergic neurons, also were found in the hypothalamus, particularly in the paraventricular and supraoptic nuclei, with axons projecting to the median eminence and neural lobe which contains large amounts of nNOS. These findings indicated that the enzyme is synthesized at all levels of the neuron from perikaryon to axon terminals.[20]

Because of this distribution in the hypothalamus in regions which contain peptidergic neurons that control pituitary hormone secretion, we decided to determine the role of this soluble gas in hypothalamic-pituitary function. The approach was to use sodium nitroprusside (NP) that spontaneously liberates NO to see if this altered the release of various hypothalamic transmitters. Hemoglobin, which scavenges NO by a reaction with the heme group on the molecule, and inhibitors of NOS, such as N^G-monomethyl-L-arginine (NMMA), a competitive inhibitor of NOS, were used to determine the effects of decreased NO. Two types of studies were performed. In the first set of experiments, medial basal hypothalamic (MBH) explants were preincubated *in vitro* and then exposed to neurotransmitters which modify the release of the various hypothalamic peptides in the presence or absence of inhibitors of the release of NO. The response to NO itself, provided by sodium NP, was also evaluated. Anterior pituitaries were incubated similarly *in vitro* and the effect of these compounds that increase or decrease the release of NO into the tissue on the release of pituitary hormones was examined.

In order to determine if the results *in vitro* also held *in vivo*, substances were microinjected into the third ventricle (3V) of the brain of conscious, freely moving animals to determine the effect on pituitary hormone release.[21]

Our most extensive studies were carried out with regard to the release of LHRH. Not only does LHRH act after its secretion into the hypophyseal-portal vessels to stimulate LH and to a lesser extent FSH release, but it also induces mating behavior in female rats and penile erection in male rats by hypothalamic action.

Our experiments showed that release of NO from sodium NP *in vitro* promoted LHRH release and that the action was blocked by hemoglobin, a scavenger of NO. NP also caused an increased release of prostaglandin E_2 (PGE_2) from the tissue, which previous experiments showed played an important role in release of LHRH. Furthermore, it caused the biosynthesis and release of prostanoids from ^{14}C-arachidonic acid. The effect was most pronounced for PGE_2, but there also was release of lipoxygenase products that have been shown to play a role in LHRH release. Inhibitors of cyclooxygenase, the responsible enzyme for prostanoid synthesis, such as indomethacin and salicylic acid, blocked the release of LHRH induced by NE. This provided further evidence for the role of NO in the control of LHRH release via the activation of cyclooxygenase-1. Needleman's group also showed that NO activates cyclooxygenase-1 and cyclooxygenase-2 in cultured fibroblasts. The action is probably mediated by combination of NO with the heme group of cyclooxygenase, altering its conformation. The action on lipoxygenase is similar; although it contains ferrous iron, the actual presence of heme in lipoxygenase has yet to be demonstrated.[21,22]

The previously accepted pathway for the physiologic action of NO is by activation of soluble guanylate cyclase by interaction of NO with the heme group of this enzyme, thereby causing conversion of guanosine triphosphate into cyclic guanosine monophosphate (cGMP), which mediates the effects on smooth muscle by decreasing the intracellular [Ca^{++}]. On the other hand, Muelam's group has shown in incubated pancreatic acinar cells that cGMP has a biphasic effect on intracellular [Ca^{++}], elevating it at low concentration and lowering it at higher concentrations. We postulate that the NO released from the NOergic neurons near the LHRH neuronal terminals increases the intracellular free calcium required to activate phospholypase A_2 (PLA_2). PLA_2 causes the conversion of membrane phospholipids in the LHRH terminal to arachidonate, which then can be converted to PGE_2 via the activated cyclooxygenase. The released PGE_2 activates adenyl cyclase causing an increase in cAMP release which then activates protein kinase-A, leading to exocytosis of LHRH secretory granules into the hypophyseal portal capillaries for transmission to the anterior pituitary gland.[23]

Norepinephrine (NE) has previously been shown to be a powerful releasor of LHRH. In the present experiments, we showed it acted by activation of the NOergic neurons since the activation of these neurons and the release of LHRH could be blocked by a competitive inhibitor of NOS, NMMA. NE acts to stimulate the release of NO from the NOergic neurons by α_1 adrenergic receptors since its action can be blocked by phentolamine, an α–receptor blocker, and prazosine, an α_1 receptor blocker. Activation of the α_1 receptors is postulated to increase intracellular [Ca^{++}] that combines with calmodulin to activate NOS, leading to generation of NO.

We measured the effect of NE on the content of NOS in the MBH explants at the end of the experiments by homogenizing the tissue and adding ^{14}C-arginine and measuring its conversion to citrulline on incubation of the homogenate. Since arginine is converted to equimolar quantities of NO and citrulline, measurement of citrulline production provides a convenient estimate of the activity of the enzyme. The NO disappears rapidly making its measurement very difficult. NE caused an increase in the apparent content of the enzyme. That we were actually measuring enzyme content was confirmed, because incubation of the homogenate with l-nitroarginine methyl ester, another inhibitor of NOS, caused a drastic decline in the conversion of arginine to citrulline. We further confirmed that we actually had increased the content of enzyme by isolating the enzyme according to the method of Bredt and Snyder[24] and then measuring the conversion of labeled arginine to citrulline. The conversion was significantly increased by NE.[25]

Glutamic acid (GA), at least in part by N-methyl-D-aspartate (NMDA) receptors, also plays a physiologically significant role in controlling the release of LHRH.

Therefore, we evaluated where GA fit into the picture. It also acted via NO to stimulate LHRH release, but we showed that the effect of GA could be completely obliterated by the α-receptor-blocker phentolamine. Consequently, we concluded that GA acted by stimulation of the noradrenergic terminals in the MBH to release NE, which then initiated NO release and stimulation of LHRH release.[26]

Oxytocin has actions within the brain to promote mating behavior in the female and penile erection in the male rat. Since LHRH mediates mating behavior, we hypothesized that oxytocin would stimulate the LHRH release that, after secretion into the hypophyseal portal vessels, mediates LH release from the pituitary. Consequently, we incubated MBH explants and demonstrated that oxytocin (10^{-7}–$10^{-10}M$) induced LHRH release via NE stimulation of nNOS. Therefore, oxytocin may be very important as a stimulator of LHRH release. Furthermore, NO acted as a negative feedback to block oxytocin release.[27]

One of the few receptors to be identified on LHRH neurons is the γ-aminobutyric acid-a (GABAa) receptor. Consequently, we evaluated the role of GABA in LHRH release and the participation of NO in this. The experiments showed that GABA blocked the response of the LHRH neurons to NP that acts directly on the LHRH terminals. We concluded that GABA suppressed LHRH release by blocking their response to NO. Additional experiments showed that NO stimulated the release of GABA, providing thereby an inhibitory feed-forward pathway to inhibit the pulsatile release of LHRH initiated by NE. As NE stimulated the release of NO, this would stimulate the release of GABA, which would then block the response of the LHRH neuron to the NO released by NE.[28]

Other studies indicated that NO would suppress the release of dopamine and NE. We have already described the ability of NE to stimulate LHRH release, and dopamine also acts as a stimulatory transmitter in the pathway. Therefore, there is an ultrashort loop negative feedback mechanism to terminate the pulsatile release of LHRH since the NO released by NE would diffuse to the noradrenergic terminals and inhibit the release of NE, thereby terminating the pulse of NE, LHRH, and finally LH.[29]

We further examined the possibility that other products from this system might have inhibitory actions. Indeed, we found that as we added increasing amounts of NP we obtained a bell-shaped dose-response curve of the release of LHRH, such that the release increased with increasing concentrations of NP up to a maximum at around 600 μm and then declined with higher concentrations. When the effect of NP on NOS content at the end of the experiment was measured, we found that high concentrations of NP lowered the NOS content. Furthermore, NP could directly decrease NOS content when incubated with MBH homogenates — results that indicate a direct effect on NOS, probably by interaction of NO with the heme group on the enzyme. Thus, when large quantities of NO are released, as could occur following induction of iNOS in the brain during infections, the release of NO would be decreased by an inhibitory action on the enzyme at these high concentrations. Furthermore, high concentrations of cGMP released by NO also acted in the explants or even in the homogenates to suppress the activation of NOS. This pathway could also be active in the presence of high concentrations of NO, such as would occur in infection by induction of inducible NOS by bacterial or viral products.[25]

14.1.2 Effect of Cytokines (IL-1 and GMCSF) on NOergic Control of LHRH Release

The cytokines that have been tested, for example IL-1 and granulocyte macrophage colony-stimulating factor (GMCSF), act within the hypothalamus to suppress the release of LHRH as revealed in both *in vivo* and *in vitro* studies. We have examined the mechanism of this effect and found that, for IL-1, it occurs by inhibition of cyclooxygenase as shown by the fact that there is blockage of the conversion of labeled arachidonate to prostanoids, particularly PGE_2, and the release of PGE_2 induced by NE also is blocked.[30]

A principle mechanism of action is by suppression of the LHRH release induced by NO donors such as NP.[30] We first believed that there were IL-1 and GMCSF receptors on the LHRH neuron which blocked the response of the neuron to NO. However, since we also had shown that GABA blocks the response to NP, and earlier work had shown that GABA receptors are present on the LHRH neurons, we evaluated the possibility that the action of cytokines could be mediated by stimulation of GABAergic neurons in the MBH. Indeed, in the case of GMCSF its inhibitory action on LHRH release can be partially reversed by the GABAa receptor blocker, bicuculine, which also blocks the inhibitory action of GABA itself, on the response of the LHRH terminals to NO. Therefore, we believe that the inhibitory action of cytokines on LHRH release is mediated by stimulation of GABA neurons.[31]

14.1.3 Role of NO in Mating Behavior

LHRH controls lordosis behavior in the female rat and is also involved in mediating male sex behavior. Studies *in vivo* have shown that NO stimulates the release of the LHRH involved in inducing sex behavior. This behavior can be stimulated by 3V injection of NP and is blocked by inhibitors of NOS. Apparently, there are two LHRH neuronal systems: one with axons terminating on the hypophyseal portal vessels, the other with axons terminating on neurons which mediate sex behavior.[32] NO is also involved in inducing penile erection by the release of NO from NOergic neurons innervating the corpora cavernosa penis. The role of NO in sexual behavior in both sexes has led us to change the name of NO to the sexual gas.[20]

14.1.4 Effect of NO in the Release of Other Hypothalamic Peptides

NO appears to act similarly to stimulate the release of corticotropin-releasing hormone (CRH),[33] growth hormone-releasing hormone (GHRH), somatostatin,[34] but not FSHRF[21] since this is not affected by inhibitors of NOS or by donors of NO. On the other hand, the release of vasopressin and oxytocin release is suppressed by NO.[27]

14.1.5 ACTION OF NO TO CONTROL RELEASE OF ANTERIOR PITUITARY HORMONES

NOS is localized in certain pituitary cells, principally the folliculostellate cells, which are modified glial cells that bear a resemblance to macrophages, and also in the LH gonadotropes as revealed by immunocytochemistry. When pituitaries are incubated *in vitro*, most pituitary hormones are secreted only in small quantities. The exception to this rule is prolactin, which is secreted in large amounts because of removal of inhibitory hypothalamic control by dopamine.[35] In the case of pituitary hormones that are secreted at low levels because of lost stimulatory hypothalamic input, NO donors have little affect on this basal release, for example, in the case of LH and GH. On the other hand, in the case of prolactin, which is released in large amounts, NO donors suppress the release of the hormone and inhibitors of NOS usually enhance the release, indicating that there is still some capability for the gland to further increase release of prolactin *in vitro*.

Dopamine is the most important prolactin-inhibiting factor by action on dopamine type 2 receptors (D_2 receptors) in the gland. The dramatic inhibitory action of dopamine can be prevented by D_2 receptor blockers and also is prevented by incubation in the presence of inhibitors of NOS. Therefore, we conclude that the primary inhibitory action of dopamine is mediated by its action to stimulate D_2 receptors on the NOS-containing cells in the pituitary gland with resultant release of NO which diffuses to the lactotropes and activates guanylate cyclase, causing the release of cGMP that mediates the inhibition of prolactin secretion. Consistent with this hypothesis is the fact that NO donors suppress prolactin release and the addition of cyclic GMP can also lower the release of the hormone from incubated pituitaries.[35]

The LH-releasing action of LHRH has long been known to be caused by an increase in intracellular free calcium, and this holds also for FSH.[2,3,36] Since NOS has been localized to gonadotropes, it occurred to us that NO might play a role in control of gonadotropin secretion. Indeed, in early work we found that cGMP, but not cAMP, would activate both FSH and LH release.[37] We have now determined that blockade of NO formation by the use of the inhibitor of NOS, N^G-monomethyl-L-arginine (NMMA), inhibits the release of FSH and LH induced by LHRH. Presumably, the NO activates GC causing the release of cGMP which as already indicated, can release both gonadotropins, presumably by an action on protein kinase-G.[38]

Certain cytokines also can effect gonadotropin secretion by the pituitary including GMCSF (Kimura, M. et al., in preparation, 1997), which increases LH release from the gland. It has not yet been determined whether this is mediated via NO.

14.2 POTENTIAL ROLE OF LEPTIN IN REPRODUCTION

The hypothesis that leptin may play an important role in reproduction stems from several findings. First, the Ob/Ob mouse, lacking the leptin gene, is infertile and has atrophic reproductive organs.[39] Gonadotropin secretion is impaired and very sensitive to negative feedback by gonadal steroids as is the case for prepubertal animals.[40] It has now been shown that treatment with leptin can recover the reproductive system in the Ob/Ob mouse by leading to growth and function of the reproductive organs and fertility[41] via secretion of gonadotropins.[42]

The critical weight hypothesis of the development of puberty states that when body fat stores have reached a certain point, puberty occurs.[43] This hypothesis in its original form does not hold, since if animals are underfed puberty is delayed, but with access to food, the rapid weight gain leads to onset of puberty at weights well below the critical weight under normal nutritional conditions.[44] We hypothesized that during this period of refeeding or at the time of the critical weight in the normally fed animal, there is increased release of leptin into the blood-stream from the adipocytes and that this acts on the hypothalamus to stimulate the release of LHRH with resultant induction of puberty. Indeed, leptin has recently been found to induce puberty.[45]

Therefore, when we were able to obtain leptin in mid-March 1996, we initiated studies on its possible effect on hypothalamic-pituitary function. We anticipated that it would also be active in adult rats and therefore studied its effect on the release of FSH and LH from hemipituitaries, and also its possible action to release LHRH from MBH explants *in vitro*. To determine if it was active *in vivo*, we used a model which we have often employed to evaluate stimulatory effects of peptides on LH release, namely, the ovariectomized, estrogen-primed rat. Since our supply of leptin was limited, we began by microinjecting it into the 3V in conscious animals bearing implanted third ventricular cannulae and also catheters in the external jugular vein extending to the right atrium, so that we could draw blood samples before and after the injection of leptin and measure the effect on plasma FSH and LH.[46]

14.2.1 EFFECT OF LEPTIN ON LH RELEASE

We found that under our conditions leptin had a bell-shaped dose-response curve to release LH from anterior pituitaries incubated *in vitro*. There was no consistent stimulation of LH release with a concentration of $10^{-5}\,M$. Results became significant with $10^{-7}\,M$ and remained on a plateau through $10^{-11}\,M$, with reduced release at a concentration of $10^{-12}\,M$ that was no longer statistically significant. The release was not significantly less than that achieved with LHRH ($4 \times 10^{-8}\,M$). Under these conditions, there was no additional release of LH when leptin was incubated ($10^{-7}\,M$) together with LHRH ($4 \times 10^{-8}\,M$). In certain other experiments, there was an additive effect when leptin was incubated with LHRH; however, this effect was not uniformly seen. The results indicate that leptin was only slightly less effective to release LH than LHRH itself.

14.2.2 EFFECT OF LEPTIN ON FSH RELEASE

In the incubates from these same glands we also measured FSH release and found that it showed a similar pattern as that of LH, except that the sensitivity in terms of FSH release was much less than that for LH. The minimal effective dose for FSH was $10^{-9}\,M$, whereas it was $10^{-11}\,M$ for LH. The responses were roughly of the same magnitude at the effective concentrations as obtained with LH and the responses were clearly equivalent to those observed with $4 \times 10^{-9}\,M$ LHRH. Combination of LHRH with a concentration of leptin that was just below significance gave a clear additive effect.

14.2.3 Effect of Leptin on Prolactin Release

The results with prolactin were in contrast to those with FSH and LH in that the maximal response (a 4.5-fold increase) was seen with the highest concentration tested ($10^{-5}\,M$) and prolactin release declined with lower concentrations, such that there was no longer a significant effect with $10^{-8}\,M$ leptin.

14.2.4 Effect of Leptin on LHRH Release

There was no significant effect of leptin in a concentration range of 10^{-6}–$10^{-12}\,M$ on LHRH release during the first 30 min of incubation; however, during the second 30 min, the highest concentration produced a borderline significant decrease in LHRH release with $10^{-6}\,M$. This was followed by a tendency to increase with lower concentrations and a significant, plateaued increase with the lowest concentrations tested (10^{-10} and $10^{-12}\,M$). The overall significance, combining the results with both effective doses, was $p < .01$.

14.2.5 Effect of Intraventricularly Injected Leptin on Plasma Gonadotropin Concentrations in Ovariectomized, Estrogen-Primed Rats

The injection of the diluent for leptin into the 3V (Krebs-Ringer bicarbonate, 5 µl) had no effect on pulsatile FSH or LH release, but the injection of leptin (10 µg) uniformly produced an increase in plasma LH with a variable time-lag ranging from 10 to 50 min, so that the maximal increase in LH from the starting value was highly significant ($p < .01$) and constituted a mean increase of 60% above the initial concentration. In contrast, leptin inhibited FSH release on comparison with the results with the diluent, but the effect was delayed and occurred mostly in the second hour. Therefore, at this dose of estrogen it appears that leptin stimulates the release of LHRH and inhibits the release of FSHRF.[46]

14.2.6 Mechanism of Action of Leptin on the Hypothalamic-Pituitary Axis

Recently, we have shown that leptin exerts its action at both the hypothalamic and pituitary level by activating NOS, since its effect to release LHRH, FSH, and LH *in vitro* is blocked by NMMA.[38]

Leptin, in essence, is a cytokine secreted by the adipocytes. Like the cytokines, it appears to reach the brain via a transport mechanism mediated by the Ob/Ob$_a$ receptors[47] in the choroid plexus.[48] These receptors have an extensive extracellular domain but a greatly truncated intracellular domain,[47] and mediate transport of the cytokine by a saturable mechanism.[49] Following uptake into the cerebrospinal fluid (CSF) through the choroid plexus, leptin is carried by the flow of CSF to the 3V, where it either diffuses into the hypothalamus through the ependymal layer lining the ventricle or combines with Ob/Ob$_b$[47] receptors on terminals of responsive neurons that extend to the ventricular wall.

The Ob/Ob$_b$ receptor has a large intracellular domain that presumably mediates the action of the protein.[48] These receptors are widespread throughout the brain,[48] but particularly localized in the region of the paraventricular (PVN) and arcuate nuclei (AN). Leptin activates Stat 3 within 30 min following its intraventricular injection.[50] Stat 3 is a protein which is important in conveying information to the nucleus to initiate DNA-directed mRNA synthesis. Following injection of bacterial lipopolysaccharide (LPS), it is also activated, but in this case the time delay is 90 min, presumably because LPS has been shown to induce IL-1β mRNA in the same areas, namely, the PVN and AN,[51] IL-1β mRNA would then cause production of IL-1β that would activate Stat 3. On entrance into the nucleus, Stat 3 would activate or inhibit DNA-directed mRNA synthesis. In the case of leptin, it activates corticotropin-releasing hormone (CRH) mRNA in the PVN, whereas in the AN it inhibits neuropeptide Y (NPY) mRNA, resulting in increased CRH synthesis and presumably release in the PVN and decreased NPY synthesis and release in the AN.[48] Presumably, a combination of leptin with these transducing receptors either increases or decreases the firing rate of that particular neuron. In the case of the AN-median eminence area, leptin may enter the median eminence by diffusion between the tanycytes or alternatively by combining with its receptors on terminals of neurons projecting to the tanycytes. Activation or inhibition of these neurons would induce LHRH release.

The complete pathway of leptin action in the MBH to stimulate LHRH release is not yet elucidated. Arcuate neurons bearing Ob/Ob receptors may project to the ME to the tanycyte/portal capillary junction. Leptin would either combine with its receptors on the terminals that transmit information to the cell bodies in the AN or diffuse to the AN to combine with its receptors on the perikarya of AN neurons. Since leptin decreases NPY mRNA, and presumably NPY biosynthesis in NPY neurons in the AN, we postulate that leptin causes a decrease in NPY release. Since NPY inhibited LH release in intact and castrated male rats,[52] we hypothesize that NPY decreases the release of LHRH by inhibiting the noradrenergic neurons which mediate pulsatile release of LHRH. Therefore, when the release of NPY is inhibited by leptin, noradrenergic impulses are generated that act on α_1 receptors on the NOergic neurons causing the release of NO which diffuses to the LHRH terminals and activates LHRH release by activating guanylate cyclase and cyclooxygenase$_1$ as shown in our prior experiments reviewed above. Leptin acts to activate NOS as indicated since its release of LHRH is blocked by inhibition of NOS.[38] The LHRH enters the portal vessels and is carried to the anterior pituitary gland where it acts to stimulate FSH and particularly LH release by combining with its receptors on the gonadotropes. The release of LH and to a lesser extent FSH is further increased by the direct action of leptin on its receptors[53] in the pituitary gland.[46]

We hypothesize that leptin may be a critical factor in induction of puberty as the animal nears the so-called critical weight. Either metabolic signals reaching the adipocytes, or signals related to their content of fat, cause the release of leptin which increases LHRH and gonadotropin release, thereby initiating puberty and finally ovulation and onset of menstrual cycles. In the male, the system would work similarly; however, there is no preovulatory LH surge brought about by the positive feedback of estradiol. Sensitivity to leptin is undoubtedly under steroid control and we are actively working to elucidate this problem.

During fasting, the leptin signal is removed and LH pulsatility and reproductive function decline quite rapidly. In women with anorexia nervosa, this causes a reversion to the prepubertal state which can be reversed by feeding. Thus, leptin would have a powerful influence on reproduction throughout the reproductive life span of the individual. The consequences to gonadotropin secretion of overproduction of leptin, as has already been demonstrated in human obesity, are not clear. There are often reproductive abnormalities in this circumstance and whether they are due to excess leptin production or other factors remains to be determined.

In conclusion, it is now clear that leptin plays an important role in control of reproduction by actions on the hypothalamus and pituitary. It may also act in the gonads since its receptors have been found there.

ACKNOWLEDGMENTS

This work was supported by NIH grants DK4390 and MH51853. We would like to thank Judy Scott and Jason Holland for their excellent secretarial assistance.

REFERENCES

1. McCann, S. M., Taleisnik, S., and Friedman, H. M., LH-releasing activity in hypothalamic extracts, *Proc. Soc. Exp. Biol. Med.*, 104, 432, 1960.
2. McCann, S. M. and Ojeda, S. R., The anterior pituitary and hypothalamus, *Textbook of Endocrine Physiology*, 3rd ed., Griffin, J. and Ojeda, S. R., Eds., Oxford University Press, New York, 1996, 101.
3. Reichlin, S., Neuroendocrinology, *Textbook of Endocrinology*, Wilson, J. D. and Foster, D. W., Eds., Williams, Philadelphia, 1992, 135.
4. McCann, S. M., Marubayashi, U., Sun, H.-Q., and Yu, W. H., Control of follicle stimulating hormone and luteinizing hormone release by hypothalamic peptides. Intraovarian Regulators and Polycystic Ovarian Syndrome: Recent Progress on Clinical and Therapeutic Aspects, *Ann. NY Acad. Sci.*, 687, 55, 1993.
5. Dees, W. L., Rettori, V., Kozlowski, J. G., and McCannm S. M., Ethanol and the pulsatile release of luteinizing hormone, follicle stimulating hormone and prolactin in ovariectomized rats, *Alcohol*, 2, 641, 1985.
6. Rettori, V, Gimeno, M. F., Karara, A., Gonzalez, M. C., and McCann, S. M., Interleukin-1α inhibits prostaglandin E_2 release to suppress pulsatile release of luteinizing hormone but not follicle-stimulating hormone, *Proc. Natl. Acad. Sci., U.S.A.*, 88, 2763, 1991.

7. McCann, S. M. and Krulich, L., Role of neurotransmitters in control of anterior pituitary hormone release, in *Endocrinology,* 2nd ed., W.B. Saunders, Philadelphia, 1989, 117.
8. Lumpkin, M. D., McDonald, J. K., Samson, W. K., and McCann, S. M., Destruction of the dorsal anterior hypothalamic region suppresses pulsatile release of follicle stimulating hormone but not luteinizing hormone, *Neuroendocrinology,* 50, 229, 1989.
9. Marubayashi, U., McCann, S. M., and Antunes-Rodrigues, J., Altered gonadotropin and prolactin release induced by median eminence (ME) lesions and pharmacological manipulation of prolactin release: further evidence for separate hypothalamic control of FSH and LH release, *Brain Res. Bull.,* 23, 193, 1989.
10. Marubayashi, U., Yu, W. H., and McCann, S. M., Median eminence lesions reveal separate hypothalamic control of pulsatile follicle-stimulating hormone and luteinizing hormone release, *Proc. Soc. Exp. Biol. Med.,* submitted 1998.
11. Lumpkin, M. D., Moltz, J. H., Yu, W., Samson, W. K., and McCann, S. M., Purification of FSH-releasing factor: its dissimilarity from LHRH of mammalian, avian, and piscian origin, *Brain Res. Bull.,* 18, 175, 1987.
12. Yu, W. H., Karanth, S., Walczewska, A., Sower, S. A., and McCann, S. M., A hypothalamic follicle-stimulating hormone-releasing decapeptide in the rat, *Proc. Natl. Acad. Sci. U.S.A.,* 94, 9499, 1997.
13. Yu, W. H., Millar, R. P., Milton, S. C. F., del Milton, R. C., and McCann, S. M., Selective FSH-releasing activity of [D-Trp^9]GAP$_{1-13}$: comparison with gonadotropin-releasing abilities of analogs of GAP and natural LHRHs, *Brain Res. Bull.,* 25, 867, 1990.
14. Yu, W. H. and McCann, S. M., Comparison of the selective FSH-releasing action of rat and human gonadotropin-releasing hormone associated peptide$_{1-13}$ *in vivo* and *in vitro* in the rat, 10th Int. Cong. Endocrinology, San Francisco, CA, June 12–15, 1996. Abstr. # P3-235, p 813.
15. Yu, W. H., McCann, S. M., and Li, C. H., Synthetic human seminal α-inhibin-92 selectively suppresses FSH release *in vivo, Proc. Natl. Acad. Sci. U.S.A.,* 85, 289, 1988.
16. Li, C. H., Hammonds, R. G., Jr., Ramasharma, K., and Chung, D., Human seminal α inhibins: isolation, characterization, and structure, *Proc. Natl. Acad. Sci. U.S.A.,* 82, 4041, 1986.
17. Yu, W. H., Riedel, M., Yamashiro, D., Ramasharma, K., and McCann, S. M., Effects of α-inhibin-92 fragments and α-inhibin-92 antiserum on the control of follicle-stimulating hormone release in male rats, *Life Sci.,* 55, 93, 1994.
18. Yu, W. H. and McCann, S. M., Feedback of follicle-stimulating hormone to inhibit luteinizing hormone and stimulate follicle-stimulating hormone release in ovariectomized rats, *Neuroendocrinology,* 53, 453, 1991.
19. Rage, F., Hill, D. F., Sena-Esteves, M., Breakefield, X. O., Coffey, R. J., Costa, M. E., McCann, S. M., and Ojeda, S. R., Targeting transforming growth factor α expression to discrete loci of the neuroendocrine brain induces female sexual precocity, *Proc. Natl. Acad. Sci. U.S.A.,* 94, 2735, 1997.
20. McCann, S. M. and Rettori, V., The role of nitric oxide in reproduction, *Proc. Soc. Exp. Biol. Med.,* 211, 7, 1996.
21. Rettori, V., Belova, N., Dees, W. L., Nyberg, C. L., Gimeno, M, and McCann, S. M., Role of nitric oxide in the control of luteinizing hormone-releasing hormone release *in vivo* and *in vitro, Proc. Natl. Acad. Sci. U.S.A.,* 90, 10130, 1993.

22. Rettori, V., Gimeno, M., Lyson, K., and McCann, S. M., Nitric oxide mediates norepinephrine-induced prostaglandin E_2 release from the hypothalamus, *Proc. Natl. Acad. Sci. U.S.A.,* 89, 11543, 1992.
23. Canteros, G., Rettori, V., Franchi, A., Genaro A., Cebral, E., Saletti, A., Gimeno, M., and McCann, S. M., Ethanol inhibits luteinizing hormone-releasing hormone (LHRH) secretion by blocking the response of LHRH neuronal terminals to nitric oxide, *Proc. Natl. Acad. Sci. U.S.A.,* 92, 3416, 1995.
24. Bredt, D. S. and Snyder, S. H., Nitric oxide mediates glutamate-linked enhancement of cGMP levels in the cerebellum, *Proc. Natl. Acad. Sci. U.S.A.,* 86, 9030, 1989.
25. Canteros, G., Rettori, V., Genaro, A., Suburo, A., Gimeno, M., and McCann, S. M., Nitric oxide synthase (NOS) content of hypothalamic explants: increased by norepinephrine and inactivated by NO and cyclic GMP, *Proc. Natl. Acad. Sci. U.S.A.,* 93, 4246, 1996.
26. Kamat, A., Yu, W. H., Rettori, V., and McCann, S. M., Glutamic acid stimulated luteinizing-hormone releasing hormone release is mediated by α adrenergic stimulation of nitric oxide release, *Brain Res. Bull.,* 37, 233, 1995.
27. Rettori, V., Canteros, G., Renoso, R., Gimeno, M., and McCann, S. M., Oxytocin stimulates the release of luteinizing hormone-releasing hormone from medial basal hypothalamic explants by releasing nitric oxide, *Proc. Natl. Acad. Sci. U.S.A.,* 94, 2741, 1997.
28. Seilicovich, A., Duvilanski, B. H., Pisera, D., Thies, S., Gimeno, M., Rettori, V., and McCann, S. M., Nitric oxide inhibits hypothalamic luteinizing hormone-releasing hormone release by releasing α-aminobutyric acid, *Proc. Natl. Acad. Sci. U.S.A.,* 92, 3421, 1995.
29. Seilicovich, A., Lasaga, M., Befumo, M., Duvilanski, B. H., del C. Dias, M., Rettori, V., and McCann, S. M., Nitric oxide inhibits the release of norepinephrine and dopamine from the medial basal hypothalamus of the rat, *Proc. Natl. Acad. Sci. U.S.A.,* 92, 11299, 1995.
30. Rettori, V., Belova, N., Kamat, A., Lyson, K., and McCann, S. M., Blockade by interleukin-1-α of the nitricoxidergic control of luteinizing hormone-releasing hormone release *in vivo* and *in vitro*, *Neuroimmunomodulation,* 1, 86, 1994.
31. Kimura, M., Yu, W. H., Rettori, V., Walczewska, A., and McCann, S. M., Granulocyte-macrophage colony stimulating factor suppresses LHRH release mediated through nitric oxide and α-aminobutyric acid, *Neuroimmunomodulation,* in press, 1997.
32. Mani, S. K., Allen, J. M. C., Rettori, V., O'Malley, B. W., Clark, J. H., and McCann, S. M., Nitric oxide mediates sexual behavior in female rats by stimulating LHRH release, *Proc. Natl. Acad. Sci. U.S.A.,* 91, 6468, 1994.
33. Karanth, S., Lyson, K., and McCann, S. M., Role of nitric oxide in interleukin 2-induced corticotropin-releasing factor release from incubated hypothalami, *Proc. Natl. Acad. Sci. U.S.A.,* 90, 3383, 1993.
34. Rettori, V., Belova, N., Yu, W. H., Gimeno, M., and McCann, S. M., Role of nitric oxide in control of growth hormone release in the rat, *Neuroimmunomodulation,* 1, 195, 1994.
35. Duvilanski, B. H., Zambruno, C., Seilicovich, A., Pisera, D., Lasaga, M., Diaz, M., del C, Belova, N., Rettori, V., and McCann, S. M., Role of nitric oxide in control of prolactin release by the adenohypophysis, *Proc. Natl. Acad. Sci. U.S.A.,* 92, 170, 1995.
36. Wakabayashi, K., Kamberi, I. A., and McCann, S. M., *In vitro* responses of the rat pituitary to gonadotrophin-releasing factors and to ions, *Endocrinology,* 85, 1046, 1969.

37. Nakano, H., Fawcett, C. P., Kimura, F., and McCann, S. M., Evidence for the involvement of guanosine 3′,5′-cyclic monophosphate in the regulation of gonadotropin release, *Endocrinology,* 103, 1527, 1978.
38. Yu, W. H., Walczewska, A., Karanth, S., and McCann, S. M., Nitric oxide mediates leptin-induced luteinizing hormone-releasing hormone (LHRH) and LHRH and leptin-induced LH release from the pituitary gland, *Endocrinology,* 138, 5055, 1997.
39. Swerdloff, R., Batt, R., and Bray, G., Reproductive hormonal function in the genetically obese (ob/ob) mouse, *Endocrinology,* 98, 1359, 1976.
40. Swerdloff, R., Peterson, M., Vera, A., Batt, R., Heber, D., and Bray, G., The hypothalamic-pituitary axis in genetically obese (ob/ob) mice: response to luteinizing hormone-releasing hormone, *Endocrinology,* 103, 542, 1978.
41. Chehab, F. F., Lim, M. E., and Ronghua, L., Correction of the sterility defect in homozygous obese female mice by treatment with the human recombinant leptin, *Nat. Genet.,* 12, 318, 1996.
42. Barash, I. A., Cheung, C. C., Weigle, D. S., Ren, H., Kabigting, E. B., Kuijper, J. L., Clifton, D. K., and Steiner, R. A., Leptin is a metabolic signal to the reproductive system, *Endocrinology,* 137, 3144, 1996.
43. Frisch, R. E. and McArthur, J. W., Menstrual cycles: fatness as a determinant of minimum weight for height necessary for their maintenance or onset, *Science,* 185, 949, 1974.
44. Ronnekleiv, O. K., Ojeda, S. R., and McCann, S. M., Undernutrition, puberty and the development of estrogen positive feedback in the female rat, *Biol. Reprod.,* 19, 414, 1978.
45. Chehab, F. F., Mounzih, K., Lu, R., and Lim, M. E., Early onset of reproductive function in normal female mice treated with leptin, *Science,* 275, 88, 1997.
46. Yu, W. H., Kimura, M., Walczewska, A., Karanth, S., and McCann, S. M., Role of leptin in hypothalamic-pituitary function, *Proc. Natl. Acad. Sci. U.S.A.,* 94, 1023, 1997.
47. Cioffi, J., Shafer, A., Zupancic, T., Smith-Gbur, J., Mikhail, A., Platika, D., and Snodgrass, H., Novel B219/ob receptor isoforms: possible role of leptin in hematopoiesis and reproduction, *Nat. Med,* 2, 585, 1996.
48. Schwartz, M. W., Seeley, R. J., Campfield, L. A., Burn, P., and Baskin, D. G., Identification of targets of leptin action in rat hypothalamus, *J. Clin. Invest.,* 98, 1101, 1996.
49. Banks, W. A., Kastin, A. J., Huang, W., Jaspan, J. B., and Maness, L. M., Leptin enters the brain by a saturable system independent of insulin, *Peptides,* 17(2), 305, 1996.
50. Vaisse, C., Halaas, J. L., Horvath, C. M., Darnell, J. E., Jr., Stoffel, M., and Friedman, J. M., Leptin activation of stat3 in the hypothalamus of wildtype and Ob/Ob mice but not Db/Db mice, *Nat. Genet.,* 14, 95, 1996.
51. Wong, M.-L., Bongiorno, P. B., Rettori, V., McCann, S. M., and Licinio, J., Interleukin 1β interleukin 1 receptor antagonist, interleukin 10, and interleukin 13 gene expression in the central nervous system and anterior pituitary during systemic inflammation: pathophysiological implications, *Proc. Natl. Acad. Sci. U.S.A.,* 94, 227, 1997.
52. Reznikov, A. G. and McCann, S. M., Effects of neuropeptide Y on gonadotropin and prolactin release in normal, castrated or flutamide-treated male rats, *Neuroendocrinology,* 57, 1148, 1993.
53. Naivar, J. S., Dyer, C. J., Matteri, R. L, and Keisler, D. H., Expression of leptin and its receptor in sheep tissues, Proc. Soc. Study Reprod., Seattle, WA, 1996. Abstr. #391, p. 154.

15 Interferon and the Central Nervous System

Nachum Dafny

CONTENTS

15.1 Introduction 221
15.2 Interferon Modulates the Peripheral Nervous System (PNS) 222
15.3 Interferon Modulates Electroencephalogram (EEG) Activity 222
15.4 IFN-α Modulates Evoked Field Potentials 223
15.5 IFN-α Modulates Neuronal Activity 223
15.6 Interferons and Opioids 224
Acknowledgments 228
References 228

Key Words: Interferon, behavior, endocrine, opiate, thermosensitive neurons, PNS, CNS, EEG, evoked potentials, single cell, glucose sensitive neurons, electrical activity

15.1 INTRODUCTION

Interferons were originally detected in immunological cells and have been shown to be produced during nonimmunological responses of both central and peripheral origins. The biological properties of interferon include antiviral activity as an inhibitor of viral proliferation and enhancement of immune function.[42] Subsequently, it has been shown to have diverse biological activities, including antitumor and biological response modifier activities.[15,16] Furthermore, it is established that a variety of stimuli act on different cells to give rise to various types of interferons. Three types of interferons are known: α, β, and γ. They are a family of proteins that appear in a large variety of vertebrates, from fish to Homo sapiens, and are comprised of complex endogenous proteins, glycoproteins, and peptides.[4,49,50,61] Initially, interferon-α (IFN-α) was thought to be produced by macrophages, leukocytes, and monocytes. *In vivo*, IFN-α is produced at a constant low "physiological" level.[16]

Immunologic therapy uses cytokines such as interferon to treat various hematologic malignancies and infectious ailments as well as autoimmune diseases, such as multiple sclerosis.[3,4,15,35,54,61] Multiple sclerosis (MS), a chronic disease of the nervous system, and experimental autoimmune encephalomyelitis (EAE), an animal model resembling MS, both exhibit a deficiency of natural killer (NK) cells, which is in turn correlated with disease severity. IFN-α administration in mice and rats inhibits EAE inducers,[1] protecting mice from EAE and death.[76]

Clinically, each of the interferon types is usually administered daily or every other day by injection, but long-term treatment is needed for the therapy to be effective. Because the cytokines used in immunologic therapy are produced naturally in the body, they were thought to be nontoxic.[35] However, several adverse effects are reported to be frequent in long-term IFN-α therapies. Besides the activation of immunity, IFN-α produces a broad spectrum of nonimmunologic host defenses in the counterreply to infection, including sensory and motor disturbance, fever, anorexia, confusion, and depression, which are considered symptoms of central nervous system (CNS) dysfunction.[2,19,38,42,50,74]

The above features indicate that IFN-α affects CNS processes. Although the effects of IFN-α have been extensively studied in a variety of systems, the role of IFN-α in the nervous system has not been studied and warrants further explanation. Until the early 1980s, only a few laboratories[8,18] using behavioral and electrophysiological procedures had investigated whether IFN-α affects CNS activity.

15.2 INTERFERON MODULATES THE PERIPHERAL NERVOUS SYSTEM (PNS)

Components of the immune system, such as the spleen, are innervated by noradrenergic sympathetic nerves originating in the hypothalamus.[5] An intracerebroventricular (i.c.v.) injection of IFN-α elicits increased splenic sympathetic activity which results in suppression of splenic cell activity.[43,77,78] Splenic denervation has been reported to completely abolish the immunosuppressive effect of an i.c.v. IFN-α injection.[5,80] This observation suggests that sympathetic nerves innervating the spleen are involved in regulating splenic NK activity. The IFN-α effects on increasing splenic activity can be abolished by splenic denervation and by the opioid antagonist naloxone, but not by adrenalectomy.[44,78] Although an i.c.v. injection of IFN-α also suppresses the hypothalamus-pituitary-adrenocortical axis, this suppression is prevented by naloxone.[44,71,72] These observations suggests that brain IFN-α stimulates opiate receptors through activation of the brain corticotrophin-releasing-factor (CRF) system which elicits increased activity of the sympathetic nerves innervating the spleen, which in turn results in the suppression of spleen NK cytotoxicity.[78] In conclusion, it was demonstrated that IFN-α modulates the PNS.

15.3 INTERFERON MODULATES ELECTROENCEPHALOGRAM (EEG) ACTIVITY

Multiple IFN-α injection therapy elicits several CNS side effects indicating that this therapy affects not only the sensory, motor, and limbic systems structures, but also the hypothalamus, thalamus, and the reticular formation.[50,55] Based on these reports, permanent electrodes were implanted in several brain sites of the rat: in the motor and sensory cortices for EEG recording; in the caudate nucleus, which is a subcortical site known to be part of the motor system; in the mesencephalic reticular formation and parafascicular nucleus of the thalamus as representatives of the nonspecific sensory system; in the hippocampus, a representative site of the limbic system; and in the hypothalamus[23] to record EEG-like activity. EEG and EEG-like activity were recorded from these seven structures bilaterally and simultaneously in freely behaving rats before and following single IFN-α (150 units per gram of body weight, i.p.) injections. In all the recording sites, single IFN-α injections elicited high-amplitude synchronized activity independently (unsynchronized in time). The initial effects were observed in the hypothalamus (280 s postinjection) followed by the somatosensory cortex, the hippocampus, the motor cortex, caudate nucleus, and the parafasciculus nucleus, respectively.[23] Others have noted identical, dose-dependent EEG slowing in rodents, without altering other aspects of sleep.[7,40,44,46,69] IFN-α-elicited sleep has been reported also in primates[63] and rabbits.[46] Similar EEG showings from patients treated with IFN-α also have been reported.[50,54,60,74] The sleep-promoting activity of IFN-α may be related to the feeling of lassitude and sleepiness that often accompanies viral disease and IFN-α therapy.[46] It has been shown that the sleep-inducing effects of IFN-α are likely to be mediated via the locus ceruleus.[30] IFN-α given daily (150 unit per gram of body weight, i.p.) to rats for 3 weeks results in an almost flat EEG, and behaviorally the animals became extremely agitated and aggressive. The EEG returned to a normal pattern 1 week after cessation of IFN-α (unpublished observation).

15.4 IFN-α MODULATES EVOKED FIELD POTENTIALS

Single injections of IFN-α in unanesthetized freely behaving rats, implanted previously with permanent electrodes, resulted in the potentiation of the evoked potential recorded from the sensory cortex and ventromedial hypothalamus (VMH) as compared with control recording.[23] The same treatment in the same animals resulted in attenuating the evoked responses recorded from the preoptic/anterior hypothalamus (PO/AH) area. Thus, in some areas of the brain IFN-α elicits increases in the evoked activity while in other structures it elicits decreases in the evoked activity. In rat hippocampal slices, IFN-α reduces the size of short-term potentiation and suppresses long-term potentiation with a dose-dependent characteristic.[20] Since short-term potentiation and long-term potentiation are synaptic events,[13,37] D'Arcangelo et al.[20] suggest that IFN-α probably affects both pre- and postsynaptic sites. IFN-α produced both: (1) excitatory effects on CA3 pyramidal cells recorded from rat hippocampal slice and cultures, and (2) a decrease in evoked inhibitory postsynaptic potential amplitude, which eventually led to epileptiform bursting.[55]

15.5 IFN-α MODULATES NEURONAL ACTIVITY

Clavet and Gresser[18] were the first to report the effect of IFN-α on single neuronal activity. They reported that IFN-α enhanced the spontaneous activity of neurons in cerebral and cerebellar cat cortices as well as rat nerve cell cultures. These firing-rate changes occurred after about 30 min posttreatment and the excitation lasted for several hours. When repetitive stimulation was applied and the responses before and after IFN-α treatment were compared, marked shortening of the latency to the initial evoked activity and an increase in firing rate were observed. In another series of experiments, other investigators used three different sources of IFN-α as well as IFN-γ in *in vivo* experiments that were recorded from four different brain areas using the multibarreled procedure of single-cell recording and microiontophoretic application of interferons and other drugs.[23,24,26,32,62-68] They reported that all three sources of IFN-α elicited increases in the cerebral cortex and hippocampal neurons in a dose-dependent manner, while the majority of the thalamic neurons failed to respond to IFN-α injection, and the VMH cells responded in a mixed pattern (i.e., some increased and others decreased their baseline firing after IFN-α applications). As the dose of IFN-α was increased, some VMH neurons changed their direction of response from an increased firing rate at low doses to a decreased firing rate after high IFN-α doses, while in the other sites higher IFN-α doses intensified the initial effects. On the other hand, interferon-γ as well as some of its fractions failed to alter the electrical activity of the same cerebral cortex, hippocampus, thalamus, and the VMH neurons.[26,63-65]

The PO/AH area was the target of several investigators. Nakashima et al.,[56-58] using brain slices recorded from the PO/AH area and from the VMH, obtained mixed responses. IFN-α caused some cells to increase activity and others to decrease activity, similar to other *in vivo* experiments.[26,62] Other studies obtained only decreased neuronal activity in PO/AH following central administration of IFN-α in conscious rats.[44,70] The reduction in PO/AH activity was accompanied with a shift in the EEG-activity pattern toward synchronization (sleep) and a decrease in basal secretion of adrenal corticosteroids. In *in vitro* recording from PO/AH neurons with IFN-α perfusion experiments,[12,57,73] combined with antidromic electrophysiologically identified hypothalamic paraventricular nucleus neuron preparation,[72] IFN-α resulted in short-latency decreased firing rates when applied i.p. and i.c.v. The rapidity of the IFN-α effect observed by these investigators suggests that the effect of this agent was mediated via membrane-bound receptors.

Interferon is a large molecule and, when given systemically, there is a question whether it crosses the blood-brain barrier (BBB). To confirm whether IFN-α affects brain activity directly or indirectly, two routes of IFN-α application were used in *in vivo* experiments: (1) a direct local injection into the brain (i.c.v.) to bypass the BBB, and (2) systemic (i.v.) application while recording from the same neuron (Figure 15.1). The responses obtained were similar, the only difference being the latency, i.e., IFN-α applied locally and systemically elicited similar effects on the same single neurons.

15.6 INTERFERONS AND OPIOIDS

The effects of opioid peptides on the secretion of interferons and other cytokines has generally been described as stimulatory and bimodal.[67] According to Brown and Van Epps,[14] β-endorphin and met-enkephalin increase IFN-γ production while Lysle et al.,[48] reported that morphine suppresses release of IFN-γ. Structural similarities between IFN-α, proopiomelanocortin, and β-endorphin which link IFN-α to opioids

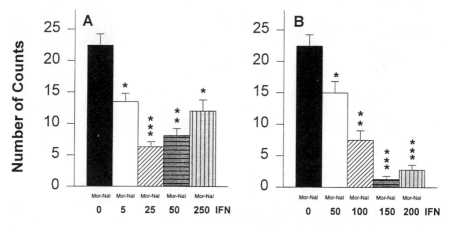

Figure 15.1 Histograms summarizing the group mean and SE of wet dog shake behavioral sign following naloxone injection (Nal; 1.0 mg/kg, i.p.) in morphine-dependent rats (100 mg pellet per animal; each group N = 8). Interferon-α (IFN) and Nal were injected 71 and 72 h after morphine pellet implantation, respectively. In A, IFN-α was given i.c.v. and the numbers under each bar represent the international unit (IU) each rat received. In B, IFN-α was given i.p., and the number under each bar represent the IU per gram of body weight. The figure demonstrate that IFN-α given i.c.v. (in A) or i.p. (in B) attenuated the severity of the morphine withdrawal sign. * = $p < .05$; ** = $p < .01$; *** $p < .001$.

have been demonstrated.[69] Other evidence linking IFN-α with the opioids is provided by experiments showing that lymphocytes stimulated IFN-α inducers to produce ACTH and endorphin-like substances. Furthermore, they exhibit a similar antigenic reaction, suggesting that these peptides have some common structural properties,[8-11,69-75] as well as an affinity for ^{3}H-morphine-binding sites in mouse brain membranes. IFN-α, but not IFN-β or IFN-γ, binds to opiate receptors and shares some pharmacological properties of opioids such as analgesia, reduced motor activity, and catatonia. These effects are reversed or prevented by the opioid antagonist naloxone.[7-9,30]

Repetitive use of opioids elicits tolerance and dependence.[41] Morphine dependence is manifested by an abstinence syndrome when opioid use is discontinued.[41] The opiate abstinence syndrome represents a fundamental feature of the addictive process. Physical dependence (addiction) on opioids can be quantified by the frequency or intensity of the abstinence (withdrawal) behaviors observed after abrupt termination of opiate intake or after injection of an opiate antagonist such as naloxone.[41] The degree of addiction, therefore, is assessed by the intensity of the withdrawal behavior. Evidence has been accumulating which suggests that tolerance to and/or dependence on opioids may be related to endogenous production of peptides and/or proteins in the CNS capable of counteracting the effects of the drug and its side effects.[79] With the finding of endogenous opioids, efforts have been directed toward the discovery of active endopeptide analogues which would prevent tolerance of and physical dependence on opioids.[51] It has been suggested that some endogenous substances are produced and released along with endogenous opioids in order to

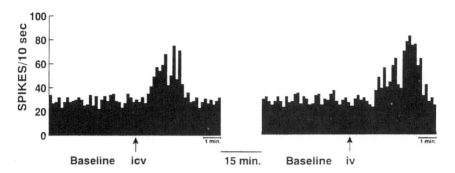

Figure 15.2 Rate meter showing the average electrical discharges (spikes each 10 s) of one neuron following i.c.v. IFN-α (25 IU per animal) and 1 h later from the same neuron following i.v. IFN-α (15 IU per gram of body weight). The i.c.v. injection of IFN-α elicits "short" latency excitatory effects and i.v. IFN-α elicits similar effects but exhibit "longer" latency excitatory effects. The figure demonstrate that IFN-α given locally within the brain (i.c.v.) or peripherally (i.v.) which cross the blood-brain barrier activate the same neurons in a similar pattern.

prevent the organism from becoming tolerant to, and dependent on, its own morphine-like peptides.[6] In mice, interaction between IFN-α and morphine has been demonstrated using single injections of morphine which resulted in reduced levels of serum IFN-α induced by poly I:C or endotoxin.[36,39] Moreover, the opiate antagonist naloxone, when given prior to IFN-α treatment, prevents the analgesia and catatonic behavior produced by IFN-α given alone.[9] This evidence suggests that the effects of IFN-α on CNS activity are mediated via a number of different mechanisms, some of which may be antagonized by naloxone.[7]

The observations that IFN-α is widely present in many systems, including the CNS, and that IFN-α receptors are localized and synthesized in the brain, form the rationale for the following hypothesis: *IFN-α is the endogenous substance which serves to prevent the development of tolerance and/or dependence on brain opioids*.[24,25,29,31,33,34] To test this hypothesis, in a series of experiments IFN-α was used to reduce the severity of the opioid abstinence syndrome and to prevent the development of physical dependence on opioids in rats.[21,22,25-28] Figure 15.2 shows the effects of single i.p. IFN-α injection (in A) and i.c.v. (in B) to morphine-dependent subjects 1 h prior to naloxone injection. IFN-α treatment remarkably decreased the severity of all opioid abstinence behaviors. Whatever the IFN-α mechanism is, the striking ability to modify the severity of opiate withdrawal behavior in a dosage comparable to that used in clinical therapy suggests that IFN-α helps maintain homeostasis in the morphine-dependent subjects. In a subsequent experiment,[26] a single IFN-α injection (i.p.) was given prior to chronic morphine treatment and similar results were obtained. The reduced severity of opioid withdrawal in these experiments indicates that IFN-α interferes with both the development of physical dependence to and withdrawal behaviors from morphine. In dose-response experiments using IFN-α and IFN-γ, it has been demonstrated that only IFN-α, but *not*

IFN-γ, remarkably reduces the severity of the opiate abstinence syndrome precipitated by naloxone in morphine-dependent rats in a dose-related manner.[25,26,29]

To test whether IFN-α exerts its effects via opiate receptors, electrical recordings were made using two preparations: (1) the guinea-pig ileum[17,66] and, (2) several neurophysiological procedures (evoked field potentials and single-cell recording) from different brain areas.[24-26,62,64-66] The guinea-pig ileum preparations have been used to study the effects of opiates on transmission between autonomic nerves and intestinal smooth muscle.[45,59] Opiate receptors are present on the presynaptic neuromuscular endings, and activation of those endings by opiate agonists at the receptor site inhibits the release of acetylcholine (ACh) at the neuromuscular junction.[59] Use of this "simple" system has allowed the characterization of various opiate receptors and the study of agonist-antagonist interactions of these receptors.[17] IFN-α elicits excitation in the guinea-pig ileum, and naloxone fails to prevent this IFN-α-induced excitation. In the neurophysiological experiments using single-unit recording, IFN-α elicited excitation in cortical neurons and in some ventromedial hypothalamic (VMH) neurons, while naloxone failed to antagonize the IFN-α effects. However, in some VMH cells IFN-α elicited the opposite effects, i.e., decreased the firing rates, and in these neurons naloxone antagonized the IFN-α effects. This observation suggests that IFN-α action on these two preparations may not be mediated via opiate receptors but rather through IFN-α receptors.[25,26] However, not all possibilities were tested; opiates can act through a number of receptors and not all of these receptors are blocked by naloxone. Thus, IFN-α may be acting on an opioid receptor that has not yet been investigated.

In brain slice preparations IFN-α elicits both an increase and a decrease in the firing rate recorded from the anterior hypothalamus (AH) and from VMH, respectively. Both responses are blocked by naloxone.[56] These studies suggest that there are at least three different functional and/or receptor sites for IFN-α within the CNS: (1) an inhibitory site — IFN-α produces reduced neuronal activity of naloxone-dependent sites possibly representing the μ-receptor type; (2) an excitatory site which is naloxone dependent, i.e., IFN-α produces excitation in neuronal activity which the opiate antagonist naloxone reverses — it is possible that this response represents the κ or δ receptor sites;[56] and (3) an excitatory site which is not naloxone dependent, i.e., IFN-α produces excitation in the neuronal activity which is not antagonized by naloxone.[26,66]

An i.c.v. injection of IFN-α suppresses the cytotoxic activity of the cells in the spleen of mice, and this effect is prevented by pretreatment with the opiate antagonist naltroxone.[77,78] Using *in vivo* rat brain membrane preparations, IFN-α was shown to inhibit the binding of ^{3}H-naloxone.[53] A number of physiological effects of IFN-α may be explained as a competition between IFN-α and naloxone for the responsible membrane binding sites. This may be the mechanism of attenuating morphine withdrawal symptoms when IFN-α is injected systemically 1 h prior to naloxone in morphine-dependent subjects.[53] In summary, cytokines include IFN-α, and opioid peptide mediation of neuroimmune interactions seems to operate via complex mechanisms involving multiple controls acting at different levels of integration.[67]

ACKNOWLEDGMENTS

The constructive help and comments of Drs. L. Laufman and S. A. Brod are appreciated and thanks to Lela Trotter for the manuscript preparation.

REFERENCES

1. Abreu, S. L., Tondreau, J., Levine, S., and Sowinski, R., Inhibition of passive localized experimental allergic encephalomyelitis by interferon, *Int. Arch. Allergy Appl. Immunol.,* 72, 30, 1983.
2. Akerman, S. K., Hochstein, H. D., Zoon, K., Browne, W., Rivera, E., and Elisberg, B., Interferon fever: absence of human leukocytic pyrogen response to recombinant α-interferon, *J. Leukocyte Biol.,* 36, 17, 1984.
3. Arnason, B. G. W., and Reder, A. T., Interferons and multiple sclerosis, *Clin. Neuropharmacol.,* 17, 495, 1984.
4. Baron, S., Tyring, S. K., Fleischmann, W. R., Jr., Coppenhaver, D. H., Niesel, D. W., Klimpel, G. R., Stanton, G. J., and Hughes, T. K., The interferons: mechanisms of action and clinical applications, *JAMA,* 266, 1375, 1991.
5. Bellinger, D. L., Felten, S. Y., Lorton, D., and Felten, D. L., Origin of noradrenergic innervation of the spleen in rats, *Brain Behav. Immunol.,* 3, 291, 1989.
6. Bertolini, A., Poggioli, R., and Fratta, W., Withdrawal symptoms in morphine-dependent rats intracerebroventrically injected with ACTH1-24 and with B-MSH, *Life Sci.,* 29, 249, 1981.
7. Birmanns, B., Saphier, D., and Abramsky, O., Interferon modifies cortical EEG activity: dose-dependence and antagonism by naloxone, *J. Neurol. Sci.,* 100, 22, 1990.
8. Blalock, J. E. and Smith, E. M., Human leukocyte interferon: structural and biological relatedness to adrenocorticotropic hormone and endorphins, *Proc. Natl. Acad. Sci. U. S. A.,* 77, 5972, 1980.
9. Blalock, J. E. and Smith, E. M., Human leukocyte interferon (Hu IFN-α): potent endorphin-like opioid activity, *Biochem. Biophys. Res. Commun.,* 101, 472, 1981.
10. Blalock, J. E. and Smith E. M., Structure and function of interferon (IFN) and neuroendocrine hormones, in *The Biology of the Interferon System,* De Maeyer, E., Galasso, G., and Schellekens, H., Eds., Elsevier, Amsterdam, 1981, 93.
11. Blalock, J. E. and Stanton, J. D., Common pathways of interferon and hormonal action, *Nature,* 283, 406, 1980.
12. Blatteis, C. M., Xin, L., and Quan, N., Neuromodulation of fever: apparent involvement of opioids, *Brain Res. Bull.,* 26, 219, 1991.
13. Bliss, T. V. P., Maintenance is presynaptic, *Nature,* 346, 698, 1990.
14. Brown, S. L. and Van Epps, D. E., Opioid peptides modulate production of interferon γ by human mononuclear cells, *Cell. Immun.,* 103, 19, 1986.
15. Bocci, V. Physiochemical and biologic properties of interferons and their potential uses in drug delivery systems, *Crit. Rev. Ther. Drug Carrier Sys.,* 9, 91, 1992.
16. Bocci, V., Paulesu, L., Muscettola, M., and Viti, A., The physiologic interferon response. VI. Interferon activity in human plasma after a meal and drinking, *Lymphokine Res.,* 4, 151, 1985.
17. Burks, T. F., Actions of drugs on gastrointestinal motility, in *Physiology of the Gastrointestinal Tract,* Vol. 1, Johnson, L. R., Christensen, J., Grossman, M., Jacobson, I. E. D., and Schultz, S. G., Eds., Raven Press, New York, 1981, 495.

18. Calvet, M. C. and Gresser, I., Interferon enhances the excitability of cultured neurons, *Nature,* 278, 558, 1979.
19. Cantell, K., Pulkkien, E., Eluoso, R., and Suominen, J., Effect of interferon on severe psychiatric diseases, *Ann. Clin. Res.,* 12, 131, 1980.
20. D'Arcangelo, G., Grassi, F., Ragozzino, D., Santoni, A., Tancredi, V., and Eusebi, F., Interferon inhibits synaptic potentiation in rat hippocampus, *Brain Res.,* 564, 245, 1991.
21. Dafny, N., Modification of morphine withdrawal by interferon, *Life Sci.,* 32, 303, 1983.
22. Dafny, N., Interferon modifies morphine withdrawal phenomena in rodents, *Neuropharmacology,* 22, 647, 1983.
23. Dafny, N., Interferon modifies EEG and EEG-like activity recorded from sensory, motor, and limbic system structures in freely behaving rats, *Neurotoxicology,* 4, 235, 1983.
24. Dafny, N., Interferon as an endocoid candidate preventing and attenuating opiate addiction, *Endocoids,* Alan R. Liss, New York, 1985, 269.
25. Dafny, N., Lee, J. R., and Dougherty, P. M., Immune response products alter CNS activity: interferon modulates central opioid function, *J. Neurosci. Res.,* 19, 130, 1988.
26. Dafny, N., Prieto Gomex, B., and Reyes Vazquez, C., Does the immune system communicate with the central nervous system? Interferon modifies central nervous system activity, *J. Neuroimmunol.,* 9, 1, 1985.
27. Dafny, N. and Reyes Vazques, C., Three different types of interferons alter naloxone induced abstinence in morphine addicted rats, *Immunopharmacology,* 9, 13, 1985.
28. Dafny, N. and Reyes Vazquez, C., Single injection of three different preparations of α-interferon modifies morphine abstinence signs for a prolonged period, *Int. J. Neurosci.,* 32, 953, 1987.
29. Dafny, N., Zielinski, M., and Reyes Vazquez, C., Alteration of morphine withdrawal to naloxone by interferon. Neuropeptides 3, 453, 1983.
30. De Sarro, G. B., Masuda, Y., Ascioti, C., Audino, M. G., and Nistico, G., Behavioural and ECG spectrum changes induced by intracerebral infusion of interferons and interleukin 2 in rats antagonized by naloxone, *Neuropharmacology,* 29, 167, 1990.
31. Doughtery, P. M., Aronowski, J., Samorajaski, T., and Dafny, N., Opiate antinociception is altered by immunomodification: the effect of interferon, cyclosporine and radiation induced immune suppression upon acute and long-term morphine activity, *Brain Res.,* 385, 401, 1986.
32. Dougherty, P. M. and Dafny, N., Interaction of immune cytokines and CNS opioids: a possible interface for stress induced immune suppression, in *Stress and Immunity,* Plotnikoff N. P., Murgo, A., Faith, R. E., and Wybran, J., Eds., Plenum Press, New York, 1991, 373.
33. Dougherty, P. M., Harper, C., and Dafny, N., The effect of α interferon, cyclosporine A and radiation immune suppression on morphine induced hypothermia and tolerance, *Life Sci.,* 39, 2191, 1986.
34. Doughtery, P. M., Pearl, J., Krajewski, K. J., Pellis, N. R., and Dafny, N., Differential modification of morphine and methadone dependence by interferon-α, *Neuropharmacology,* 26, 1595, 1987.
35. Goldstein, D. and Laszlo, J., The role of interferon in cancer therapy: a current perspective, *Cancer J. Clin.,* 38, 258, 1988.
36. Gober, W. F., Lefkowitz, S. S., and Hung, C. Y., Effect of morphine, hydromorphine, methadone, mescaline, trypan blue, vitamin A, sodium salicylate, and caffeine on the serum interferon level in response to viral infection, *Arch. Int. Pharmacodyn.,* 214, 322, 1975.

37. Gustafsson, B. and Wingströ, H., Physiological mechanisms underlying long-term potentiation, *Trends Neurosci.,* 11, 156, 1988.
38. Hori, T., Nakashima, T., Take, S., Kaizuka, Y., Mori, T., and Katafuchi, T., Immune cytokines and regulation of body temperature, food intake and cellular immunity, *Brain Res. Bull.,* 27, 309, 1991.
39. Hung, C. Y., Lefkowitz, S. S., and Geberg, W. F., Interferon inhibition by narcotic analgesics, *Proc. Soc. Exp. Biol. Med.,* 142, 106, 1973.
40. Isaacs, A. and Lindenmann, J., Virus interference. I. The interferon, *Proc. R. Soc. London,* Ser. B 147, 258, 1957.
41. Jaffe, J. H., Drug addition and drug abuse, in *The Pharmacological Basis of Therapeutics,* 8th ed., Gilman, A. G., Ralls, T. W., Nies, A. S., and Taylor, P., Eds., Macmillan, New York, 1990, 522.
42. Iivanainen, M., Laaksonen, R., Niemi, M. L., Färkkilä, M., Bergström, L., Mattson, K., Niiranen, A., and Cantell, K., Memory and psychomotor impairment following high dose interferon treatment in amyotrophic lateral sclerosis, *Acta Neurol. Scand.,* 72, 475, 1985.
43. Katafuchi, T., Take, S., and Hori, T., Roles of sympathetic nervous system in the suppression of cytotoxicity of splenic natural killer cells in the rat, *J. Physiol. (London),* 465, 343, 1993.
44. Kidron, D., Saphier, D., Ovadia, H., Weidenfeld, J., and Abramsky, O., Central administration of immunomodulatory factors alters neural activity and adrenocortical secretion, *Brain Behav. Immun.,* 3, 15, 1989.
45. Kosterlitz, H. W. and Waterfield, A. A., An analysis of the phenomenon of acute tolerance to morphine in the guinea-pig isolated brain, *Br. J. Pharmacol.,* 53, 131, 1975.
46. Krueger, J. M., Dinarello, C. A., Shoham, S., Davenne, D., Walter, J., and Kubillus, S., Interferon α-2 enhances slow-wave sleep in rabbits, *Int. J. Immunopharmacol.,* 9, 23, 1987.
47. Larsson, I., Landstrom, L. E., Larner, E., Lundgren, E., Miorner, H., and Strannegard, O., Interferon production in glia and glioma cell lines, *Infect. Immun.,* 22, 786, 1978.
48. Lysle, D. T., Coussons, M. E., Watts, V. J., and Dykstra, L. A., Morphine-induced alteratins of immune status: dose dependency, compartment specificity and antagonism by naltrexone, *J. Pharmacol. Exp. Ther.,* 265, 1071, 1993.
49. Marcovitz, R., Tsiang, H., and Hovannesiam, A. G., Production and action of interferon in mice affected with rabies virus, *Ann. Virol.,* 135E, 19, 1984.
50. Mattson, K., Niiranen, A., Iivanainen, M., Färkkilä, M., Bergström, L., Holsti, L. R., and Cantell, K., Neurotoxicity of interferon, *Cancer Treat. Rep.,* 67, 958, 1983.
51. McCain, H. W., Lamster, I. B., Bozzone, J. M., and Grbic, J. T., Endorphin modulates human immune activity via non-opiate receptor mechanisms, *Life Sci.,* 31, 1619, 1982.
52. McDonald, E. M., Mann, A. H., and Thomas, H. C., Interferons as mediators of psychiatric morbidity: an investigation in a trial of recombinant α-interferon in hepatitis-B carriers, *Lancet,* November 21, 1175, 1987.
53. Menzies, R. A., Patel, R., Hall, N. R. S., O'Grady, M. P., and Rier, S. E., Human recombinant interferon α inhibits naloxone binding to rat brain membranes, *Life Sci.,* 50, PL227, 1992.
54. Meyers, C. A. and Valentine, A. D., Neurological and psychiatric adverse effects of immunological therapy, *CNS Drugs,* 3, 56, 1995.

55. Muller, M., Fontana, A., Zbinden, G., and Ghwiler, B. H., Effects of interferons and hydrogen peroxide on CA3 pyramidal cells in rat hippocampal slice cultures, *Brain Res.*, 619, 157, 1993.
56. Nakashima, T., Hori, T., Kuriyama, K., and Kiyohara, T., Naloxone blocks the interferon-α induced changes in hypothalamic neuronal activity, *Neurosci. Lett.*, 82, 332, 1987.
57. Nakashima, T., Hori, T., Kuriyama, K., and Matsuda, T., Effects of interferon-α on the activity of preoptic thermosensitive neurons in tissue slices, *Brain Res.*, 454, 361, 1988.
58. Nakayama, T., Yamamoto, K., Ishikawa, Y., and Imai, K., Effects of preoptic thermal stimulation on the ventromedial hypothalamic neurons in rats, *Neurosci. Lett.*, 26, 177, 1981.
59. Paton, W. D. M., The action of morphine and related substances on contraction and on acetylcholine output of coaxially stimulated guinea-pig ileum, *Br. J. Pharmacol.*, 11, 119, 1957.
60. Pavol, M. A., Meyers, C. A., Rexer, J. L., Valentine, A. D., Mattis, P. J., and Talpaz, M., Pattern of neurobehavioral deficits associated with interferon α therapy for leukemia, *Neurology*, 45, 947, 1995.
61. Pestka, S., Langer, J. A., Zoon, K. C., and Samuel, C. E., Interferons and their actions, *Annu. Rev. Biochem.*, 56, 727, 1987.
62. Prieto-Gomez, B., Reyes-Vazquez, C., and Dafny, N., Differential effects of interferon on ventromedial hypothalamus and dorsal hippocampus, *J. Neurosci. Res.*, 10, 273, 1983.
63. Reite, M., Laudenslager, M., Jones, J., Crnic, L., and Kaemingk, K., Interferon decreases REM latency, *Biol. Psychiatry*, 22, 104, 1987.
64. Reyes-Vazquez, C., Prieto-Gomez, B., and Dafny, N., Novel effect of interferon on the brain: microiontophoretic application and single cell recording in the rat, *Neurosci. Lett.*, 34, 201, 1982.
65. Reyes-Vazquez, C., Prieto-Gomez, B., Georgiades, J. A., and Dafny, N., α and γ interferons effects on cortical and hippocampal neurons: microiontophoretic application and single cell recording, *Int. J. Neurosci.*, 25, 113, 1984.
66. Reyes-Vazquez, C., Weisbrodt, N., and Dafny, N., Does interferon exert its action through opiate receptors?, *Life Sci.*, 35, 1015, 1984.
67. Roda, G., Bongiorno, L., Trani, E., Urgani, A., and Marini M., Positive and negative immunomodulation by opioid peptides, *Int. J. Immunopharmacol.*, 18, 1, 1996.
68. Root-Bernstein, R. S., 'Molecular Sandwiches' as a basis for structural and functional similarities of interferons, MSH, ACTH, LHRH, myelin basic protein, and albumins, *FEBS Lett.*, 168, 208, 1984.
69. Saphier, D., Kidron, D., Abramsky, O., Trainin, N., Pecht, M., Burstein, Y., and Ovadia, H., Neurophysiological changes in the brain following central administration of immunomodulatory factors, *Isr. J. Med. Sci.*, 24, 261, 1988.
70. Saphier, D., Kidron, D., Ovadia, H., Weidenfeld, J., Abramsky, O., Burstein, Y., Pecht, M., and Trainin, N., Preoptic area (POA) multiunit activity (MUA) and cortical EEG changes following intracerebroventricular (ICV) administration of α-interferon (INF), thymic humoral factor (THF), histamine (HIS), and interleukin-1 (IL-1), *Rev. Clin. Basic Pharmacol.*, 6, 265, 1987.
71. Saphier, D., Welch, J. E., and Chuluyan, H. E., α-Interferon inhibits adrenocortical secretion via μ_1-opioid receptors in the rat, *Eur. J. Pharmacol.*, 236, 186, 1993.

72. Saphier, D., Roerig, S. C., Ito, C., Vlasak, W. R., Farrar, G. E., Broyles, J. E., and Welch, J. E., Inhibition of neural and neuroendocrine activity by α-interferon: neuroendocrine, electrophysiological, and biochemical studies in the rat, *Brain Behav. Immunol.,* 8, 37, 1994.
73. Shibata, M. and Blatteis, C. M., Differential effects of cytokines on thermosensitive neurons in guinea pig preoptic area slices, *Am. J. Physiol.,* 261, R1096, 1992.
74. Smedley, H., Katrak, M., Sikora, K., and Wheeler, T., Neurological effects of recombinant human interferon, *Br. Med. J.,* 286, 262, 1983.
75. Smith, E. M. and Blalock, J. E., Human lymphocyte production of corticotropin and endorphin-like substances: association with leukocyte interferon, *Proc. Natl. Acad. Sci. U.S.A.,* 78, 7530, 1981.
76. Stanton, G., Hughes, T., Heard, H., Georgiades, J., and Whorton, E., Modulation of a natural virus defense system by low concentration of interferons at mucosal surfaces, *J. Interferon. Res.,* 10, (Suppl. 1), S99.
77. Take, S., Mori, T., Kaizuka, Y., Katafuchi, T., and Hori, T., Central interferon-α suppresses the cytotoxic activity of natural killer cells in the mouse spleen, *Ann. NY Acad. Sci.,* 650, 46, 1992.
78. Take, S., Mori, T., Katafuchi, T., and Hori, T., Central interferon-α inhibits natural killer cytotoxicity through sympathetic innervation, *Am J. Physiol.,* 265, R453, 1993.
79. Zimmerman, E. and Krivoy, W., Antagonism between morphine and the polypeptides ACTH, $ACTH_{1-24}$, and B-MSH in the nervous system, *Prog. Brain Res.,* 39, 383, 1973.

16 Opioid Systems: Cytokines and Immunity

Hemendra N. Bhargava

CONTENTS

16.1 Introduction 233
16.2 Investigative Procedure 234
    16.2.1 Methods 235
    16.2.2 Chemicals 235
    16.2.3 *In Vitro* Studies 235
    16.2.4 *In Vivo* Studies 236
16.3 Induction and Assessment of Tolerance to Physical Dependence on Morphine 236
16.4 Immune Function Studies 237
16.5 Results and Discussion 239
    16.5.1 *In Vitro* Effects of Opioid Drugs on Immune Response Parameters 239
        16.5.1.1 Morphine and Its Metabolites 239
        16.5.1.2 Heroin and Methadone 239
        16.5.1.3 Fentanyl and Meperidine 240
        16.5.1.4 δ-Opioid Receptor Agonist and Antagonists 240
        16.5.1.5 Effects of Chronic Administration Morphine and Its Subsequent Withdrawal on Immune Parameters 241
References 242

16.1 INTRODUCTION

Drug abuse is a sociomedical problem which initially was confined to few countries but is now a worldwide phenomenon. Besides the sociomedical problem, the abuse of licit and illicit drugs also has socioeconomic problems. Although a wide variety of drugs with diverse chemical nuclei are being abused[1] the discussion in this chapter will be confined to opiates only. The medical complications of opiates involve alterations of the immune system, resulting in immunotoxicity.[2] Opiate abuse in humans, particularly of morphine or heroin, is associated with modification of immunological parameters and with an increased incidence of infectious diseases.[3-7] Opiate abuse has been linked to the emergence of the acquired immunodeficiency syndrome (AIDS), a retroviral infection spread either by sexual contact or by the sharing of needles among intravenous drug users. A plethora of studies have been described in the literature which indicate that morphine or heroin causes immuno-suppression as measured by the traditional methodology using splenic and thymic cells (natural killer (NK) cell activity, mitogen-induced proliferation of blood lymphocytes, or antibody response as measured by the plaque-forming cell assay). These effects appeared to be centrally mediated because intraperitoneal administration of N-methyl morphine, a quaternary analog of morphine, had no effect on lymphocyte proliferation, plasma corticosterone concentration, or analgesic response.[8] The immuno-suppressive activity appears to be regulated indirectly via glucocorticoid receptors.[9,10] On the other hand, morphine-induced decreases *in vivo* antibody responses were not correlated with elevated levels of corticosterone. It is possible that corticosteroids may be involved in regulation of some immune parameters and not others. Not only does exogenously administered morphine affect the immune function, but there is increasing evidence that morphine and codeine are endogenous compounds and may be immunoregulators. Intraperitoneal administration of muramyl dipeptide (MDP), a product of bacterial cell-wall degradation and a potent immunomodulatory agent, produced a significant increase in tissue (spleen, brain, small intestine, and heart) morphine levels.[12] MDP stimulates the production of interleukin-1 (IL-1) and tumor necrosis faction-α (TNF-α). It is thus possible that MDP may signal the cell to modify the immune system through a link to the endogenous opiate alkaloid morphine.[12]

Although the mechanisms by which morphine and related compounds suppress various parameters of the immune system are not known, recent studies have implicated the possible role of cytokines. The immune response appears to involve the participation of a number of cell types and is generally believed to be coordinated by CD4+ T cells. At least three functionally different types of CD4+ T-cell clones can be discriminated. In the mouse, Th1 cells are characterized by the secretion of IL-2, interferon-γ (IFN-γ), and lymphotoxin, whereas Th2 cells are characterized by the secretion of IL-4, IL-5, and IL-10. The cells have unrestricted cytokine profiles.[13] Morphine treatment was shown *in vitro* to decrease the production of IFN-γ by human peripheral blood cells.[14] Morphine-induced decrease in primary antibody response *in vitro*, as measured by the plaque-forming cell assay, was attenuated by IL-6, IL-1β, or IFN-γ in a concentration-dependent manner, but was unaffected by the addition of IL-2, IL-4, or IL-5.[15] The cytokines that reversed morphine-induced suppression of antibody response are either produced by IL-1β, IL-6 or directly stimulate IFN-γ macrophages, and suggest that macrophages may be an important target for the *in vivo* effects of morphine.

16.2 INVESTIGATIVE PROCEDURE

Based on the above findings, studies were undertaken to determine the changes in the production of various cytokines by the immune cells following *in vitro* and *in vivo* exposure of splenic lymphocytes to morphine and its derivatives and metabolites. Studies were also undertaken to determine the effects of *in vitro* exposure to murine splenic lymphocytes from B6C3F1 (C57BL/6 × C3H) hybrid mice. The parameters of immune function included measurement of IL-2, IL-4, IL-6, TNF,

Opioid Systems: Cytokines and Immunity

NK-cell activity, and evaluation of the *in vitro* induction of cytotoxic T lymphocytes (CTLs), a population of T cells bearing the CD3/CD8 surface antigens. These cells are capable of exhibiting cytotoxicity towards specific target cells after prior exposure to antigen, and thus represent a central effector mechanism of cell-mediated immunity and host resistance.[16]

16.2.1 METHODS

Both *in vitro* and *in vivo* studies were undertaken using 6- to 8-week-old female B6C3F1 mice obtained from the National Cancer Institute, (Frederick, MD) and housed five per cage in polypropylene cages over hardwood bedding. The animals were provided with rodent chow and tap water *ad libitum* and were maintained on a 12-h light/dark cycle. For *in vivo* studies, to lessen the potential for morphine-induced mortality the mice were allowed to reach approximately 22 g prior to treatment with morphine.

16.2.2 CHEMICALS

All test drugs were obtained from the National Institute on Drug Abuse (NIDA), Rockville, MD and were dissolved in saline for injection. Implantable pellets were obtained from NIDA and contained either 75 mg morphine in an inert carrier or inert carrier only. RPMI-1640 medium, Hank's balanced salt solution (HBSS), and Dulbecco's phosphate-buffered saline (PBS) were all purchased from JRH Biosciences (Lenexa, KS), and Ultraculture was from Biowhittaker (Walkersville, MD). Brewer thioglycolate was from Fisher Scientific (Chicago, IL), anti-murine IgM antibody was from Jackson ImmunoResearch (West Grove, PA), bacterial lipopolysaccharide (LPS) was from List Biologicals (Campbell, CA), Eagle's minimal essential medium (E-MEM) was from Gibco (Grand Island, NY), and fetal bovine serum (FBS) was from Hyclone (Logan, UT). Mercaptoethanol, mitomycin C, and phenazine methosulfate were all obtained from Sigma Chemical Co. (St. Louis, MO). Antimurine CD3 monoclonal antibody was from Pharmingen (San Diego, CA), interleukin-2 (IL-2) and interleukin-4 (IL-4) were both form PeproTech (Rocky Hill, NJ), and interleukin-6 (IL-6) was purchased from R&D Systems (Minneapolis, MN). ^{51}Cr was from ICN (Costa Mesa, CA), and 2,3-bis [2-methoxy-4-nitro-5-sulfophenyl]-5- [(phenyl-amino)carbonyl]-2H-tetrazolium hydroxide (XTT) was obtained from Polysciences (Warrenton, PA).

16.2.3 IN VITRO STUDIES

The animals were euthanized, the spleens removed, and single-cell suspensions were prepared by rubbing the spleens through sterile nylon mesh (Spectrum Medical, Los Angeles, CA). Five individual splenocyte pools consisting of three to five animals per pool were utilized for each assay. The drugs were dissolved in culture medium consisting of RPMI-1640 (Biowhittaker, Walkersville, MD) supplemented with 10% heat-inactivated fetal bovine serum (FBS; Hyclone, Logan, UT), *l*-glutamine, and gentamicin. Sterile stock solutions of 1 mM were prepared, and log dilutions of each stock were subsequently prepared in culture medium.

16.2.4 *In Vivo* Studies

The effects of morphine tolerance, physical dependence, and abstinence on the immune function were studied.

16.3 INDUCTION AND ASSESSMENT OF TOLERANCE TO PHYSICAL DEPENDENCE ON MORPHINE

Mice were rendered tolerant-dependent on morphine by a modified procedure described earlier.[17] Mice were divided into two groups: one was injected in the morning with vehicle, and the other group was injected with morphine sulfate (20 mg/kg, i.p.). The injections were repeated in the afternoon of day 1. On day 2, the vehicle-injected mice were implanted s.c. with a placebo pellet, while morphine-injected mice were implanted with a morphine pellet. On day 5, the pellets were removed and 8 h later the dose-response relationships with respect to the analgesic and hypothermic effects of morphine were determined.

The analgesic effect of morphine (5, 10, and 20 mg/kg, s.c.) was measured by using a tail-flick apparatus.[17] The tail-flick latencies to thermal stimulation were determined before and at various times up to 300 min after injection of an appropriate dose of morphine. The basal latencies were found to be 2 s in duration. A cut-off time of 10 s was used to prevent any injury to the tail. The basal response was subtracted from the effect induced by morphine, and the area under the time-response curve, $AUC_{0-300\,min}$, was calculated for each mouse. The data were expressed as mean $AUC_{0-300\,min} \pm$ S.E.M. Eight mice were used for each treatment group. The analgesic response to morphine in placebo and morphine pellet-implanted mice was compared by using Student's *t*-test. A value of $p \leq .05$ was considered to be statistically significant.

Tolerance to the hypothermic response to morphine was determined by measuring the changes in colonic temperature of placebo and morphine pellet-implanted mice after the administration of varying doses of morphine (5, 10, and 20 mg/kg, s.c.). Upon injection of an appropriate dose of morphine, the colonic temperature was recorded at various time intervals for a period of 300 min. Colonic temperature was recorded using a Cole-Palmer digital thermometer (Model 8502-20) and a thermistor probe. The animals were lightly hand-restrained, and temperature was recorded when a constant reading was obtained. The data were expressed as mean $AUC_{0-300\,min} \pm$ S.E.M. Eight mice were used for each dose of morphine. The data were analyzed as described for the analgesic response.

Both abrupt and antagonist (naltrexone)-induced withdrawal responses were used to determine the degree of physical dependence. For abrupt withdrawal responses, mice were made tolerant-dependent as described above. The body weight and colonic temperature were determined at the time of pellet implantation, pellet removal, and at various intervals after pellet removal. The differences in the body weight and body temperature of the placebo- and morphine-treated mice were determined by analysis of variance (ANOVA) followed by Scheffe's *S*-test. A value of $p \leq .05$ was considered statistically significant.

In precipitated withdrawal, naltrexone (10–200 μg/kg) was injected s.c. and the mice were placed on a circular platform. The total number of jumps per group as well as the number of mice exhibiting stereotypic jumping out of a total number tested was determined. The ED_{50} value of naltrexone and its 95% confidence limits were determined by the method described earlier.[18]

16.4 IMMUNE FUNCTION STUDIES

For immune function studies, the animals were treated with vehicle or morphine and subsequently implanted with placebo or morphine pellets as described above. Two sets of mice were used. In the first group, the animals were sham-operated and the pellets were left intact at the time of immune function studies; these mice were labeled as morphine-tolerant animals. In the second group, the pellets were removed and the animals were used for immunological evaluation 8 h later; these mice were labeled as morphine-abstinent animals. Animal sacrifice for both groups was performed 8 h following sham operation or pellet removal. All immune function data were analyzed for statistical difference from respective placebo controls by analysis of variance (ANOVA) and Dunnett's multicomparison test. A p value of $\leq .05$ was considered significant.

Immunopathology — To assess changes in body weight, mice were weighed prior to pellet implantation, prior to removal of the pellets, and at sacrifice. In addition, the spleen weight and cellularity and the thymus weight and cellularity were quantitated.

Preparation of splenic lymphocytes for *in vitro* studies — Animals were euthanized by cervical dislocation, and the spleens were removed aseptically. The individual spleens were transferred to Petri dishes containing approximately 1 ml of culture medium and 1 square of nylon macromesh (Spectrum Medical, Los Angeles, CA). A single-cell suspension was prepared by cutting the capsule several times and then gently expelling the contents by rubbing through the nylon mesh with a sterile syringe plunger. The spleens were gently pipetted several times through a syringe fitted with a 23-gauge needle, and transferred to sterile tubes. The preparations were allowed to settle at room temperature for approximately 5 min before transfer to clean tubes. The cells were washed once by centrifugation, and the cell pellet was resuspended in fresh medium; the number of cells were determined by Coulter counting, and cell viability was assessed by vital dye exclusion. All *in vitro* immune functional assays were performed on cells from individual animals.

***In vitro* proliferation of B lymphocytes** — Evaluation of B-lymphocyte function was performed using a previously described method.[19] Lymphocytes suspended in RPMI-1640/10% FBS were added to 96-well, flat-bottom microculture plates at 1×10^5 cells per well. Recombinant murine IL-4 and antimurine IgM antibody were added to each well at final concentrations of 400 pg/ml and 1 μg/ml, respectively. The plates were incubated at 37°C for 68 h. XTT was added to all wells 4 h prior to assay termination and optical density was determined as below for the cytokine bioassays.

Production of T-cell derived cytokines — Isolated lymphocytes were adjusted to a concentration to 2×10^6 viable cells per milliliter in culture medium and added to 24-well culture plates (Costar) in a volume of 1 ml per well. Antimurine CD3 antibody was added to the cell cultures at a final concentration of 20 ng/ml, and the plates were incubated at 37°C for 48 h. At the end of the incubation period the cultures were harvested by centrifugation to pelletize the cells and debris in the cultures, and the resultant supernatant fluids were stored at –70°C until assayed.

Measurement of interleukin-2 — IL-2 production was quantitated by the method outlined earlier.[20] Lymphocyte culture supernatants were cultured for 20 h with log-phase CTLL-2 indicator cells. A calorimetric indicator reagent was prepared with 1 mg/ml XTT and 0.25 mM phenazine methosulfate in PBS, and this reagent was added to all wells. The plates were incubated for an additional 4 h, and the optical density of the solution in each well was measured with a microplate spectrophotometer at 450 nm. The concentration of IL-2 in the samples was extrapolated from a reference curve constructed with recombinant murine IL-2.

Measurement of interleukin-4 — IL-4 production was quantitated using a modification of the method described earlier.[21] Lymphocyte culture supernatants were cultured for 44 h with log-phase CT.4S cells, XTT reagent was added, and the optical density was measured as described above. The concentration of IL-4 in the test samples was extrapolated from a reference curve constructed with recombinant murine IL-4.

Basal and augmented NK activity — Basal and augmented NK cell activity was measured by a modification of the method described before.[22] Splenocytes were cultured in duplicate in E-MEM/10% FBS for 24 h both with and without rhIL-2 and various concentrations of drugs. The cells were then harvested, adjusted to 5×10^6 viable cells per milliliter in RPMI-1640/10% FBS, and two 1:3 dilutions of the cell suspension were prepared. These dilutions were added to quadruplicate wells of 96-well, round-bottom microculture plates in a volume of 100 µl per well. YAC-1 cells were radiolabeled with ^{51}Cr and added to all wells at a concentration of 5×10^3 viable cells per well. The plates were incubated at 37°C for 4 h, harvested with a supernatant collection system (Skatron, Sterling, VA), and radiolabel release was quantitated in a gamma counter. Percent cytotoxicity was calculated by the formula:

$$\text{Percent cytotoxicity} = \frac{ER - SR \times 100}{TR - SR}$$

where ER represents experimental release, SR represents spontaneous release, and TR represents total releasable counts.

***In vitro* induction of cytotoxic T lymphocytes (CTL)** — CTL function was evaluated by a modification of the method described earlier.[23] Splenocytes were cultured for 5 days in serum-free Ultraculture medium containing mitomycin C-inactivated P815 stimulator cells. After this induction phase the effector cells were harvested and resuspended to 5×10^6 viable cells per milliliter, and two serial dilutions of the cell suspension were prepared. The dilutions were added to quadruplicate wells of 96-well, round-bottom microculture plates, and radiolabeled P815 cells were added to all wells at a concentration of 2×10^4 viable cells per well. The plates were incubated at 37°C for 4 h, and the supernatant fluids were collected

(Skatron, Sterling, VA). Percent specific cytotoxicity was calculated by the formula as described above for the NK cell assay.

Preparation of macrophage-derived cytokines — Mice were injected i.p. with Brewer thioglycolate medium to elicit peritoneal macrophages on the day of pellet implantation. Following sacrifice, peritoneal exudate cells from individual animals were harvested by peritoneal lavage with HBSS, and the macrophages were enriched by adherence to plastic for 2 h. Adherent monolayers were removed with a cell scraper and washed once in PBS. Macrophages were adjusted to a concentration of 1×10^5 viable cells per milliliter in culture medium supplemented with 10 µg/ml LPS and the cultures were incubated at 37°C for 48 h. Culture supernatants were harvested by centrifugation and stored at –70°C until assay.

Measurement of tumor necrosis factor (TNF) — TNF-α production by thioglycolate-elicited peritoneal macrophages was assessed as described previously.[24] Macrophage culture supernatants were incubated for 24 h with L929 fibroblasts in the presence of actinomycin D. Then 4 h prior to harvest, XTT reagent was added and the optical density was measured as described above. The concentration of TNF in the test samples was determined from a reference curve constructed with recombinant TNF-α.

16.5 RESULTS AND DISCUSSION

16.5.1 IN VITRO EFFECTS OF OPIOID DRUGS ON IMMUNE RESPONSE PARAMETERS

16.5.1.1 Morphine and Its Metabolites

Murine B6C3FI splenic lymphocytes or peritoneal macrophages were cultured *in vitro* at concentrations of 0.0001–100 µmol/l morphine sulfate, morphine-3-glucuronide, morphine-6-glucuronide, or normorphine. B-cell proliferation was significantly suppressed following exposure to all drugs. Production of IL-2, IL-4, and IL-6 was affected only moderately by all drugs except morphine-6-glucuronide, which produced a marked suppression at 100 µmol/l. Both basal and augmented NK cell function were unaffected by any drug except morphine-6-glucuronide, which enhanced NK cell activity at concentrations between 0.0001 and 1.0 µmol/l. In contrast, both morphine-3-glucuronide and morphine-6-glucuronide significantly inhibited cytotoxic T-lymphocyte induction at concentrations between 0.0001 and 100 µmol/l, whereas morphine and normorphine were inactive in this assay. In summary, in the absence of direct cellular cytotoxicity, a differential immunomodulation was observed following *in vitro* exposure to morphine and its metabolites.[25]

16.5.1.2 Heroin and Methadone

In vitro exposure to heroin resulted in decreased B-cell proliferation at concentrations of 1–100 µ*M* and methadone had a similar effect at concentrations of 0.1–100 µ*M*.

Production of IL-2 was suppressed by 0.1–100 μM of heroin, whereas exposure to methadone appeared to result in a generalized modulation, with suppression of IL-2 at most concentrations. In contrast, IL-4 production was only affected at the 100-μM concentration of both drugs. CTL was suppressed by exposure to 100 μM heroin, whereas NK cell activity was suppressed at high concentrations of both heroin and methadone, and macrophage function also was differentially affected. The results presented here indicate that both drugs display some immunomodulatory potential.[27]

These studies are especially important in patients exposed to HIV. Since morphine and heroin suppress the immune function directly and also enhance the replication of HIV-1,[28] the combined effects in AIDS patients could be disastrous.

16.5.1.3 Fentanyl and Meperidine

Exposure to fentanyl and meperidine was associated with a differential suppression of IL-4 production by T cells, and a more generalized suppression of cytokine production by macrophages. In addition, T-cell cytolytic activity was suppressed at high drug concentrations. B-cell proliferation and NK cell activity were also inhibited, but to a lesser degree than noted with T-cell function. Addition of naltrexone to the cultures did not reverse these alterations in immune function, suggesting that these changes are not mediated via opioid receptors.[26]

16.5.1.4 δ-Opioid Receptor Agonists and Antagonists

16.5.1.4.1 δ-Opioid Receptor Agonists

Peptides examined in this study included DSLET (H-Tyr-D-Ser-Gly-Phe-Leu-Thr-OH), DTLET (H-Tyr-D-Thr-Gly-Phe-Leu-Thr-OH), DPDPE (H-Tyr-D-Pen-Gly-Phe-D-Pen-OH), DPDPE-trifluoroacetate (TFA-Tyr-D-Pen-Gly-Phe-D-Pen-OH), DALA-HCl (HCl-Tyr-D-Ala-Gly-Phe-Met-NH$_2$), DADLE (H-Tyr-D-Ala-Gly-Phe-D-Leu-OH), and deltorphin-1 (H-Tyr-D-Ala-Phe-Asp-Val-Val-Gly-NH). Murine splenic lymphocytes and peritoneal macrophages were cultured *in vitro* with peptides at concentrations of 0.00001–10 μM. These peptides had minimal effects on B-cell proliferation at any concentration examined. In comparison, enhancement of cytokine production by T-helper cells occurred following exposure to several of the compounds — to a significant extent with DPDPE-trifluoroacetate, or deltorphin-1, and most pronounced at concentrations between 0.00001 and 0.1 μM. Likewise, IL-6 production by macrophages was significantly augmented by exposure to these peptides. NK cell function was significantly enhanced by *in vitro* exposure to several of the peptides, with enhancement generally noted at concentrations between 0.00001 and 0.01 μM. However, some of the peptides (most notably DADLE) greatly suppressed NK cell activity. These data suggest that δ-opioid agonists are broadly immunostimulatory.[29] IL-2 secretion from CD4 T-cells was also shown to be suppressed by δ-opioid receptor agonists.[30]

16.5.1.4.2 δ-Opioid Receptor Antagonists

Drugs examined in this study included TIPP (Tyr-Tic-Phe-Phe-OH), D-TIPP (Tyr-D-Tic-Phe-Phe-NH$_2$), and ICI 174864 HCl (N, N-diallyl-Tyr-Aib-Aib-Phe-Leu-OH

HCl; Aib = α-aminoisobutyric acid). In addition, the effects of nonpeptidic δ-opioid receptor antagonists like 7-benzylidene-naltrexone (BNTX), naltriben (NTB), and naltrindole (NTI) were also studied.

None of the peptides examined significantly affected B-cell proliferation. Production of IL-2 by T cells was not consistently affected by exposure to either TIPP or D-TIPP, but was significantly suppressed at 10 μM ICI 174864. Production of IL-4, however, was significantly suppressed by low concentrations of either TIPP or D-TIPP, and by cells 10 μM exposed to ICI 174864. IL-6 production by macrophages was unaffected except for sporadic incidents of enhanced production in cell exposure to ICI 174864. NK cell function exhibited a differential pattern of suppression, with the greatest degree of suppression observed following exposure to TIPP and only slight suppression in cells exposed to either D-TIPP or ICI 174864. These data suggest that peptidic δ-opioid receptor antagonists do not exhibit the same pattern or degree of immunosuppressive activity as the nonpeptidic antagonists at equivalent *in vitro* concentrations.[31]

In vitro exposure to BNTX or NTI resulted in immunosuppression of B-cell proliferation, cytokine production, and NK cell activity at concentrations between 0.1 and 10 μM. However, NTB had limited effect on immune parameters.[32] These results are consistent with those of Arakawa et al.[33,34]

16.5.1.5 Effects of Chronic Administration Morphine and Its Subsequent Withdrawal on Immune Parameters

Female B6C3FI mice were rendered tolerant-dependent on morphine by a combination of injections and pellet implantation. Mice were injected with morphine sulfate (20 mg/kg, s.c.) twice a day on day 1. On day 2, they were implanted s.c. with a 75-mg morphine pellet for 3 days. On day 5, the pellets were either left intact (tolerant) or removed 8 h prior (abstinent) to carrying out the immune function tests. A high degree of tolerance to the analgesic and hypothermic effects of morphine developed as a result of this procedure. Similarly, physical dependence also developed as evidenced by the signs of the abrupt and naltrexone-precipitated abstinence syndrome. Implantation with morphine pellets resulted in a profound, statistically significant reduction in spleen and thymus weight and cellularities, with the greatest degree of reduction noted in abstinent animals. Morphine tolerance was associated with suppressed B-cell proliferation following *in vitro* stimulation, as well as IL-2 and IL-4 production by T cells. NK cell activity was significantly reduced in morphine-tolerant, but not in morphine-abstinent, mice following a 24-h incubation in the presence or absence of IL-2. In comparison, the *in vitro* induction of cytotoxic T cells was significantly depressed in morphine-abstinent, but not morphine-tolerant, animals. Exposure to morphine apparently had limited effect on macrophage function as assessed by production of tumor necrosis factor. These studies demonstrate a differential effect on immune effector and regulatory mechanisms in morphine tolerance and abstinence processes.[35]

The effects of naltrexone on tolerance/dependence, as well as alterations in cellular immune function induced by morphine administration, were determined.

Mice were rendered tolerant to and physically dependent on morphine by subcutaneous implantation of pellets containing 75 mg of morphine. Implantation of naltrexone pellets (10 mg) blocked the development of tolerance to the analgesic action of morphine, as well as the development of physical dependence. Morphine suppressed lymphoid organ weights and cellularities, and this suppression was blocked by naltrexone. B-cell proliferation was suppressed in morphine-tolerant but not in morphine-abstinent mice, and this suppression was exacerbated by naltrexone. Morphine tolerance and abstinence were associated with suppression of IL-2 production, which was completely blocked by naltrexone. NK cell activity was not significantly affected by either morphine or naltrexone exposure. The results suggest that the effects of morphine on the immune system are at least partially mediated through opioid receptors.[36]

Endogenous opioid endorphins also appear to modulate cytokine production, which in turn may affect the immune status of AIDS patients. β-endorphin and its fragments β-endorphin$_{6-31}$ and β-endorphin$_{18-31}$ enhanced IL-2 and IL-4 production. Similarly, the production of IL-6 and IFN-γ was increased by these peptides. Thus, at least two distinct sites of β-endorphin were found to exert stimulatory effects on cytokine production.[37] Interestingly, in patients with HIV infection and basal CD4$^+$ counts above 200 cells per milliliter, intermittent infusions of IL-2 were shown to produce substantial and sustained increases in CD4$^+$ counts and yet there was no increase in plasma HIV RNA levels.[38] Thus, there appears to be a clear connection between opioid drugs and immunity that may be mediated by alterations in the production of cytokines.

REFERENCES

1. House, R. V., Thomas P. T., and Bhargava,, H. N. *In vitro* exposure to peptidic delta opioid receptor antagonists results in limited immunosuppression, *Neuropeptides*, 31, 89, 1997.
2. Pillai, R. M. and Watson, R. R., *In vitro* immunotoxicology and immunopharmacology: studies on drugs of abuse, *Toxicol. Lett.*, 53, 269, 1990.
3. Louria, D. B., Hensle, T., and Rose J., The major medical complications of heroin addiction, *Ann. Intern. Med.*, 67, 1, 1967.
4. McDonough, R. J., Madden, J. J., Falek, A., Shafer, D. A., Pline, M., Gordon, D., Bokos, P., Kuehnle, J. C., and Mendelson, J., Alteration of T and null lymphocyte frequencies in the peripheral blood of human opioid addicts: *In vivo* evidence for opiate receptor sites on T lymphocytes, *J. Immunol.*, 125, 2539, 1980.
5. Tubaro, E., Borelli, G., Croce, C., Cavallo, G., and Santiangeli, C., Effects of morphine on resistance to infection, *J. Infect. Dis.*, 148, 656, 1983.
6. Tubaro, E., Avico, U., Santiangeli, C., Zuccaro, P., Cavallo, G., Pacifici, R., Croce, C., and Borelli, G., Morphine and methadone impact on human phagocyte physiology, *Int. J. Immunopharmacol.*, 7, 865, 1985.
7. Bhargava, H. N., Opioid peptides, receptors and immune function, in *Drugs of Abuse: Chemistry, Pharmacology, Immunology, and AIDS*, Pham, P. T. K. and Rice, K., Eds., NIDA Monogr. 96, National Institute on Drug Abuse, Rockville, MD, 1990.
8. Hernandez, M. C., Flores, L. R., and Bayer, B. M., Immunosuppression by morphine is mediated by central pathways, *J. Pharmacol. Exp. Ther.*, 267, 1336, 1993.

9. Bryant, H. V., Berton, E. W. Kenner, J. R., and Holaday, J. W., Role of adrenal cortical activation in the immuno-suppressive effects of chronic morphine treatment, *Endocrinology,* 128, 3253, 1991.
10. Pruett, S. B., Han, Y.-C., and Fuchs, B. A., Morphine suppresses primary humoral immune response by a predominantly indirect mechanism, *J. Pharmacol. Exp. Ther.,* 262, 923, 1992.
11. Lockwood, L. L., Silbert, L. H., Fleshner, M. Laudenslager, M. L., Watkins, L. R., and Maier, S. F., Morphine-induced decreases in *in vivo* antibody response, *Brain Behav. Immunol.,* 8, 24, 1994.
12. Horak, P., Haberman, F., and Spectgor, S., Endogenous morphine and codeine in mice — effect of muramyl dipeptide, *Life Sci.,* 52, PL255, 1993.
13. Mosmann, T. R., Schumacher, J. H., Street, N. F., Budd, R., O'Garra, A., Fong, T. A. T., Bond, M. W., Moore, K. W. M., Sher, A., and Fiorentino, D. F. Diversity of cytokine synthesis and function of mouse CD4$^+$ T cells, *Immunol. Rev.,* 123, 209, 1991.
14. Peterson, P. K., Sharp, B., Gekker, G., Brummitt, C., and Keane, W. F., Opioid-mediated suppression of interferon-γ production by cultured peripheral blood mononuclear cells, *J. Clin Invest.,* 80, 824, 1987.
15. Bussiere, J. L., Adler, M. W., Rogers, T. J., and Eisenstein, T. K., Cytokine reversal of morphine-induced suppression of the antibody response, *J. Pharmacol. Exp. Ther.,* 264, 591, 1993.
16. Berke, G., The cytolytic T lymphocyte and its mode of action, *Immunol. Lett.,* 20, 169, 1989.
17. Bhargava, H. N., Dissociation of tolerance to the analgesic and hypothermic effects of morphine by using thyrotropin releasing hormone, *Life Sci.,* 29, 1015, 1981.
18. Litchfield, J. T. and Wilcoxon, F. A., Simplified method of evaluating dose effect experiments, *J. Pharmacol. Exp. Ther.,* 96, 99, 1949.
19. Abbas, A. K., Urioste, S., Collins, T. L., and Boom, W. H., Heterogeneity of helper/inducer T-lymphocytes. IV. Stimulation of resting and activated B cells by Th1 and Th2 clones, *J. Immunol.,* 144, 2031, 1990.
20. Gillis, S., Ferm, M. M., Ou, W., and Smith, K. A., T-cell growth factor: parameters of production and a quantitative microassay for activity, *J. Immunol.,* 120, 2027, 1978.
21. Hu-Li, J., Ohara, J., Watson, C., Tsang, W., and Paul, W. E., Derivation of a T-cell line that is highly responsive to IL-4 and IL-2 (CT. 4R) and an IL-2 hyporesponsive mutant of that line (CT. 4S), *J. Immunol.,* 142, 800, 1989.
22. Talmadge, J. E., Immunoregulation and immunostimulation of murine lymphocytes by recombinant human interleukin-2, *J. Biol. Res. Modif.,* 4, 18, 1985.
23. House, R. V., Lauer, L. D., Murray, M. J., and Dean, J. H., Suppression of T-helper cell function in mice following exposure to the carcinogen 7, 12-dimethylbenza[a]anthracene and its restoration by interleukin-2, *Int. J. Immunopharmacol.,* 9, 89, 1987.
24. Meager, A., Leung, H., and Woolley, J., Assays for tumor necrosis factor, *J. Immunol. Meth.,* 141, 1, 1989.
25. Thomas, P. T., Bhargava, H. N., and House, R. V., Immunomodulatory effects of *in vitro* exposure to morphine and its metabolites, *Pharmacology,* 50, 51, 1995.
26. House, R. V., Thomas P. T., and Bhargava, H. N., *In vitro* evaluation of fentanyl and meperidine for immunomodulatory activity, *Immunol. Lett.,* 46, 117, 1995.
27. Thomas, P. T., House, R. V., and Bhargava, H. N., Direct cellular immunomodulation produced by diacetylmorphine (heroin) or methadone, *Gen. Pharmacol.,* 26, 123, 1995.

28. Peterson, P. K., Gekker, G., Schut, R., Hu, S., Balfour, H. H., Jr., and Chao, C. C., Enhancement of HIV-1 replication by opiates and cocaine: the cytokine connection, in *Drugs of Abuse, Immunity and AIDS*, Friedman, H., et al., Eds., Plenum Press, New York, 1993, 181.
29. House, R. V., Thomas P. T., and Bhargava, H. N., A comparative study of immunomodulation produced by *in vitro* exposure to delta opioid receptor agonist peptides, *Peptides*, 17, 75, 1996.
30. Shahabi, N. A. and Sharp, B. M., Antiproliferative effects of δ-opioids on highly purified $CD4^+$ and $CD8^+$ murine T cells, *J. Pharmacol. Exp. Ther.*, 273, 1105, 1995.
31. House, R. V., Thomas P. T., and Bhargava, H. N., Immuno-toxicology of opioids, inhalants and other drugs of abuse, *NIDA Monogr.*, 173, 175, 1997.
32. House, R. V., Thomas, P. T., Kozak, J. T., and Bhargava, H. N., Suppression of immune function by non-peptidic delta opioid receptor antagonists, *Neurosci. Lett.*, 198, 119, 1995.
33. Arakawa, K., Akarmi, T., Okamoto, M., et al., Immunosuppressive effect of δ-opioid receptor antagonist on xenogeneic mixed lymphocyte response. *Transplant Proc.*, 24, 696, 1992.
34. Arakawa, K., Akarmi, T., Okamoto, M., et al., Immunosuppression by delta opioid receptor antagonists, *Transplant Proc.*, 25, 738, 1993.
35. Bhargava, H. N., Thomas, P. T., Thorat, S. N., and House, R. V., Effect of morphine tolerance and abstinence on cellular immune function. *Brain Res.* 642, 1, 1994.
36. Bhargava, H. N., House, R. V., Thorat, S. N., and Thomas, P. T., Effect of naltrexone on morphine-induced tolerance and physical dependence and changes in cellular immune function in mice, *Brain Res.*, 690, 121, 1995.
37. Van den Bergh, P., Rozing, J., and Nagelkerken, L., Identification of distinct sites of β-endorphin that stimulate lymphokine production by murine $CD4^+$ T cells, *Lymphokine Cytokine Res.*, 13, 63, 1994.
38. Kovacs, J. A., Vogel, S., Albert, J. M., Falloon, J., Davey, R. T., Jr., Walker, R. E., Polis, M. A., Spooner, K., Metcalf, J. A., Baseler, M., Fyfe, G., and Lane, H. C., Controlled trial of interleukin-2 infusions in patients infected with the human immuno-deficiency virus, *N. Engl. J. Med.*, 335, 1350, 1996.

17 Opioid Growth Factor, [Met⁵]-Enkephalin, and the Etiology, Pathogenesis, and Treatment of Gastrointestinal Cancer

Ian S. Zagon, Jill P. Smith, and Patricia J. McLaughlin

CONTENTS

17.1 Introduction 245
17.2 Hypothesis 247
17.3 Human Colon Cancer 248
    17.3.1 Does OGF Function in Human Colon Cancer? 248
    17.3.2 Does OGF Modulate Human Colon Cancer Xenografts? 249
    17.3.3 Is There Evidence That a Growth-Related Opioid Peptide, OGF, and Its Receptor, ζ, Are Present in Human Colon Cancer? 251
    17.3.4 What Is the Nature of the Opioid Receptor Related to OGF? 251
    17.3.5 Is There Evidence for the Autocrine Production of OGF? 251
    17.3.6 Are OGF and the ζ Receptor Involved in Cellular Renewal of the Normal Colon? 253
17.4 Human Pancreatic Cancer 253
17.5 Conclusion 254
Acknowledgments 256
References 256

17.1 INTRODUCTION

Cancer of the gastrointestinal tract, including the esophagus, stomach, colon, and rectum, occurs in more than 250,000 Americans each year, and claims the lives of well over 100,000 individuals annually.[1-8] The incidence of some of these neoplasias (i.e., colorectal) places it as the second leading cause of cancer-related deaths in the U.S., and by the year 2000 it is estimated that this will be the most common neoplasia. The average survival rate from these cancers depends on the type of neoplasia encountered. In the case of colorectal cancer, for example, the overall 5-year survival rate is 45%. However, in patients with pancreatic carcinoma the median survival after diagnosis is 3 to 6 months, with a 5-year survival of approximately 2%. Although treatment modalities include surgery, chemotherapy, radiation therapy, immunotherapy, and hormonal therapy, in many types of gastrointestinal cancer (e.g., pancreatic) these are ineffective. Indeed, the survival rate has not changed for pancreatic patients in the past 40 years, but the incidence of this cancer has tripled in this time period.

Recently, a number of naturally occurring growth factors have been identified that may be of importance in tumorigenesis,[9] including neoplasias of the gastrointestinal tract.[9-28] Alterations (e.g., quantity, integrity) in these growth factors and/or their receptors has been postulated to lead to disease states such as cancer, and may offer important clues in the treatment of oncogenesis. Some factors associated with gastrointestinal cancer include somatostatin, gastrin, bombesin (gastrin-releasing peptide), colon mitosis inhibitor, transforming growth factor-β, transforming growth factor-α, epidermal growth factor, basic fibroblast growth factor, and neurotensin. Interestingly, a number of growth factors found to be important in other neoplastic cells appear to have no function in some gastrointestinal cancers. For instance, insulin-like growth factors I and II have little influence on the growth of HT-29 human colon cancer cells. One growth factor that has emerged as potentially important in basic science and clinical medicine is gastrin and related antagonists.[12,13,29-33] Administration of gastrin has been shown to stimulate the growth of carcinogen-induced colon cancers in animals, human colon cancer cells in culture, and human tumors xenografted to nude mice. Gastrin works by way of a direct mechanism, and does so through a CCK-B/gastrin-like receptor.

In addition to effects on such modalities as analgesia and behavior, endogenous opioid peptides serve to regulate the growth of developing, neoplastic, renewing, and healing tissues, and function in both prokaryotes and eukaryotes.[34-64] One native opioid peptide, [Met^5]-enkephalin, has emerged as a receptor-mediated growth factor; this peptide is encoded by the preproenkephalin gene (PPE). Native peptides related to [Met^5]-enkephalin, but not synthetic opioids, have shown some growth regulatory properties to a lesser extent. To distinguish the role of this peptide as a growth factor in neural and nonneural cells and tissues, and in prokaryotes and eukaryotes, [Met^5]-enkephalin has been termed opioid growth factor (OGF). OGF is a potent, reversible, species-unspecific, tissue-unspecific peptide that is a negative growth regulator. This neuropeptide is autocrine and possibly paracrine produced and secreted, and effective at concentrations consistent with the binding affinity of its receptor. OGF is rapid in biological action, not cytotoxic, obedient to the intrinsic rhythms of the cell (e.g., circadian rhythm), and acts stereospecifically and in a receptor-mediated fashion. The peptide is targeted to cell proliferation, but also appears to influence cell migration, differentiation, survival, and tissue organization. OGF is tonically active, and interference with peptide-receptor interaction accelerates cell growth by removing inhibitory signaling.

OGF interacts with an opioid receptor, zeta (ζ), to influence growth. Unlike other opioid receptors, the function (e.g., growth), distribution (neural and nonneural), transient appearance during ontogeny, ligand specificity (i.e., [Met⁵]-enkephalin), competitive inhibition profile, subcellular location (i.e., nucleus), and the fact that ligands for other known opioid receptors (e.g., μ, δ, and κ) do not influence growth have provided a unique set of characteristics that distinguish the ζ receptor from other opioid receptors. The ζ receptor has 4 binding subunits: 32, 30, 17, and 16 kDa that are of basic pH (i.e., 7.3_8.3). Chymotryptic peptide mapping revealed that the 32- and 30-kDa subunits had strong homology, but not with the 17- and 16-kDa polypeptides. The 17- and 16-kDa subunits had only a partial homology to each other. Polyclonal and monoclonal antibodies raised against the binding subunits recognized all four binding polypeptides in developing brain by Western blotting; no reaction was recorded in postmitotic tissues (e.g., adult brain).

Enkephalins and endorphins have been reported in the gastrointestinal tract, and are believed to be involved in neurotransmission and motility.[65,66] A centrally mediated sympathethic connection between opioid peptides and glucose homeostasis has been documented, as well as a direct effect on islet insulin secretion and liver glucose production.[67] In the gastrointestinal tract of the adult rat, μ, δ, and κ opioid receptors, as well as immunoreactive [Met⁵]-enkephalin and β-endorphin, have been observed.[67] Acid extracts prepared from human tissues revealed [Met⁵]-enkephalin and [Leu⁵]-enkephalin, and these eluates displaced radiolabeled [Met⁵]-enkephalin from rat brain membranes in an opioid radioreceptor assay.[68] β-endorphin also has been described in the human gastrointestinal system.[69-71]

A number of studies have documented opioids in human gastrointestinal cancer.[15,40,41,51,54,71,72] Opioid peptides and opioid receptors have been found in human colon cancer tissues.[15,40,41,51,69-72] Bostwick et al.[72] detected the expression of opioid peptides and their precursors in cancer of the large bowel by immunocytochemistry, and Tari et al.[70,71] using radioimmunoassays reported β-endorphin-like immunoreactivity in normal mucosa and the muscle layer of the normal colon, adenocarcinomas derived from the colon mucosa, and colon polyps which were histologically confirmed to be adenoma without a focus of carcinoma. β-Endorphin has been found to be normal in the plasma of patients with colorectal cancer.[69] Gustin and co-workers[73] mention (but do not present data) examining a colon adenocarcinoma in a study exploring opioid peptides by radioimmunoassay and immunocytochemistry.

17.2 HYPOTHESIS

We have been aware for 10 years that endogenous opioids are associated with human gastrointestinal cancer. Following our discovery that opioids are involved in neoplasia, we performed a study that screened a series of diverse human and animal cancers for the presence of opioid peptides and receptors.[51] Amongst the tumors examined, we studied moderately differentiated and well-differentiated adenocarcinomas of the rectum. Using ligands that recognized neurotransmitter-like opioid receptors, we detected δ- and κ-like receptors, but not μ-like, opioid receptors. We also found both [Met⁵]-enkephalin and β-endorphin in these tumor tissues. One interpretation of the results is that opioid receptors and peptides known to be present in the normal human gastrointestinal tract and presumably acting in a neurotransmitter role, are retained and expressed in tumor cells. Alternatively, and possibly concurrently, one or more of these peptides and receptors could have been involved in growth-related activities as reported for other cancers (e.g., neuroblastoma, pancreatic).

In recent years, intrigued by the discovery that OGF, [Met^5]-enkephalin, functions as a growth factor in developing, renewing, and carcinogenic cells and/or tissues, a concerted effort was initiated to examine the hypothesis that endogenous opioids inhibit human gastrointestinal cancer by interaction with a unique opioid receptor. The following is an account of the evidence and findings supporting this hypothesis.

17.3 HUMAN COLON CANCER

17.3.1 Does OGF Function in Human Colon Cancer?

Given the presence of OGF in human colon cancer tissue,[51] the function of this endogenous opioid in such neoplasias required elucidation. Using a tissue culture model of HT-29 human colon cancer cells, addition of OGF depressed growth by 17 to 41% at 12 to 72 h. OGF action exhibited a dose-response relationship, had a reversible and noncytotoxic effect, was active at physiological concentrations, and was opioid receptor mediated. If OGF was inhibitory and tonically active, as well as autocrine produced, then blockade of OGF-ζ receptor interaction with an antibody to OGF would be predicted to increase cell number. Such experiments indicated that the number of cells in cultures treated with antibodies to OGF was almost twofold greater than control cultures, but no differences in cell number were observed between nonimmune IgG-treated and control cultures. Likewise, blockade of OGF-receptor interfacing with a potent and long-acting opioid antagonist such as NTX should result in accelerated growth. Persistent opioid receptor blockade with NTX yielded a 38% increase in cell number from control levels within 12 h of the initiation of experimentation. At 24, 48, and 72 h after addition of NTX, the HT-29 cells exhibited increases in number ranging from 38 to 55% of control values. To ascertain dependence of OGF action on serum-related factors, the cultures of HT-29 cells were adapted to serum-free conditions over a 4-week period by slowly reducing serum from 10%, 2.5%, 1.25%, to 0%. Despite some small loss in plating efficiency (about 10% less than in serum-containing cultures) cell growth in serum-free and serum-containing media was comparable (e.g., at 48 h, serum-free cultures contained 94% of the number of cells observed in cultures incubated in serum-containing media). Growth inhibition by OGF was prominent in the serum-free media, reducing cell number by one-third of control values by 48 h after addition of drug treatment (Figure 17.1). From a wide variety of opioid peptides screened (both synthetic and natural, and selective for μ, δ, κ, ε, and σ receptors) only OGF, [Met^5]-enkephalin, influenced growth. The modulatory capability of this endogenous opioid was not restricted to HT-29 cells, but also was noted in two other human colon cancer cell lines examined: WiDr and COLO 205. OGF continually repressed growth because an increase in cell number was noted when cells were exposed to the potent and long-acting opioid antagonist naltrexone (NTX) or an antibody to OGF. These results led to the suggestion that OGF functions in a direct, tonic, and inhibitory manner on the growth of human colon cancer cells, and that exogenous application of OGF can influence cell growth in a tissue culture setting.

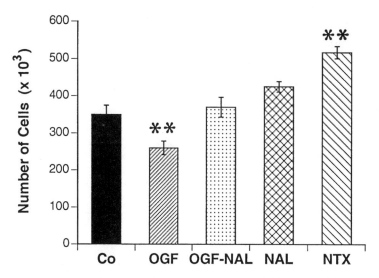

Figure 17.1 Effects of OGF and NTX on the number of HT-29 human colon cancer cells grown in serum-free media after 48 h. Twenty-four hours after seeding of HT-29 cells acclimated to serum-free conditions, cultures were exposed to OGF, OGF-NAL, NAL alone, or NTX at a concentration of 10^{-6} M; media and compounds were changed daily. Data represent means ± SE for at least two aliquots per well from two wells per group. **Significantly different at $p < .01$. (From Zagon, I. S., Hytrek S. D., and McLaughlin, P. J., *Am. J. Physiol.*, 271, R511, 1996. With permission.)

17.3.2 DOES OGF MODULATE HUMAN COLON CANCER XENOGRAFTS?

Although OGF had a marked influence on colon cancer cell growth *in vitro*, it may not reflect OGF's function *in vivo*. Thus, experiments were designed to subject nude mice inoculated with a dosage of tumor cells that would produce maximal tumor incidence to OGF. Administration of the peptide on a daily basis at dosages of 0.5, 5, or 25 mg/kg to nude mice inoculated with HT-29 colon cancer cells showed that more than 80% of the mice receiving OGF beginning at the time of tumor cell injection did not exhibit neoplasias within 3 weeks in comparison to a tumor incidence of 93% in control subjects (Table 17.1). Even 7 weeks after cancer cell inoculation, 57% of the mice given OGF did not display a tumor. OGF delayed tumor appearance and growth in animals developing colon cancer with respect to the control group. The suppressive effects of OGF on oncogenicity were opioid receptor mediated. These results show that exogenous administration of a naturally

TABLE 17.1
Tumor Latency and Incidence in Nude Mice Inoculated With 1×10^6 HT-29 Human Colon Cancer Cells and Treated With Various Doses of OGF ([Met^5]-enkephalin) and/or Naloxone or an Equivalent Volume of Sterile Water (Control)

	Control	0.5 mg/kg OGF	5 mg/kg OGF	25 mg/kg OGF	5 mg/kg OGF + 10 mg/kg Nal	10 mg/kg Nal
Latency for palpable tumor, days	10.2 ± 1.6	18.6 ± 3.2[b]	19.2 ± 2.2	22.0 ± 4.5[b]	13.7 ± 1.6	9.3 ± 1.4
Latency for visible tumor, days	11.2 ± 1.6	22.4 ± 2.6[a]	20.2 ± 2.2[b]	23.0 ± 4.5[a]	14.5 ± 0.9	10.3 ± 1.4
Latency for measurable tumor appearance, days	12.4 ± 1.4	28.0 ± 0.0[a]	22.2 ± 2.3[a]	24.0 ± 4.5[a]	15.6 ± 0.9	12.8 ± 1.3
	6–19	(28)	13–22	(12–32)	(13–20)	(10–16)
No. of mice with tumors 21 days after inoculation	13/14	2/14[a]	4/14[a]	2/14a	7/8	4/5
No. of mice with tumors 50 days after inoculation	13/14	7/14[b]	6/14[a]	5/14[a]	7/8	4/5

Note: Data for latencies represent means ± SE. Range shown in parentheses. OGF, opioid growth factor; Nal, naloxone. n = 14 for OGF alone, 8 for OGF + Nal, 5 for Nal alone, and 14 for control.

[a] Significantly different from controls at $p < .01$.
[b] Significantly different from controls at $p < .05$.

From Zagon, I. S., Hytrek, S. D., Lang, C. M., McGarrity, T. J., Wu, Y., and McLaughlin, P. J., *Am. J. Physiol.*, 271, R780, 1996. With permission.

occurring opioid peptide acts as a potent negative regulator of human gastrointestinal cancer under *in vivo* conditions.

17.3.3 Is There Evidence That a Growth-Related Opioid Peptide, OGF, and Its Receptor, ζ, Are Present in Human Colon Cancer?

Before proceeding with any further functional and mechanistic questions about opioids and human colon cancer, it was incumbent to address the question of whether an opioid-related growth factor (i.e., OGF) and its receptor, ζ, were associated with these neoplasias. In an earlier report[51] we had found [Met⁵]-enkephalin in human rectal adenocarcinomas by radioimmunoassay and immunocytochemistry. Subsequent examination revealed that both [Met⁵]-enkephalin and the ζ opioid receptor were present in human colon cancers obtained at surgery, human colon cancer cell lines growing in tissue culture (Figure 17.2), and human colon cancer xenografts in nude mice.[40,41] The immunocytochemical staining was cytoplasmic and perinuclear; although receptor binding would later be localized to the nuclear fraction. The immunoreactivity detected in the cytoplasm was related to the site of synthesis and location of function. These studies suggest that OGF ([Met⁵]-enkephalin) and the ζ opioid receptor appear to be ubiquitous components in neoplasia of the human colon.

17.3.4 What Is the Nature of the Opioid Receptor Related to OGF?

Although we found that OGF had a marked inhibitory influence on the course of colon neoplasia in nude mice, and that this antitumor activity was receptor mediated, little was known about the opioid receptor(s) involved in growth regulation. Ligand-binding assays using HT-29 human colon cancer tissue and [³H]-[Met⁵]-enkephalin were performed to characterize this receptor. Specific and saturable binding was detected in nuclear homogenates, and Scatchard analysis revealed that the data were consistent for a single binding site with a binding affinity (K_d) of 15.4 ± 2.0 nM and a binding capacity (B_{max}) of 364.8 ± 25.7 fmol/mg protein (Figure 17.3). Subcellular fractionation studies indicated that binding was restricted to the nuclear fraction. Competition experiments showed that cold [Met⁵]-enkephalin was the most effective ligand at displacing [³H]-[Met⁵]-enkephalin. Binding to radiolabeled [Met5]-enkephalin also was detected in colon cancers obtained from surgical resections. The function, pharmacological and biochemical characteristics, distribution, and subcellular location of this OGF receptor in human colon cancer were consistent with the ζ opioid receptor.

17.3.5 Is There Evidence for the Autocrine Production of OGF?

Two aspects of experimentation may suggest that OGF is made by the colon cancer cells/tissues. First, as mentioned above, OGF can be detected in human colon tumor

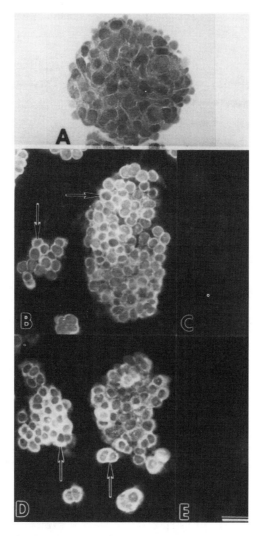

Figure 17.2 Photomicrographs of HT-29 colon cancer cells growing in culture for 3 days. Cultures were stained with hematoxylin and eosin (A) or antibodies to OGF [Met^5]-enkephalin (1:100) (B) or ζ-opioid receptor (D); rhodamine-conjugated IgG (1:100) served as the secondary antibody. Note immunoreactivity in cytoplasm of epithelial cells (arrows); cell nuclei were not stained. No immunoreactivity was detected in preparations stained with antibodies preabsorbed with either the OGF (C) or the 17-kDa polypeptide of the ζ-opioid receptor (E). Scale bar, 75 μm. (From Zagon, I. S., Hytrek S. D., and McLaughlin, P. J., *Am. J. Physiol.*, 271, R511, 1996. With permission.)

cells and specimens at biopsy. Second, using tissue culture (albeit with the inclusion of serum-containing media), OGF was detected in the media and in the cells. These data could indicate that OGF is produced by the colon cancer cells either in response to the process of tumorigenesis and/or as a compromised defense mechanism to cancer.

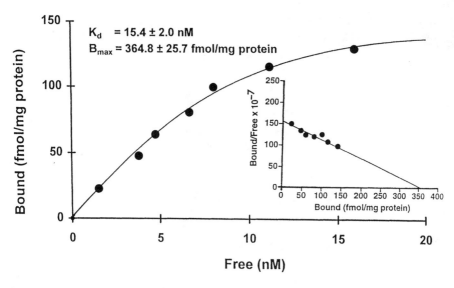

Figure 17.3 Representative saturation isotherm of specific binding of [^{3}H]-[Met^5]-enkephalin to HT-29 nuclear homogenates. Mean ± SE binding affinity (K_d) and maximal binding capacity (B_{max}) values from 16 experiments performed in duplicate are shown. Representative Scatchard plot (inset) of specific binding of radiolabeled [Met^5]-enkephalin to HT-29 nuclear homogenates revealed a one-site model of binding. (From Hytrek, S. D., Smith, J. P., McGarrity, T. J., McLaughlin, P. J., Lang, C. M., and Zagon, I. S., *Am. J. Physiol.*, 271, R115, 1996. With permission.)

17.3.6 ARE OGF AND THE ζ RECEPTOR INVOLVED IN CELLULAR RENEWAL OF THE NORMAL COLON?

Because OGF and the ζ receptor function to modulate cell growth in neoplastic cells, what is the relationship of these elements to normal cell renewal? Especially pertinent in the case of colon is that the cells which are neoplastic normally undergo cellular replenishment. In a recent publication[53] we have discovered that OGF plays a role in homeostatic cellular renewal of epithelial cells in the esophagi of mice (Figure 17.4). Both OGF and the ζ opioid receptor were present in mouse and human esophagi, as well as in murine colon. Although more work is needed, these results indicate that OGF and the ζ receptor are components involved in cell turnover of normal colon epithelial cells. If this is the case, one might speculate that defects in peptide and/or receptor could alter the process of cellular renewal in the colon, thereby contributing to the etiology of these cancers.

17.4 HUMAN PANCREATIC CANCER

In order to address the question of whether OGF modulates other gastrointestinal cancer, experiments were conducted to examine if OGF influences human pancreatic cancer neoplasia. Nude mice inoculated with human pancreatic cancer (BxPC-3) cells and receiving 5 mg/kg of opioid growth factor ([Met^5]-enkephalin; OGF) three

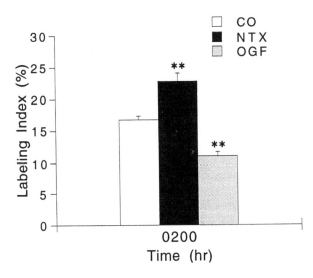

Figure 17.4 Labeling indexes of cells (i.e., percentage of labeled cell nuclei) in the basal epithelial layer of esophagus from adult mice examined at 0200 h. NTX, OGF, or an equivalent volume of sterile water (CO) was injected 2 h before mice were killed. [^{3}H]thymidine was given 30 min before death. In comparison to control levels, NTX increased the labeling index by 36%, whereas OGF decreased DNA synthesis by 34%. Values are means ±SE; at least 2000 cells per group were analyzed. **Significantly different from CO labeling index by $p < .01$. (From Zagon, I. S., Wu, Y., and McLaughlin, P. J., *Am. J. Physiol.*, 272, R1094, 1997. With permission.)

times daily exhibited a marked retardation in tumorigenicity compared to animals injected with sterile water (controls) (Table 17.2). OGF-treated animals had a delay of 43% in initial tumor appearance compared to control subjects (10.6 days). At the time when all of the control mice had tumors, 62% of the mice in the OGF group had no signs of neoplasia. Tumor tissue excised from mice after 30 days was assayed for levels of [Met^5]-enkephalin and ζ opioid receptors. Tumor tissue levels of [Met^5]-enkephalin were 24-fold greater in OGF-treated mice than controls, but plasma levels of OGF were 8.6-fold lower in animals receiving OGF. Specific and saturable binding of radiolabeled [Met^5]-enkephalin to nuclear homogenates of pancreatic tumor tissue was recorded, with a binding affinity (K_d) of 10 nM and a binding capacity (B_{max}) of 46.8 fmol/mg protein. Binding capacity, but not affinity, of [^{3}H]-[Met^5]-enkephalin was reduced by 58% of control levels in tumor tissue from mice of the OGF group. OGF and the zeta (ζ) opioid receptor were detected in human pancreatic tumor cells by immunocytochemistry. These results demonstrate that an endogenous opioid and its receptor are present in human pancreatic cancer, and act as a negative regulator of tumorigenesis *in vivo*.

17.5 CONCLUSION

Endogenous opioid peptides serve as the body's substance for regulating cell proliferation. We have identified an opioid peptide — [Met^5]-enkephalin — that

TABLE 17.2
Latency and Incidence of Tumor Appearance in Nude Mice Inoculated With BxPC-3 Human Pancreatic Tumor Cells and Treated With 5 mg/kg OGF Three Times Daily or Sterile Water

	Control	OGF
Latency (days) for visible tumor[a]	9.4 ± 1.1	12.7 ± 1.4[d]
Latency (days) for initial tumor appearance[a]	10.6 ± 1.1	15.2 ± 1.2[d]
Latency days for tumors of 62.5 mm³ in size[a]	12.2 ± 0.8	17.2 ± 1.1[d]
Number of mice with tumors 7 days after tumor cell inoculation[b]	3/10[c]	0/13[c]
Number of mice with tumors 14 days after tumor cell inoculation[b]	8/10	5/13[c]
Number of mice with tumors 15 days after tumor cell inoculation[b]	10/10	8/13[c]
Number of mice with tumors 22 days after tumor cell inoculation[b]	10/10	12/13
Number of mice with tumors 30 days after tumor cell inoculation[b]	10/10	12/13

[a]Data are means ± SE. Values were analyzed using one-way analysis of variance with subsequent comparisons made with Newman-Keuls tests.
[b]All tumors were ≥.5 mm³ Data were analyzed using chi-square tests.
[c]Significantly different from control group at $p < .05$.
[d]Significantly different from control group at $p < .01$.

From Zagon, I. S., Hytrek, S. D., Smith, J. P., and McLaughlin, P. J., *Cancer Lett.*, 112, 167, 1997. With permission.

represses the synthesis of molecules involved in cellular replication. This peptide, termed opioid growth factor (OGF) to signify its relationship to growth, is normally in the gastrointestinal tract and functions to control the cellular renewal processes required to replace the lining of the esophagus, stomach, colon, and rectum. Normally, this peptide regulates the delicate balance of cell generation in relationship to cell death. We now find that OGF is involved in cancers of the gastrointestinal area. Thus, supplementation of OGF to animals with these types of cancers suppresses the initiation and progress of neoplasia. These data would suggest that the normal levels of OGF are altered in individuals with gastrointestinal cancer, and that restoration (and perhaps excess) of this peptide to more homeostatic levels, has a profound effect on cellular replication. Thus, one may postulate that problems in transcription for preproenkephalin, the gene encoding OGF, and/or translation of OGF, may be defective in these forms of cancer. This would give rise to a situation of a deficit of OGF, and because this peptide is in a constant equilibrium in the cell in order to govern processes of cell renewal, cells would escape the mechanism of OGF regulation. Presumably, this could contribute to the etiology and pathogenesis of these cancers.

One should not forget that OGF interacts with the ζ opioid receptor, and that problems with transcription and/or translation of this receptor also could be associated with the origins — and in the treatment — of gastrointestinal cancer. Therefore, defective ζ receptors (e.g., affinity, concentration) would compromise the action of OGF, and permit cells to replicate in an abnormal fashion.

Thus, the work presented may speak to the issue of the treatment of at least colorectal and pancreatic cancers in humans. If, indeed, dysfunction of OGF levels does occur, then replacement of OGF (e.g., systemically or locally by implants of matrix material containing OGF) possibly could be of import to at least delaying the early stages of growth or as adjuvant therapy following resection. In a related scenario, if alteration in ζ receptor occurs, rectifying this receptor to its normal status (perhaps by replacement through genetic mechanisms) would be in order.

ACKNOWLEDGMENT

This work was supported by NIH grant CA-66783.

REFERENCES

1. Steele, G., Tepper, J., Motwani, B. T., and Bruckner, H. W., Adenocarcinoma of the colon and rectum, in *Cancer Medicine*, 3rd ed., Holland, J. F., Free, E., Bast, R. C., Kufe, D. W., Morton, D. L., and Weichselbaum, R. R., Eds., Lea and Febiger, Philadelphia, 1993, 1493.
2. Neugut, A. I., Jacobson, J. S., and Rella, V. A., Prevalence and incidence of colorectal adenomas and cancer in asymptomatic persons, *Gastrointest. Endosc. Clin. N. Am.*, 7, 387, 1997.
3. Warshaw, A. L. and Fernandez-del Castillo, F., Pancreatic carcinoma, *N. Engl. J. Med.*, 326, 455, 1992.
4. Boland, C. R., The biology of colorectal cancer, *Cancer*, Suppl. 71, 4180, 1993.
5. Boring, C. C., Squires, T. S., and Tong, T., Cancer statistics, *CA*, 42, 19, 1992.
6. MacLennan, R., Diet and colorectal cancer, *Int. J. Cancer*, Suppl. 10, 10, 1997.
7. Lynch, J., The genetics and natural history of hereditary colon cancer, *Semin. Oncol. Nursing*, 13, 91, 1997.
8. Johnston, P. G. and Allegra, C. J., Colorectal cancer biology: clinical implications, *Semin. Oncol.*, 22, 418, 1995.
9. Aaronson, S. A., Growth factors and cancer, *Science*, 254, 1146, 1991.
10. Anzano, M. A., Rieman, D., Prichett, W., Bowen-Pope, D. F., and Grieg, R., Growth factor production by human colon carcinoma cell lines, *Cancer Res.*, 49, 2898, 1989.
11. Thorup, I., Histomorphological and immunohistochemical characterization of colonic aberrant crypt foci in rats: relationship to growth factor expression, *Carcinogenesis*, 18, 465, 1997.
12. Smith, J. P. and Solomon, T. E., Effects of gastrin, proglumide, and somatostatin on growth of human colon cancer, *Gastroenterology*, 95, 1541, 1988.
13. Smith, J. P., Stock, E. A., Wotring, M. G., McLaughlin, P. J., and Zagon, I. S., Characterization of the CCK-B/gastrin-like receptor in human colon cancer, *Am. J. Physiol.*, 271, R797, 1996.
14. Markowitz, S. D., Molentin, K., Gerbic, C., Jackson, J., Stellato, T., and Willson, J. K. V., Growth stimulation by coexpression of transforming growth factor-α and epidermal growth factor-receptor in normal and adenomatous human colon epithelium, *J. Clin. Invest.*, 86, 356, 1990.
15. Hytrek, S. D., Smith, J. P., McGarrity, T. J., McLaughlin, P. J., Lang, C. M., and Zagon, I. S., Identification and characterization of ζ-opioid receptor in human colon cancer, *Am. J. Physiol.*, 271, R115, 1996.

16. Cendan, J. C., Souba, W. W., Copeland, E. M., and Lind, D. S., Characterization and growth factor stimulation of L-arginine transport in a human colon cancer cell line, *Ann. Surg. Oncol.*, 2, 257, 1995.
17. Baghdiguian, S., Verrier, B., Gerard, C., and Fantini, J., Insulin like growth factor I is an autocrine regulator of human colon cancer cell differentiation and growth, *Cancer Lett.*, 62, 23, 1992.
18. Barnard, J. A. and Warwick, G., Butyrate rapidly induces growth inhibition and differentiation in HT-29 cells, *Cell Growth Differ.*, 4, 4915, 1993.
19. Ciardiello, F., Kim, N., Saeki, T., Dono, R., Persico, M. G., Plowman, G. D., Garrigues, J., Radke, S., Todaro, G. J., and Salomon, D. S., Differential expression of epidermal growth factor-related proteins in human colorectal tumors, *Proc. Natl. Acad. Sci. U.S.A.*, 99, 7792, 1991.
20. Coffey, R. J., Shipley, G. D., and Moses, H. L., Production of transforming growth factors by human colon cancer lines, *Cancer Res.*, 46, 1164, 1986.
21. Culouscou, J.-M., Remacle-Bonnet, M., Garrouste, F., Fantini, J., Marvaldi, J., and Pommier, G., Production of insulin-like growth factor II (IGF-II) and different forms of IGF-binding proteins by HT-29 human colon carcinoma cell line, *J. Cell. Physiol.*, 143, 405, 1990.
22. Huang, S., Truijillo, J. M., and Chakrabarty, S., Proliferation of human colon cancer cells: role of epidermal growth factor and transforming growth factor α, *Int. J. Cancer*, 52, 978, 1992.
23. Iishi, H., Tatsuta, M., Baba, M., Yamamoto, R., and Taniguchi, H., Enhancement by bombesin of colon carcinogenesis and metastasis induced by azoxymethane in Wistar rats, *Int. J. Cancer*, 50, 834, 1992.
24. Lahm, H., Suardet, L., Laurent, P. L., Fischer, J. R., Ceyhan, A., Givel, J. C., and Odartchenko, N., Growth regulation and co-stimulation of human colorectal cancer cell lines by insulin-like growth factor I, II and transforming growth factor α, *Br. J. Cancer*, 65, 341, 1992.
25. Radulovic, S., Miller, G., and Schally, A. V., Inhibition of growth of HT-29 human colon cancer xenografts in nude mice by treatment with bombesin/gastrin releasing peptide antagonist (RC-3095), *Cancer Res.*, 51, 6006, 1991.
26. Roberts, A. B., Anzano, M. A., Wakefield, L. M., Rocke, N. S., Stern, D. F., and Sporn, M. B., Type β transforming growth factor: a bifunctional regulator of cellular growth, *Proc. Natl. Acad. Sci. U.S.A.*, 82, 119, 1985.
27. Skraastand, O., and Reichelt, K. L., An endogenous colon mitosis inhibitor reduces proliferation of colon carcinoma (HT-29) in serum-restricted medium, *Virchows Arch. B Cell Pathol.*, 56, 393, 1989.
28. Wan, C.-W., McKnight, M. K., Brattain, D. E., and Yeoman, L. C., Different epidermal growth factor responses and receptor levels in human colon carcinoma cell lines, *Cancer Lett.*, 43, 139, 1988.
29. Smith, J. P., Kramer, S. T., and Demers, L. M., Effects of gastrin and difluoromethylornithine on growth of human colon cancer, *Dig. Dis. Sci.*, 38, 520, 1993.
30. Smith, J. P., Rickabaugh, C. A, McLaughlin, P. J., and Zagon, I. S., Cholecystokinin receptors and PANC-1 human pancreatic cancer cells, *Am. J. Physiol.*, 265, G149, 1993.
31. Smith, J. P., Liu, G., Soundararajan, V., McLaughlin, P. J., and Zagon, I. S., Identification and characterization of CCK-B/gastrin receptors in human pancreatic colon cancer cell lines, *Am. J. Physiol.*, 266, R277, 1994.
32. Smith, J. P., Fantaskey, A. P, Liu, G., and Zagon, I. S., Identification of gastrin as a growth peptide in human pancreatic cancer, *Am. J. Physiol.*, 268, R135, 1995.

33. Smith, J. P., Shih, A, Wu, Y., McLaughlin, P. J., and Zagon, I. S., Gastrin regulates growth of human pancreatic cancer in a tonic and autocrine fashion, *Am. J. Physiol.*, 270, R1078, 1996.
34. McLaughlin, P. J., Opioid antagonist modulation of rat heart development, *Life Sci.*, 54, 1423, 1994.
35. McLaughlin, P. J. and Zagon, I. S., Modulation of human neuroblastoma transplanted into nude mice by endogenous opioid systems, *Life Sci.*, 41, 1465, 1987.
36. Murgo, A. J., Inhibition of B16-BL6 melanoma growth in mice by methionine-enkephalin, *J. Natl. Cancer Inst.*, 75, 341, 1985.
37. Zagon, I. S., Endogenous opioids and neural cancer: an immunoelectron microscopic study, *Brain Res. Bull.*, 22, 1023, 1989.
38. Zagon, I. S., Goodman, S. R., and McLaughlin, P. J., Demonstration and characterization of zeta (ζ), a growth-related opioid receptor, in a neuroblastoma cell line, *Brain Res.*, 511, 181, 1990.
39. Zagon, I. S., Goodman, S. R., and McLaughlin, P. J., Zeta (ζ), the opioid growth factor receptor: identification and characterization of binding subunits, *Brain Res.*, 605, 50, 1993.
40. Zagon, I. S., Hytrek, S. D., Lang, C. M., Smith, J. P., McGarrity, T. J., Wu, Y., and McLaughlin, P. J., Opioid growth factor ([Met^5]-enkephalin) prevents the incidence and retards the growth of human colon cancer, *Am. J. Physiol.*, 271, R780, 1996.
41. Zagon, I. S., Hytrek S. D., and McLaughlin, P. J., Opioid growth factor, [Met^5]-enkephalin: a tonically active negative regulator of human colon cancer cell proliferation in tissue culture, *Am. J. Physiol.*, 271, R511, 1996.
42. Zagon, I. S., Isayama, T., and McLaughlin, P. J., Preproenkephalin mRNA expression in the developing and adult rat brain, *Mol. Brain Res.*, 21, 85, 1994.
43. Zagon, I. S. and McLaughlin, P. J., Naltrexone modulates tumor response in mice with neuroblastoma, *Science,* 221, 671, 1983.
44. Zagon, I. S. and McLaughlin, P. J., Endogenous opioid systems, stress, and cancer, in *Enkephalins-Endorphins: Stress and the Immune System*, Plotnikoff, N. P., Murgo, A. J., Faith, R. E., and Good, R. A., Eds., Plenum Press, New York, 1986, 81.
45. Zagon, I. S. and McLaughlin, P. J., Endogenous opioids and the growth regulation of a neural tumor, *Life Sci.*, 43, 1313, 1988.
46. Zagon, I. S. and McLaughlin, P. J., Endogenous opioid systems regulate growth of neural tumor cells in culture, *Brain Res.*, 490, 14, 1989.
47. Zagon, I. S. and McLaughlin, P. J., Opioid antagonist modulation of murine neuroblastoma: a profile of cell proliferation and opioid peptides and receptors, *Brain Res.*, 480, 16, 1989.
48. Zagon, I. S., and McLaughlin, P. J., The role of endogenous opioids and opioid receptors in human and animal cancers, in *Stress and Immunity*, Plotnikoff, N. P., Murgo, A. J., Faith, R. E., and Wybran, J., CRC Press, Boca Raton, FL, 1991, 343.
49. Zagon, I. S. and McLaughlin, P. J., Opioid growth factor receptor in the developing nervous system, in *Receptors in the Developing Nervous System,* Vol. 1, Growth Factors and Hormones, Zagon, I. S. and McLaughlin, P. J., Eds., Chapman and Hall, London, 1993, 39.
50. Zagon, I. S. and McLaughlin, P. J., Production and characterization of polyclonal and monoclonal antibodies to the zeta (ζ) opioid receptor, *Brain Res.*, 630, 295, 1993.
51. Zagon, I. S., McLaughlin, P. J., Goodman, S. R., and Rhodes, R. E., Opioid receptors and endogenous opioids are present in diverse human and animal cancers, *J. Natl. Cancer Inst.*, 79, 1059, 1987.

52. Zagon, I. S., Wu, Y., and McLaughlin, P. J., Opioid growth factor (OGF) inhibits DNA synthesis in mouse tongue epithelium in a circadian rhythm-dependent manner, *Am. J. Physiol.*, 267, R645, 1994.
53. Zagon, I. S., Wu, Y., and McLaughlin, P. J., The opioid growth factor, [Met^5]-enkephalin, and the zeta (ζ) opioid receptor are present in human and mouse gastrointestinal tract and inhibit DNA synthesis in the esophagus, *Am. J. Physiol.*, 272, R1094, 1997.
54. Zagon, I. S., Hytrek, S. D., Smith, J. P., and McLaughlin, P. J., Opioid growth factor (OGF) inhibits human pancreatic cancer transplanted into nude mice, *Cancer Lett.*, 112, 167, 1997.
55. Bartolome, J. V., Bartolome, M. V., Lorber, B. A., Dileo, S. J., and Schanberg, S. M., Effects of central administration of β-endorphin on brain and liver DNA synthesis in preweaning rats, *Neuroscience*, 40, 289, 1991.
56. McLaughlin, P. J., Regulation of DNA synthesis of myocardial and epicardial cells in the developing rat heart by [Met^5]-enkephalin, *Am. J. Physiol.*, 271, R122, 1996.
57. Villiger, P. M. and Lotz, M., Expression of prepro-enkephalin in human articular chondrocytes is linked to cell proliferation, *EMBO J.*, 11, 135, 1992.
58. Davila-Garcia, M. I. and Azmitia, E. C., Effects of acute and chronic administration of leu-enkephalin on cultured serotonergic neurons: evidence for opioids as inhibitory neuronal growth factors, *Dev. Brain Res.*, 56, 35, 1990.
59. Hauser, K. F., McLaughlin, P. J., and Zagon, I. S., Endogenous opioid systems and the regulation of dendritic growth and spine formation, *J. Comp. Neurol.*, 281, 13, 1989.
60. Steine-Martin, A. and Hauser, K. F., Opioid-dependent growth of glial cultures: suppression of astrocyte DNA synthesis by met-enkephalin, *Life Sci.*, 46, 91, 1990.
61. Isayama, T., McLaughlin, P. J., and Zagon, I. S., Endogenous opioids regulate cell proliferation in the retina of developing rat, *Brain Res.*, 544, 79, 1991.
62. Vertes, Z., Melegh, G., Vertes, M., and Kovacs, S., Effect of naloxone and D-Met^2-pro^5-enkephalinamide treatment on the DNA synthesis in the developing rat brain, *Life Sci.*, 31,119, 1982.
63. Barg, J., Belcheva, M., McHale, R., Levy, R., Vogel, Z., and Coscia, C. J., β-Endorphin is a potent inhibitor of thymidine incorporation into DNA via μ- and κ-opioid receptors in fetal rat brain cell aggregates in culture, *J. Neurochem.*, 60, 765, 1993.
64. Meriney, S. D., Ford, M. J., Oliva, D., and Pilar, G., Endogenous opioids modulate neuronal survival in the developing avian ciliary ganglion, *J. Neurosci.*, 11, 3705, 1991.
65. Konturek, S. J., Opiates and the gastrointestinal tract, *Am. J. Gastroenterol.*, 74, 285, 1980.
66. North, R. A. and Egan, T. M., Actions and distributions of opioid peptides in peripheral tissues, *Brit. Med. Bull.*, 39, 71, 1983.
67. Khawaja, X. Z., Green, I. C., Thorpe, J. R., and Titheradge, M. A., The occurrence and receptor specificity of endogenous opioid peptides within the colon and liver of the rat, *Biochem. J.*, 267, 233, 1990.
68. Feurle, G. E., Helmstaedter, V., and Weber, U., Met- and leu-enkephalin immuno- and bio-reactivity in human stomach and colon, *Life Sci.*, 31, 2961, 1982.
69. Esposti, D., Lissoni, P., Tancini, G., Barni, S., Crispino, S., Paolorossi, F., Rovelli, F., Ferri, L., Cattaneo, G., Esposti, G., Lucini, V., and Fraschini, F., A study on the relationship between the pineal gland and the opioid system in patients with cancer, *Cancer*, 62, 494, 1988.

70. Tari, A., Miyachi, Y., Hide, M., Sumii, K., Kajiyama, G., Tahara, E., Tanaka, G., and Miyoshi, A., β-Endorphin-like immunoreactivity and somatostatin-like immunoreactivity in normal gastric mucosa, muscle layer, and adenocarcinoma, *Gastroenterology*, 88, 670, 1985.
71. Tari, A., Miyachi, Y., Sumii, K., Haruma, K., Yoshihara, M., Kajiyama, G., and Miyoshi, A., β-Endorphin-like immunoreactivity in normal mucosa, muscle layer, adenocarcinoma, and polyp of the colon, *Dig. Dis. Sci.*, 33, 429, 1988.
72. Bostwick, D. G., Null, W. E., Holmes, D., Weber, E., Barchas, J. D., and Bensch, K. G., Expression of opioid peptides in tumors, *N. Engl. J. Med.*, 317, 1439, 1987.
73. Gustin, T., Bachelot, T., Verna, J. -M. G., Molin, L. F., Brunet, J. -F. M., Berger, F. R., and Benabid, A. L., Immunodetection of endogenous opioid peptides in human brain tumors and associated cyst fluids, *Cancer Res.*, 53, 4715, 1993.

18 The Control of Pain in Peripheral Tissue by Cytokines and Neuropeptides

Michael Schäfer and Christoph Stein

CONTENTS

18.1 Introduction 261
18.2 Neuroimmune Interactions in the Generation of Pain..262
    18.2.1 Inflammatory Response and Nociceptors...262
    18.2.2 Nerve Growth Factor...262
    18.2.3 Substance P and CGRP...263
    18.2.4 Cytokines...263
18.3 Neuroimmune Interactions in the Control of Pain 263
    18.3.1 Opioid Receptors Expressed In Primary Afferent Neurons 263
    18.3.2 Opioid Peptides Expressed in Immune Cells 264
    18.3.3 CRF and IL-1 Release Opioid Peptides..265
    18.3.4 Exogenous CRF and IL-1 in Nociception 265
    18.3.5 Local Expression of CRF and Pain Control...266
18.4 Clinical Relevance...266
References..267

18.1 INTRODUCTION

Our knowledge about changes within the central nervous system following a painful stimulus has considerably increased over the last several years. Peripheral nociceptors (mainly thinly myelinated Aδ- and unmyelinated C-fibers) transduce the noxious stimulus into an electrical excitation of the cell membrane, leading to an increased firing of primary afferent neurons.[1] In the dorsal horn of the spinal cord, where the central terminals of these neurons project, synaptic transmission onto second-order sensory neurons occurs. Here the propagation of the sensory input from the periphery is subject to multiple modulations by other incoming stimuli ("central plasticity").[2,3] Depending on the predominance of excitatory or inhibitory impulses and/or neurotransmitters that are released, this modulation can lead to either an enhanced ("central sensitization") or a decreased sensitivity of dorsal horn neurons which determines the final sensory output from the spinal cord to higher pain centers.

Recent research has shown that, in addition to these mechanisms in the central nervous system, intrinsic modulation of nociception can occur at the peripheral terminals of afferent nerves. The results of this research have changed our view about the interaction of the neuroendocrine and immune systems. While both systems have long been regarded as two separate entities, evidence now seems to emerge that the immune system can interact with peripheral sensory-nerve endings and vice versa.[4,5] This chapter focuses on the neuroimmune cross-talk in peripheral tissue triggered by injury and pain and outlines how this contributes to both the generation as well as the control of pain.

18.2 NEUROIMMUNE INTERACTIONS IN THE GENERATION OF PAIN

18.2.1 INFLAMMATORY RESPONSE AND NOCICEPTORS

Pain is thought to serve as a physiological warning to guard the integrity of the organism and to prevent injury. If tissue damage occurs, an inflammatory response develops triggering local mechanisms that contribute to the maintenance of the painful state. The inflammatory response is characterized in the early phase by a migration of polymorphonuclear leukocytes (PMNL) and in the later phase by lymphocytes, macrophages, and monocytes in the affected area.[6] Products from cell breakdown (prostaglandins, protons), plasma leakage (bradykinin), and inflammatory cells (histamine, serotonin) either directly stimulate (e.g., bradykinin, protons)[7,8] and/or indirectly sensitize (e.g., prostaglandins) the nerve endings of primary afferent neurons.[9] Recent evidence suggests that this sensitization is caused by activation of a specific subtype of sodium channels, the tetrodotoxin-resistant voltage-gated Na^+ channel.[10,11] As a consequence, the nociceptive thresholds of C- and Aδ-primary afferent nerve endings (nociceptors) are reduced. This leads to an enhanced sensitivity of the injured (primary hyperalgesia) and adjacent areas (secondary hyperalgesia) to noxious as well as innocuous stimuli (allodynia) (for review, see References 2 and 3). A subpopulation of nociceptors ("silent nociceptors") which are unresponsive under normal conditions may be activated, leading to an expansion of their peripheral receptive fields.[3]

18.2.2 NERVE GROWTH FACTOR

Nerve growth factor (NGF), a neurotrophic factor that is critical for the development and maintenance of sensory and sympathetic neurons, seems to play a key role in the development of inflammatory pain. Local administration of NGF produces hyperalgesia,[12] whereas neutralization of endogenous NGF by specific antisera diminishes hyperalgesia.[13,14] NGF accumulates in inflamed tissue[14,15] and the high-affinity NGF receptor, trkA, is selectively expressed in primary sensory neurons.[16] NGF is thought to bind to its receptor on sensory nerve terminals and to be transported retrogradely together with its receptor to the cell body. In the dorsal root ganglion, NGF induces an upregulation in the synthesis of substance P (SP) and calcitonin gene-related peptide (CGRP), two proinflammatory neuropeptides.[17] Furthermore, NGF enhances the axonal transport of both peptides from the dorsal root ganglion towards the peripheral tissue.[13] The elevated levels of these neuropeptides probably result in their increased release from peripheral nerve terminals.[13]

18.2.3 SUBSTANCE P AND CGRP

Upon release from peripheral sensory nerve endings, SP and CGRP can modulate the synthesis and secretion of inflammatory mediators and cytokines from resident immune cells which may contribute to an increased inflammatory response. SP, for instance, stimulates the generation of arachidonic acid-derived mediators[18] and the synthesis of tumor necrosis factor α (TNF-α) in mast cells,[19] alters phagocytosis and the release of lysosomal enzymes in leukocytes, and enhances the synthesis and release of immunoglobins and B lymphocytes.[18,20] CGRP stimulates adenylate cyclase in macrophages.[21] Receptors for both SP[18,20] and CGRP[21,22] have been demonstrated on lymphocytes and/or macrophages. Thus, not only do immune cell-derived substances act on peripheral sensory nerve endings (see above), but also vice versa — neuropeptides released from peripheral sensory nerve endings act on resident immune cells. This closes the vicious circle of mutually reinforcing effects between immune cells and primary afferent nerve terminals, resulting in enhanced hyperalgesia and pain.

18.2.4 CYTOKINES

The interaction of cytokines and nociceptors in inflammatory pain is less clear. A few studies have examined the effects of interleukin-1 (IL-1) in noninflamed tissue. They have shown that injection of exogenous IL-1 into noninflamed tissue produces hyperalgesia.[14,23] They have suggested an indirect interaction either by a release of prostaglandins[23] or by an increase in local NGF levels.[14] None of these studies could show direct interactions between immune cell-derived IL-1 and nociceptors. This may be partly explained by a lack of IL-1 receptors, since they could not be demonstrated on peripheral nerve endings of primary afferent neurons.[24]

18.3 NEUROIMMUNE INTERACTIONS IN THE CONTROL OF PAIN

18.3.1 OPIOID RECEPTORS EXPRESSED IN PRIMARY AFFERENT NEURONS

Concurrent to the generation and maintenance of the inflammatory painful state, counteractive endogenous mechanisms are established to inhibit inflammatory pain at the site of the injury. These mechanisms are also based on an interaction between the immune and nervous systems. Opioid receptors have been identified on peripheral terminals of primary afferent neurons, namely, on thinly myelinated and unmyelinated cutaneous nerves, by immunocytochemistry[25] and by autoradiography.[26]

Recent cloning of the opioid receptors has made it possible to demonstrate that primary afferent neurons express mRNA specific for μ-, δ-, and κ-opioid receptors indicating their synthesis in dorsal root ganglia.[27,28] By use of specific antibodies against the C- or N-terminal region of the cloned receptors, μ-, δ-, and κ-opioid receptor proteins have been identified on peripheral nerve terminals of primary afferent neurons by immunofluorescence.[29,30] Apparently, these receptor proteins are transported from the site of their synthesis, the dorsal root ganglion, along the axon to the peripheral nerve terminal.[25,26] These findings are in line with functional studies indicating that capsaicin-sensitive C-fiber neurons mediate the peripheral antinociceptive effects of morphine.[31]

Occupation of peripheral neuronal opioid receptors by an opioid agonist leads to an inhibition of the excitability of the nociceptive input terminal and/or the propagation of action potentials induced by painful stimuli.[32,33] In addition, the release of excitatory neuropeptides (e.g., substance P) from peripheral nerve endings of primary afferent neurons is attenuated.[34] These mechanisms may account for the potent antinociceptive effects of opioids in peripheral tissues — similar to the activation of opioid receptors within the central nervous system, but without the typical central side effects.[5,35] Moreover, they may account for anti-inflammatory actions of opioids in peripheral tissues.[5,35]

Following an injury, the number of opioid receptors at the peripheral nerve terminal increases in parallel to the duration and development of the inflammatory response.[26,27] This is due to an enhanced axonal transport of opioid receptors to the peripheral sensory nerve terminal.[26] It may partly explain why peripheral antinociceptive effects of exogenous opioids are enhanced under inflammatory conditions. Other mechanisms also contribute to this enhanced efficacy. The inflammatory process leads to a disruption of the perineural sheath of sensory nerve endings, a physical barrier which preserves the microenvironment of the nerve,[36] thus facilitating the access of opioids to the nerve terminal.[37] In addition, the inflammatory milieu leads to an enhanced coupling of G proteins and opioid receptors which change previously inactive opioid receptors at the nerve terminal into an active state.[38]

18.3.2 OPIOID PEPTIDES EXPRESSED IN IMMUNE CELLS

In close proximity to the opioid receptors on peripheral sensory nerve terminals, opioid peptides, the natural ligands of opioid receptors, have been identified in resident immune cells of inflamed tissue.[25,39,40] Three families of opioid peptides have been characterized in the central nervous and neuroendocrine systems, each deriving from a distinct gene and precursor protein, namely, proopiomelanocortin (POMC), proenkephalin (PENK), and prodynorphin. POMC, PENK, and prodynorphin mRNA and the corresponding opioid peptides (β-endorphin, enkephalin, dynorphin) have been demonstrated within resident B and T lymphocytes, monocytes, and macrophages.[25,40] Both mRNA and peptide levels are increased under inflammatory conditions. This suggests that the opioid peptides may be released from immune cells, occupy opioid receptors present on peripheral sensory nerve terminals, and lead to an inhibition of painful stimuli similar to the exogenous administration of opioids.

18.3.3 CRF AND IL-1 RELEASE OPIOID PEPTIDES

Corticotropin-releasing factor (CRF), a neuropeptide, is a major physiological secretagogue for POMC-derived peptides in the pituitary. IL-1, a cytokine, potentiates the CRF releasing effect[41] and also directly stimulates β-endorphin release in the pituitary.[42] Receptors for each substance are present on immune cells[24,43,44] and are increased in number during inflammation.[24] Short-term (5 to 10 min) incubation with CRF or IL-1 releases β-endorphin *in vitro* in immune cell suspensions prepared from lymph nodes of inflamed tissue.[45] This is consistent with other *in vitro* studies showing a release of β-endorphin from plasma-derived lymphocytes and macrophages.[46] In these studies, the immune cells were stimulated *in vitro* and required a long-term incubation (up to 10 to 20 h) with CRF or IL-1,[46] while in the other studies the immune cells were stimulated by an inflammation (Freund's adjuvant) *in vivo*, which resembles the clinical situation much more closely.[45] Consequently, a short-term incubation (5 to 10 min) was sufficient. Further experiments showed that the CRF- or IL-1-induced release of β-endorphin from immune cells is specifically mediated by activation of the respective CRF or IL-1 receptors on immune cells.[45,47] This release is calcium dependent and mimicked by increasing potassium concentrations,[47] suggesting a vesicular release via the regulated secretory pathway.

18.3.4 EXOGENOUS CRF AND IL-1 IN NOCICEPTION

Earlier reports showed that CRF inhibits the abdominal constrictor response to intraperitoneal injection of phenylbenzoquinone in mice,[48] the paw withdrawal response of anesthetized rats to hot water,[49] and the paw-lick response of rats in the hot-plate test.[50] In the rat carrageenan model of inflammation, peripherally administered CRF significantly inhibited hyperalgesia,[51] suggesting a peripheral site of action within inflamed subcutaneous tissue. In addition, CRF attenuates the heat-evoked firing response of single units recorded from the spinal trigeminal nucleus in urethane-anesthetized rats.[52] This suggests that CRF may interact with peripheral sensory nerves to decrease afferent transmission of nociceptive stimuli. However, CRF binding sites appear to be absent on peripheral nerve endings of subcutaneous sensory nerves.[24] Instead, abundant CRF binding sites are detectable on inflammatory cells (lymphocytes and monocytes/macrophages) within subcutaneous tissue[24,53] with a similar pharmacological profile[24] to CRF receptors within the anterior pituitary gland.[54,55] The number of these CRF receptors is markedly upregulated within inflamed rat hindpaws inoculated with Freund's adjuvant.[24] Thus, the peripheral antinociceptive effect of CRF seems to be best explained by an indirect mechanism, most likely by a release of opioids from immune cells.

On the other hand, reports about the properties of IL-1 in nociception are controversial. After intracerebroventricular injection of IL-1, both analgesic[56-58] and hyperalgesic effects[58,59] have been described. After peripheral injection of IL-1, hyperalgesic effects have been shown in healthy tissue,[14,23] while analgesic effects have been determined in fully inflamed tissue in which opioid peptide-containing immune cells are present.[45]

In animals with a localized Freund's adjuvant inflammation, exogenous application of CRF and IL-1[45] or other cytokines[60] into the inflamed paw produces potent analgesic effects. They are dose-dependent, receptor-specific, and critically dependent on the functional integrity of the immune cells because immunosuppression by cyclosporin A abolishes these effects.[45] The CRF- and IL-1-induced analgesia is inhibited by immunoneutralization of β-endorphin in the inflamed paw[45] and by antagonists selective for μ- and δ-opioid receptors and naloxone.[45] Therefore, β-endorphin seems to be the predominant opioid peptide which is released by CRF and IL-1 from immune cells of inflamed tissue, activating μ- and δ-opioid receptors on peripheral sensory nerve terminals and resulting in the inhibition of pain. In addition to β-endorphin, [Met]-enkephalin and dynorphin A seem to contribute to the analgesic effect of CRF and IL-1, respectively.[45]

18.3.5 Local Expression of CRF and Pain Control

In inflamed, painful tissue, the expression of CRF is increased[53,61] and the number of CRF receptors on immune cells is upregulated.[24] Eventually, local endogenous CRF may play a role in nociception. This has been investigated by means of a specific environmental stress paradigm that produces potent localized analgesia in peripheral inflamed tissue without the involvement of central pathways.[61,62] The stress-induced analgesia was attenuated by application of μ- and δ-opioid receptor antagonists and by β-endorphin antisera into the inflamed tissue,[62] indicating the involvement of local β-endorphin acting on peripheral μ- and δ-opioid receptors. In addition, local, but not systemic injection of a CRF-selective receptor antagonist (α-helical CRF) and of a CRF-specific antiserum abolishes the stress-induced analgesic effect.[61] This indicates activation of CRF receptors in the inflamed paw by local but not by circulating CRF. When blocking CRF expression in the peripheral inflamed tissue by a CRF-antisense oligodeoxynucleotide targeted against the translation initiation site of the rat CRF mRNA, the stress-induced analgesic effect is abolished.[61] The same treatment significantly reduces the amount of CRF extracted from inflamed paws and the number of CRF-immunoreactive cells.[61] Thus, endogenous CRF, expressed within inflamed tissue, can release opioid peptides from immune cells by a paracrine action and produces intrinsic inhibition of pain. IL-1 does not seem to be involved in these mechanisms,[61] which is consistent with studies demonstrating that the IL-1 gene expression peaks very early (1 to 2 h) after the onset of an inflammation[63] and is probably present in insufficient amounts at later stages of this inflammation.

18.4 CLINICAL RELEVANCE

Application of exogenous opioids has already been successfully used for the treatment of pain in humans.[5] A growing body of controlled clinical trials demonstrate the analgesic efficacy of small doses of morphine, applied intraarticularly, during knee surgery, without the occurrence of systemic opioid effects (reviewed in Reference 5). These effects are long lasting and reversible by naloxone, indicating specificity for peripheral opioid receptors.[5] Recently, opioid peptides (mainly β-endorphin and [Met]-enkephalin) have also been identified in synovial lining cells and in immune cells of inflamed synovial tissue in patients undergoing knee surgery.[64] Blocking of intraarticular opioid receptors by the local administration of naloxone resulted in significantly increased postoperative pain.[64] This suggests that opioids are tonically released from inflamed tissue and activate peripheral opioid receptors to inhibit clinically significant pain. However, they do not interfere with exogenous agonists since intraarticular morphine has equally potent analgesic effects in patients with and without opioid-containing synovitis.[65] Thus, opioid tolerance may develop to a lesser degree in the peripheral than in the central nervous system. In view of the role of CRF in intrinsic opioid pain inhibition in animals (see above), it may well play a similar role in painful inflammatory states of humans to counteract ongoing pain. These findings may eventually lead to the development of a novel generation of analgesics which may act in a manner similar to CRF by releasing endogenous opioids within injured tissue.

REFERENCES

1. Willis, W. D. and Coggeshall, R. E., *Sensory Mechanisms of the Spinal Cord*, Plenum Press, New York, 1991.
2. Coderre, T. J., Katz, J., Vaccarino, A. L., and Melzack, R., Contribution of central neuroplasticity to pathological pain: review of clinical and experimental evidence, *Pain*, 52, 259, 1993.
3. Schaible, H. G. and Grubb, B. D., Afferent and spinal mechanisms of joint pain, *Pain*, 55, 5, 1993.
4. Weihe, E., Büchler, M., Müller, S., Friess, H., Zentel, H. J., and Yanaihara, N., Peptidergic innervation in chronic pancreatitis, in *Chronic Pancreatitis*, Beger, H. G., Büchler, M., Ditschuneit, H., and Malfertheiner, P., Eds., Springer, Berlin, 1990, 83.
5. Stein, C., Mechanisms of disease: the control of pain in peripheral tissue by opioids, *N. Engl. J. Med.*, 332, 1685, 1995.
6. Roitt, I., *Essential Immunology*, Blackwell Scientific, Oxford, 1991, 11.
7. Dray, A. and Perkins, M., Bradykinin and inflammatory pain, *Trends Neurosci.*, 16, 99, 1993.
8. Steen, K. H., Reeh, P. W., Anton, F., and Handwerker, H. O., Protons selectively induce lasting excitation and sensitization to mechanical stimulation of nociceptors in rat skin, *in vitro*, *J. Neurosci.*, 12, 86, 1992.
9. Rueff, A. and Dray A., Pharmacological characterization of the effects of 5-hydroxytryptamine and different prostaglandins on peripheral sensory neurons *in vitro*, *Agents Actions*, 38, C13, 1993.
10. Gold, M. S., Reichling, D. B., Shuster, M. J., and Levine, J. D., Hyperalgesic agents increase a tetrodotoxin-resistant Na^+ current in nociceptors, *Proc. Natl. Acad. Sci. U.S.A.*, 93, 1108, 1996.
11. Akopian, A. N., Sivilotti, L., and Wood, J. N., A tetrodotoxin-resistant voltage-gated sodium channel expressed by sensory neurons, *Nature*, 379, 257, 1996.
12. Andreev, N. Y., Dimitrieva, N., Koltzenburg, M., and McMahon, S. B., Peripheral administration of nerve growth factor in the adult rat produces a thermal hyperalgesia that requires the presence of sympathetic post-ganglionic neurones, *Pain*, 63, 109, 1995.

13. Donnerer, J., Schuligoi, R., and Stein, C., Increased content and transport of substance P and calcitonin gene-related peptide in sensory nerves innervating inflamed tissue: evidence for a regulatory function of nerve growth factor *in vivo*, *Neuroscience*, 49, 693, 1992.
14. Safieh-Garabedian, B., Poole, S., Allchorne, A., Winter, J., and Woolf, C. J., Contribution of interleukin-1 β to the inflammation-induced increase in nerve growth factor levels and inflammatory hyperalgesia, *Br. J. Pharmacol.*, 115, 1265, 1995.
15. Westkamp, G. and Otten, U., An enzyme-linked immunoassay for nerve-growth factor (NGF): a tool for studying regulatory mechanisms involved in NGF production in brain and peripheral tissues, *J. Neurochem.*, 48, 1779, 1987.
16. Carrol, S. L., Silos-Santiago, I., Frese, S. E., Ruit, K. G., Milbrandt, J., and Snider, W. D., Dorsal root ganglion neurons expressing trk are selectively sensitive to NGF deprivation in utero, *Neuron*, 9, 779, 1992.
17. Lindsay, R. M. and Harmer, A. J., Nerve growth factor regulates expression of neuropeptide genes in adult sensory neurons, *Nature*, 337, 362, 1989.
18. McGillis, J. P., Organist, M. L., and Payan, D. J., Substance P and immunoregulation, *Fed. Proc.*, 46, 196, 1987.
19. Ansel, J. C., Brown, J. R., Payan, D. G., and Brown, M. A., Substance P selectively activates TNF-α gene expression in murine mastcells, *J. Immunol.*, 150, 4478, 1993.
20. Bost, K. L. and Pascual, D. W., Substance P: a late acting B lymphocyte differentiation cofactor, *Am. J. Physiol.*, 262, C537, 1992.
21. Abello, J., Kaiserlian, D., Cuber, J. C., Revillard, J. P., and Chayvialle, J. A., Characterization of calcitonin gene-related peptide receptors and adenylate cyclase response in the murine macrophage cell line P388 D1, *Neuropeptides*, 19, 43, 1991.
22. Gates, T. S., Zimmerman, R. P., Mantyh, C. R., Vigna, S. R., and Matyh, P. W., Calcitonin gene-related peptide-α receptor binding sites in the gastrointestinal tract, *Neuroscience*, 31, 757, 1987.
23. Ferreira, S. H., Lorenzetti, B. B., Bristow, A. F., and Poole, S., Interleukin-1 β as a potent hyperalgesic agent antagonized by a tripeptide analogue, *Nature*, 334, 698, 1988.
24. Mousa, S. A., Schäfer, M., Mitchel, W. M., Hassan, A. H. S., and Stein, C., Local upregulation of corticotropin-releasing hormone and interleukin-1 receptors in rats with painful hindlimb inflammation, *Eur. J. Pharmacol.*, 311, 221, 1996.
25. Stein, C., Hassan, A. H. S., Przewlocki, R., Gramsch, C., Peter, K., and Herz, A., Opioids from immunocytes interact with receptors on sensory nerves to inhibit nociception in inflammation, *Proc. Natl. Acad. Sci. U.S.A.*, 87, 5935, 1990.
26. Hassan, A. H. S., Ableitner, A., Stein, C., and Herz, A., Inflammation of the rat paw enhances axonal transport of opioid receptors in the sciatic nerve and increases their density in the inflamed tissue, *Neuroscience*, 55, 185, 1993.
27. Schäfer, M., Imai, Y., Uhl, G. R., and Stein, C., Inflammation enhances peripheral µ-opioid analgesia, but not µ-opioid receptor transcription in dorsal root ganglia, *Eur. J. Pharmacol.*, 279, 165, 1995.
28. Maekawa, K., Minami, M., Yabuuchi, K., Toya, T., Katao, Y., Hosoi, Y., Onogi, T., and Satoh, M., *In situ* hybridization study of µ- and κ-opioid receptor mRNAs in the rat spinal cord and dorsal root ganglia, *Neurosci. Lett.*, 168, 97, 1994.
29. Dado, R. J., Law, P. Y., Loh, H. H., and Elde, R., Immunofluorescent identification of a delta-opioid receptor on primary afferent nerve terminals, *NeuroReport*, 5, 341, 1993.
30. Ji, R.-R., Zhang, Q., Law, P.-Y., Low, H. H., Elde, R., and Hökfelt, T., Expression of mu-, delta-, and kappa-opioid receptor-like immunoreactivities in rat dorsal root ganglia after carrageenan-induced inflammation, *J. Neurosci.*, 15, 8156, 1995.

31. Barthó, L., Stein, C., and Herz, A., Involvement of capsaicin-sensitive neurons in hyperalgesia and enhanced opioid antinociception in inflammation, *Naunyn-Schmiedebergs Arch. Pharmacol.*, 342, 666, 1990.
32. Andreev, N., Urban, L., and Dray, A., Opioids suppress spontaneous activity of polymodal nociceptors in rat paw skin induced by ultraviolet irradiation, *Neuroscience*, 58, 793, 1994.
33. Russel, N. J. W., Schaible H. G., and Schmidt R. F., Opiates inhibit the discharges of fine afferent units from inflamed knee joint of the cat, *Neurosci. Lett.*, 76, 107, 1987.
34. Yaksh, T. L., Substance P release from knee joint afferent terminals: modulation by opioids, *Brain Res.*, 458, 319, 1988.
35. Barber, A. and Gottschlich, R., Opioid agonists and antagonists: an evaluation of their peripheral actions in inflammation, *Med. Res. Rev.*, 12, 525, 1992.
36. Olsson, Y., Microenvironment of the peripheral nervous system under normal and pathological conditions, *Crit. Rev. Neurobiol.*, 5, 265, 1990.
37. Antonijevic, I., Mousa, S. A., Schäfer, M., and Stein, C., Perineurial defect and peripheral opioid analgesia in inflammation, *J. Neurosci.*, 15, 165, 1995.
38. Selley, D. E., Breivogel, C. S., and Childers, S. R., Modification of G protein-coupled functions by low-pH pretreatment of membranes from NG108-15 cells: increase in opioid efficacy by decreased inactivation of G proteins, *Mol. Pharmacol.*, 44, 731, 1993.
39. Carr, D. J. J., The role of endogenous opioids and their receptors in the immune system, *Proc. Soc. Exp. Biol. Med.*, 198, 710, 1991.
40. Przewlocki, R., Hassan, A. H. S., Lason, W., Epplen, C., Herz, A., and Stein, C., Gene expression and localization of opioid peptides in immune cells of inflamed tissue. Functional role in antinociception, *Neuroscience*, 48, 491, 1992.
41. Fagarasan, M. O., Eskay, R., and Axelrod, J., Interleukin-1 potentiates the secretion of β-endorphin induced secretagogues in a mouse pituitary cell line, *Proc. Natl. Acad. Sci. U.S.A.*, 86, 2070, 1989.
42. Bernton, E. W., Beach, J. E., Holaday, J. W., Smallridge, R. C., and Fein, H. G., Release of multiple hormones by a direct action of interleukin-1 on pituitary cells, *Science*, 238, 519, 1987.
43. Webster, E. L., Tracey, D. E., Jutila, M. A., Wolfe, S. A., and De Souza, E. B., Corticotropin-releasing factor receptors in mouse spleen: identification of receptor-bearing cells as resident macrophages, *Endocrinology*, 127, 440, 1990.
44. Dinarello, C. A., The interleukin-1 family: 10 years of discovery, *FASEB J.*, 8, 1314, 1994.
45. Schäfer, M., Carter, L., and Stein, C., Interleukin-1β and corticotropin-releasing factor inhibit pain by releasing opioids from immune cells in inflamed tissue, *Proc. Natl. Acad. Sci. U.S.A.*, 91, 4219, 1994.
46. Heijnen, C. J., Kavelaars, A., and Ballieux, R. E., β-Endorphin: cytokine and neuropeptide, *Immunol. Rev.*, 119, 41, 1991.
47. Cabot, P. J., Carter, L., Gaiddon, C., Zhang, Q., Schäfer, M., Loeffler, J. P., and Stein, C., Immune cell-derived β-endorphin: production, release, and control of inflammatory pain in rats, *J. Clin. Invest.*, 100, 142, 1997.
48. Wei, E. T., Kiang, J. G., Buchan, P., and Smith, T. W., Corticotropin-releasing factor inhibits neurogenic plasma extravasation in the rat paw, *J. Pharmacol. Exp. Ther.*, 238, 783, 1986.
49. Kiang, J. G. and Wei, E. T., Corticotropin-releasing factor inhibits thermal injury, *J. Pharmacol. Exp. Ther.*, 243, 517, 1987.

50. Hargreaves, K. M., Mueller, G. P., Dubner, R., Goldstein, D., and Dionne, R. A., Corticotropin-releasing factor (CRF) produces analgesia in humans and rats, *Brain Res.*, 422, 154, 1987.
51. Hargreaves, K. M., Dubner, R., and Costello, A. H., Corticotropin-releasing factor (CRF) has a peripheral site of action for antinociception, *Eur. J. Pharmacol.*, 170, 275, 1989.
52. Poree, L. R., Dickenson, A. H., and Wei, E. T., Corticotropin-releasing factor inhibits the response of trigeminal neurons to noxious heat, *Brain Res.*, 502, 349, 1989.
53. Crofford, L. J., Sano, H., Karalis, K., Webster, L. E., Goldmuntz, E. A., Chrousos, G. P., and Wilder, R. L., Local secretion of corticotropin-releasing hormone in the joints of Lewis rats with inflammatory arthritis, *J. Clin. Invest.*, 90, 2555, 1992.
54. Wynn, P. C., Aguilera, G., Morell, J., and Catt, K. J., Properties and regulation of high-affinity pituitary receptors for corticotropin-releasing factor, *Biochem. Biophys. Res. Commun.*, 110, 602, 1983.
55. De Souza, E. B., Perrin, M. H., Rivier, J., Vale W., and Kuhar, M. J., Corticotropin-releasing factor receptors in rat pituitary gland: autoradiographic localization, *Brain Res.*, 296, 202, 1984.
56. Nakamura, H., Nakanishi, K., Kita, A., and Kadokawa, T., Interleukin-1 induces analgesia in mice by a central action, *Eur. J. Pharmacol.*, 149, 49, 1988.
57. Bianchi, M., Sacerdote, P., Ricciardi-Castagnoli, P., Mantegazza, P., and Panerai, A. E., Central effects of tumor necrosis factor α on nociceptive thresholds and spontaneous locomotor activity, *Neurosci. Lett.*, 148, 76, 1992.
58. Oka, T., Oka, K., Hosoi, M., Aou, S., and Hori, T., The opposing effects of interleukin-1 β microinjected into the preoptic hypothalamus and the ventromedial hypothalamus on nociceptive behavior in rats, *Brain Res.*, 700, 271, 1995.
59. Adams, J. U., Bussiere, J. L., Geller, E. B., and Adler, M. W., Pyrogenic doses of intracerebroventricular interleukin-1 did not induce analgesia in the rat hot-plate or cold-water tail-flick tests, *Life Sci.*, 53, 1401, 1993.
60. Czlonkowski, A., Stein, C., and Herz, A., Peripheral mechanisms of opioid antinociception in inflammation: involvement of cytokines, *Eur. J. Pharmacol.*, 242, 229, 1993.
61. Schäfer, M., Mousa, S. A., Zhang, Q., Carter, L., and Stein, C., Expression of corticotropin-releasing factor in inflamed tissue is required for intrinsic peripheral opioid analgesia, *Proc. Natl. Acad. Sci. U.S.A.*, 93, 6096, 1996.
62. Stein, C., Gramsch, C., and Herz, A., Intrinsic mechanisms of antinociception in inflammation. Local opioid receptors and β-endorphin, *J. Neurosci.*, 10, 1292, 1990.
63. Schindler, R., Clark, B. D., and Dinarello, C. A., Dissociation between interleukin-1 β mRNA and protein synthesis in human peripheral blood mononuclear cells, *J. Biol. Chem.*, 265, 10232, 1990.
64. Stein, C., Hassan, A. H. S., Lehrberger, K., Giefing, J., and Yassouridis, A., Local analgesic effect of endogenous opioid peptides, *Lancet*, 342, 321, 1993.
65. Stein, C., Pflüger, M., Yassouridis, A., Hoelzl, J., Lehrberger, K., Welte, C., and Hassan, A. H. S., No tolerance to peripheral morphine analgesia in presence of opioid expression in inflamed synovia, *J. Clin. Invest.*, 98, 793, 1996.

19 Interleukins and Immunocyte β-Endorphin

Paola Sacerdote

CONTENTS

19.1 Introduction 271
19.2 BE Production by Immunocytes 272
19.3 Central IL-1 and Immunocyte BE 272
   19.3.1 Modulation of IL-1-Induced BE Increase 273
19.4 Footshock and Immunocyte BE 275
19.5 Involvement of Immunocyte BE Increase in IL-1 and
   Footshock-Induced Immunosuppression 276
19.6 Conclusions 277
Acknowledgment 278
References 279

19.1 INTRODUCTION

The cytokine interleukin-1 (IL-1) has an important role not only in modulating the immune responses, but also in eliciting a variety of centrally mediated adaptive responses during infections, inflammation, and injury.[1] IL-1 influences complex neural circuits involved in thermoregulation, food intake, sleeping patterns, and behavior.

Moreover, this cytokine affects neuroendocrine functions, and its ability to activate the hypothalamus-pituitary-adrenal axis (HPA) is now well recognized. IL-1 acts at the hypothalamic level as a potent activator of corticotropin-releasing hormone (CRH); it increases CRH secretion and stimulates hypothalamic CRH messenger RNA.[2,3] CRH is the main mediator of stress responses;[4] it initiates series of behavioral and physiological responses that are adaptive during stressful situations.

Consistently, IL-1 seems to activate the same biological responses of stress. Stress-related effects of IL-1 include activation of the HPA,[1-3] inhibition of the reproductive axis,[5] immunosuppression,[6] as well as behavioral modifications.[7]

In this chapter, we have outlined another interesting effect shared by the central administration of IL-1 and by a classic experimental model of stress, footshock, on the concentrations of the opioid peptide β-endorphin (BE) in rat splenocytes, lymph node cells, and peripheral blood mononuclear cells (PBMC).[8,9]

Finally, evidence will be provided for the involvement of immunocyte-derived BE in central IL-1 and footshock-induced immunosuppression.

19.1 BE PRODUCTION BY IMMUNOCYTES

BE is a 31-aminoacid opioid peptide known in the last 20 years for its action in the central nervous system. It is produced as a larger precursor molecule, proopiomelanocortin (POMC). The POMC precursor molecule is posttranslationally processed by a series of enzymatic events which give rise to a number of peptide hormones, e.g., ACTH, βLPH, αMSH, and BE. In the CNS, BE is synthesized mainly in the arcuate nucleus of the hypothalamus and projects into several brain areas. In the periphery the main source of BE is the intermediate pituitary or its vestige.

BE is also constitutively synthesized in the cells of the immune system, which, upon appropriate stimuli, release the opioid peptide.[10,11] The hypothalamic peptide CRH, which is considered the main signal for BE synthesis and release in the brain and in the pituitary, can also induce the synthesis and the secretion of the POMC-derived peptides from PBMC and splenocytes.[12] Our laboratory also demonstrated that BE is similarly controlled in lymphocytes and brain by the same neurotransmitters; in fact BE is under a dopaminergic and GABAergic inhibitory tonic control, and a serotoninergic tonic stimulatory input.[13]

Moreover, a dysfunction of the brain and the pituitary BE systems has been involved in many different neurological pathologies and disorders; a wide literature has reported on the modifications of cerebral spinal fluid and plasma BE concentrations in the presence of chronic pain and in patients affected with migraine or anorexia nervosa. In these disorders we found that BE concentrations were similarly modified in cells of the immune system, showing the existence of an unique system of control of BE in the body.[14,15]

19.3 CENTRAL IL-1 AND IMMUNOCYTE BE

In order to define the effects of central IL-1 on BE levels in cells of the immune system, recombinant human IL-1α was injected intracerebroventricularly (i.c.v.) in Sprague-Dawley male rats at the dose of 1ng per rat. At 2 and 24 h after the cytokine administration, the concentrations of BE in splenocytes, lymph node cells, and in PBMC were evaluated by radioimmunoassay. As depicted in Figure 19.1, the BE concentrations were increased in all the tissues considered. In order to evaluate whether the effect of the cytokine was mediated by specific brain IL-1 receptors, the animals were i.c.v. pretreated with the IL-1 receptor antagonist (IL-1RA) at a dose 2000 times higher than IL-1. The complete block of the IL-1-induced increase achieved by i.c.v. administration with IL-1RA indicates that IL-1 seems to act in the brain by binding to its specific receptors. Although *in vitro* studies have shown that IL-1 is able to directly stimulate POMC in immunocytes, the effects on BE concentrations observed probably are not due to a direct effect of IL-1 on the cells, since it is unlikely that the cytokine injected in the brain could reach relevant peripheral concentrations, especially considering that IL-1 does not seem to easily

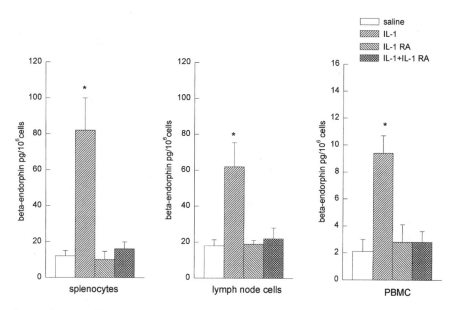

Figure 19.1 BE concentrations in splenocytes, lymph node cells, and PBMC 2 h after i.c.v. IL-1 administration. IL-1RA was injected i.c.v. 15 min before IL-1. * $p < .01$ vs. saline (ANOVA).

cross the blood-brain barrier. Furthermore, when IL-1 RA was injected intravenously, it did not affect the BE increase induced by i.c.v. IL-1.

Analogous results were obtained when the cytokine was injected i.c.v. in Swiss albino mice.

19.3.1 MODULATION OF IL-1-INDUCED BE INCREASE

Since CRH is a potent stimulator of the POMC gene, and IL-1 activates CRH in the paraventricular nucleus of the hypothalamus, this hormone was an obvious candidate as a mediator of the effects of IL-1 on immunocyte BE. As expected, when animals were pretreated with the CRH antagonist (α-helical CRH 9-41, 20 μg per rat, i.c.v.), the effects of IL-1 on immunocyte BE disappeared. These results are in agreement with what has been reported for many other effects of IL-1. CRH is involved in the pyrogenic and thermogenic effects of IL-1, in IL-1 modulation of food intake, and in the suppression of cellular immune responses induced by central IL-1.[1,16]

Many biological effects of IL-1 are also mediated through prostaglandins,[1] but this does not seem to be the case for the effects on BE, since the cyclooxygenase blocker indomethacin did not affect the increase of immunocyte BE concentrations induced by IL-1.

IL-1 has been shown to activate brain neurotransmitters. A stimulation of the central noradrenergic system has been observed in the hypothalamus, hippocampus, brainstem, and spinal cord.[17,18] The chemical disruption of the central catecholamin-

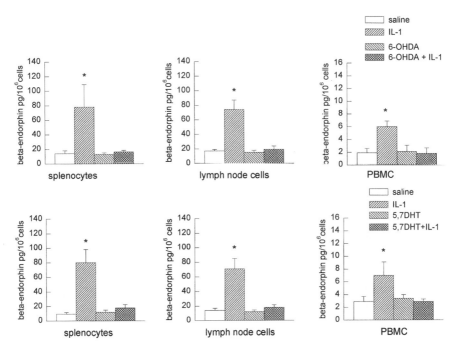

Figure 19.2 BE concentrations in splenocytes, lymph node cells, and PBMC 2 h after i.c.v. IL-1 in catecholamine- and serotonin-depleted rats. **Upper panels.** For the depletion of brain catecholamines rats were injected i.c.v. twice with 250 μg 6-OHDA, with a 24-h interval between doses. **Lower panels.** For the depletion of brain 5-HT, animals were given an i.c.v. injection of 200 μg/ml 5-7DHT. Before 5-7DHT treatment, animals were treated i.p. with 25 mg/kg desimipramine to protect catecholaminergic terminals. * $p < .01$ vs. saline (vehicle) (ANOVA).

ergic system achieved by the i.c.v. central administration of 6-hydroxydopamine (6-OHDA) completely abolished the IL-1-induced increase of immunocyte BE (Figure 19.2, upper panels). Since it has been demonstrated that noradrenergic innervation of the paraventricular nucleus mediates the effect of IL-1 on CRH,[18] the depletion of this innervation could prevent CRH stimulation by IL-1 and, as a consequence, the rise of BE. In addition, it also is well known that a positive feedback loop exists between CRH and noradrenergic cells of the locus ceruleus, a main noradrenergic system in the brain involved in stress and behavioral responses. The application of CRH into the locus ceruleus increases its firing rate, activating descending catecholaminergic pathways.[4] Therefore, the depletion of NA content could also inhibit the response of noradrenergic neurons to CRH.

As reported in Section 19.2, the serotoninergic system exerts an important positive control on immunocyte BE.[13] Since IL-1 activates 5-HT turnover in the hypothalamus,[19] the role of this neurotransmitter system could be relevant in the IL-1-induced increase of immunocyte BE. In fact, in animals where the serotoninergic system had been depleted throughout the i.c.v. administration of the specific neuro-

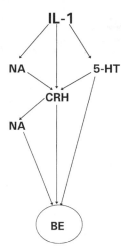

Figure 19.3 Scheme of the brain neurotransmitter network involved in the stimulation of immunocyte BE induced by central IL-1.

toxin 5,7-dihydroxytryptamine, the effects of IL-1 on immunocyte BE concentrations were completely lost (Figure 19.2, lower panels). As previously reported, IL-1 is a strong activator of the HPA axis, causing a rise in plasma ACTH and subsequently of glucocorticoids. However, the increase of immunocyte BE concentrations induced by IL-1 is still present in adrenalectomized as well as in hypophysectomized animals. The BE increase is therefore independent of HPA activation or of other pituitary-dependent hormonal pathways.

In conclusion, for the spleen and the lymph nodes, which are directly innervated, it can be suggested that neural rather than hormonal pathways are involved in the increase in immunocyte BE concentrations induced by IL-1. However, it is more difficult to explain the role of the central neurotransmitter systems on PBMC BE concentrations.

A summary of the network of multiple regulatory neurotransmitters which participate in mediating the effect of IL-1 on immunocyte BE is depicted in Figure 19.3.

19.4 FOOTSHOCK AND IMMUNOCYTE BE

In consideration of the main role played by CRH in the stress responses, and of the involvement of CRH in the increase of immunocyte BE induced by IL-1, it could be expected that stress also could affect immunocyte BE concentrations. In order to address this problem we chose a classical experimental stress paradigm, inescapable footshock, which has been widely characterized for its ability to activate the brain endogenous opioid system.[20,21] Rats were exposed to intermittent footshock, receiving a shock of 10 mA intensity for a duration of 1 s every 5 s for 20 min. Similar to that observed after IL-1 treatment, BE concentrations in splenocytes, lymph node cells, and PBMC were significantly increased after footshock (Figure 19.4). The effect became evident 30 min after the end of the stress, and the BE levels were still elevated 24 h later. This footshock paradigm was previously shown to activate the opiatergic

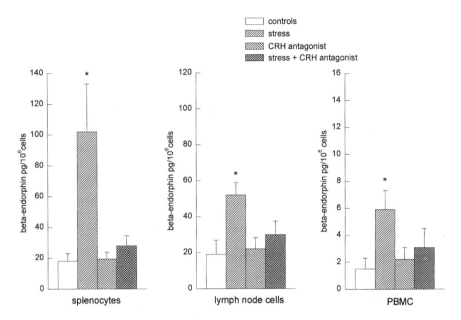

Figure 19.4 BE concentrations in splenocytes, lymph node cells, and PBMC 2 h after intermittent footshock. CRH antagonist (20 µg) was administered i.c.v. 15 min before footshock. * $p < .01$ vs. control animals (ANOVA).

system and to increase BE levels in the brain;[21] therefore, the footshock-induced BE increase further confirms that BE is controlled in a parallel way in the brain and in the immune system, and that the peptide in the brain and in the cells of the immune system are similarly modified in different physiological or pathological conditions.

As expected, when animals were i.c.v. treated with the CRH antagonist 15 min before footshock, the rise of immunocyte BE concentrations was prevented (Figure 19.4). On the contrary, as already observed for the immunocyte BE increase induced by central IL-1, the stress-induced BE increase was not affected by either adrenalectomy or hypophysectomy. The action of CRH on BE is expressed within the brain, independent of the activation of the HPA axis. CRH seems, therefore, to be the key molecule on which both IL-1 and footshock converge for the subsequent effects on BE.

19.5 INVOLVEMENT OF IMMUNOCYTE BE INCREASE IN IL-1 AND FOOTSHOCK-INDUCED IMMUNOSUPPRESSION

From the data reported, it can be concluded that the increase in immunocyte BE concentrations is an effect shared by central IL-1 and physical stress throughout the activation of CRH. Another common effect of central IL-1, physical stress, and central CRH is the induction of peripheral immunosuppression.[4-6,8]

Evidence for the involvement of BE in the regulation of immune function is continuing to accumulate, and most data indicate that BE *in vivo* possesses immunosuppressive properties.[22,23] It has been shown that BE exerts a physiological inhibitory effect on immune function, since the administration of the opiate receptor antagonist naloxone, or of immunoglobulins which neutralize the activity of BE, induced an increase of natural killer activity (NK) and of mitogen-induced splenocyte proliferation within minutes.[22,23] Immunoglobulins do not cross the blood-brain barrier, therefore when injected by the intravenous route their effect is exerted only peripherally. Interestingly, the increase induced by the intravenous administration of the anti-BE immunoglobulins also was present in hypophysectomized rats, i.e., in the absence of pituitary cells, the main source of plasma BE.[22] The alternative source of the opioid is therefore likely to be the immune cells such as PBMC and splenocytes. The inhibitory effect of BE could be exerted by the opioid on the same cells that have synthesized and released it, or on a cell nearby, i.e., through an autocrine or paracrine effect.

Both intermittent footshock and the same doses of IL-1 that increased immunocyte BE induced a decrease of splenocyte proliferation and NK activity, and this immunosuppressive effect was prevented by pretreatment with the CRH antagonist.[6,8,16] It can, therefore, be suggested that the increase of immunocyte BE can have a role in the decrease of immune responses observed after footshock and IL-1.

In order to test the involvement of the increase of immunocyte BE in footshock and central IL-1-induced immunosuppression, animals were injected intravenously with an anti-BE antibody before stress or IL-1 injection. As reported in Figure 19.5, 2 h after footshock and central IL-1 injections, splenocyte proliferation to phytohemoagglutinin and NK activity were significantly reduced. The pretreatment with anti-BE immunoglobulins 1 h before IL-1 or footshock prevented the development of immunosuppression. In this situation, the effect was also present in hypophysectomized rats, suggesting a direct involvement of immunocyte BE in the modulation of immune system induced by footshock and central IL-1.

19.6 CONCLUSIONS

Both central IL-1 and footshock stress induce an increase of BE concentrations in immunocytes obtained by different immune organs, showing that this effect is a generalized response of all the cells of the immune system. The activation of central CRH neurons is essential for this effect, independently of the stimulation of the HPA axis. Moreover, multiple neurotransmitters participate in mediating the increase of immunocyte BE induced by IL-1 and stress.

On consideration of the immunosuppressive properties of immunocyte BE these results are particularly interesting, since they point to BE as one of the final effector molecules of the modulation of the immune system induced by a stressful situation perceived centrally. In other words, in the complex network of neuroimmunomodulation the BE released from cells of the immune system could be a link between the brain and immune system.

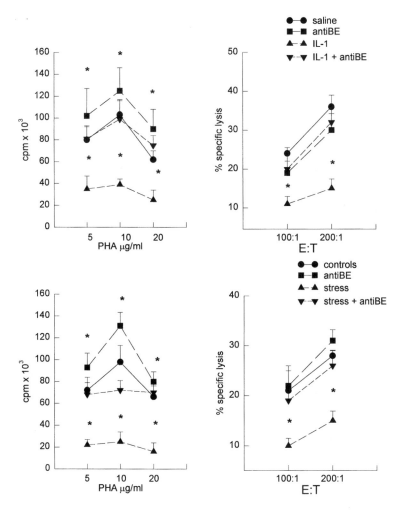

Figure 19.5 Effect of the intravenous administration of anti-BE immunoglobulins on immunosuppression induced by central IL-1 (upper panels) and footshock (lower panels). The immune parameters evaluated were PHA-induced splenocyte proliferation and NK activity. Splenocytes were collected 2 h after treatment. In order to evaluate proliferation, splenocytes were cultured for 48 h in presence of PHA, and ^{3}H-thymidine incorporation (c.p.m.) was evaluated (for details, see Reference 23). NK activity was evaluated by a 4-h ^{51}Cr release assay. Splenocytes (effector cells) and mouse lymphoblastoma ^{51}Cr YAC-1 (target cells) were incubated together at an Effector:Target ratio of 200:1 and 100:1 (for details, see Reference 8). * $p < .01$ vs. saline/controls (two-way ANOVA).

ACKNOWLEDGMENT

I am grateful to Professor A. E. Panerai for useful discussions and helpful suggestions in the preparation of this manuscript.

REFERENCES

1. Rothwell, N. J., Function and mechanisms of interleukin-1 in the brain, *TIPS*, 12, 30, 1991.
2. Sapolsky, R., Rivier, C., Yamamoto, G., Plotsky, P., and Vale, W., Interleukin-1 stimulates the secretion of hypothalamic corticotropin releasing factor, *Science*, 238, 522, 1987.
3. Suda, T., Tozawa, F., Ushiyama, T., Sumitomo, T., Yamada, H., and Demura, H., Interleukin-1 stimulates corticotropin releasing factor gene expression in rat hypothalamus, *Endocrinology*, 126,1223, 1990.
4. Johnson, E. O., Kamilaris, C. T., Chrousos, G. P., and Gold, P. W., Mechanism of stress: a dynamic overview of hormonal and behavioral homeostasis. *Neurosci. Biobehav. Rev*, 16, 115, 1992.
5. Rivest, S. and Rivier, C., Influence of the paraventricular nucleus of the hypothalamus in the alteration of neuroendocrine functions induced by intermittent footshock or interleukin, *Endocrinology*, 129, 2049, 1991.
6. Sundar, S. K., Becker, K. J., Cierpial, M. A., Carpenter, M. D., Rankin, L. A., Fleener, S. L., Ritchie, J. C., Simson, P. E., and Weiss, J. M., Intracerebroventricular infusion of interleukin-1 rapidly decreases peripheral cellular immune responses, *Proc. Natl. Acad. Sci. U.S.A.*, 86, 6398, 1989.
7. Bianchi, M., Sacerdote, P., Locatelli, L., and Panerai, A. E., Corticotropin releasing hormone, interleukin-1 and tumor necrosis factor share characteristics of stress mediators, *Brain Res.*, 546, 139, 1991.
8. Sacerdote, P., Manfredi, B., Bianchi, M., and Panerai, A. E., Intermittent but not continuous inescapable footshock stress affects immune responses and immunocyte β-endorphin concentrations in the rat, *Brain Behav. Immunol.*, 8, 251, 1994.
9. Sacerdote, P., Bianchi, M., Manfredi, B., and Panerai, A. E., Intracerebroventricular interleukin-1 α increases immunocyte β-endorphin concentrations in the rat: involvement of corticotropin releasing hormone and neurotransmitters, *Endocrinology*, 135, 1346, 1994.
10. Sacerdote, P., Breda, M., Barcellini, W., Meroni, P. L., and Panerai, A. E., Age-related changes of β-endorphin and cholecystokinin in human and rat mononuclear cells, *Peptides*, 12, 1353, 1991.
11. Manfredi, B., Clementi, E., Sacerdote, P., Bassetti, M., and Panerai, A. E., Age-related changes in mitogen induced β-endorphin release from human peripheral blood mononuclear cells, *Peptides*, 16, 699, 1995.
12. Heijnen, C. J., Kavelaars, A., and Ballieux, R. E., β-Endorphin, cytokine and neuropeptide, *Immunol. Rev.*, 119, 41, 1991.
13. Sacerdote, P., Rubboli, F., Locatelli, L., Ciciliato, I., Mantegazza, P., and Panerai, A. E., Pharmacological modulation of neuropeptides in peripheral mononuclear cells, *J. Neuroimmunol.*, 32, 35, 1991.
14. Leone, M., Sacerdote, P., D'Amico, D., Panerai, A. E., and Bussone, G., β-endorphin concentrations in the peripheral blood mononuclear cells of migraine and tension-type headache patients, *Cephalalgia*, 12, 155, 1992.
15. Panerai A. E. and Sacerdote, P., β-endorphin in the immune system: a role at last? *Immunol. Today*, 18, 317, 1997.
16. Saperstein, A., Brand, H., Audhya, T., Nabrinsky, D., Hutchinson, B., Rosenweig, S., and Hollander, C. S., Interleukin-1-β mediates stress-induced immunosuppression via corticotropin-releasing factor, *Endocrinology*, 130, 1552, 1992.

17. Shintani, F., Kamba, S., Nakaki, T., Nibuya, M., Kinoshita, N., Suzuki E., Yagi, G., Kato, R., and Asai, M., Interleukin-1β augments release of norepinephrine, dopamine and serotonin in the rat anterior hypothalamus, *J. Neurosci.*, 13, 3574, 1993.
18. Chuluyan, H. E., Saphier, D., Rohn, W. M., and Dunn, A. J., Noradrenergic innervation of the hypothalamus participates in adrenocortical responses to IL-1, *Neuroendocrinology*, 56, 106, 1992.
19. Gemma, C., Ghezzi, P., and De Simoni, M. G., Activation of the hypothalamic serotoninergic system by central IL-1, *Eur. J. Pharmacol.*, 209, 139, 1991.
20. Lewis, J. W., Cannon, J. T., and Liebeskind, J. C., Opioid and nonopioid mechanisms of stress-induced analgesia, *Science*, 208, 623, 1980.
21. Rossier, J., French, E. D., Rivier, C., Ling, N., Guillemin, R., and Bloom F. E., Foot-shock induced stress increases β-endorphin levels in brain, *Nature*, 270, 618, 1977.
22. Panerai, A. E., Manfredi, B., Granucci, F., and Sacerdote, P., The β-endorphin inhibition of mitogen-induced splenocytes proliferation is mediated by central and peripheral paracrine/autocrine effects of the opioid, *J. Neuroimmunol.*, 58, 71, 1995.
23. Manfredi, B., Sacerdote, P., Bianchi, M., Veljic-Radulovic, J., and Panerai, A. E., Evidence for an opioid inhibitory effect on T cell proliferation, *J. Neuroimmunol.*, 44, 43, 1993.

20 Opioids, Opioid Receptors, and the Immune System

Ricardo Gomez-Flores and Richard J. Weber

CONTENTS

Abstract 281
20.1 Introduction 282
20.2 Opioid Actions on Lymphocytes 284
    20.2.1 Effect of Opioids on T Cell Proliferation 284
    20.2.2 Effect of Opioids on T Cell-Mediated Cytotoxicity 284
    20.2.3 Opioid-Mediated Cytokine Production by T Lymphocytes 286
    20.2.4 Modulation of B Cell Functions by Opioids 287
    20.2.5 Natural Killer Cell Activity 289
20.3 Effect of Opioids on Macrophage Functions 289
    20.3.1 Opioid Action on Macrophage Proliferation
           and Differentiation 289
    20.3.2 Opioid-Mediated Cytokine Production by Macrophages 293
    20.3.3 Effect of Opioids on Macrophage Respiratory Burst 293
    20.3.4 Opioid Modulation of Nitric Oxide Production by Macrophages ..293
    20.3.5 Opioid Action on Phagocytosis by Macrophages 297
    20.3.6 Effect of Opioids on Macrophage Tumoricidal Activity 297
    20.3.7 Additional Opioid-Regulated Macrophage Functions 297
20.4 Opioid Receptors on Cells of the Immune System 298
20.5 Modulation of Signal Transduction Pathways by Opioid Action 298
20.6 Discussion 300
Acknowledgement 301
References 301

ABSTRACT

This review describes the role of opioids on lymphocyte, natural killer (NK) cell, and macrophage functions. Opioid agonists have been demonstrated to bind to μ, κ, and δ opioid receptor types on the surface of these cells. Morphine, the most studied opioid, as well as endorphin, enkephalin, deltorphin, dynorphin A, fentanyl, methadone, and U50,488 have been reported to affect lymphocyte, NK cell, and macrophage properties including proliferation and differentiation, cytotoxicity, cytokine and antibody production, phagocytosis, chemotaxis, and signal transduction pathways. The biological functions of these cell populations were found to be enhanced, inhibited, or unaffected by opioid action; such differences may be related to cell source, opioid type and dose, route of opioid administration, and duration of treatment. Most of the studies of opioid action on immune responses has been developed *in vitro* and a few of them *ex vivo*, however, it is necessary to prompt *in vivo* studies to elucidate the role of opioid agonists on immune function against infectious diseases and cancer.

Key Words: opioids, opioid action, macrophages, macrophage functions, macrophage immunoregulation

20.1 INTRODUCTION

The association of the central nervous system (CNS) and the immune system is now well recognized. During stress the brain induces the release of hormones such as glucocorticoids that produce analgesia, but that also provoke immunosuppression.[1-4] As a feedback, immune cells release cytokines that may affect the CNS.[5-7] Because of the interaction between CNS and the immune system, it may be inferred that a drug acting on the CNS may affect immune cell functions, and vice versa. Opioids can affect the immune system through their effects on the hypothalamo-pituitary-adrenal (HPA) axis, leading to the production of corticosteroids or catecholamines, or by binding to and acting through opioid receptors on the surface of immune cells.[4]

Opioid agonists represent a group of natural, semisynthetic, or synthetic drugs with the ability to relieve pain but with the potential risk to provoke physical dependence. Structurally, opioids are phenanthrene derivatives (morphine), phenypiperidine derivatives (fentanyl), diphenylheptane derivatives (methadone),[8] and peptide-related compounds (endorphin, enkephalin, and dynorphin).[9] Opioids exert their major pharmacologic effect on the CNS; however, they also interact with the immune system by altering macrophage, NK cell, and lymphocyte functions, thus impairing immunity against infectious diseases and cancer.[4,10-13] Opioid agonists activities depend on binding to high-affinity receptors named μ, κ, and δ[9,14] (Table 20.1). Such opioid receptor types have been found on cells of the immune system.[1,2]

The cells of the immune system are normally present in the blood and lymph, as well as in lymphoid organs, and as scattered cells in virtually all tissues of the body. The immune system has to be able to respond to a large number of foreign molecules (antigens) introduced at any site in the body, and lymphocytes will specifically recognize and respond to them. Lymphocytes differentiate from precursor cells in bone marrow and mature there (B cells) or in the Fabricius bursa (B cells in birds), or migrate to the thymus (T cells) for complete maturation. When their differentiation and maturation are complete lymphocytes migrate to and populate the peripheral lymphoid organs (blood, spleen, lymph nodes) and tissues (Peyer's patches, mucose-associated lymphoid tissue, etc.). Lymphocytes consist of two distinct subsets that differ in their functions but share morphological similarities. One class of lymphocytes is named B lymphocytes, since their existence was first recognized in the bird's Fabricius bursa. The B cell's major function is to produce antibodies to antigens. The second main class of lymphocytes are named T lymphocytes since they mature in the thymus. They are subdivided into helper T cells, and cytotoxic/suppressor T cells. In response to antigen challenge, helper T cells secrete

TABLE 20.1
Opioid-Selective Receptor Types[a]

Opioid-receptor type	Agonists		Antagonists	
	Endogenous	Exogenous	Reversible	Irreversible
μ	β-END	Morphine	CTOP	β-FNA
	β-NEND	DAMGO		BIT[b]
		DAMEA		
		DALEA		
		Fentanyl		
		Methadone		
δ	LENK	DPDPE	Naltrindole	SUPERFIT
	MENK	DADLE		FIT[c]
	Deltorphin			DALCE[d]
				UPHIT
κ	DYN-A	U50,488	nor-BNI	UPHIT
		U69,593		

[a] Reviewed in Reference 235.
[b] BIT = 2-(4-ethoxybenzyl)-1-diethylaminoethyl-5-isothiocyanatobenzimidazole.
[c] FIT = N-phenyl-N-[1-(2-(4-isothiocyanato)phenylethyl)4-piperidinyl]pro-panamide.
[d] DALCE = [D-Ala$_2$, Leu$_5$, Cys$_6$]enkephalin.

protein hormones called cytokines, or more specifically, lymphokines, whose function is to promote the proliferation and differentiation of T and B lymphocytes, and macrophages. Certain lymphokines can also recruit and activate inflammatory leukocytes (macrophages and granulocytes).[15-17] T cells recognize peptide antigens attached to proteins that are encoded by the major histocompatibility complex and expressed on the surface of other cells such as the macrophages.

Macrophages play a central role in orchestrating humoral and cellular immunity.[18] Their major function is to protect host cells against microbial diseases,[19,20] and cancer.[21] Macrophages have the ability to secrete a variety of active substances (including cytokines, complement factors, and prostaglandins)[22,23] that regulate immune system and other biological functions of the body.[24] Macrophage immunoregulatory functions depend on their activation by priming or sensitizing signals mainly provided by cytokines such as interferon-γ (IFN-γ), and by triggering signals such as the endotoxin lipopolysaccharide (LPS).[25,26] Macrophage activation will lead to the production of mediators of cytotoxicity against intracellular parasites,[27] and tumor cells;[28,29] these cytotoxic factors include the reactive oxygen intermediates (ROI) such as superoxide and hydrogen peroxide, and the reactive nitrogen intermediates (RNI) such as nitric oxide. ROI have been associated with antimicrobial function of macrophages against *Toxoplasma, Trypanosoma, Leishmania, Candida,* and *Mycobacteria.*[30-32] ROI cause peroxidation of membrane fatty acid unsaturated chains, thus affecting their functions and inactivating proteins.[33] Nitric oxide is the most studied RNI and has a relevant role against intramacrophage infections caused by *Cryptococcus, Schistosoma, Leishmania, Francisella, Listeria,* and *Mycobacte-*

ria.[27,34-40] Nitric oxide's major effects are inhibiting the production of ATP[41] and inhibiting DNA synthesis.[42] Nitric oxide production by macrophages depends on activation of the inducible enzyme nitric oxide synthase (NOS) which can be mediated by interferons, tumor necrosis factor-α (TNF-α), or LPS.[43-47]

Macrophage activation facilitates the process of phagocytosis of microbial pathogens which are eventually killed by intracellular mechanisms, including lysosomal enzymes, and ROI or RNI. Another essential function of activated macrophages is to serve as antigen-presenting cells. Macrophage presentation of antigens to T helper cells induces production of IFN-γ by these lymphocytes; IFN-γ then activates macrophage cytotoxic/cytocidal activities, thus potentiating immune responses against infections and tumor cells.[48-50]

This chapter will review the effect of opioid agonists on lymphocyte, NK cell, and macrophage functions.

20.2 OPIOID ACTIONS ON LYMPHOCYTES

20.2.1 Effect of Opioids on T Cell Proliferation

T lymphocytes respond to antigen challenge by proliferating and expanding the antigen-specific lymphocyte clones, thus amplifying immune responses. Functional T cell proliferating activity can be studied by the use of polyclonal mitogens such as concanavalin A (Con A) and phytohemagglutinin (PHA), which bind to certain sugar residues on T cell surface glycoproteins, including the T cell receptor and CD3 protein, and stimulate T cell proliferative response. Morphine,[51] β-endorphin (β-END, Tyr-Gly-Gly-Phe-Met-Thr-Ser-Glu-Lys-Ser-Gln-Thr-Pro-Leu-Val-Thr-Leu-Phe-Lys-Asn-Ala-Ile-Val-Lys-Asn-Ala-His-Lys-Gly-Gln-OH),[52-57] dynorphin-A (DYN-A, Try-Gly-Gly-Phe-Leu-Arg-Arg-Ile-Arg-Pro-Lys-Leu-Lys-Trp-Asp-Asn-Gln-OH),[58] and Met-enkephalin (MENK, Try-Gly-Gly-Phe-Met-OH)[59,60] have been reported to increase mitogen-mediated T cell proliferation; MENK and DPDPE (D-Pen$_2$,D-Pen$_5$-enkephalin) also increased T cell proliferation in the absence of mitogen (Table 20.2). This effect was reversed by the δ-selective antagonist ICI 174,864;[61] in some cases, this effect was reversed by naloxone.[57,59,60] In contrast, morphine,[51,62-75] β-END,[59,62,76] MENK,[77] cocaine,[69,78] and 2-N-pentiloxy-2-phenyl-4-methyl-morpholine (PM)[79] have been shown to suppress mitogen-induced T cell proliferation. This response could be suppressed by naloxone,[57,76,77,79] naltrexone,[63-66,68,72,75] or by the nitric oxide inhibitor L-N^G-monomethyl-L-arginine.[71] Naloxone itself has also been reported to inhibit T cell proliferation.[80] Absence of opioid effect on T cell proliferative response was observed with morphine,[73,80] and MENK[52,80] (Table 20.2).

20.2.2 Effect of Opioids on T Cell-Mediated Cytotoxicity

Cytolytic T lymphocytes (CTLs) are a subpopulation of T cells that kill target cells expressing specific antigen. CTLs are essential effector cells against intracellular infections of nonphagocytic cells, or infections not eliminated by phagocytosis (viral infections, or infection by *Listeria monocytogenes*); other CTLs functions are related

TABLE 20.2
Effect of Opioids on T Cell Proliferation

Opioid Agonist	Cell Source	Mitogen	Proliferative Response	Reverted by Naloxone	Ref.
Morphine (10^{-3} or 10^{-8} M)[a]	Human PBMC[b]	PHA	▼ 10^{-3} M, ▲ 10^{-8} M	ND[c]	51
β-END (10^{-10} M), α-END (10^{-8} M), MENK (10^{-8} M)	Rat spleen cells	Con A/PHA	▲ β-END, ↔ α-END, MENK	No	52
β-END (10^{-13} M)	Human PBMC	Con A	▲	No	53
β-END (10^{-8} M)	Murine spleen cells	Con A	▲	No	54
β-END (10^{-6} M)	Rat lymph node cells	PHA	▲	No	55
β-END (10^{-12} M)	Rat spleen cells	Con A	▲	No	56
β-END (10^{-13} M)	Rat and murine spleen cells	PHA	▲ Murine, ▼ rat	Yes	57
Dyn-A (10^{-12} M)	Human PBMC	PHA	▲	No	58
MENK (10^{-12} M)	Human PBMC	Con A	▲	Yes	59
MENK (10^{-14} M)	Murine lymph node cells	Con A	▲	Yes	60
MENK (10^{-8} M), DPDPE (10^{-8} M)	Human PBMC	No mitogen	▲	ND[d]	61
β-END (10^{-8} M), morphine (10^{-3} M)	Human PBMC	PHA	▼	No	62
Morphine (25 mg/kg, sc)	Rat PBMC	Con A	▼	ND[d]	63
Morphine (10^{-4} M, 5 mg/kg, sc)	Rat PBMC	Con A	▼	ND[d]	64
Morphine (1 mg/kg, im)	*Rhesus* monkey PBMC	PHA, Con A, PWM[e]	▼ Mitogen, ▲ No mitogen	ND[d]	65
Morphine (5 mg/kg, sc)	Rat spleen cells and PBMC	Con A	▼	ND[d]	66
Morphine (10 mg/kg, sc)	Rat PBMC	Con A	▼	ND[d]	67
Morphine (7 mg/kg, sc)	Rat PBMC	Con A	▼	ND[d]	68
Morphine (10 mg/kg, sc), cocaine (5 mg/kg, iv)	Rat PBMC	Con A	▼	ND	69, 70

[a] Values in parentheses represent minimal active dose. *Ex vivo* experiments are in bold letters; ▲, increase; ▼, decrease; ↔, no effect.
[b] PBMC = peripheral blood mononuclear cells.
[c] ND = not determined.
[d] Reversed by ICI-174864,[61] naltrexone,[63-66,68] or RU486.[67]
[e] PWM = pokeweed mitogen.

TABLE 20.2 (CONTINUED)
Effect of Opioids on T Cell Proliferation

Opioid Agonist	Cell Source	Mitogen	Proliferative Response	Reverted by Naloxone	Ref.
DAMGO (20 μg/ml, ICV)[a]	Rat spleen cells	TSST-1[b]	▼	ND[c,d]	71
Morphine (75 mg, pellet)	Murine lymph node cells	No mitogen[e]	▼	ND[d]	72
Morphine (4 μg, PAG/ICV)	Rat spleen cells	Con A, PHA	▼ ICV, ⇔ PAG	ND	73
Morphine (10^{-4} M)	Rat spleen cells	Con A	▼	ND	74
Morphine (10 mg/kg, sc)	Rat spleen cells	Con A, PHA	▼	ND[d]	75
β-END (1 ng, ICV)	Rat spleen cells	PHA	▼	Yes	76
MENK (10^{-7} M)	Murine spleen cells	Con A	▼	Yes	77
Cocaine (5 mg/kg, iv)	Rat spleen cells and PBMC[f]	Con A	▼	ND	78
PM[g] (5 μg/ml)	Rat spleen cells	Con A	▼	Yes	79, 93
Morphine (10^{-8} M)	Human PBMC	Con A, PHA	⇔	No[h]	80

[a] Values in parentheses represent minimal active dose. *Ex vivo* experiments are indicated in bold letters ▲, increase; ▼, decrease; ⇔, no effect.
[b] TSST-1 = toxic shock syndrome toxin.
[c] ND = not determined.
[d] Reversed by L-N^G-monomethyl-L-arginine,[71] or naltrexone.[72,75]
[e] Response was measured by IUdR uptake using fluorodeoxyuridine as a potentiator.
[f] PBMC = peripheral blood mononuclear cells.
[g] PM = 2-n-pentiloxy-2-phenyl-4-methyl-morpholine.
[h] Naloxone itself inhibited lymphocyte proliferation.

to acute allograft rejection, and antitumor activity. MENK and β-END,[81] and camphor odor[82] have been reported to increase CTL activity; this response was reversed by naloxone[81] or naltrexone[82] (Table 20.3). In contrast, morphine,[83-87] and fentanyl (N-phenyl-N- [1- (2-phenylethyl)-4-piperidinyl]propanamide), and meperidine[88] were shown to inhibit CTL activity. This effect was blocked by β-funaltrexone (β-FNA)[85] or naltrexone,[86,87] but not by naltrindole[84] (Table 20.3).

20.2.3 Opioid-Mediated Cytokine Production by T Lymphocytes

Cytokines are glycoproteins involved in signaling between cells during immune responses. They are essential in limiting viral infections (interferons), and stimulat-

TABLE 20.3
Effect of Opioids on T Cell Mediated Cytotoxicity

Opioid Agonist	Cell Source	Assay/Target Cells	CTLs Activity	Reverted by Naloxone	Ref.
β-END (10^{-14} M), MENK (10^{-12} M)[a]	Murine spleen cells	MLC[b]	▲	Yes	81
Camphor odor	Murine spleen cells	Alloimmunization/ EL-4 mouse lymphoma cell line	▲	ND[c,d]	82
Morphine (50 mg/kg, sc)	Murine peritoneal and spleen cells	Alloimmunization/ EL-4 cell line	▼	ND[d]	83–86
Morphine (10 mg/kg, sc)	Murine spleen cells	Alloimmunization/ EL-4 cell line	▼	ND[d]	87
Fentanyl, meperidine (10^{-4} M)	Murine spleen cells	P815 cell line	▼	ND	88

[a] Values in parentheses represent minimal active dose. *Ex vivo* experiments are indicated in bold letters; ▲, increase; ▼, decrease; ↠, no effect.
[b] MLC = mixed lymphocyte culture.
[c] ND = not determined.
[d] Reversed by naltrexone,[82,86,87] or β-funaltrexone.[85]
[e] Not reverted by naltrindole.[84]

ing cell population differentiation or mediating inflammatory responses (interleukins). IFN-γ production has been shown to be enhanced by treatment with β-END or MENK;[89] this effect was not reversed by naloxone.[89] In contrast, morphine[74,90] and β-END[90] were reported to decrease IFN-γ production in a naloxone-reversible fashion.[90] β-END and MENK were shown to suppress the production of T lymphocyte chemotactic factor (LCF)[91] while β-END increased the production of leukocyte migration inhibitor factor (LIF).[92] In addition, IL-1 production was suppressed by PM and reversed by naloxone[93] (Table 20.4). IL-2 production has been reported to be enhanced by morphine,[94] and deltorphin and DPDPE.[95] On the contrary, D-Ala$_2$-Met$_5$-enkephalinamide (DAMEA) and PM[93] were shown to inhibit IL-2 production, while fentanyl and meperidine did not affect it.[88] However, fentanyl and meperidine suppressed IL-4 production,[88] whereas DAMEA and DPDPE enhanced it.[95]

20.2.4 MODULATION OF B CELL FUNCTIONS BY OPIOIDS

In birds, B cells mature and differentiate in the bursa of Fabricius, while in mammals, B cells differentiate in the fetal liver and in the adult bone marrow. B cells play an essential role in adaptive immune responses by producing immunoglobulins (IgM, IgG, IgD, IgE, and IgA isotypes) which specifically recognize, bind to, and neutralize

TABLE 20.4
Effect of Opioids on T Cell Cytokine Production

Opioid Agonist	Cell Source	Cytokine Studied	Effect	Reverted by Naloxone	Ref.
Morphine (15 mg/kg, sc)[a]	Rat spleen cells	IFN-γ	▼	ND[b]	74
Morphine (56 mg/kg, sc)	Murine spleen cells	IFN-γ	▼	ND	87
Fentanyl, meperidine (10^{-4} M)	Murine spleen cells	IL-2, IL-4	⇸ IL-2, ▼ IL-4	ND	88
β-END, MENK (10^{-14} M)	Human PBMC[c]	IFN-γ	▲	No	89
β-END (10^{-12} M), morphine (10^{-10} M)	Human PBMC	IFN-γ	▼	Yes	90
β-END, MENK (10^{-11} M)	Human PBMC	LCF[d]	▼	ND	91
β-END (10^{-10} M)	Human PBMC	LIF[e]	▲	ND	92
PM[f] (10^{-6} M)	Murine spleen cells	IL-1, IL-2	▼	Yes	93
Morphine (0.1 ml/kg/day)	*Rhesus* monkey PBMC	IL-2	▲	ND	94
Deltorphin, DPDPE (10^{-10} M), DAMEA (10^{-6} M)	Murine spleen cells	IL-2, IL-4	▲ IL-2 deltorphin, DPDPE ▼ IL-2 DAMEA ▲ IL-4 DAMEA, DPDPE	ND	95

[a] Values in parentheses represent minimal active dose. *Ex vivo* experiments are indicated in bold letters; ▲, increase; ▼, decrease; ⇸, no effect.
[b] ND = not determined.
[c] PBMC = peripheral blood mononuclear cells.
[d] LCF = T lymphocyte chemotactic factor.
[e] LIF = leukocyte migration inhibitory factor.
[f] PM = 2-n-pentiloxy-2-phenyl-4-methyl-morpholine.

antigens. Although B cells can bind to free antigens and become activated, B cell recognition and response to antigens mostly depend on interacting with T lymphocytes responding to antigen challenge. B cell proliferation has been reported to be enhanced by MENK,[77] and suppressed by fentanyl and meperidine,[88] U50,488 (*trans*-3,4-dichloro-*N*-methyl-*N*-[2(1-pyrrolidinyl)-cyclohexyl]-benzene-acetamide),[96] morphine,[73,74] and DSLET ([D-Ser$_2$,Leu$_5$]enkephalin-Thr$_6$) and DPDPE[95] (Table 20.5). Absence of an effect by morphine on B cell proliferative response was also reported.[73,75] Opioid agonists have been also shown to affect immunoglobulin secretion. IgM production has been reported to be suppressed by MENK,[93,97,98] and naltrindole;[99] IgG secretion was inhibited by treatment with DPDPE and U50,488,[96] MENK,[97] β-END,[98] and PM,[93] whereas IgA production was suppressed by β-END,[98] and PM.[93] In addition, total plaque-forming cell responses to sheep red blood cells were shown to be suppressed by β-END[57] and U50,488.[100] Suppression of immunoglobulin production by opioids was reverted by naloxone,[57,93,97] or by [D'Tic]cTAP, norBNI, and ICI 174,864 (μ-, κ-, and δ-opioid receptor-selective antagonists, respectively).[96] In contrast, enhancement of IgM or IgG production was induced by morphine[101] or naloxone[102] (Table 20.5).

20.2.5 NATURAL KILLER CELL ACTIVITY

Natural killer (NK) cells are cytotoxic lymphocytes differing from CTLs in that NK cells lack the specific T cell receptor for antigen recognition. They are not T or B lymphocytes as demonstrated by surface phenotype and lineage; in addition, NK cells do not undergo thymic maturation, or immunoglobulin or T cell receptor gene rearrangements. NK cells are essential effector cells in natural rather than specific immunity; they are cytotoxic against tumor cells and normal cells infected by virus, and activate macrophage cytotoxic activity by secreting IFN-γ. It has been reported that β-END,[103-106] nonopioid fragments of β-END,[107] Leu-enkephalin (LENK, Try-Gly-Gly-Phe-Leu-OH),[103,108] MENK,[104,108,109] DYN-A, DSLET and DAMGO (D-Ala$_2$-N-MePhe$_4$,Gly-ol$_5$-enkephalin),[108] the fentanyl derivative OHM3295,[110,111] fentanyl,[112] naloxone,[113] DAMEA, DPDPE, and deltorphin,[95] and camphor odor plus poly I:C[114] are able to increase NK cell activity. This effect was blocked by naloxone,[103,106-108] naltrexone,[66,110,111] β-FNA,[110,111] or antibodies against β-END[105] (Table 20.6). Conversely, morphine,[66,73,74,87,101,115-119] MENK, LENK, DYN-A, DSLET, and DAMGO,[108] fentanyl,[88,110,111] naloxone,[102,113,120] and camphor odor[82] were reported to suppress NK cell activity; this response was inhibited by naloxone,[108] and naltrexone[66,110,111,118,119] (Table 20.6). In addition, absence of an opioid effect on NK cell activity has been shown for morphine,[83-86,103] cocaine,[69,78] the fentanyl derivative OHM3507,[121] and naloxone.[113] Central opioid action has been observed to suppress (periaqueductal gray [PAG] matter opioid injection),[73,119] or enhance (intracerebroventricular [i.c.v.] opioid injection)[106] NK cell activity. However, central injection (i.c.v.) of IFN-α in rats has been reported to suppress NK cell cytotoxicity by activating the splenic sympathetic nerve through brain opioid receptors; this effect was blocked by i.c.v. injection of naltrexone,[122] α-helical CRF$_{(9-41)}$ (antagonist of corticotropin-releasing factor),[122,123] i.v. injection of naloxone,[124] splenic denervation,[122] and by i.v. injection of the β-adrenergic antagonist nadolol,[124] thus suggesting a mechanism involving β-adrenergic receptors on splenic cells.[122-124]

20.3 EFFECT OF OPIOIDS ON MACROPHAGE FUNCTIONS

20.3.1 OPIOID ACTION ON MACROPHAGE PROLIFERATION AND DIFFERENTIATION

Immune responses depend on cellular proliferation and differentiation leading to cell-mediated immunity and antibody production. Morphine has been reported to

TABLE 20.5
Effect of Opioids on B Cell Functions

Opioid Agonist	Cell Source	Stimulus	Function Studied and Effect	Reverted by Naloxone	Ref.
β-END (10^{-5} M)[a]	Murine spleen cells	SRBC[b]	▼ PFC[c]	Yes	57
Morphine (40 μg, PAG/ICV)	Rat spleen cells	LPS	▼ Proliferation$_{i.c.v.}$ ↔ Proliferation$_{PAG}$	ND	73
Morphine (10^{-4} M)	Rat spleen cells	LPS	▼ Proliferation	ND	74
Morphine (10 mg/kg, sc)	Rat spleen cells	LPS	↔ Proliferation	ND	75
MENK (10^{-7} M)	Murine spleen cells	LPS	▲ Proliferation	ND	77
Fentanyl, meperidine (10^{-10} M)	Murine spleen cells	IL-4 + anti-IgM	▼ Proliferation	ND	88
PM[d] (10^{-6} M)	Murine spleen cells	SRBC, LPS	▼ IgM, IgG, IgA	Yes	93
DSLET, DPDPE (10^{-7} M)	Murine spleen cells	IL-4 + anti-IgM	▼ Proliferation	ND	95
DAMGO, DPDPE (10^{-10} M), U50,488 (10^{-7} M)	Human PBMC	SAC[e]	▼ Proliferation$_{U50,488}$ ▼ IgG$_{DAMGO,DPDPE,U50488H}$	ND[f]	96
MENK (10^{-12} M)	Murine spleen cells	LPS, dextran sulfate	▼ IgM, IgG$_3$, IgG$_{2a}$	Yes	97
β-END (10^{-8} M)	Murine Peyer's patch cells	Con A	▼ IgM, IgG, IgA	No	98
Naltrindole (10^{-7} M)	Murine spleen cells	LPS	▼ IgM	ND[f]	99
U50,488 (10^{-6} M)	Murine spleen cells	SRBC	▼ PFC	ND	100
Morphine (3.2 mg/kg/day)	*Rhesus* monkey PBMC	PWM[g]	▲ IgM, IgG	ND	101
Naloxone (0.1 mg/kg)	Murine spleen cells	SRBC	▲ IgM	ND	102

[a] Values in parentheses represent minimal active dose. *Ex vivo* experiments are indicated in bold letters; ▲, increase; ▼, decrease; ↔, no effect.
[b] SRBC = sheep red blood cells.
[c] PFC = plaque-forming cells.
[d] PM = 2-n-pentiloxy-2-phenyl-4-methyl-morpholine.
[e] SAC = *Staphylococcus aureus* Cohen strain I.
[f] Reversed by [D'Tic]cTAP, ICI 174864, nor-BNI,[96] or oxymorphindole.[99]
[g] PWM = pokeweed mitogen.

TABLE 20.6
Effect of Opioids on NK Cell Activity

Opioid Agonist	Cell Source	Cr[51] Release Assay	NK Cell Activity	Reverted by Naloxone	Ref.
Morphine (5 mg/kg, sc)[a]	Rat spleen cells	YAC-1 cell line	▼	ND[b]	66
Cocaine (5 mg/kg, iv)	Rat spleen cells	YAC-1 cell line	↔	ND	69, 78
Morphine (4 μg, i.c.v./ PAG)	Rat spleen cells	YAC-1 cell line	▼	ND	73
Morphine (15 mg/kg, sc)	Rat spleen cells	YAC-1 cell line	▼	ND	74
Camphor odor	Murine spleen cells	YAC-1 cell line	▼	ND	82
Morphine (50 mg/kg, sc)	Murine spleen cells	YAC-1 cell line	↔	ND	83–86
Morphine (56 mg/kg, sc)	Murine spleen cells	YAC-1 cell line	▼	ND	87
Fentanyl, meperidine (10^{-4} M)	Murine spleen cells	YAC-1 cell line	▼	ND	88
DAMEA (10^{-9} M), DPDPE, deltorphin (10^{-11} M)	Murine spleen cells	YAC-1 cell line	▲	ND	95
Morphine (3.2 mg/kg/day)	*Rhesus* monkey PBMC[c]	K562 cell line	▼	ND	101
Naloxone (0.1 mg/kg),	Murine spleen cells	YAC-1 cell line	▼	ND	102
β-END (10^{-8} M), LENK, morphine (10^{-7} M)	Human PBMC	K562 cell line	▲ β-END, LENK ▼ morphine	Yes	103
β-END (10^{-10} M), MENK (10^{-8} M)	Human PBMC	K562 cell line	▲	ND	104
β-END (0.02 μg, cisterna magna)	Murine spleen cells	YAC-1 cell line	▲	ND[b]	105
β-END (120 μg in 6 days, i.c.v.)	Rat spleen cells	YAC-1 cell line	▲	Yes	106
β-END (10^{-15}M, non-opioid fragments)	Human PBMC	K562 cell line	▲	Yes	107

[a] Values in parentheses represent minimal active dose. *Ex vivo* experiments are indicated in bold letters; ▲, increase; ▼, decrease; ↔, no effect.
[b] Reversed by naloxone,[66] or antibodies to β-END injected in the cisterna magna.[105]
[c] PBMC = peripheral blood mononuclear cells.

TABLE 20.6 (CONTINUED)
Effect of Opioids on NK Cell Activity

Opioid Agonist	Cell Source	Cr^{51} Release Assay	NK Cell Activity	Reverted by Naloxone	Ref.
MENK, LENK, DYN-A (10^{-10} M), DSLET, DAMGO (10^{-8} M)^a	Human PBMC^b	K562 cell line	▲ Low NK cell responders, ▼ High NK cell responders	Yes	108
MENK (10^{-8} M)	Human PBMC	K562 cell line	▲	ND	109
Fentanyl (0.25 mg/kg, sc), OHM3295 (3.2 mg/kg, sc)	Murine spleen cells	YAC-1 cell line	▲ OHM3295 ▼ Fentanyl	Yes^c	110, 111
Fentanyl (5 mg/kg, sc)	Human PBMC	K562 cell line	▲	ND	112
Naloxone (10^{-10} M)	Human PBMC	K562 cell line	▲ ▼ ↔	ND	113
Camphor odor	Murine spleen cells	YAC-1 cell line	▲	ND	114
Morphine (25 mg/kg, sc)	Murine spleen cells	YAC-1 cell line	▼	ND	115
Morphine (50 mg/kg, sc)	Murine spleen cells	YAC-1 cell line	▼	ND	116
Morphine (10 mg, iv)	Human PBMC	K562 cell line	▼	ND	117
Morphine (32 mg/kg, sc)	Murine spleen cells	YAC-1 cell line	▼	ND^c	118
Morphine (6.6 nmol, PAG)	Rat spleen cells	YAC-1 cell line	▼	ND^c	119
Naloxone (10^{-3} M)	Human PBMC	K562 cell line	▼	ND	120
OHM3295 (1 mg/kg, sc)	Murine spleen cells	YAC-1 cell line	↔	ND	121

^{a} Values in parentheses represent minimal active dose. *Ex vivo* experiments are indicated in bold letters; ▲, increase; ▼, decrease; ↔, no effect.
^{b} PBMC = peripheral blood mononuclear cells.
^{c} Reversed by β-FNA,[110,111] naltrexone.[118,119]

inhibit proliferation and differentiation of macrophage-colony stimulating factor (M-CSF)-dependent macrophage progenitor cells;[125,126] this event was reversed by naloxone.[125,126] Similarly, β-END, MENK, and LENK have been shown to reduce macrophage[126] or granulocyte-macrophage[127,128] colony formation (Table 20.7); the effect of β-END was not reversed by naloxone,[126] whereas this antagonist itself was found to inhibit granulocyte-macrophage colony formation.[127] In contrast, β-END alone or in combination with IL-1, was reported to enhance M-CSF-dependent bone-marrow macrophage differentiation (Table 20.7); this effect was completely abolished by naloxone.[129]

20.3.2 OPIOID-MEDIATED CYTOKINE PRODUCTION BY MACROPHAGES

Opioid agonists have been largely associated with altering cytokine production by macrophages. Monokines such as IL-1, IL-6, and TNF-α are essential mediators of host inflammatory responses in natural immunity.[130,131] MENK,[132-134] LENK, and β-neoendorphine (β-NEND, Tyr-Gly-Gly-Phe-Leu-Arg-Lys-Try-Pro)[135] and β-END[135-137] have been shown to potentiate LPS-induced IL-1 production by macrophages. MENK effect was not blocked[132] or partially suppressed[134] by naloxone, whereas β-END effect was abrogated by naloxone[135,136] (Table 20.7). On the contrary, morphine,[138] the opioid U50,488,[139,140] and PM[93] have been shown to inhibit LPS-mediated IL-1 production. This effect was not blocked by naloxone,[138] or the effect was reversed by this antagonist or by the κ-selective opioid receptor antagonist norbinaltorphimine (norBNI).[93,139,140] In addition, LPS-induced TNF-α production by macrophages has been reported to be inhibited by morphine,[138] MENK,[133] and U50,488[139,140] (Table 20.7). In contrast, morphine[141,142] and DYN-A[143] were shown to activate LPS-induced TNF-α production by microglial cells (Table 20.7); such an effect was abrogated by naloxone or by β-FNA[141,142] and by norBNI.[143] However, microinjection of morphine into the PAG did not affect LPS-mediated TNF-α and IL-6 production by macrophages.[144] In addition, β-END,[138] DPDPE,[95,145] deltorphin, DPDPE,[95] and MENK[95a,95b] were reported to potentiate LPS-induced IL-6 production by macrophages, while U50,488[139] inhibited this response (Table 20.7).

20.3.3 EFFECT OF OPIOIDS ON MACROPHAGE RESPIRATORY BURST

Production of reactive oxygen intermediates by activated macrophages is largely associated with the killing of intracellular parasites[30-33] and tumor cells.[29] Morphine, β-END, DYN-A, and MENK have been reported to increase superoxide and/or hydrogen peroxide production by macrophages of different sources.[142,146-150] On the contrary, [D-Ala$_2$-Met$_5$]-enkephalinamide (DAMEA) was shown to inhibit macrophage respiratory burst[151] (Table 20.7); enhancement or inhibition of respiratory burst by opioid agonists was reversed by naloxone.[146-149,151]

20.3.4 OPIOID MODULATION OF NITRIC OXIDE PRODUCTION BY MACROPHAGES

Nitric oxide production is catalyzed by the enzyme NOS which is induced after macrophage activation by cytokines, bacteria, and tumor cells.[152-154] Nitric oxide release is associated with resistance to infection[27,34-40] (this has been clearly observed in mice lacking inducible NOS which are more susceptible to infection[155]) and with antitumor activity.[156] However, toxic levels of nitric oxide may impair macrophage functions.[157,158] Morphine[159-161] and the agonist DAMGO[162] have been reported to

TABLE 20.7
Effect of Opioids on Macrophage Functions

Opioid Agonist	Target Cells	Function Studied and Effect	Reverted by Naloxone	Ref.
Morphine (10^{-5} M)[a]	Murine macrophage cell line Bac 1.2F5	▼ Proliferation and differentiation	Yes	125
Morphine (75 mg. pellets, sc, or 10^{-5} M), β-END (10^{-7} M)	Murine bone marrow cells	▼ Macrophage colony formation	Yes[b]	126
MENK and LENK (10^{-15} M)	Murine bone marrow cells	▼ Granulocyte-macrophage colony formation	ND[b]	127, 128
β-END (10^{-13} M)	Murine bone marrow cells	▲ Macrophage colony formation	Yes	129
DAMEA (10^{-10} M), deltorphin, DPDPE (10^{-8} M)	Murine spleen cells	▲ IL-6	ND	95
MENK (10^{-12} M)	Murine peritoneal macrophages	▲ IL-1 production	No	132
MENK (10^{-10} M)	Murine peritoneal macrophages	▼ IL-1 t TNF-α production	ND	133
MENK (10^{-8} M)	Murine microglial cells	▲ IL-1-β production	Yes	134
β-END and LENK (10^{-14} M), β-NEND (10^{-12} M)	Murine bone marrow cells	▲ IL-1 production	Yes	135, 136
β-END (10^{-11} M)	Murine splenic and peritoneal macrophages	▲ IL-1 and IL-6 production	ND	137
Morphine (10^{-8} M for IL-1); (10^{-3} M for TNF-α)	Murine peritoneal macrophages	▼ IL-1 and TNF-α production	No	138
U50,488 (10^{-10} M for IL-1); (10^{-8} M for TNF-α)	Murine macrophage cell line P388D1	▼ IL-1 and TNF-α production	Yes[b]	139
U50,488 (10^{-9} M for IL-1; 10^{-7} M for IL-6; 10^{-9} M for TNF-α)	Murine peritoneal macrophages	▼ IL-1, IL-6 and TNF-α production	Yes[b]	140
Morphine (10^{-10} M)	Human and murine microglial cells	▲ TNF-α production	Yes[b]	141, 142
DYN-A (10^{-13} M)	Human microglial cells	▲ IL-6 and TNF-α production	Yes[b]	143
Morphine (3672 ng in PAG)	Rat splenic macrophages	↔ IL-6 and TNF-α production	ND	144
DPDPE (10^{-8} M)	Murine peritoneal macrophages	▲ IL-6 production	ND	145

[a] Values in parentheses represent minimal active dose. *Ex vivo* experiments are indicated in bold letters ▲, increase; ▼, decrease; ↔, no effect.

[b] The effect of β-END was not reversed by naloxone,[126] while naloxone itself inhibited granulocyte-macrophage colony formation.[127] The effect of U50,488 was also reversed by norBNI.[139,140] Morphine effect was also reversed by β-FNA,[141,142] while the effect of DYN-A was also reverted by norBNI.[143]

Opioid Agonist	Target Cells	Function Studied and Effect	Reverted by Naloxone	Ref.
Morphine (10^{-10} M)	Murine microglial cells	▲ O_2^- release	Yes	142
β-END and DYN-A (10^{-12} M)	Murine macrophage cell line J774	▲ O_2^- and H_2O_2 release	Yes	146
β-END and DYN-A (10^{-14} M)	Human peritoneal macrophages	▲ O_2^- release	Yes	147
β-END (10^{-12} M)	Human monocytes	▲ O_2^- release	Yes	148
MENK (10^{-14} M)[a]	DA and AO rat peritoneal macrophages	▲ (DA), t (AO) H_2O_2 release	Yes	149
MENK (10^{-9} M)	Rat peritoneal macrophages	▲ O_2^- release	ND	150
DAMEA (10^{-14} M)	Murine peritoneal macrophages	▼ O_2^- and H_2O_2 release	Yes	151
Morphine (20 mg/kg, sc)	Murine peritoneal macrophages	▲ Nitric oxide release	ND[b]	159
Morphine (15 mg/kg, sc)	Rat splenic macrophages	▲ Nitric oxide release	ND	160
Morphine (10^{-7} M)	Human monocytes	▲ Nitric oxide release	Yes	161
DAMGO (0.1 μg in ICV)	Rat splenic macrophages	▲ Nitric oxide release	ND	162
Morphine, DAMGO, and U50,488 (10^{-8} M)	Murine macrophage cell line J774	▼ Nitric oxide release	Yes	163
Morphine (50 mg. pellet, sc)	Murine peritoneal macrophages	▼ Phagocytosis of *C. albicans*	No	11
Morphine (170 mg/day)	Human monocytes	▼ Phagocytosis of *C. albicans*	ND	167
Morphine (75 mg/kg, sc)	Murine peritoneal macrophages	▼ Phagocytosis of *C. albicans*	ND	168
Morphine (75 mg. pellet, sc, or 10^{-8} M)	Murine peritoneal macrophages	▼ Phagocytosis of *C. albicans*	ND	169
Morphine (10^{-8} M)	Murine peritoneal macrophages	▼ Phagocytosis of SRBC[c]	Yes[b]	170
Morphine (10^{-10} M), DAMGO and DPDPE (10^{-11} M), U50,488 (10^{-9} M)	Murine peritoneal macrophages	▼ Phagocytosis of *C. albicans*	ND[b]	171
MENK (10^{-14} M)	Murine peritoneal macrophages	▼ Phagocytosis of SRBC	ND	133

[a] Values in parentheses represent minimal active dose. *Ex vivo* experiments are indicated in bold letters; ▲, increase; ▼, decrease; ↔, no effect.
[b] Reversed by naltrexone,[159,170] naltrindole and norBNI,[171] or β-FNA.[174]
[c] SRBC = sheep red blood cells.

TABLE 20.7 (CONTINUED)
Effect of Opioids on Macrophage Functions

Opioid Agonist	Target Cells	Function Studied and Effect	Reverted by Naloxone	Ref.
MENK (10^{-9} M)	Rat peritoneal macrophages	▼ Phagocytosis of SRBC	Yes	172
Alfentanyl (100 μg/kg, iv)	Human monocytes	▼ Phagocytosis of latex particles	Yes	173
Morphine (10^{-8} M)	Human microglial cells	▲ Phagocytosis of *Mycobacterium tuberculosis*	Yes	174
DYN-A (10^{-9} M)	Murine peritoneal macrophages	▲ Phagocytosis of latex particles	No	175
MENK (10^{-10} M), LENK (10^{-12} M)	Murine peritoneal macrophages	▲ Phagocytosis of SRBC	ND	176
Morphine (20 mg/kg, sc)	Murine peritoneal macrophages	▲ (20–40 min posttreatment), t (24 h posttreatment) antitumor activity vs. L1210 cell line	ND	159
Morphine (20 mg/kg, sc)[a], **methadone (12.5 mg/kg, sc)**	Murine peritoneal macrophages	▲ (Morphine), ↔ (methadone) antitumor activity vs. L1210 cell line	ND	180
MENK and LENK (10^{-8} M), α-END (10^{-12} M)	Human monocytes	▲ Antitumor activity vs. MA-160 cell line	Yes	181
DYN-A (10^{-9} M), MENK (10^{-7} M), LENK (10^{-10} M), β-END (10^{-12} M)	Murine peritoneal macrophages	▲ Antitumor activity vs. EL-4 cell line	Yes	182
DYN-A (10^{-10} M)	Murine peritoneal macrophages	▲ Antitumor activity vs. P815 cell line	Yes	183
α-END, β-END, MENK and LENK (10^{-5} M)	Murine peritoneal macrophages	↔ Antitumor activity vs. B16- F10 cell line	ND	184

[a] Values in parentheses represent minimal active dose. *Ex vivo* experiments are indicated in bold letters; ▲, increase; ▼, decrease; ↔, no effect.
[b] Reversed by naltrexone.[159]

activate nitrite release by primary cultures of macrophages from different sources; this effect was observed to be reverted by naltrexone[159] or naloxone.[161] In contrast, morphine, DAMGO, and U50,488, but not DPDPE and deltorphin, were shown to inhibit LPS-mediated nitrite production by the cell line J774[163] (Table 20.7); this effect was abrogated by naloxone. We have observed that microinjection of morphine into PAG significantly decreases LPS- or IFN-γ-mediated nitric oxide production by splenic, but not peritoneal, macrophages (unpublished observations). It is now recognized that opioids can regulate immune responses by acting on the PAG which has also been implicated in regulating functions such as pain, aggression, vocalization, and reproductive behavior.[119,164]

20.3.5 Opioid Action on Phagocytosis by Macrophages

Phagocytosis or endocytosis of parasites by macrophages is the first stage of a series of events leading to intracellular pathogen death. Once inside the cell, macrophages kill pathogens by generating ROI, RNI, or TNF-α.[27,28,165,166] Morphine,[11,167-171] the agonists DAMEA, LENK, and [D-Ala$_2$-Leu$_5$]-enkephalinamide (DALEA),[170] MENK,[133,171,172] as well as DAMGO, DPDPE, and U50,488[171] and alfentanyl[173] have been reported to inhibit *in vitro* and/or *in vivo* phagocytosis by macrophages of different sources (Table 20.7); this effect could be blocked by naloxone,[170,172,173] naltrexone,[170] and H-D-Phe-Cys-Tyr-D-Trp-Arg-Thr-Pen-Thr-NH$_2$, naltrindole, and norBNI.[171] However, it was also shown that the morphine inhibitory effect on macrophage phagocytosis was not reversed by naloxone.[11] In addition, we have observed that microinjection of morphine into PAG significantly decreases phagocytosis of *Candida albicans* by spleen macrophages (unpublished observations). In contrast, phagocytosis was shown to be enhanced by morphine,[174] DYN-A,[175] and MENK and LENK[176] (Table 20.7). Naloxone or β-FNA abrogated the morphine[174] but not DYN-A[175] enhancing effect on macrophage phagocytosis.

20.3.6 Effect of Opioids on Macrophage Tumoricidal Activity

Macrophages play an essential role in the host defense against neoplasia.[177-179] Morphine,[159,180] α-END, MENK, and LENK,[181] and DYN-A, LENK, and β-END[182,183] have been reported to potentiate macrophage tumoricidal activity (Table 20.7); this effect was reversed by naltrexone[159] or naloxone.[181-183] On the contrary, methadone (6-dimethylamino-4,4-diphenyl-3-heptanone),[180] α-END, MENK, and LENK[184] have shown no effect against a variety of tumor cells (Table 20.7).

20.3.7 Additional Opioid-Regulated Macrophage Functions

Opioid agonists also have been reported to affect a diversity of macrophage functions. Morphine (10^{-6} M) was shown, in a naloxone-reversible fashion, to induce conformational changes of human monocytes by stimulating nitric oxide production.[185] Spreading of murine macrophage cell line RAW264 has been shown to be stimulated by MENK (10^{-8} M).[186] Morphine (75 mg pellets, s.c.) was reported to enhance prostaglandin E$_2$ synthesis by rat peritoneal macrophages in a naloxone-independent fashion.[187] In addition, morphine, heroin, DAMGO, and DPDPE were reported to inhibit human monocyte chemotaxis to endotoxin-activated serum;[188] this effect was abrogated by naloxone. In contrast, β-END,[189,190] MENK,[189] LENK and DYN-A,[190] and the undecapeptide substance P[191] were shown to stimulate human monocyte chemotaxis.

20.4 OPIOID RECEPTORS ON CELLS OF THE IMMUNE SYSTEM

Exogenous administration and endogenous opioid action affect not only behavior, and the neuroendocrine and autonomic systems, but also the immune system. Opioids effects depend on the presence of opioid-selective receptors (μ, δ, and κ) (Table 20.1) present on the surface of immunocompetent cells. These receptors exhibit different patterns of ligand selectivity, stereoselectivity, saturability, and nanomolar affinity for opioids.[4,9,192,193] Presence of opioid receptors on cells of the immune system has been indirectly inferred by the use of μ, δ, and κ class receptor-selective antagonists. However, opioid binding sites on lymphocytes and macrophages have been identified by opioids labeled with radioisotopes or fluorochromes, or by identifying the genes responsible for opioid-receptor expression. The use of competitive binding assays with radiolabeled ligands such as [^{3}H]naloxone,[194-196] [^{3}H]U69,593,[196,197] [^{3}H]cis-(+)-3-methylfentanylisothio-cyanate (SUPERFIT),[197,198] and [^{3}H]dihydromorphine,[199] or fluorochromes such as naltrexone-fluorescein isothiocyanate (FITC)[200] and 1-(N)-fluoresceinyl naloxone thiosemicarbazone (6-FN),[201] has allowed the identification of opioid binding sites on the surface of human lymphocytes[194,195,198] and monocytes,[198,199] murine lymphocytes[197,198,201] and macrophages,[201] rat lymphocytes, monocytes, and neutrophils,[200] and in murine lymphoid[196,197] and myeloid[197] cell lines (Table 20.8). The expression of genes for opioid receptors has been demonstrated in human lymphocytes[202-204] and monocytes,[203] monkey lymphocytes,[202] human lymphoid cell lines,[203,204] murine lymphocytes,[203,205,206] murine lymphoid cell lines,[203] and rat macrophages[207] by RNA amplification using the reverse transcriptase-polymerase chain reaction (PCR) technique (Table 20.8).

20.5 MODULATION OF SIGNAL TRANSDUCTION PATHWAYS BY OPIOID ACTION

Opioids produce their effects in leukocytes via interactions with μ, δ, and κ opioid receptors. Coupling of opioids with their receptors will activate membrane G proteins which transmit signals through two major pathways, the adenylate cyclase and the phospholipase C systems (this one can also activate the guanylate cyclase system). The activation of these pathways induce protein phosphorylation which enable cells to respond rapidly to diverse signals in the extracellular environment. Protein phosphorylation is crucial for a number of cellular processes such as intermediate metabolism, cytoskeletal architecture, cell adhesion, and cell cycle progression. Protein phosphorylation induces second messengers such as inositol phosphates, diacylglycerol, Ca^{2+}, phosphatidic acid (arachidonic acid), and transcription of new proteins.

Opioid action on leukocytes activate signal transduction pathways.[208] Chronic morphine treatment has been reported to inhibit forskolin-mediated cAMP production in *Rhesus* monkey peripheral blood mononuclear cells[101] and in murine CD8$^+$ lymphocytes.[83] Similarly, Sharp et al. reported that treatment of a T cell line with DADLE ([D-Ala$_2$,D-Leu$_5$]enkephalin) decreased forskolin-mediated intracellular

TABLE 20.8
Presence of Opioid Receptors on Cells of the Immune System

Target Cells	Assay	Ligand/mRNA	Receptor Type Identified	Ref.
Human lymphocytes and platelets	Competitive binding with [^{3}H]naloxone	Morphine and naloxone	μ	194
Human T lymphocytes	Competitive binding with [^{3}H]naloxone	Morphine, β-END, MENK, and LENK	μ, δ	195
Murine lymphoma cell line	Competitive binding with [^{3}H]naloxone and [^{3}H]U69,593	U69,593	κ	196
Murine T and B cell populations, murine T, B, and macrophage cell lines	Competitive binding with [^{3}H]U69,593 and [^{3}H]SUPERFIT	U69,593 and SUPERFIT	κ, δ	197
Murine T and B cell populations, murine lymphocytes, and human lymphocytes and monocytes	Competitive binding with [^{3}H]SUPERFIT	SUPERFIT	δ	198
Human monocytes	Competitive binding with [^{3}H]dihydromorphine	Morphine, dihydromorphine, naloxone, naltrexone	μ	199
Rat lymphocytes, monocytes, and neutrophils	Competitive binding with naltrexone-FITC	Naltrexone, DADLE, and DAMGO	μ, δ	200
Murine T and B cells and macrophages	Competitive binding with 6-FN	Naloxone, morphine, and methadone	μ	201
Human and monkey lymphocytes	Reverse transcriptase-PCR	mRNA	δ	202
Human lymphocytes and monocytes, human T and B cell lines, murine spleen cells, and murine T and B cell lines	Reverse transcriptase-PCR	mRNA	κ, δ (human) δ (mouse)	203
Human PBMC[a], and human T and B cell lines	Reverse transcriptase-PCR	mRNA	Orphan opioid receptor	204
Murine CD4$^+$CD8$^+$, and CD4$^-$CD8$^-$ lymphocytes	Reverse transcriptase-PCR	mRNA	Orphan opioid receptor	205
Murine CD4$^+$ lymphocytes	Reverse transcriptase-PCR	mRNA	δ	206
Rat peritoneal macrophages	Reverse transcriptase-PCR	mRNA	μ	207

[a] PBMC = peripheral blood mononuclear cells.

cAMP levels, but increased intracellular Ca^{2+} levels (i[Ca^{2+}]).[209] It was also shown that treatment of murine splenic T cells[209,210] or T cell lines[211] with β-END increased i [Ca^{2+}]. In addition, stimulation of human B cell lines with DAMGO and U50,488 was reported to enhance i [Ca^{2+}].[212] On the other hand, β-END has been shown to inhibit the production of phosphatidylinositol in human lymphocytes,[213] to alter the phosphorylation of the CD3 γ chain of human T lymphocytes,[214] and to increase *c-myc* mRNA levels in Jurkat cells.[211] In the macrophage, opioid-mediated changes in second messengers have been also implicated. MENK has been reported to activate Ca^{2+} influx, and to increase cAMP and cGMP intracellular levels in rat peritoneal macrophages;[150] MENK-mediated increases in Ca^{2+} influx and cGMP levels, but not cAMP levels, were reversed by naloxone.

20.6 DISCUSSION

It is well recognized that lymphocytes, NK cells, and macrophages are very sensitive to opioid action. The role of opioids on regulating immune responses has become more significant because of the implications of drug abuse on immunity against infectious diseases and cancer.[4,10-13,215,216] Opioids may interact directly with opioid receptors on cells of the immune system,[57,59,60,76,77,79,81,93,97,103,105,106,108,110,111,126,135,140,149,159,163,169,183,188] or on receptors within the CNS,[119,144,162,217,218] (and our unpublished observations on *in vivo* acute injection of morphine into the PAG). However, there is no clear evidence of a direct involvement of morphine on lymphocyte, NK cell, or macrophage opioid receptors *in vivo*. Since the peripheral[11,66-70,83-87,110-112,219] or central[71,73,76,106,119,122,162] administration of opioids appears to interact with brain opioid receptors, other neuroendocrine systems are hypothesized to regulate immune function: the HPA and the sympathetic nervous system (SNS).[119] The activation of the HPA axis elicits the production of adrenocorticotropin hormone from the pituitary, which in turn elicits the release of the glucocorticoids[220] which either suppress[221] or have no effect[222] on immune functions. On the other hand, the activation of the SNS by opioids through innervation of primary and secondary lymphoid organs,[223] elicits the release of catecholamines,[224] which have been demonstrated to suppress lymphocyte,[225] NK-cell,[226] and macrophage[227] functions. In our laboratory, Hall et al.[228] have shown that microinjection of morphine in the PAG increases splenic norepinephrine and serotonin levels (unpublished observations), and suppresses anti-CD3/TcR plus IL-2 or Con A-induced rat splenic and thymic T cell proliferation and splenic NK cell cytotoxic activity which was not reversed by the glucocorticoid antagonist RU486.[229] These findings suggest that the SNS, but not the HPA axis, is involved in morphine-mediated immunosuppression. Numerous studies have provided evidence that the interactions between immune, endocrine, and neural systems maintains homeostasis. Immune, endocrine, and neural cells express receptors for cytokines, hormones, neuropeptides, and transmitters, and coexist in lymphoid, endocrine, and neural tissue.[230] Investigating the cellular mechanisms of opioid action in regulating immune functions, must be complemented by studying the molecular pathways of opioid-mediated immunosuppression through the CNS. Changes in the intracellular levels of cAMP or Ca^{2+} have been associated with opioid-mediated immunoregulation,[83,101,150,209-212,231-234] however, scarce information exists about the involvement of other second messengers or gene regulation after *ex vivo* opioid action.

Lymphocyte, NK cell, and macrophage functions have been reported to be enhanced, suppressed, or not to be affected by opioid action (Tables 20.2 to 20.7). Such differences might be related to several factors including: (1) cell source; (2) opioid type and dose, and probably opioid receptor type; (3) route of opioid administration; and (4) duration of treatment. Most of the research of opioid action on leukocyte functions is based on *in vitro* or *ex vivo* studies. Few studies have shown the *in vivo* impact of opioids in the development of disease.[11] It is then necessary to prompt *in vivo* studies to elucidate the role of opioid agonists on leukocyte function against infectious diseases and cancer.

ACKNOWLEDGEMENT

This work was supported by NIDA Grant DA/AI 08988.

REFERENCES

1. Carr, D. J. J., Rogers, T. J., and Weber, R. J., The relevance of opioid receptors on immunocompetence and immune homeostasis, *Proc. Soc. Exp. Biol. Med.*, 213, 248, 1996.
2. Rouveix, B., Opiates and immune function. Consequences on infectious diseases with special reference to AIDS, *Therapie*, 47, 503, 1992.
3. Adler, M. W., Geller, E. B., and Rogers, T. J., Opioids, receptors, and immunity, in *Drugs of Abuse, Immunity, and AIDS*, Friedman, H., Ed., Plenum Press, New York, 1993.
4. Roy, S. and Loh, H. H., Effects of opioids on the immune system, *Neurochem. Res.*, 21, 1375, 1996.
5. Blatteis, C. M., Xin, L., and Quan, N., Neuromodulation of fever: apparent involvement of opioids, *Brain Res. Bull.*, 26, 219, 1991.
6. Dafny, N., Prieto-Gomez, B., and Reyes-Vazquez, C., Does the immune system communicate with the central nervous system? Interferon modifies central nervous activity, *J. Neuroimmunol.*, 9, 1, 1985.
7. Felten, D. L., Cohen, N., Ader, R., Felten, S. Y., Carlson, S. L., and Roszman, T. L., Central neural circuits involved in neural-immune interactions, in *Psychoneuroimmunology*, Ader, R. and Cohen, N., Eds., Academic Press, San Diego, 1991.
8. McEvoy, G. K., Opiate agonists, in *American Hospital Formulary Service Drug Information*, McEvoy, G. K., Ed., American Society of Health-System Pharmacists, Bethesda, MD, 1995.
9. Carr, D. J. J., The role of endogenous opioids and their receptors in the immune system, *Proc. Soc. Exp. Biol. Med.*, 198, 710, 1991.
10. Guan, L., Towsend, R., Eisenstein, T. K., Adler, M. W., and Rogers, T. J., The cellular basis for opioid-induced immunosuppression, *Adv. Exp. Med. Biol.*, 373, 57, 1995.
11. Tubaro, E., Borelli, G., Croce, C., Cavallo, G., and Santiangeli, C., Effect of morphine on resistance to infection, *J. Infect. Dis.*, 148, 656, 1983.
12. Watson, R. R., Prabhala, R. H., Darban, H. R., Yahya, M. D., and Smith, T. L., Changes in lymphocyte and macrophage subsets due to morphine and ethanol treatment during a retrovirus infection causing murine AIDS, *Life Sci.*, 43, v, 1988.

13. Yeager, M. P. and Colacchio, T. A., Effect of morphine on growth of metastatic colon cancer *in vivo*, *Arch. Surg.*, 126, 454, 1991.
14. Lord, J. A., Waterfield, A. A., Hughes, J., and Kosterlitz, H. W., Endogenous opioid peptides: multiple agonists and receptors, *Nature*, 267, 495, 1977.
15. Ikuta, K., Uchida, N., Friedman, J., and Weissman, I. L., Lymphocyte development from stem cells, *Ann. Rev. Immunol.*, 10, 759, 1992.
16. Mackay, C. R., Immunological memory, *Adv. Immunol.*, 53, 217, 1993.
17. Abbas, A. K., Lichtman, A. H., and Pober, J. S., Cells and tissues of the immune system, in *Cellular and Molecular Immunology*, Abbas, A. K., Lichtman, A. H., and Pober, J. S., Eds., W. B. Saunder, Philadelphia, 1994.
18. Gorczynski, R. M., Control of the immune response: role of macrophages in regulation of antibody and cell-mediated immune responses, *Scand. J. Immunol.*, 5, 1031, 1976.
19. Mackaness, G. B., Cellular resistance to infection, *J. Exp. Med.*, 116, 381, 1962.
20. Steigbigel, R. T., Lambert, L. H., and Remington, J. S., Phagocytic and bactericidal properties of normal human monocytes, *J. Clin. Invest.*, 53, 131, 1974.
21. Adams, D. O., Johnson, W. J., and Marino, P. A., Mechanisms of target destruction in macrophage-mediated tumor cytotoxicity, *Fed. Proc.*, 41, 2212, 1982.
22. Adams, D. O., Macrophage activation and secretion, *Fed. Proc.*, 41, 2193, 1982.
23. Nathan, C. F., Murray, H. W., and Cohn, Z. A., The macrophage as an effector cell, *N. Engl. J. Med.*, 303, 622, 1980.
24. Douglas, S. D., Mononuclear phagocytes and tissue regulatory mechanisms, *Dev. Comp. Immunol.*, 4, 7, 1980.
25. Meltzer, M. S., Occhionero, M., and Ruco, L. P., Macrophage activation for tumor cytotoxicity: regulatory mechanisms for induction and control of cytotoxic activity, *Fed. Proc.*, 41, 2198, 1982.
26. Ruco, L. P. and Meltzer, M. S., Macrophage activation for tumor cytotoxicity: development of macrophage cytotoxic activity requires completion of a sequence of short-lived intermediary reactions, *J. Immunol.*, 121, 2035, 1978.
27. Nathan, C. and Hibb, J. B., Jr., Role of nitric oxide synthesis in macrophage antimicrobial activity, *Curr. Opin. Immunol.*, 3, 65, 1991.
28. Hibbs, J. B., Jr., Taintor, R. R., Vavrin, Z., and Rachlin, E. M., Nitric oxide: a cytotoxic activated macrophage effector molecule, *Biochem. Biophys. Res. Commun.*, 157, 87, 1988.
29. Flescher, E., Gonen, P., and Keisari, Y., Oxidative burst-dependent tumoricidal and tumorostatic activities of paraffin oil-elicited mouse macrophages, *J. Natl. Cancer Inst.*, 72, 1341, 1984.
30. Nathan, C. F., Macrophage microbicidal mechanisms, *Trans. R. Soc. Trop. Med. Hyg.*, 77, 620, 1983.
31. Gangadharam, P. R. J. and Edwards, C. K., III, Release of superoxide anion from resident and activated mouse peritoneal macrophages infected with *Mycobacterium intracellulare*, *Am. Rev. Respir. Dis.*, 130, 834, 1984.
32. Gangadharam, P. R. J. and Pratt, P. F., Susceptibility of *Mycobacterium intracellulare* to hydrogen peroxide, *Am. Rev. Respir. Dis.*, 130, 309, 1984.
33. Vladimirov, Y. A., Olenev, V. I., Suslova, T. B., and Cheremisina, Z. P., Lipid peroxidation in mitochondrial membrane, *Adv. Lipid. Res.*, 17, 173, 1980.
34. Fortier, A. H., Polsinelli, T., Green, S. J., and Nacy, C. A., Activation of macrophages for destruction of *Francisella tularensis*: identification of cytokines, effector cells, and effector molecules, *Infect. Immun.*, 60, 817, 1992.

35. Becjerman, K. P., Rogers, H. W., Corbett, J. A., Schreiber, R. D., McDaniel, M. L., and Unanue, E. R., Release of nitric oxide during T cell-independent pathway of macrophage activation. Its role in resistance to *Listeria monocytogenes*, *J. Immunol.*, 150, 888, 1993.
36. Denis, M., Interferon-γ-treated murine macrophages inhibit growth of tubercle bacilli via the generation of reactive nitrogen intermediates, *Cell. Immunol.*, 132, 150, 1991.
37. Denis, M., Tumor necrosis factor and granulocyte macrophage-colony stimulating factor stimulate human macrophages to restrict growth of virulent *Mycobacterium avium* and to kill avirulent *M. avium*: killing effector mechanism depends on the generation of reactive nitrogen intermediates, *J. Leukocyte Biol.*, 49, 380, 1991.
38. Alspaugh, J. A. and Granger, D. L., Inhibition of *Cryptococcus neoformans* replication by nitrogen oxides supports the role of these molecules as effectors of macrophage-mediated cytostasis, *Infect. Immun.*, 59, 2291, 1991.
39. Mauel, J., Ransijn, A., and Buchmuller-Rouiller, Y., Killing of Leishmania parasites in activated murine macrophages is based on an L-arginine-dependent process that produces nitrogen derivatives, *J. Leukocyte Biol.*, 49, 73, 1991.
40. Lepoivre, M., Fieschi, F., Coves, J., Thelander, L., and Fontecave, M., Inactivation of ribonucleotide reductase by nitric oxide, *Biochem. Biophys. Res. Commun.*, 179, 442, 1991.
41. Lowenstein, C. J., Dinerman, J. L., and Snyder, S. H., Nitric oxide: a physiologic messenger, *Ann. Intern. Med.*, 120, 227, 1994.
42. Kwon, N. S., Stuehr, D. J., and Nathan, C. F., Inhibition of tumor cell ribonucleotide reductase by macrophage derived nitric oxide, *J. Exp. Med.*, 174, 761, 1991.
43. Moncada, S., Palmer, R. M. J., and Higgs, E. A., Nitric oxide: physiology, pathophysiology, and pharmacology, *Pharmacol. Rev.*, 43, 109, 1991.
44. Ding, A., Nathan, C. F., and Stuehr, D. J., Release of reactive nitrogen intermediates from mouse peritoneal macrophages: comparison of activating cytokines and evidence for independent production, *J. Immunol.*, 141, 2407, 1988.
45. Drapier, J. C., Wietzerbin, J., and Hibbs, J. B., Jr., Interferon-γ and tumor necrosis factor induce the L-arginine-dependent cytotoxic effector mechanism in murine macrophages, *Eur. J. Immunol.*, 18, 1587, 1988.
46. Hibbs, J. B., Jr., Granger, D. L., Krahenbuhl, J. L., and Adams, L. B., Synthesis of nitric oxide from L-arginine: a cytokine inducible pathway with antimicrobial activity, in *Mononuclear Phagocytes*, van Furth, R., Ed., Kluwer Academic, Boston, 1992.
47. Lowestein, C. J., Alley, E., Raval, P., Snyder, S. H., Russel, S. W., and Murphy, W., Nitric oxide synthase gene: two upstream regions mediate its induction by interferon-γ and lipopolysaccharide, *Proc. Natl. Acad. Sci. U.S.A.*, 90, 9730, 1993.
48. Germain, R. H. and Marguiles, D. M., The biochemistry and cell biology of antigen processing and presentation, *Ann. Rev. Immunol.*, 11, 403, 1993.
49. Unanue, E. R. and Allen, P. M., The basis for the immunoregulatory role of macrophages and other accessory cells, *Science*, 236, 551, 1987.
50. Zinkernagel, R. M. and Doherty, P. C., Activity of sensitized thymus-derived lymphocytes in lymphocytic choriomeningitis reflects immunological surveillance against self components, *Nature*, 251, 547, 1974.
51. Palm, S., Mignar, C., Kuhn, K., Umbreit, H., Barth, J., Stuber, F., and Maier, C., Phytohemagglutinin-dependent T-cell proliferation is not impaired by morphine, *Methods Find. Exp. Clin. Pharmacol.*, 18, 159, 1996.
52. Gilman, S. C., Schwartz, J. M., Milner, R. J., Bloom, F. E., and Feldman, J. D., β-Endorphin enhances lymphocytes proliferative responses, *Proc. Natl. Acad. Sci. U.S.A.*, 79, 4226, 1982.

53. Fontana, L., Fattorossi, A., D'Amelio, R., Migliorati, A., and Perricone, R., Modulation of human concanavalin A-induced lymphocyte proliferative response by physiological concentrations of β-endorphin, *Immunopharmacology*, 13, 111, 1987.
54. Gilmore, W. and Weiner, L. P., The opioid specificity of β-endorphin enhancement of murine lymphocyte proliferation, *Immunopharmacology*, 17, 19, 1989.
55. Hemmick, L. M. and Bidlack, J. M., β-Endorphin stimulates rat T lymphocyte proliferation, *J. Neuroimmunol.*, 29, 239, 1990.
56. Van den Bergh, P., Rozing, J., and Nagelkerken, L., Two opposing modes of action of β-endorphin on lymphocyte function, *Immunology*, 72, 537, 1991.
57. Jiayou, L., Gang, L., and Jiayu, W., Effects of β-endorphin on phytohemagglutinin-induced lymphocyte proliferation and mouse plaque-forming cell response via an opioid receptor mechanism, *Chin. Med. Sci. J.*, 9, 245, 1994.
58. Barreca, T., Di Benedetto, G., Corsini, G., Lenzi, G., and Puppo, F., Effects of dynorphin on the PHA-induced lymphocyte proliferation *in vitro*, *Immunopharmacol. Immunotoxicol.*, 9, 467, 1987.
59. Hucklebridge, F. H., Hudspith, B. N., Muhamed, J., Lydyard, P. M., and Brostoff, J., Methionine-enkephalin stimulates *in vitro* proliferation of human peripheral lymphocytes via δ-opioid receptors, *Brain Behav. Immunol.*, 3, 183, 1989.
60. Dubinin, K. V., Zakharova, L. A., Khegai, L. A., and Zaitsev, S. V., Immunomodulating effect of met-enkephalin on different stages of lymphocyte proliferation induced with concanavalin A *in vitro*, *Immunopharmacol. Immunotoxicol.*, 16, 463, 1994.
61. Hucklebridge, F. H., Hudspith, B. N., Lydyard, P. M., and Brostoff, J., Stimulation of human peripheral lymphocytes by methionine enkephalin and δ-selective opioid analogues, *Immunopharmacology*, 19, 87, 1990.
62. Deitch, E. A., Xu, Dazhong, and Bridges, M., Opioids modulate human neutrophil and lymphocyte function: thermal injury alters plasma β-endorphin levels, *Surgery*, 104, 41, 1988.
63. Bayer, B. M., Daussin, S., Hernandez, M., and Irvin, L., Morphine inhibition of lymphocyte activity is mediated by an opioid dependent mechanism, *Neuropharmacology*, 29, 369, 1990.
64. Bayer, B. M., Gastonguay, M. R., and Hernandez, M. C., Distinction between the *in vitro* and *in vivo* inhibitory effects of morphine on lymphocyte proliferation based on agonist sensitivity and naltrexone reversibility, *Immunopharmacology*, 23, 117, 1992.
65. Chuang, L. F., Killam, K. F., and Chuang, R. Y., Opioid dependency and T-helper cell functions in Rhesus monkey, *in vivo*, 7, 159, 1993.
66. Lysle, D. T., Coussons, M. E., Watts, V. J., Bennett, E. H., and Dykstra, L. A., Morphine-induced alterations of immune status: dose dependency, compartment specificity and antagonism by naltrexone, *J. Pharmacol. Exp. Ther.*, 265, 1071, 1993.
67. Flores, L. R., Hernandez, M. C., and Bayer, B. M., Acute immunosuppressive effects of morphine: lack of involvement of pituitary and adrenal factors, *J. Pharmacol. Exp. Ther.*, 268, 1129, 1994.
68. Flores, L. R., Wahl, S. M., and Bayer, B. M., Mechanisms of morphine-induced immunosuppression: effect of acute morphine administration on lymphocyte trafficking, *J. Pharmacol. Exp. Ther.*, 272, 1246, 1995.
69. Bayer, B. M., Mulroney, S. E., Hernandez, M. C., and Ding, X. Z., Acute infusions of cocaine result in time- and dose-dependent effects on lymphocyte responses and corticosterone secretion in rats, *Immunopharmacology*, 29, 19, 1995.
70. Flores, L. R., Dretchen, K. L., and Bayer, B. M., Potential role of the autonomic nervous system in the immunosuppressive effects of the acute morphine administration, *Eur. J. Pharmacol.*, 318, 437, 1996.

71. Schneider, G. M. and Lysle, D. T., Effects of centrally administered opioid agonists on macrophage nitric oxide production and splenic lymphocyte proliferation, in *AIDS, Drugs of Abuse and the Neuroimmune Axis*, Friedman, H., Ed., Plenum Press, New York, 1996.
72. Bryant, H. U. and Roudebush, R. E., Suppressive effects of morphine pellet implants on *in vivo* parameters of immune function, *J. Pharmacol. Exp. Ther.*, 255, 410, 1990.
73. Lysle, D. T., Hoffman, K. E., and Dykstra, L. A., Evidence for the involvement of the caudal region of the periaqueductal gray in a subset of morphine-induced alterations of immune status, *J. Pharmacol. Exp. Ther.*, 277, 1533, 1996.
74. Fecho, K., Maslonek, K. A., Dykstra, L. A., and Lysle, D. T., Assessment of the involvement of central nervous system and peripheral opioid receptors in the immunomodulatory effects of acute morphine treatment in rats, *J. Pharmacol. Exp. Ther.*, 276, 626, 1996.
75. Hamra, J. G. and Yaksh, T. L., Equianalgesic doses of subcutaneous but not intrathecal morphine alter phenotypic expression of cell surface markers and mitogen-induced proliferation in rat lymphocytes, *Anesthesiology*, 85, 355, 1996.
76. Panerai, A. E., Manfredi, B., Granucci, F., and Sacerdote, P., The β-endorphin inhibition of mitogen-induced splenocytes proliferation is mediated by central and peripheral paracrine/autocrine effects of the opioid, *J. Neuroimmunol.*, 58, 71, 1994.
77. Gang, L. and Fraker, P. J., Methionine-enkephalin alteration of mitogenic and mixed lymphocyte culture responses in zinc-deficient mice, *Acta Pharmacol. Sin.*, 10, 216, 1989.
78. Bayer, B. M., Hernandez, M. C., and Ding, X. Z., Tolerance and cross-tolerance to the suppressive effects of cocaine and morphine on lymphocyte proliferation, *Pharmacol. Biochem. Behav.*, 53, 227, 1996.
79. Karagouni, E. E. and Hadjipetrou-Kourounakis, L., Interleukins counteract opioid agonists immunosuppression, *Int. J. Neurosci.*, 54, 157, 1990.
80. Zbrog, Z., Luciak, M., Tchorzewski, H., and Pokoca, L., Modification of some lymphocyte functions *in vitro* by opioid receptor agonists and antagonist in chronic uremic patients and healthy subjets, *J. Immunopharmacol.*, 13, 475, 1991.
81. Carr, D. J. J. and Klimpel, G. R., Enhancement of the generation of cytotoxic T cells by endogenous opiates, *J. Neuroimmunol.*, 12, 75, 1986.
82. Hiramoto, R. N., Hsueh, C.-M., Rogers, C. E., Demissie, S., Hiramoto, N. S., Soong, S.-J., and Ghanta, V. K., Conditioning of the allogeneic cytotoxic lymphocyte response, *Pharmacol. Biochem. Behav.*, 44, 275, 1993.
83. Carpenter, G. W., Garza, H. H., Jr., Gebhardt, B. M., and Carr, D. J. J., Chronic morphine treatment suppresses CTL-mediated cytolysis, granulation, and cAMP responses to alloantigen, *Brain Behav. Immunol.*, 8, 185, 1994.
84. Carr, D. J. J. and Carpenter, G. W., Morphine-induced suppression of cytotoxic T lymphocyte activity in alloimmunized mice is not mediated through a naltrindole-sensitive delta opioid receptor, *Neuroimmunomodulation*, 2, 44, 1995.
85. Carpenter, G. W. and Carr, D. J. J., Pretreatment with β-funaltrexamine blocks morphine-mediated suppression of CTL activity in alloimmunized mice, *Immunopharmacology*, 29, 129, 1995.
86. Carpenter, G. W., Breeden, L., and Carr, D. J. J., Acute exposure to morphine suppresses cytotoxic T-lymphocyte activity, *Int. J. Immunopharmacol.*, 17, 1001, 1995.
87. Scott, M. and Carr, D. J. J., Morphine suppresses the alloantigen-driven CTL response in a dose-dependent and naltrexone reversible manner, *J. Pharmacol. Exp. Ther.*, 278, 980, 1996.

88. House, R. V., Thomas, P. T., and Bhargava, H. N., *In vitro* evaluation of fentanyl and meperidine for immunomodulatory activity, *Immunol. Lett.*, 46, 117, 1995.
89. Brown, S. L. and van Epps, D. E., Opioid peptides modulate production of interferon-γ by human mononuclear cells, *Cell. Immunol.*, 103, 19, 1986.
90. Peterson, P. K., Sharp, B., Gekker, G., Brummitt, C., and Keane, W. F., Opioid-mediated suppression of interferon-γ production by cultured peripheral blood mononuclear cells, *J. Clin. Invest.*, 80, 824, 1987.
91. Brown, S. L. and van Epps, D., Suppression of T lymphocyte chemotactic factor production by the opioid peptides β-endorphin and met-enkephalin, *J. Immunol.*, 134, 3384, 1985.
92. Wolf, G. T. and Peterson, K. A., β endorphin enhances *in vitro* lymphokine production in patients with squamous carcinoma of the head and neck, *Otolaryngol. Head Neck Surg.*, 94, 224, 1986.
93. Hadjipetrou-Kourounakis, L., Karagounis, E., Rekka, E., and Kourounakis, P., Immunosuppression by a novel analgesic-opioid agonist, *Scand. J. Immunol.*, 29, 449, 1989.
94. Carr, D. J. J., Carpenter, G. W., and Garza, H. H., Jr., Chronic and infrequent opioid exposure suppresses IL-2r expression on Rhesus monkey peripheral blood mononuclear cells following stimulation with pokeweed mitogen, *Int. J. Neurosci.*, 81, 137, 1995.
95. House, R. V., Thomas, P. T., and Bhargava, H. M., A comparative study of immunomodulation produced by *in vitro* exposure to delta opioid receptor agonist peptides, *Peptides*, 17, 75, 1996.
95a. Zhong, F., Li, X. Y., Yang, S. L., Stefano, G. B., Fimiani, C., and Bilfinger, T. V., Methionine-enkephalin stimulates interleukin-6 mRNA expression: human plasma levels in coronary artery bypass grafting, *Int. J. Cardiol.*, 164, S53, 1998.
95b. Roy, S., Cain, K. J., Chapin, R. B., Charboneau, R. G., and Barke, R. A., Morphine modulates NF κ B activation in macrophages, *Biochem. Biophys. Res. Commun.*, 245, 392, 1998.
96. Morgan, E. L., Regulation of human B lymphocyte activation by opioid peptide hormones: inhibition of IgG production by opioid receptor class (μ-, κ-, and δ-) selective agonists, *J. Neuroimmunol.*, 65, 21, 1996.
97. Das, K. P., Hong, J. S., and Sanders, V. M., Ultralow concentrations of proenkephalin and [met^5]-enkephalin differentially affect IgM and IgG production by B cells, *J. Neuroimmunol.*, 73, 37, 1997.
98. Carr, D. J. J., Radulescu, R. T., deCosta, B. R., Rice, K. C., and Blalock, J. E., Differential effect of opioids on immunoglobulin production by lymphocytes isolated from Peyer's Patches and spleen, *Life Sci.*, 47, 1059, 1990.
99. Carr, D. J. J., Radulescu, R. T., deCosta, B. R., Rice, K. C., and Blalock, J. E., Opioid modulation of immunoglobulin production by lymphocytes isolated from Peyer's patches and spleen, *Ann. NY Acad. Sci.*, 650, 125, 1992.
100. Guan, L., Townsend, R., Eisenstein, T. K., Adler, M. W., and Rogers, T. J., Both T cells and macrophages are targets of κ-opioid-induced immunosuppression, *Brain Behav. Immunol.*, 8, 229, 1994.
101. Carr, D. J. J. and France, C. P., Immune alterations in morphine-treated *Rhesus* monkeys, *J. Pharmacol. Exp. Ther.*, 267, 9, 1993.
102. Carr, D. J. J. and Blalock, J. E., Naloxone administration *in vivo* stereoselectively alters antigen-dependent and antigen-independent immune responses, *Psychoneuroendocrinology*, 16, 407, 1991.
103. Kay, N., Allen, J., and Morley, J. E., Endorphins stimulate normal human peripheral blood lymphocyte natural killer activity, *Life Sci.*, 35, 35, 1984.

104. Puente, J., Maturana, P., Miranda, D., Navarro, C., Wolf, M. E., and Mosnaim, A. D., Enhancement of human natural killer cell activity by opioid peptides: similar response to methionine-enkephalin and β-endorphin, *Brain Behav. Immun.*, 6, 32, 1992.
105. Hsueh, C. M., Chen, S. F., Tyring, S. K., Ghanta, V. K., and Hiramoto, R. N., Expression of the conditioned NK cell activity is β-endorphin dependent, *Brain Res.*, 678, 76, 1995.
106. Jonsdottir, I. H., Johansson, C., Asea, A., Hellstrand, K., Thorén, P., and Hoffmann, P., Chronic intracerebroventricular administration of β-endorphin augments natural killer cell cytotoxicity in rats, *Regul. Pept.*, 62, 113, 1996.
107. Kay, N., Morley, J. E., and van Ree, J. M., Enhancement of human lymphocyte natural killing function by non-opioid fragments of β-endorphin, *Life Sci.*, 40, 1083, 1987.
108. Oleson, D. R. and Johnson, D. R., Regulation of human natural cytotoxicity by enkephalins and selective opiate agonists, *Brain Behav. Immunol.*, 2, 171, 1988.
109. Bajpal, K., Singh, V. K., Agarwal, S. S., Dhawan, V. C., Naqvi, T., Haq, W., and Mathur, K. B., Immunomodulatory activity of met-enkephalin and its two potent analogs, *Int. J. Immunopharmacol.*, 17, 207, 1995.
110. Carr, D. J. J., Baker, M. L., Holmes, C., Brockunier, L. L., Bagley, J. R., and France, C. P., OHM3295: a fentanyl-related 4-heteroanilido piperidine with analgesic effects but not suppressive effects on splenic NK activity in mice, *Int. J. Immunopharmacol.*, 16, 835, 1994.
111. Baker, M. L., Brockunier, L. L., Bagley, J. R., France, C. P., and Carr, D. J. J., Fentanyl-related 4-heteroanilido piperidine OHM3295 augments splenic natural killer activity and induces analgesia through opioid receptor pathways, *J. Pharmacol. Exp. Ther.*, 274, 1285, 1995.
112. Beilin, B., Shavit, Y., Hart, J., Mordashov, B., Cohn, S., Notti, I., and Bessler, H., Effects of anesthesia based on large versus small doses of fentanyl on natural killer cell cytotoxicity in the perioperative period, *Anesth. Analg.*, 82, 492, 1996.
113. Martin-Kleiner, I. and Gabrilovac, J., Naloxone modulates NK-cell activity of human peripheral blood lymphocytes like an opioid agonist, *Immunopharmacol. Immunotoxicol.*, 15, 179, 1993.
114. Hsueh, C. M., Tyring, S. K., Hiramoto, R. N., and Ghanta, V. K., Efferent signal (s) responsible for the conditioned augmentation of natural killer cell activity, *Neuroimmunomodulation*, 1, 74, 1994.
115. Carr, D. J. J., Gebhardt, B. M., and Paul, D., α adrenergic and μ-2 opioid receptors are involved in morphine-induced suppression of splenocyte natural killer activity, *J. Pharmacol. Exp. Ther.*, 264, 1179, 1993.
116. Carr, D. J. J, Mayo, S., Gebhardt, B., and Porter, J., Central α-adrenergic involvement in morphine-mediated suppression of splenic natural killer activity, *J. Neuroimmunol.*, 53, 53, 1994.
117. Provinciali, M., Di Stefano, G., Stronati, S., Raffaeli, W., Pari, G., and Fabris, N., Role of prolactin in the modulatoin of NK and LAK cell activity after short- or long-term morphine administration in neoplastic patients, *Int. J. Immunopharmacol.*, 18, 577, 1996.
118. Carr, D. J. J., Gerak, L. R., and France, C. P., Naltrexone antagonizes the analgesic and immunosuppressive effects of morphine in mice, *J. Pharmacol. Exp. Ther.*, 269, 693, 1994.
119. Weber, R. J. and Pert, A., The periaqueductal gray matter mediates opiate-induced immunosuppression, *Science*, 245, 188, 1989.

120. Adelson, M. O., Novick, D. M., Khuri, E., Albeck, H., Hahn, E. F., and Kreek, M. J., Effects of the opioid antagonist naloxone on human natural killer cell activity *in vitro*, *Isr. J. Med. Sci.*, 30, 679, 1994.
121. Carr, D. J. J., Scott, M., Brockunier, L. L., Bagely, J. R., and France, C. P., The effect of novel opioids on natural killer activity and tumor surveillance *in vivo*, in *AIDS, Drugs of Abuse and the Neuroimmune Axis*, Friedman, H., Ed., Plenum Press, New York, 1996.
122. Take, S., Mori, T., Katafuchi, T., and Hori, T., Central interferon-α inhibits natural killer cytotoxicity through sympathetic innervation, *Am. J. Physiol.*, 265, R453, 1993.
123. Perez, L., and Lysle, D. T., Corticotropin-releasing hormone is involved in conditioned stimulus-induced reduction of natural killer cell activity but not in conditioned alterations in cytokine production or proliferation responses, *J. Neuroimmunol.*, 63, 1, 1995.
124. Katafuchi, T., Take, S., and Hori, T., Roles of sympathetic nervous system in the suppression of cytotoxicity of splenic natural killer cells in the rat, *J. Physiol.*, 465, 343, 1993.
125. Roy, S., Sedqi, M., Ramakrishnan, S., Barke, R. A., and Loh, H. H., Differential effects of opioids on the proliferation of a macrophage cell line, Bac 1.2F5, *Cell. Immunol.*, 169, 271, 1996.
126. Roy, S., Ramakrishnan, S., Loh, H. H., and Lee, N. M., Chronic morphine treatment selectively suppresses macrophage colony formation in bone marrow, *Eur. J. Pharmacol.*, 195, 359, 1991.
127. Krizanac-Bengez, L., Boranic, M., Testa, N. G., and Marotti, T., Effect of enkephalins on bone marrow cells, *Biomed. Pharmacother.*, 46, 367, 1992.
128. Krizanac-Bengez, L. J., Breljak, D., and Boranic, M., Suppressive effect of met-enkephalin on bone marrow cell proliferation *in vitro* shows circadian pattern and depends on the presence of adherent accessory cells, *Biomed. Pharmacother.*, 50, 85, 1996.
129. Hagi, K., Inaba, K., Sakuta, H., and Muramatsu, S., Enhancement of murine bone marrow macrophage differentiation by β-endorphin, *Blood*, 86, 1316, 1995.
130. Bonta, I. L. and Ben-Efraim S., Involvement of inflammatory mediators in macrophage antitumor activity, *J. Leukocyte Biol.*, 54, 613, 1993.
131. Mukaida, N., Inflammation and pro-inflammatory cytokines, *Jpn. J. Clin. Med.*, 50, 1724, 1992.
132. Si-Xun, Y. and Xiao-Yu, L., Enhancement of interleukin-1 production in mouse peritoneal macrophages by methionine-enkephalin, *Acta Pharmacol. Sin.*, 10, 266, 1989.
133. Marotti, T., Burek, B., Rabatic, S., Balog, T., and Hrsak, I., Modulation of lipopolysaccharide-induced production of cytokines by methionine-enkephalin, *Immunol. Lett.*, 40, 43, 1994.
134. Das, K. P., McMillian, M. K., Bing, G., and Hong, J. S., Modulatory effects of [Met^5]-enkephalin on interleukin-1 β secretion from microglia in mixed brain cell cultures, *J. Neuroimmunol.*, 62, 9, 1995.
135. Apte, R. N., Durum, S. K., and Oppenheim, J. J., Opioids modulate interleukin-1 production and secretion by bone-marrow macrophages, *Immunol. Lett.*, 24, 141, 1990.
136. Apte, R. N., Oppenheim, J. J., and Durum, S. K., β-Endorphin regulates interleukin 1 production and release by murine bone marrow macrophages, *Int. Immunol.*, 1, 465, 1989.

137. Van den Bergh, P., Rozing, J., and Nagelkerken, L., β-Endorphin stimulates Ia expression on mouse B cells by inducing interleukin-4 secretion by CD4+ T cells, *Cell. Immunol.*, 149, 180, 1993.
138. Bian, T. H., Wang, X. F., and Li, X. Y., Effect of morphine on interleukin-1 and tumor necrosis factor α production from mouse peritoneal macrophages *in vitro*, *Acta Pharmacol. Sin.*, 16, 449, 1995.
139. Belkowski, S. M., Alicea, C., Eisenstein, T. K., Adler, M. W., and Rogers, T. J., Inhibition of interleukin-1 and tumor necrosis factor-α synthesis following treatment of macrophages with the *kappa* opioid agonist U50, 488H, *J. Pharmacol. Exp. Ther.*, 273, 1491, 1995.
140. Alicea, C., Belkowski, S., Eisenstein, T. K., Adler, M. W., and Rogers, T. J., Inhibition of primary murine macrophage cytokine production *in vitro* following treatment with the κ-opioid agonist U50, 488H, *J. Neuroimmunol.*, 64, 83, 1996.
141. Peterson, P. K., Gekker, G., Hu, S., Anderson, W. R., Kravitz, F., Portoghese, P. S., Balfour, H. H., Jr., and Chao, C. C., Morphine amplifies HIV-1 expression in chronically infected promonocytes cocultured with human brain cells, *J. Neuroimmunol.*, 50, 167, 1994.
142. Chao, C. C., Gekker, G., Sheng, W. S., Hu, S., Tsang, M., and Peterson, P. K., Priming effect of morphine on the production of tumor necrosis factor-α by microglia: implications in respiratory burst activity and human immunodeficiency virus-1 expression, *J. Pharmacol. Exp. Ther.*, 269, 198, 1994.
143. Chao, C. C., Gekker, G., Hu, S., Sheng, W. S., Portoghese, P. S., and Peterson, P. K., Upregulation of HIV-1 expression in cocultures of chronically infected promonocytes and human brain cells by dynorphin, *Biochem. Pharmacol.*, 50, 715, 1995.
144. Bian, T. H. and Li, X. Y., Immunomodulatory effects of morphine microinjected into periaqueductal gray, *Acta Pharmacol. Sin.*, 16, 121, 1995.
145. House, R. V., Thomas, P. T., and Bhargava, H. N., A comparative study of immunomodulation produced by *in vitro* exposure to delta opioid receptor agonist peptides, *Peptides*, 17, 75, 1996.
146. Tosk, J. M., Grim, J. R., Kinback, K. M., Sale, E. J., Bozzetti, L. P., and Will, D., Modulation of chemiluminescence in a murine macrophage cell line by neuroendocrine hormones, *Int. J. Immunopharmacol.*, 15, 615, 1993.
147. Sharp, B. M., Keane, W. F., Suh, H. J., Gekker, G., Tsukayama, D., and Peterson, P. K., Opioid peptides rapidly stimulate superoxide production by human polymoprhonuclear leukocytes and macrophages, *Endocrinology*, 117, 793, 1985.
148. Peterson, P. K., Sharp, B. M., Gekker, G., Brummitt, C., and Keane, W. F., Opioid-mediated suppression of cultured peripheral blood mononuclear cell respiratory burst activity, *J. Immunol.*, 138, 3907, 1987.
149. Radulovic, J., Dimitrijevic, M., Laban, O., Stanojevic, S., Vasiljevic, T., Kovacevic-Jovanovic, V., and Markovic, B. M., Effect of met-enkephalin and opioid antagonists on rat macrophages, *Peptides*, 16, 1209, 1995.
150. Fóris, G., Medgyesi, G. A., and Hauck, M., Bidirectional effect of met-enkephalin on macrophage effector functions, *Mol. Cell. Biochem.*, 69, 127, 1986.
151. Efanov, A. M., Koshkin, A. A., Sazanov, L. A., Borodulina, O. I., Varfolomeev, S. D., and Zaitsev, S. V., Inhibition of the respiratory burst in mouse macrophages by ultra-low doses of an opioid peptide is consistent with a possible adaptation mechanism, *FEBS Lett.*, 355, 114, 1994.
152. Henry, Y., Lepoivre, M., Drapier, J. C., Ducrocq, C., Boucher, J. L., and Guissani, A., EPR characterization of molecular targets for NO in mammalian cells and organelles, *FASEB J.*, 7, 1124, 1993.

153. Moncada, S., The L-arginine: nitric oxide pathway, *Acta Physiol. Scand.*, 145, 201, 1992.
154. Stuehr, D. J. and Marletta, M. A., Mammalian nitrate biosynthesis: mouse macrophages produce nitrite and nitrate in response to *Escherichia coli* lipopolysaccharide, *Proc. Natl. Acad. Sci. U.S.A.*, 82, 7738, 1985.
155. McMicking, J. D., Nathan, C., Hom, G., Chartain, N., Fletcher, D. S., Trumbauer, M., Stevens, K., Xie, Q. W., Sokol, K., Hutchinson, N., Chen, H., and Mudgett, J. S., Altered responses to bacterial infection and endotoxin shock in mice lacking inducible nitric oxide synthase, *Cell*, 81, 641, 1995.
156. Stuehr, D. J. and Nathan, C. F., Nitric oxide. A macrophage product responsible for cytostasis and respiratory inhibition in tumor target cells, *J. Exp. Med.*, 169, 1543, 1989.
157. Albina, J. E., Mills, C. D., Henry, W. L., and Caldwell, M. D., Regulation of macrophage physiology by L-arginine: role of the oxidative L-arginine deiminase pathway, *J. Immunol.*, 143, 3641, 1986.
158. Doi, T., Ando, M., Akaike, T., Suga, M., Sato, K., and Maeda, H., Resistance to nitric oxide in *Mycobacterium avium* complex and its implication in pathogenesis, *Infect. Immun.*, 61, 1980, 1993.
159. Pacifici, R., Minetti, M., Zuccaro, P., and Pietraforte, D., Morphine affects cytostatic activity of macrophages by the modulation of nitric oxide release, *Int. J. Immunopharmacol.*, 17, 771, 1995.
160. Fecho, K., Maslonek, K. A., Coussons-Read, M. E., Dykstra, L. A., and Lysle, D. T., Macrophage-derived nitric oxide is involved in the depressed concanavalin A responsiveness of splenic lymphocytes from rats administered morphine *in vivo*, *J. Immunol.*, 152, 5845, 1994.
161. Magazine, H. I., Liu, Y., Bilfinger, T. V., Fricchione, G. L., and Stefano, G. B., Morphine-induced conformational changes in human monocytes, granulocytes, and endothelial cells and in invertebrate immunocytes and microglia are mediated by nitric oxide, *J. Immunol.*, 156, 4845, 1996.
162. Schneider, G. M. and Lysle, D. T., Evidence for the involvement of CNS in the modulation of splenic nitric oxide production, *J. Neuroimmunol.*, 69, 25, 1996.
163. Iuvone, T., Capasso, A., D'Acquisto, F., and Carnuccio, R., Opioids inhibit the induction of nitric oxide synthase in J774 macrophages, *Biochem. Biophys. Res. Commun.*, 212, 975, 1995.
164. Weber, R. J. and Pert, A., Immune system, in *Encyclopedia of Neuroscience*, Smith, B. and Adelman, G., Eds., Birkhauser, Boston, 1997.
165. Adams, D. O. and Hamilton, T. A., The cell biology of macrophage activation, *Ann. Rev. Immunol.*, 2, 283, 1984.
166. Appelberg, R., Castro, A. G., Pedrosa, J., Silva, R. A., Orme, I. M., and Minóprio, P., Role of γ interferon and tumor necrosis factor α during T-cell-independent and -dependent phases of *Mycobacterium avium* infection, *Infect. Immun.*, 62, 3962, 1994.
167. Tubaro, E., Avico, U., Santiangeli, C., Zuccaro, P., Cavallo, G., Pacifici, R., Croce, C., and Borelli, G., Morphine and methadone impact on human phagocytic physiology, *Int. J. Immunopharmacol.*, 7, 865, 1985.
168. Tubaro, E., Santiangeli, C., Belogi, L., Borelli, G., Cavallo, G., Croce, C., and Avico, U., Methadone vs morphine: comparison of their effect on phagocytic functions, *Int. J. Immunopharmacol.*, 9, 79, 1987.
169. Rojavin, M., Szabo, I., Bussiere, J. L., Rogers, T. J., Adler, M. W., and Eisenstein, T. K., Morphine treatment *in vitro* or *in vivo* decreases phagocytic functions of murine macrophages, *Life Sci.*, 53, 997, 1993.

170. Casellas, A. M., Guardiola, H., and Renaud, F. L., Inhibition by opioids of phagocytosis in peritoneal macrophages, *Neuropeptides*, 18, 35, 1991.
171. Szabo, I., Rojavin, M., Bussiere, J. L., Eisenstein, T. K., Adler, M. W., and Rogers, T. J., Suppression of peritoneal macrophage phagocytosis of *Candida albicans* by opioids, *J. Pharmacol. Exp. Ther.*, 267, 703, 1993.
172. Fóris, G., Medgyesi, G. A., Gyimesi, E., and Hauck, M., Met-enkephalin induced alterations of macrophage functions, *Mol. Immunol.*, 21, 747, 1984.
173. Carrera, J., Catala, J. C., Monedero, P., Carrascosa, F., Arroyo, J. L., and Subira, M. L., Depression of the mononuclear phagocyte system caused by high doses of narcotics, *Rev. Med. Univ. Navarra*, 37, 119, 1992.
174. Peterson, P. K., Gekker, G., Hu, S. X., Sheng, W. S., Molitor, T. W., and Chao, C. C., Morphine stimulates phagocytosis of mycobacterium tuberculosis by human microglial cells: involvement of a G protein-coupled opiate receptor, *Adv. Neuroimmunol.*, 5, 299, 1995.
175. Ichinose, M., Asai, M., and Sawada, M., Enhancement of phagocytosis by dynorphin A in mouse peritoneal macrophages, *J. Neuroimmunol.*, 60, 37, 1995.
176. Lin, J., Ageing suppresses the enhancement of T cell mitogenesis by opioid peptides and enkephalins increase phagocytosis of murine macrophages, *Acta Acad. Med. Sin.*, 14, 233, 1992.
177. Alexander, P., The functions of the macrophage in malignant disease, *Annu. Rev. Med.*, 27, 207, 1976.
178. Nathan, C. F., Murray, H. W., and Cohn, Z. A., The macrophage as an effector cell, *N. Eng. J. Med.*, 303, 622, 1980.
179. Hibbs, J. B., Jr., Lambert, L. H., and Remington, J. S., Possible role of macrophage mediated nonspecific cytotoxicity in tumor resistance, *Nature*, 235, 48, 1972.
180. Pacifici, R., Patrini, G., Venier, I., Parolaro, D., Zuccaro, P., and Gori, E., Effect of morphine and methadone acute treatment on immunological activity in mice: pharmacokinetic and pharmacodynamic correlates, *J. Pharmacol. Exp. Ther.*, 269, 1112, 1994.
181. Cameron, D. J., Effect of neuropeptides on macrophage mediated cytotoxicity in normal donors and cancer patients, *Jpn. J. Exp. Med.*, 57, 31, 1987.
182. Hagi, K., Uno, K., Inaba, K., and Muramatsu, S., Augmenting effect of opioid peptides on murine macrophage activation, *J. Neuroimmunol.*, 50, 71, 1994.
183. Foster, J. S. and Moore, R. N., Dynorphin and related opioid peptides enhance tumoricidal activity mediated by murine peritoneal macrophages, *J. Leukocyte Biol.*, 42, 171, 1987.
184. Koff, W. C. and Dunegan, M. A., Modulation of macrophage-mediated tumoricidal activity by neuropeptides and neurohormones, *J. Immunol.*, 135, 350, 1985.
185. Magazine, H. I., Liu, Y., Bilfinger, T. V., Fricchione, G. L., and Stefano, G. B., Morphine-induced conformational changes in human monocytes, granulocytes, and endothelial cells and in invertebrate immunocytes and microglia are mediated by nitric oxide, *J. Immunol.*, 156, 4845, 1996.
186. Petty, H. R. and Martin, S. M., Combinative ligand-receptor interactions: effects of cAMP, epinephrine, and met-enkephalin on RAW264 macrophage morphology, speading, adherence, and microfilaments, *J. Cell. Physiol.*, 138, 247, 1989.
187. Elliott, G. R., van Batenburg, M. J., and Dzoljic, M. R., Enhanced prostaglandin E_2 and thromboxane B_2 release from resident peritoneal macrophages isolated from morphine-dependent rats, *FEBS Lett.*, 217, 6, 1987.
188. Pérez-Castrillón, J. L., Pérez-Arellano, J. L., Garc'a-Palomo, J. D., Jiménez-López, A., De Castro, S., Opioids depress *in vitro* human monocyte chemotaxis, *Immunopharmacology*, 23, 57, 1992.

189. van Epps, D. E. and Daland, L., β-endorphin and met-enkephalin stimulate human peripheral blood mononuclear cell chemotaxis, *J. Immunol.*, 132, 3046, 1984.
190. Ruff, M. R., Wahl, S. M., Mergenhagen, S., and Pert, C. B., Opiate receptor-mediated chemotaxis of human monocytes, *Neuropeptides*, 5, 363, 1985.
191. Ruff, M. R., Wahl, S. M., and Pert, C. B., Substance P receptor-mediated chemotaxis of human monocytes, *Peptides*, 6, 107, 1985.
192. Sibinga, N. E. S. and Goldstein, A., Opioid peptides and opioid receptors in cells of the immune system, *Annu. Rev. Immunol.*, 6, 219, 1988.
193. Weber, R. J., and Pert, C. B., Opiatergic modulation of the immune system, in *Central and Peripheral Endorphins: Basic and Clinical Aspects,* Muller, E. E. and Genazzani, A. R., Eds., Raven Press, New York, 1984.
194. Mehrishi, J. N. and Mills, I. H., Opiate receptors on lymphocytes and platelets in man, *Clin. Immunol. Immunopathol.*, 27, 240, 1983.
195. Madden, J. J., Donahoe, R. M., Zwemer-Collins, J., Shafer, D. A., and Falek, A., Binding of naloxone to human T lymphocytes, *Biochem. Pharmacol.*, 36, 4103, 1987.
196. Bidlack, J. M., Saripalli, L. D., and Lawrence, D. M. P., κ-Opioid binding sites on a murine lymphoma cell line, *Eur. J. Pharmacol.*, 227, 257, 1992.
197. Carr, D. J. J., DeCosta, B. R., Kim, C.-H., Jacobson, A. E., Guarcello, V., Rice, K. C., and Blalock, J. E., Opioid receptors on cells of the immune system: evidence for δ- and κ-classes, *J. Endocrinol.*, 122, 161, 1989.
198. Carr, D. J. J., DeCosta, B. R., Kim, C.-H., Jacobson, V., Rice, K. C., and Blalock, J. E., Evidence for a δ-class opioid receptor on cells of the immune system, *Cell. Immunol.*, 122, 161, 1989.
199. Stefano, G. B., Digenis, A., Spector, S., Leung, M. K., Bilfinger, T., Makman, M. H., Scharrer, B., and Abumrad, N. N., Opiate-like substances in an invertebrate, an opiate receptor on invertebrate and human immunocytes, and a role in immunosuppression, *Proc. Natl. Acad. Sci. U.S.A.*, 90, 11099, 1993.
200. Lang, M. E., Jourd'Heuil, D., Meddings, J. B., and Swain, M. G., Increased opioid binding to peripheral white blood cells in a rat model of acute cholestasis, *Gastroenterology*, 108, 1479, 1995.
201. Patrini, G., Massi, P., Ricevuti, G., Mazzone, A., Fossati, G., Mazzucchelli, I., Gori, E., and Parolaro, D., Changes in opioid receptor density on murine splenocytes induced by *in vivo* teatment with morphine and methadone, *J. Pharmacol. Exp. Ther.*, 279, 172, 1996.
202. Chuang, L. F., Chuang, T. K., Killam, K. F., Jr., Chuang, A. J., Kung, H., Yu, L., and Chuang, R. Y., Delta opioid receptor gene expression in lymphocytes, *Biochem. Biophys. Res. Commun.*, 202, 1291, 1994.
203. Gaveriaux, C., Peluso, J., Simonin, F., Laforet, J., and Kieffer, B., Identification of κ- and δ-opioid receptor transcripts in immune cells, *FEBS Lett.*, 369, 272, 1995.
204. Wick, M. J., Minnerath, S. R., Roy, S., Ramakrishnan, S., and Loh, H. H., *Mol. Brain Res.*, 32, 342, 1995.
205. Halford, W. P., Gebhardt, B. M., and Carr, D. J. J., Functional role and sequence analysis of a lymphocyte orphan opioid receptor, *J. Neuroimmunol.*, 59, 91, 1995.
206. Miller, B., δ Opioid receptor expression is induced by concanavalin A in CD4+ T cells, *J. Immunol.*, 157, 5324, 1996.
207. Sedqi, M., Roy, S., Ramakrishnan, S., Elde, R., and Loh, H. H., Complementary DNA cloning of a μ-opioid receptor from rat peritoneal macrophages, *Biochem. Biophys. Res. Commun.*, 209, 563, 1995.
208. Reith, M. E. A., Intracellular messengers and drug addiction, in *Signal Transduction, An Introduction,* Ari, S., Ed., Birkhauser, Boston, 1997.

209. Sharp, B. M., Shahabi, N. A., Heagy, W., McAllen, K., Bell, M., Huntoon, C., and McKean, D. J., Dual signal transduction through delta opioid receptors in a transfected human T-cell line, *Proc. Natl. Acad. Sci. U.S.A.*, 93, 8294, 1996.
210. Shahabi, N. A., Heagy, W., and Sharp, B. M., β-endorphin enhances concanavalin-A-stimulated calcium mobilization by murine splenic T cells, *Endocrinology*, 137, 3386, 1996.
211. Hough, C. J., Halperin, J. I., Mazorow, D. L., Yeandle, S. L., and Millar, D. B., β-Endorphin modulates T-cell intracellular calcium flux and *c-myc* expression via a potassium channel, *J. Neuroimmunol.*, 27, 163, 1990.
212. Heagy, W., Shipp, M. A., and Finberg, R. W., Opioid receptor agonists and Ca^{2+} modulation in human B cell lines, *J. Immunol.*, 149, 4074, 1992.
213. Chiappelli, F., Nguyen, L., Bullington, R., and Fahey, J. L., β-Endorphin blunts phosphatidylinositol formation during *in vitro* activation of isolated human lymphocytes: preliminary report, *Brain Behav. Immun.*, 6, 1, 1992.
214. Kavelaars, A., Eggen, B. J. L., De Graan, P. N. E., Gispen, W. H., and Heijnen, C. J., The phosphorylation of the CD3 γ chain of T lymphocytes is modulated by β-endorphin, *Eur. J. Immunol.*, 20, 943, 1990.
215. Bryant, H. V. and Roudebush, R. E., Suppressive effects of morphine pellet implants on *in vivo* parameters of immune functions, *J. Pharmacol. Exp. Ther.*, 255, 410, 1990.
216. Bayer, B. M. and Flores, C. M., Effects of morphine on lymphocyte function: possible mechanisms of interaction, in *Drugs of Abuse and Immune Function,* Watson, R. R., Ed., CRC Press, Boca Raton, FL, 1990.
217. Carr, D. J. J., Rogers, T. J., and Weber, R. J., The relevance of opioid receptors on immunocompetence and immune homeostasis, *Proc. Soc. Exp. Biol. Med.*, 213, 248, 1996.
218. Makman, M. H., Morphine receptors in immunocytes and neurons, *Adv. Neuroimmunol.*, 4, 69, 1994.
219. Shavit, Y., Depaulis, A., Martin, F. C., Terman, G. W., Pechnick, R. N., Zane, C. J., Gale, R. P., and Liebeskind, J. C., Involvement of brain opiate receptors in the immune-suppressive effect of morphine, *Proc. Natl. Acad. Sci. U.S.A.,* 83, 7114, 1986.
220. Bryant, H. V., Bernton, E. W., Kenner, J. R., and Holaday, J. W., Role of adrenal cortical activation in the immunosuppressive effects of chronic morphine treatment, *Endocrinology*, 128, 3253, 1991.
221. Freier, D. O. and Fuchs, B. A., A mechanism of action for morphine-induced immunosuppression: corticosterone mediates morphine-induced suppression of natural killer cell activity, *J. Pharmacol. Exp. Ther.*, 270, 1127, 1994.
222. Flores, L. R., Hernández, M. C., and Bayer, B. M., Acute immunosuppressive effects of morphine: lack of involvement of pituitary and adrenal factors, *J. Pharmacol. Exp. Ther.*, 268, 1129, 1994.
223. Felten, D. L., Felten, S. Y., Carlson, S. Y., Olschowka, J. A., and Livnat, S., Noradrenergic and peptidergic innervation of lymphoid tissue, *J. Immunol.*, 135, 755s, 1985.
224. Conway, E. L., Brown, M. J., and Dollery, C. T., Plasma catecholamine and cardiovascular responses to morphine and D-Ala-D-Leu-enkephalin in conscious rats, *Arch. Int. Pharmacodyn.*, 265, 244, 1983.
225. Flores, L. R., Dretchen, K. L., and Bayer, B. M., Potential role of the autonomic nervous system in the immunosuppressive effects of acute morphine administration, *Eur. J. Pharmacol.*, 318, 437, 1996.
226. Katafuchi, T., Take, S., and Hori, T., Roles of sympathetic nervous system in the suppression of cytotoxicity of splenic natural killer cells in the rat, *J. Physiol.*, 465, 343, 1993.

227. Spengler, R. N., Chensue, S. W., Giacherio, D. A., Blenk, N., and Kunkel, S. L., Endogenous norepinephrine regulates tumor necrosis factor-α production from macrophages *in vitro*, *J. Immunol.*, 152, 3024, 1994.
228. Hall, D. M., Suo, J.-L., and Weber, R. J., Opioid mediated effects on the immune system: sympathetic nervous system involvement, *J. Neuroimmunol.*, 83, 29, 1998.
229. Suo, J.-L. and Weber, R. J., Immunomodulation mediated by microinjection of morphine into the periaqueductal gray matter of the mesencephalon, in *Drugs of Abuse, Immunomodulation, and AIDS*, Friedman, H., Madden, J., and Klein, T., Eds., Plenum Press, New York, 1998.
230. Tomaszewska, D. and Przekop, F., The immune-neuro-endocrine interactions, *J. Physiol. Pharmacol.*, 48, 139, 1997.
231. Bartik, M. M., Brooks, W. H., and Roszman, T. L., Modulation of T cell proliferation by stimulation of the β-adrenergic receptor: lack of correlation between inhibition of T cell proliferation and cAMP accumulation, *Cell. Immunol.*, 148, 408, 1993.
232. Carlson, S. L., Brooks, W. H., and Roszman, T. L., Neurotransmitter-lymphocyte interactions: dual receptor modulation of lymphocyte proliferation and cAMP production, *J. Neuroimmunol.*, 24, 155; 1989.
233. Chou, R. C., Stinson, M. W., Noble, B. K., and Spengler, R. N., β-adrenergic receptor regulation of macrophage-derived tumor necrosis factor-α production from rats with experimental arthritis, *J. Neuroimmunol.*, 67, 7, 1996.
234. Feinstein, D. L., Galea, E., and Reis, D. J., Norepinephrine suppresses inducible nitric oxide synthase activity in rat astroglial cultures, *J. Neurochem.*, 60, 1945, 1993.

21 Methionine Enkephalin-Induced Immune Modulation and Cytokines

Rebecca Bowden-Elkes and Steven C. Specter

CONTENTS

21.1 Introduction 315
    21.1.1 Methionine Enkephalin as an Immunoregulator 315
    21.1.2 Cytokines and Viral Infections 316
21.2 Cytokine Modulation in FLV Infection 317
    21.2.1 Methods 317
21.3 Conclusions 318
Acknowledgment 321
References 321

21.1 INTRODUCTION

21.1.1 METHIONINE ENKEPHALIN AS AN IMMUNOREGULATOR

The pentapeptide methionine enkephalin (MENK) is an endogenously produced opioid that has varied effects on the immune system. Activated T-helper cells produce MENK and the presence of MENK increases active T-cell rosette formation.[1,2] MENK has also been shown to be chemotactic for T cells.[3] Chukwuocha and colleagues found that MENK has dual immunoregulatory effects *in vitro*, either it upregulates or downregulate the primary antibody response depending on the dose of the antigen.[4] High concentrations (millimolar) of MENK have been demonstrated to suppress both humoral and cell-mediated responses while low concentrations (micromolar) enhance immune responses, particularly those immune responses relating to the cell-mediated branch of the immune system.[5] MENK has also been shown to enhance natural killer (NK) cell activity both *in vivo* and *in vitro*.[6,7] Therefore, MENK appears to function as a cytokine in mediating homeostasis of the immune system.[8]

    The modulation of immune activity by MENK is believed to occur through interaction with opioid receptors present on cells of the immune system.[9] MENK interacts with the δ and the μ opioid receptors on T cells. Mu and κ opioid receptor ligands primarily downregulate the immune system while δ and ε receptor ligands act to upregulate the immune system, accounting for the observed dichotomous effect of MENK *in vitro* and *in vivo*.[10] T-cell opioid receptor binding of MENK and other opioids can be competitively inhibited by the opioid antagonist naloxone.[9] Nyland et al. showed that treatment using MENK or μ and δ receptor agonists increased the amount of syncytia formation in human immunodeficiency virus (HIV) or human T-cell leukemia virus type 1 (HTLV-1)-infected SupT1 cells, a human CD4+ T cell line.[11] This response appears to be concentration-dependent with higher concentrations of MENK and/or agonist causing the greatest enhancement of syncytia formation. The ability of MENK to enhance viral replication as measured by syncytia formation demonstrates the need for MENK to be used in combination with antivirals such as azidothymidine (AZT) when considered for use in a chemotherapeutic approach for the treatment of human and mouse retroviral diseases. For these reasons, the use of MENK as an immunostimulant in conjunction with antivirals is being investigated in retroviral diseases.

HIV-infected patients demonstrated increased CD4+ cell counts after treatment with MENK although clinical effects were not demonstrated in the long term.[12] Burger et al. reported that MENK treatment enhanced NK cell cytotoxicity and virus-specific cytotoxic T lymphocyte activity against influenza A virus in mice.[13] Our initial studies examined the combination of AZT and MENK as treatment for Friend leukemia virus (FLV) infection in the susceptible BALB/c mouse.[14] FLV murine retroviral disease is characterized by establishment of an erythroleukemia in mice leading to advanced splenomegaly, susceptibility to opportunistic infections, and death.[15] There are several advantages to using the FLV model of immunosuppression for screening potential therapeutic protocols for retroviral disease: it is rapid in onset and progression, affects an animal model that is known in great detail, and has analogies with HIV.[16] We observed that when compared to AZT treatment alone, the combination of MENK and AZT increased survival by approximately twofold to 50% and decreased morbidity as measured by splenomegaly.[14] All the infected untreated control mice or those treated only with MENK experienced 100% mortality. More recently, Sin et al. demonstrated that MENK works in a immunostimulatory manner to reduce infectivity when combined with AZT in an FLV infection of *Mus dunni* cells *in vitro*.[17]

21.1.2 CYTOKINES AND VIRAL INFECTIONS

The concept of type 1 vs. type 2 T helper cytokine responses in infections is not new. Type 1 cytokine responses tend to favor protection from viral infections while type 2 responses are associated with disease progression. Clerici and Shearer postulated that cytokine dysregulation from type 1 to type 2 response accompanied the progression of HIV to AIDS.[18] Thus, therapy that enhances a type 1 response relative to type 2 may provide antiretroviral activity. These cytokines may be therapeutically modulated to prevent or limit the progression of the infection. The alteration of IFNγ levels in FLV-infected splenocyte cultures raised the question of the importance of cytokines in MENK-mediated immune modulation.[19] Incubation of human peripheral blood mononuclear cells (PBMC) with MENK and the mitogen phytohemagglutinin (PHA) increased IL-2 production.[20] Similar results have been reported on the effect of MENK on IFNγ production.[21]

Sin and Specter recently investigated the role of IFNγ in the observed antiretroviral activity of MENK and AZT used in FLV-infected *Mus dunni* cell culture.[22] The combination of AZT and MENK was shown to reduce viral replication and stimulate IFNγ production compared to either treatment alone. Antibodies to IFNγ reversed the antiretroviral effect produced by treatment with MENK and AZT, lending support to the hypothesis that the MENK antiretroviral effect is mediated through the induction of cytokine production, predominantly the type 1 cytokine IFNγ. This led us to further investigate the ability of MENK to modulate cytokine production in FLV-infected mice. It must be noted that immune activation has been reported to enhance HIV replication. Our observations suggest that mice treated with MENK may experience this phenomenon.[14] Thus, the use of an immune stimulant like MENK only in combination with antiviral agents is critical to deriving benefit from the immune stimulation.

21.2 CYTOKINE MODULATION IN FLV INFECTION

21.1.1 Methods

Mice were infected intraperitoneally (i.p.) with approximately 50 infectious doses of FLV and then either treated or not with 0.0125 mg/ml AZT *ad libitum*, 60 μg MENK i.p. or a combination of the two, initiation beginning 3 days postinfection. Mice were killed and spleens were harvested at 7, 14, 21, or 28 days posttreatment. Splenocytes were cultured aseptically in the presence of concanavalin A for 72 h to measure induction of release of the cytokines interleukin (IL)-2, IL-4, IL-10, IL-12, and IFNγ by ELISA.

At 7 days posttreatment, type 1 cytokines IL-2 and IL-12 were not significantly changed compared to noninfected controls (Figures 21.1 and 21.2). IFNγ was significantly decreased in the infected mice and the single-treatment groups (Figure 21.3). The type 2 cytokines IL-4 and IL-10 were not significantly changed (Figures 21.4 and 21.5). By day 14 there was a significant decrease in all cytokines in the infected control mice. To this point in time, the type 1 and type 2 cytokines were not maintained by the single- or combination-treatment groups. It is interesting to note that by day 21 type 1 cytokines were still significantly decreased in the infected controls but the combination therapy partly restored cytokine production. By day 28 type 1 cytokine production again decreased in the combination-treatment group compared to day 21, but was still significantly higher than the infected controls. The type 2 cytokines did not seem to be significantly influenced by the treatment groups over all time periods when compared to infected controls. IL-4 was slightly decreased overall but IL-10 seemed to be unaffected. The time course of the establishment of the infection related to cytokine production tells an interesting story. During FLV establishment and probable viremia, cytokine production is decreased overall (day 7, Figures 21.1 through 21.5). As the infection progresses, the immune system responds to the treatment protocol and type 1 cytokine production is partly restored,

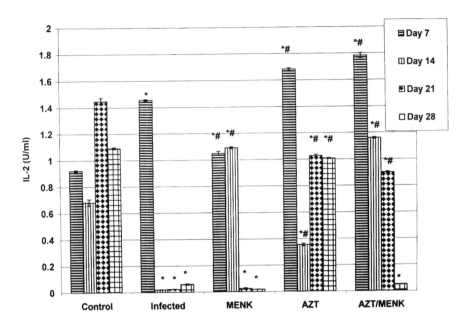

Figure 21.1 Induction of IL-2 production on days 7, 14, 21, and 28 as measured by ELISA. * $p < .005$ as compared to uninfected control values; # $p < .005$ as compared to infected control values.

particularly IL-2 and IFNγ (Figures 21.1 and 21.3). The therapeutic protocol of AZT and MENK resulted in protection as measured by survival for 44% of the mice as compared to infected controls by day 28 (data not shown). Thus, partial modulation of the amount and type of cytokines produced may be related to protective effects of the cytokines against FLV-induced retroviral disease *in vivo*. The induction of IL-2 and IFNγ *in vivo* support the findings of Sin and Specter on the antiviral role of IFNγ *in vitro*.[22] Further efforts to enhance survival are in progress, such as examining drug concentrations, timing of administration, and expanding the drug combination to therapy including a second antiviral agent.

21.3 CONCLUSIONS

Immunoregulatory cytokines may modify the resistance or susceptibility of a host to retroviral infections. The relevance of the ratio of type 1 vs. type 2 cytokines in HIV disease progression has been well documented.[18] The use of antiviral compounds with immunomodulatory agents has been seriously considered as a chemotherapeutic approach in the treatment of retroviral infections. Our studies indicate that a treatment protocol of MENK and AZT can modulate cytokines during an *in vivo* infection by FLV. These studies showed that combination treatment resulted in survival of 44% of infected mice. In previous studies, approximately 50% of the mice on this therapy resulted in survival well beyond the time when infected control mice had died.[14] Therefore, a correlation can be drawn between a treatment protocol

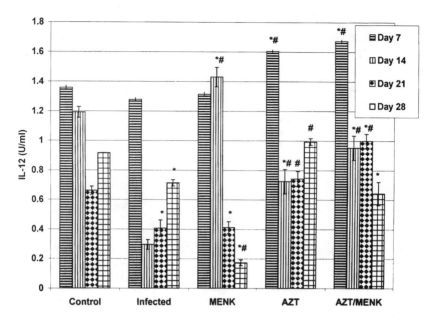

Figure 21.2 Induction of IL-12 production on days 7, 14, 21, and 28 as measured by ELISA. * $p < .005$ as compared to uninfected control values; # $p < .005$ as compared to infected control values.

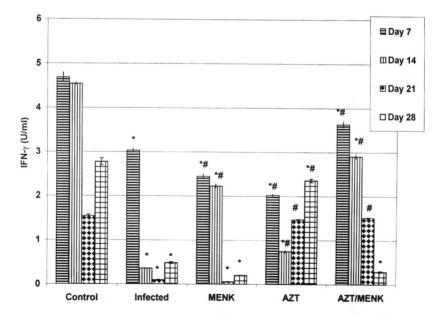

Figure 21.3 Induction of IFNγ production on days 7, 14, 21, and 28 as measured by ELISA. * $p < .005$ as compared to uninfected control values; # $p < .005$ as compared to infected controls.

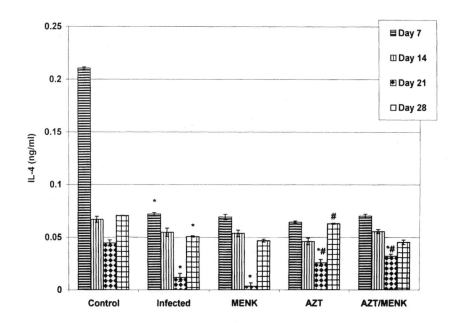

Figure 21.4 Induction of IL-4 production on days 7, 14, 21, and 28 as measured by ELISA. * $p < .005$ as compared to uninfected control values; # $p < .005$ as compared to infected controls.

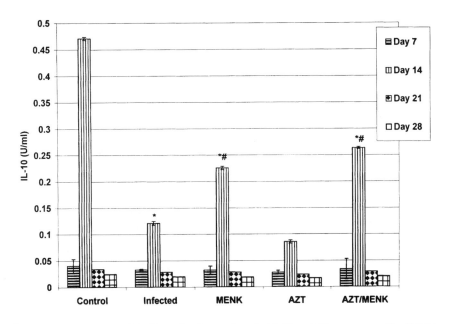

Figure 21.5 Induction of IL-10 production on days 7, 14, 21, and 28 as measured by ELISA. * $p < .005$ as compared to uninfected control values; # $p < .005$ as compared to infected controls.

that leads to diminished morbidity and mortality due to FLV infection and elevated type 1 cytokines.[14] Altered cytokine production has been associated with FLV infection. Lopez-Cepero et al. demonstrated that IL-2 production in FLV-infected splenocytes is suppressed when stimulated with mitogen in culture.[23] Sin and Specter demonstrated that the anti-FLV effects of spleen cells treated with MENK and used in combination with AZT is mediated, in part, by IFNγ secreted by lymphocytes.[22] Long-term studies need to be performed to determine if type 1 cytokines remain elevated in mice that do not exhibit characteristics of progressive FLV disease. Moreover, it will be interesting to determine if combination-treated mice that survive have type 1 cytokine profiles that differ from those mice who do not survive (approximately 50%).

The observation that certain cytokines are suppressed while others are stimulated can be partially explained by other studies with MENK. Recent reports by Marotti et al.[24] and Sin[22] indicated that MENK is capable of stimulating production of IL-1, IL-2, IL-4, and IFNγ and increasing expression of IL-2 receptors. Presumably it is through opioid receptors on lymphocytes that MENK stimulates the production IFNγ and possibly other cytokines.[25] Studies using both opioid antagonists and agonists are being performed in this laboratory to assess the importance of opioid receptors.

Our hypothesis is that MENK, or other mild stimulants of type 1 cytokines, is an improvement over using a single cytokine because it tips the balance of immune homeostasis to favor type 1, but does not grossly disrupt homeostasis. Trials with the combination therapy of AZT and MENK have been done in a small number of patients infected with HIV. The combination therapy resulted in increased numbers of CD3+, CD4+, CD8+, NK, and cytotoxic T cells but the researchers did not follow patients for a sufficient period of time to determine if there was any clinical benefit.[12] Whether the combination therapy with AZT and MENK will generate the same cytokine responses in humans as it did in mice, and whether this will correlate with clinical efficacy, will be an interesting and revealing study. *In vitro* studies by Nyland and colleagues suggest that MENK can inhibit replication of HIV and HTLV-I, suggesting further that this is an important area to investigate (personal communication).

ACKNOWLEDGMENT

This study was supported in part by PHS grants DA10161 and DA07245.

REFERENCES

1. Miller, G. C., Murgo, A. J., and Plotnikoff, N. P., Neuropeptides increase T cell activity *in vitro*, *Clin. Immunol. Immunopathol.*, 31, 132, 1994.
2. Zurawski, G., Benedik, M., Kamb, B. J., Abrams, J. S., Zurawski, S. M., and Lee, F. D., Activation of mouse T-helper cells induces abundant preproenkephalin mRNA synthesis, *Science*, 232, 772, 1986.
3. Heagy, W., Laurance, M., Cohen, E., and Finberg, R., Neurohormones regulate T cell function, *J. Exp. Med.*, 171, 1625, 1990.

4. Chukwuocha, R. U., Reyes, E., and Tokuda, S., The *in vivo* effects of opioid peptides on the murine immune response, *Int. J. Immunopharmacol.*, 16, 205, 1994.
5. Bryant, H. U. and Holaday, J. W., Opioids in immunological processes, in *Handbook of Experimental Pharmacology*, Herz, A., Ed., Springer-Verlag, Berlin, 1993, 361.
6. Matthews, P. M., Froelich, C. J., Sibbitt, W. L., and Bankhurst, A. D., Enhancement of natural cytotoxicity by β-endorphin, *J. Immunol.*, 152, 2172, 1983.
7. Shavit, Y., Lewis, J. W., Treman, G. W., Gale, R. P., and Leibeskind, J. C., Opioid peptides mediate the suppressive effect of stress on natural killer cell cytotoxicity, *Science*, 223, 188, 1984.
8. Plotnikoff, N. P., Faith, R. E., Murgo, A., Herberman, R. B., and Good, R. A., Methionine enkephalin: a new cytokine–human studies, *Clin. Immunol. Immunopathol.*, 82 (1), 93, 1997.
9. Wybran, J., Appelboom, T., Famaey, J., and Govaerts, A., Suggestive evidence for receptors for morphine and methionine-enkephalin on normal blood T lymphocytes, *J. Immunol.*, 123, 1068, 1979.
10. Bhargava, N. P., Opioid peptide immunomodulation, in *Stress and Immunity*, Plotnikoff, N. P., Murgo, A. J., Faith, R. E., and Wybran, J., Eds., CRC Press, Boca Raton, FL, 1991.
11. Nyland, S., Specter, S., Sin, J. I., and Ugen, K., Opiate effects on *in vitro* human retroviral infection, in *Advances in Experimental Medicine and Biology*, Plenum Press, New York, 1993.
12. Bihari, B., Plotnikoff, N. P., Freeman, K. D., Cort, S., Dowling, J., Diuguid, C., Smith, B., Altmann, E., and Gum, S., Methionine enkephalin in the treatment of AIDS related complex, Proc. 7th Int. Conf. AIDS, Florence, Italy, 1991.
13. Burger, R. A., Warren, R. P., Huffman, J. H., and Sidwell, R. W., Effect of methionine enkephalin on natural killer cell and cytotoxic T lymphocyte activity in mice infected with influenza A virus, *Immunopharmacol. Immunotoxicol.*, 17, 323, 1995.
14. Specter, S., Plotnikoff, N. P., Bradley, W. G., and Goodfellow, D., Methionine enkephalin combined with AZT therapy reduce murine retrovirus-induced disease, *Int. J. Immunopharmacol.*, 16, 911, 1994.
15. Soldaini, E., Matteucci, D., Lopez-Cepero, M., Specter, S., Friedman, H., and Bendinelli, M., Friend leukemia complex infection of mice as an experimental model for AIDS studies, *Vet. Immunol. Immunopathol.*, 21, 97, 1989.
16. Metcalf, D., Furth, J., and Buffett, R. F., Pathogenesis of mouse leukemia caused by Friend virus, *Cancer Res.*, 19, 52, 1959.
17. Sin, J. I., Plotnikoff, N., and Specter, S., Anti-retroviral activity of methionine enkephalin and AZT in a murine cell culture, *Int. J. Immunopharmacol.*, 18, 305, 1996.
18. Clerici, M. and Shearer, G. M., The Th1-Th2 hypothesis of HIV infection: new insights, *Immunol. Today*, 15, 575, 1994.
19. Liew, F. Y. and O'Donnell, C. A., Immunology of leishmaniasis, *Adv. Parasitol.*, 32, 161, 1993.
20. Wybran, J., Enkephalins as molecules of lymphocyte activation and modifiers of the biological response, in *Enkephalins-Endorphin Stress and the Immune System*, Plotnikoff, N., Murgo, A. J., and Faith, R. E., Eds., Plenum Press, New York, 1986, 253.
21. Brown, S. L., Tokuda, S., Saland, L. C., and Van Epps, D. E., Opioid peptides effects on leukocyte migration, in *Enkephalins-Endorphin Stress and the Immune System*, Plotnikoff, N., Murgo, A. J., and Faith, R. E., Eds., Plenum Press, New York, 1986, 367.

22. Sin, J. I. and Specter, S., The role of interferon-γ in antiretroviral activity of methionine enkephalin and AZT in a murine cell culture, *J. Pharmacol. Exp. Ther.*, 279, 1268, 1996.
23. Lopez-Cepero, M., Specter, S., Matteucii, D., Friedman, H., and Bendinelli, M., Altered interleukin production during Friend leukemia virus infection, *Proc. Soc. Exp. Biol. Med.*, 188, 353, 1988.
24. Marotti, T., Haberstock, H., Sverko, V., and Hrsak, I., Met- and Leu-enkephalin modulate superoxide anion release for human polymorphonuclear cell, *Ann. NY Acad. Sci.*, 650, 146, 1992.
25. Mazumder, S., Nath, I., and Dhar, M. M., Immunomodulation of human T cell responses with receptor selective enkephalins, *Immunol. Lett.*, 35 (1), 33, 1993.

22 The Immune System and the Hypothalamus-Pituitary-Adrenal (HPA) Axis

P. Falaschi, A. Martocchia, A. Proietti, and R. D'Urso

CONTENTS

22.1 Introduction..325
22.2 HPA Axis and the Immune System Share Signal Molecules
 and Receptors 325
22.3 HPA Axis Products Can Modulate Immune System Functions 327
22.4 Immune System Products Can Modulate HPA Axis Activity 329
22.5 Conclusions...332
References..333

22.1 INTRODUCTION

Evidence of a bidirectional communication between neuroendocrine and immune systems has recently been described. Elucidating these interactions, shared common signal molecules and receptors of both systems seems to be "common words for a single language." In this chapter we will describe these interactions at the level of the HPA axis, and their implications in physiopathological situations.

22.2 HPA AXIS AND THE IMMUNE SYSTEM SHARE SIGNAL MOLECULES AND RECEPTORS

The presence of glucocorticoid, adrenocorticotropin hormone (ACTH), and corticotropin-releasing hormone (CRH) receptors has been demonstrated in immune system cells, respectively: (1) glucocorticoid receptors with a capacity of 7000 sites per cell and a Kd of 4.8 nM; (2) ACTH receptors with high affinity (Kd = 0.04 nM and a capacity of 2000 sites per cell) and low affinity (Kd = 3.4 nM and a capacity of 41,600 sites per cell) binding sites; (3) affinity and capacity not yet determined for CRH receptors.[1-3] Conversely, specific cytokine (IL-1, IL-2, IL-6) binding sites have been found within the HPA axis (Table 22.1).[4-9] IL-1 receptors (IL-1Rs) have been localized in the hippocampus at the highest density and have characteristics similar to immune system IL-1Rs.[4,5,7] It is noteworthy that the hippocampus contains a high density of glucocorticoid receptors (type I and II) involved in the negative feedback on the HPA axis.[11] IL-1Rs on pituitary cells are upregulated by CRH, so that any mild psychological or physical stress able to increase CRH activates the pituitary gland to inflammatory IL-1 action, producing an amplified glucocorticoid-induced immune suppression.[12]

TABLE 22.1
Cytokine Receptors on HPA Axis

Cytokines	Receptors Affinity and Capacity[a]	Level
IL-1	0.2 nM	Hippocampus,
	3.05 fmol/mg protein	hypothalamus (rat), pituitary
IL-2	0.2 nM	Hippocampus[5]
	0.63 fmol/mg protein	hypothalamus (rat), pituitary
IL-3	0.2 nM	Hippocampus[5]
	1.70 fmol/mg protein	
IL-6		Pituitary

[a] Binding of radioligands (^{125}I ILs) to human hippocampal homogenates assessed using a single concentration (0.2 nM) within the range of Kd values reported for ILs in the peripheral immune system.[5]

Lymphoid cells can locally produce several hormones and neuropeptides that may exert an autocrine or paracrine action on leukocytes themselves. In particular, CRH and ACTH have been detected in leukocytes.[13,14] Human peripheral leukocytes infected by viruses synthesize a peptide biologically similar to pituitary ACTH, showing a peak concentration after 6 to 8 h;[14] leukocyte-derived ACTH probably originates from B cells.[12] Hypophysectomized mice infected by Newcastle disease virus show an increase in corticosterone and IFN-α production 8 h after infection (this effect is blocked by prior treatment with dexamethasone).[15] Usually, patients with primary adrenocortical insufficiency require more cortisol replacement than patients with secondary symptons. Corticosteroid response to stress is usually measured 2 h after the induction of stress, whereas the lymphocyte-mediated corticosteroid response is measurable 2 to 8 h afterwards.[15] The production of ACTH and endorphins by leukocytes *in vitro* is induced by CRH and IL-1, whereas it is suppressed by dexamethasone.[16,17] ACTH production and inhibition from B cells are probably mediated by macrophage-derived IL-1 action.[12] The concentrations of CRH (0.7 to 70 nM) required to induce immunoreactive ACTH production in leukocytes are within the range of those required by pituitary cells, but while pituitary ACTH responds within minutes after CRH stimulation, leukocyte-derived ACTH responds within 24 to 48 h owing to the requirement of the *de novo* expression.[16] Human unstimulated leukocytes normally express the mRNA for proopiomelanocortin (POMC) (800 nucleotides long),[18] probably representing a RNA species lacking both exon 1 and exon 2 sequences and part of the 5' region of exon 3. In inflamed tissue, leukocyte-derived β-endorphin may act on opiod receptors of sensory neurons to induce analgesia.[12] Moreover, sensory fibers signal the CNS about the site of inflammation and secrete proinflammatory or anti-inflammatory neuropeptides (such as substance P or somatostatin) that may act locally.

Furthermore, there is evidence of cytokine presence (IL-1, IL-2, IL-3, and IL-6) in human CNS, suggesting their role as locally synthesized neuromodulators. With regard to the HPA axis, IL-1α and IL-1β immunoreactive fibers were found in the hippocampus, hypothalamus (including the paraventricular nucleus), and pituitary gland. The presence of IL-1 has been demonstrated in neurons, astrocytes, microglial cells, and brain macrophages. In pituitary cells, IL-1β is colocalized with TSH in thyrotropes. Astroglial and microglial cells produce IL-1; these cells are numerous in the median eminence region of the hypothalamus — a favorable anatomic site to modify hypothalamic CRH secretion and pituitary ACTH release. The neuroanatomical site of astroglial IL-1 would allow it to act as a hypothalamic releasing or inhibiting hormone and/or to modulate classically defined releasing hormone action on target cells in the pituitary gland. IL-2 is localized in the hippocampus, whereas IL-3 is present in the hippocampus and in the hypothalamus.[4] The basomedial hypothalamus produces small amount of IL-6, probably derived from nonneuronal cells. The presence of IL-6 has been ascertained in folliculostellate cells of the pituitary gland; IL-6 release is stimulated by IL-1, PACAP38 (38-residue pituitary adenylate cyclase activating polypeptide), and CGRP (calcitonin gene-related peptide).[19-21] Probably, IL-1 and IL-6 in the pituitary have paracrine actions. Cytokines (IL-1, IL-6) can also be found in the adrenals; IL-6 has been found in the zona glomerulosa of the adrenal cortex and IL-1 in the medulla.[22,23]

22.3 HPA AXIS PRODUCTS CAN MODULATE IMMUNE SYSTEM FUNCTIONS

Glucocorticoids may be considered as "controllers" of cells traffic though blood vessels (from inside to outside and vice versa). It is noteworthy that migration of leukocytes is an active process that involves chemotactic factors (i.e., leukotrienes, IL-8, MCP, and TGF-β) and adhesion molecules; some effects of glucocorticoids might be explained though their inhibiting effects on IL-3 and IL-8. Glucocorticoids typically cause neutrophil leukocytosis; on the contrary, lymphocyte, monocyte and eosinophil counts decrease. A single administration of corticosteroids in humans produces a decrease in circulating lymphocytes and monocytes of 70% and 90%, respectively.[1] Within 4 to 6 h after corticosteroid administration, T lymphocytes rather than B lymphocytes and CD4 rather than CD8 cells are depleted from the circulation. A similar phenomenon occurs during chonic steroid administration.

Kinetic studies of recirculating CD4 cells after glucocorticoid infusion show that the return of CD4 cells to the bloodstream is greatly influenced while their exit from blood is only mildly affected.[24] The circadian rhythm of cortisol is inversely related to peripheral blood lymphocytes (T, B, and NK cells).[1] T lymphocytes are generally more sensitive than B lymphocytes to the *in vivo* corticosteroids immunosuppressive effects. In animal models, corticosteroids inhibit the production of IL-2 and the cell-mediated response (Th1 response), whereas they induce the production of IL-4 and the humoral response (Th2 response).[25] Table 22.2 shows the predominance of Th1 and Th2 immune responses in human diseases.[26,27] However, IgG, IgA, and IgM serum levels are suppressed by high-dose methylprednisolone treatment, which determines an initial increase in immunoglobulin catabolism and, subsequently, a decreased production.[28]

TABLE 22.2
TH1 and TH2 Response

Agent/Disease/Status	Response
HIV	Th1-Th2 switch
Measles	Th2
Sézary syndrome	Th2
Mycosis fungoides	Th1
Mycobacterium tuberculosis	Th1
Psoriasis vulgaris	Th1
Multiple sclerosis	Th1
IDDM	Th1?
Graves' disease	Th2?
Rheumatoid arthritis	Th1
Allergy	Th2
Atopic asthma	Th2
Pregnancy	Th1 suppressed

Hydrocortisone blocks resting (G0) B cells from enlarging, from acquiring activation antigens, and from progressing to the G1 cell cycle phase. Intermediate events in the B cell cycle are less sensitive to *in vitro* glucocorticoid suppressive effects, whereas late events, such as differentiation to immunoglobulin secreting cells, are resistant.

Once B lymphocytes are activated, steroids have only minimal effects on subsequent events. It has been recently shown that zinc (Zn) and bioactive thymulin (Zn-FTS) circulating levels are low in hypercortisolemic patients.[29] Corticosteroids inhibit macrophage class II antigen expression and IL-1 production.[28] Glucocorticoids do not inhibit early T lymphocyte activation events, but they inhibit cells at the G1a phase, blocking the completion of RNA synthesis for proliferation; IL-2 reverses this inhibitory effect. IL-2 production is blocked by glucocorticoids. Dexamethasone inhibits IL-2 and IL-2R p55 chain mRNA accumulation, as well as their product. In addition, glucocorticoids inhibit the production of other lymphokines such as IL-3, IL-6, TNF, IFN-γ and GM-CSF. Although their effects on T suppressor cells are variable, generally corticosteroids either have no effect or enhance suppressor-cell function: T suppressor cells, once generated, are resistant to the inhibitory effect of steroids. Therefore, their overall effect at the cellular functional level is immunosuppressive.

Cortisol, at physiological concentrations, is a potent *in vitro* inhibitor of NK cell activity.[30] Patients with Cushing's syndrome have significantly lower NK activity than controls.[31] Recently it was shown that two human corticostatins-defensins, HP-1 and HP-4, are negative modulators of NK activity.[32] Since resting immunological cells are more sensitive to glucocorticoids than activated cells, cortisol can prevent the excessive expansion of cells with low affinity for the antigen and of cells which are recruited under the polyclonal influence of lymphokines. On the other hand, glucocorticoids at increased but still physiological concentrations inhibit IL-1 and IL-2 production and control the clonal expansion of committed cells with high affinity for the antigen.[33] In this way, glucocorticoids suppress aspecific reactions and enhance specific immune response. Chonic activation of the HPA axis may also cause a relative decrease in the production of δ-5 androgens (like DHEAS) by the adrenals,[34] resulting in a predominance of Th2 immune response.[35]

In addition to cytokines, anti-inflammatory actions of glucorticoids also involve different humoral mediators, like vasoactive amines (histamine, serotonin), metabolites of arachidonic acid (leukotrienes, prostaglandins), proteases, and enzymes (collagenases) (some of these actions being mediated by the induction of lipocortin-I from macrophages). Moreover, glucocorticoids enhance hepatic response to cytokines released into blood from inflamed sites, increasing the synthesis of IL-1- and IL-6-responsive acute phase proteins (respectively called, APP type I and II) (Figure 22.1).[36] Particularly in conditions of inflammation, there is a reduction (IL-6-mediated) of corticosteroid-binding globulin, with a modulation of the amount of cortisol reaching target tissues.[37]

In addition to their classical neuroendocrine actions, neuropeptides may also function as direct immunoregulatory agents. ACTH reduce the number of antibody-producing cells[38] and the IFN-γ production,[39] whereas it is a positive modulator of NK activity.[40] *In vitro* CRH suppresses NK activity,[41] blocks IL-2-induced proliferation, and increases cAMP production.[42]

22.4 IMMUNE SYSTEM PRODUCTS CAN MODULATE HPA AXIS ACTIVITY

The presence and the physiological significance of detectable cytokine levels in plasma basal conditions is still controversial, though elevated cytokine levels during immune activation have been demonstrated both in plasma and locally, with paracrine effects. Various cytokines are reported to stimulate the HPA axis *in vivo* and *in vitro* at different sites of action (hypothalamus, pituitary and adrenal gland). The *in vivo* administration (i.v., i.p., or i.c.v.) of IL-1α or IL-1β increases peripheral blood ACTH and corticosterone, portal blood CRH1-41, and vasopressin (AVP)[43] (an effect medi-

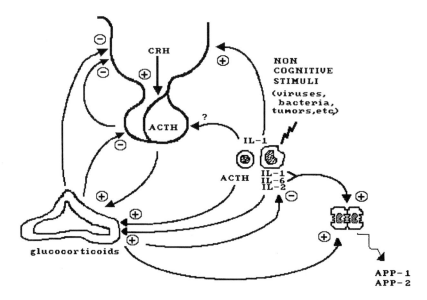

Figure 22.1 "Feedback loop" between the immune system and HPA axis.

ated by prostaglandins and blocked by anti-CRH serum).[44,45] Human IL-1α is about one-tenth as potent as human IL-1β in increasing ACTH levels in rats. Evidence shows that IL-1β increases corticosterone levels in hypophysectomized rats, probably though a direct adrenal effect.[46] The i.v. administration of IL-2[47] increases ACTH and glucocorticoid levels. About 4 h after rhIL-2 infusion in patients with cancer (30,000 to 100,000 U/kg), plasma ACTH and cortisol levels increase, and then they progressively decrease during the first treatment cycle. Interestingly, after a rest period during the second treatment cycle, hormonal levels are higher than those observed during the first one. The marked increase in these hormones suggests an anamnestic response to IL-2 by the immune system and the HPA axis. A similar effect has been observed after i.v. IFN-α administration (with a higher increase in cortisol production at the beginning of each treatment cycle), but not after i.m. injection of IFN-α administered on alternate days.

An increase in ACTH and cortisol levels has been observed after i.v. injection of rhIL-6 in rats[48] and in patients with cancer[49] (an effect blocked by anti-CRH serum).[48]

At high doses, IL-6 causes elevation of plasma AVP, indicating a role of IL-6 in the inappropriate secretion of antidiuretic hormone that can occur during inflammatory disease.[50] IL-6 may inhibit hepatic CBG (corticosteroid-binding globulin) synthesis, thus increasing bioavailability of free cortisol concentrations during HPA hyperactivity.[51] Evidences indicate that IL-1β is stronger than IL-6 in activating the HPA axis, so the HPA axis activation during immune responses is thought to be largely dependent on IL-1.

In our study the administration of rhIFN-α 2a (3 million UI im.) in patients with chonic active B and C hepatitis induced a significant rise in ACTH and in both plasma and urinary free cortisol levels, in comparison with the spontaneous hormone profiles obtained in a control day.[52] Since ACTH and cortisol levels after rhIFN-α treatment increased significantly, with peaks after 6 h, night administration of the drug is advisable to preserve the physiological circadian rhythm of ACTH and cortisol. Since glucocorticoids are powerful immunomodulators, a cortisol increase after rhIFN-α administration could contribute to the immune response control. A reduced or abolished cortisol response to IFN-α administration was observed after week 6 of therapy; a similar result was recently confirmed by other authors.[53] *In vivo* administration of IFN-γ and TNF-α also stimulates the HPA axis.[54,55] The IFN-α administration in humans induces an increase of IFN-γ and IL-6, whereas IL-1 blood levels remain unchanged; after endotoxin administration, TNF-α usually appears first, followed by IL-1 and IL-6.[56-58] This evidence supports the hypothesis of a different and well-defined sequence in the cascade of cytokines able to activate the HPA axis.

The *in vitro* effects of different cytokines on the HPA axis were studied at different levels. At the adrenal level, IL-1α (3 to 300 U/ml), IL-1β (30 to 100 U/ml), IL-2 (3 to 300 U/ml) and IL-6 (300 U/ml) stimulate glucocorticoid release from cultured adrenocortical cells in a dose- and time-dependent manner (an effect blocked by cyclooxygenase inhibitors, e.g., aspirin, but not by lipooxygenase inhibitors, e.g., AA-861 and NDGA).[59] ACTH acts synergically with interleukins in stimulating steroidogenesis. Using a rat organ culture system it was shown that IFN-α also stimulates the adrenal secretion[60] and some authors noted a structural and biological relatedness of IFN to ACTH.

Evidence in the literature suggests that the interleukins have at least two sites of action in stimulating the HPA axis and glucocorticoid production. One is an early effect through activation of hypothalamic CRH, and the other is a delayed effect which requires a 12-h latency period and could be the result of direct stimulation of the adrenal glands. The controversial effect on pituitary corticotrophs would also require a 12 to 24-h latency period.[61] An intraadrenal CRH/ACTH system can mediate the stimulatory effect of cytokines on the adrenal glands.[62] In fact, rhIL-1β administration in rats increased CRH-like immunoreactivity in both hypothalamus and adrenal glands.[63] On examination of the paracrine role of lymphocyte-derived peptides, human monocytes significantly increased cortisol production from adrenocortical cells after 24, 48, and 72 h of coculture.[64] Data on the effects of cytokines at pituitary level are still controversial. In fact, using different *in vitro* systems (dispersed anterior pituitary cells, pituitary pieces, or hemipituitaries), both increased and unchanged ACTH levels were found after the administration of IL-1, IL-2, IL-6, TNF, IFN-α, and IFN-γ.[65] In the authors' experience, IL-1 and IL-6 failed to modify ACTH release from isolated and dispersed rat pituitary cells in perfusion. At the hypothalamic level, IL-1 and IL-6 incubations of rat hypothalami (including the paraventricular nucleus) produced a marked dose-dependent stimulation of CRH1-41 in the dose range of 1 to 100 U/ml and 10 to 100 U/ml, respectively (an effect blocked by inhibiting prostaglandin synthesis via the cyclo-oxygenase pathway).[66] In this study, other cytokines (IL-2, IL-8, TNF-α, IFN-α2, and IFN-γ) failed to modify hypothalamic CRH secretion. On the contrary, another study showed that

IL-2 increases CRH release from continuously perfused hypothalami in a dose-dependent manner (25 to 100 U/ml).[67]

22.5 CONCLUSIONS

The immune system can be considered as a sensory organ,[68] able to recognize "noncognitive stimuli" (i.e., bacteria, viruses, and tumors), whereas the central nervous system recognizes classic "cognitive stimuli" (physical, emotional, and chemical). The immune, central nervous, and the neuroendocrine systems interact though the release of cytokines and hormones, with a "feedback loop" (Figure 22.1). In this circuit the afferent limb is represented by the immune system that informs the CNS and stimulates the HPA axis though the release of activated lymphocyte products. Moreover, the immune system also induces typical physiological changes, different from those observed after stress activation by "cognitive stimuli" (i.e., fever, fatigue, sleep, eating behaviour, and mood state changes). The efferent limb is represented by CRH activation that induces ACTH and glucocorticoid secretion. On the other hand, glucocorticoids counterregulate the immune system and the HPA axis.

When the inflammatory stress becomes chonic, CRH neurons are mildly suppressed, whereas AVP neurons become active. When the inflammatory stress subsides, AVP secretion returns to normality, but the HPA axis may become transiently hypofunctional. During these periods the immune system may be hyperactive.

Immune system and HPA axis interactions may be involved in pathological conditions. Examining the above-described model, we expect an excessive HPA axis activity to induce resistance to autoimmune/inflammatory diseases (and susceptibility to infectious agents/tumors). Conversely, a defective HPA axis activity induces susceptibility to autoimmune/inflammatory diseases (and resistance to infectious agents/tumors).

Two animal models of decreased (Lewis rats) and increased (Fischer rats) HPA axis activity exist; they are susceptible and resistant to inflammatory diseases, respectively.[58]

In chonic fatigue syndrome, reduced pituitary and adrenal responses to CRH were described and this hypofunctional HPA axis may play a role in immune system hyperactivity and clinical aspects of the disease.[69] Patients with rheumatoid arthitis have lower blood cortisol levels than individuals with a control inflammatory disorder, especially in the morning when the symptoms are generally more severe, despite similar increases in IL-1β and IL-6 blood levels.[70]

However, the beneficial and permissive actions of glucocorticoids during acute stress reactions and, in particular, during severe bacterial infections cannot be disregarded. In these situations if adrenocortical insufficiency is quite apparent (low urinary free cortisol excretion, low response to ACTH test), replacement doses of glucocorticoids appear appropriate.[71]

In situations of hyperactivity of the HPA axis (mental deterioration, depression, aging, and chonic stress), high glucocorticoid levels induce a significant immune suppression, characterized by alterations of humoral and cellular immune mechanisms with increased susceptibility to infectious diseases and neoplasms.[65] During the last trimester of pregnancy, the placenta secretes CRH, producing a mild hypercortisolism with immune suppression and a prevalence of Th2 immune response. During post partum there is a rebound of the Th1 immune response with reactivation of autoimmune diseases (i.e. chonic thyroiditis and rheumatoid arthitis).[58]

An enhanced immune system activity may also derive from glucocorticoid resistance in target tissues,[72] a syndrome characterized by a defective glucocorticoid receptor or a post-receptorial cascade (the glucocorticoid receptor gene has been localized on chomosome 5). The glucocorticoid resistance syndrome must necessarily be partial, even if it is generalized; the complete resistance is not consistent with life. The altered negative feedback of glucocorticoids on ACTH secretion increases the production of ACTH itself, with subsequent high levels of cortisol, mineralcorticoids (corticosterone and deoxycorticosterone), and androgens (androstenedione and DHEAS); the classical clinical feature is represented by hypokalemia, hypertension, and hyperandrogenism. The response to the low-dose dexamethasone suppression test is absent, but the response to a high dose is present. Recently, it has been stated that the glucorticoid resistance syndrome may be acquired and not generalized (for example, it may be limited to lymphocytes).

Alterations of glucocorticoid receptors have been demonstrated in hematological diseases (leukemia, lymphoma, and myeloma), nephotic syndrome, rheumatoid arthitis, steroid-resistant asthma, AIDS, degenerative osteoarthitis and central nervous system disorders (mental deterioration, depression, aging, and chonic stress).[58]

Hypo- and hyperresponsiveness of the HPA axis could be important to examine immunoneuroendocrine interactions. Some authors proposed the endotoxin test as an appropriate and safe tool to explore the adaptation of the HPA axis to a "noncognitive stimuli", thus providing complementary information to the CRH test.[73]

REFERENCES

1. Berczi, I., The influence of pituitary-adrenal axis on the immune system, in *Pituitary Function and Immunity,* Berczi, I., Ed., CRC Press, Boca Raton, Florida, 1986, 49.
2. Smith, E. M., Brosman, P., Meyer, W. J., and Blalock, J. E., A corticotropin receptor on human mononuclear lymphocytes: correlation with adrenal ACTH receptor activity, *N. Engl. J. Med.,* 317, 1266, 1987.
3. Audhya, T., Jain, R., and Hollander, C. S., Receptor-mediated immunomodulation by corticotropin-releasing factor, *Cell. Immunol.,*134, 77, 1991.
4. Koenig, J. I., Presence of cytokines in the hypothalamic- pituitary axis, *Prog. Neuroendocrinimmunol.,* 4,143, 1991.
5. Araujo, D. M., and Lapchak P. A., Induction of immune system mediators in the hippocampal formation in Alzheimer's and Parkinson's diseases: selective effects on specific interleukins and interleukin receptors, *Neuroscience,* 61, 745, 1994.
6. Arzt, E., Stelzer, G., Renner, U., Lange, M., Muller, O. A., and Stalla, G. K., Interleukin-2 and interleukin-2 receptor expression in human corticotropic adenoma and murine pituitary cell cultures, *J. Clin. Invest.,* 90, 1944, 1992.
7. Marguette, C., Ban, E., Fillon, G., and Haourb, F., Receptors for IL-1 α and β, IL-2, IL-6 in mouse, rat and human pituitary, *Neuroendocrinology,* 52 (Suppl.1), 48, 1990.

8. Ohmichi, M., Hirota, K., Koike, K., Kurachi, H., Ohtsuka, S., and Matsuzaki, N., Binding sites for interleukin-6 in the anterior pituitary gland, *Neuroendocrinology,* 55, 199, 1992.
9. Hanisch, U. K. and Quirion, R., Interleukin-2 as a neuroregulatory cytokine, *Brain Res. Rev.,* 21, 246, 1996.
10. Schettini, G., Interleukin 1 in the neuroendocrine system: from gene to function, *Prog. Neuroendocrinimmunol.,* 3, 157, 1990.
11. Reul, J. and De Kloet, E. R., Two receptor systems for corticosterone in rat brain microdistribution and differential occupation, *Endocrinology,* 117, 2505, 1985.
12. Blalock, J. E., The syntax of immune-neuroendocrine communication, *Immunol. Today,* 15, 504, 1994.
13. Stephanou, A., Jessop, D. S., Knight, R. A., and Lightman, S. L., Corticotrophin-releasing factor-like immunoreactivity and mRNA in human leukocytes, *Brain Behav. Immun.,* 4, 67,1990.
14. Blalock, J. E. and Smith, E. M., Human leukocyte interferon: structural and biological relatedness to adrenocorticotropic hormone and endorphines, *Proc. Natl. Acad. Sci. U.S.A.,* 77, 5972, 1980.
15. Smith, E. M., Meyer, W. J., and Blalock, J. E., Virus-induced corticosterone in hypophysectomized mice: a possible lymphoid axis, *Science,* 218, 1311, 1982.
16. Smith, E. M., Morrill, A. C., Meyer, W. J., and Blalock, J. E., Corticotropin releasing factor induction of leukocyte- derived immunoreactive ACTH and endorphins, *Nature,* 321, 881, 1986.
17 Buzzetti, R., Ciancio, A., Ciucci, E., Celli, V., and Giovannini, C., ACTH di origine linfocitaria in condizioni normali e patologiche, *Ann. Ital. Med. Int.,* 6, 357, 1991.
18. Buzzetti, R., McLoughlin, L., Lavender, P. M., Clark, A. J., and Rees, L. R., Expression of pro-opiomelanocortin gene and quantification of adrenocorticotropic hormone-like immunoreactivity in human normal peripheral mononuclear cells and lymphoid and myeloid malignancies, *J. Clin. Invest.,* 83, 733, 1989.
19. Vankelecom, H., Carmeliet, P., Van Damme, J., Billiau, A., and Denef, C., Production of interleukin-6 by folliculo-stellate cells of the anterior pituitary gland in a histiotypic cell aggregate culture system, *Neuroendocrinology,* 49, 102, 1989.
20. Spangelo, B. L., Judd, A. M., Isakson P. C., and MacLeod R. M., Interleukin 1 stimulates interleukin-6 release from rat anterior pituitary cells *in vitro, Endocrinology,* 128, 2685, 1991.
21. Tatsuno, I., Somogyvari-Vigh, A., Mizuno, K., Gottschall, P. E., Hidaka, H., and Arimura, A., Neuropeptide regulation of interleukin-6 production from the pituitary: stimulation by pituitary adenylate cyclase activating polypeptide and calcitonin gene-related peptide, *Endocrinology,* 129, 1797, 1991.
22. Achultberg, M., Anderson, C., Unden, A., Troye-Blowberg, M., Svenson, S. B., and Bartfai, T., Interleukin-1 in adrenal chromaffin cells, *Neuroscience,* 30, 805, 1989.
23. Judd, A. M., Spangelo, B. L., and MacLeod, R. M., Rat adrenal glomerulosa cells produce interleukin-6, *Prog. Neuroendocrinimmunol.,* 3, 282, 1992.
24. Ottaway, C. A. and Husband, A. J., The influence of neuroendocrine pathways on lymphocyte migration, *Immunol. Today,* 15, 511, 1994.
25. Daynes, R. A. and Araneo, B. A., Contrasting effects of glucocorticoid on the capacity of T cells to produce the growth factors interleukin 2 and interleukin 4, *Eur. J. Immunol.,* 19, 2319, 1989.
26. Mosmann, T. R. and Sad, S., The expanding universe of T-cell subsets: Th1, Th2 and more, *Immunol. Today,* 17, 138, 1996.

27. Song, Y. H., Li, Y., and MacLaren, N. K., The nature of autoantigens targeted in autoimmune endocrine diseases, *Immunol. Today,* 17, 232, 1996.
28. Boumpas, D. T., Paliogianni, S., Anastassiou, E. D., and Balow, J. E., Glucocorticosteroid action on the immune system: molecular and cellular aspects, *Clin. Exp. Rheumatol.,* 9, 413, 1991.
29. Travaglini, P., De Min, C., Mocchegiani, E., Re, T., Colombo, P., and Fabris, N., Thymulin and zinc circulating levels in hypercortisolemic patients: preliminary observations, *J. Endocrinol. Invest.,* 15, 187, 1992.
30. Masera, R., Gatti, G., Sartori, M. L., Carignola, R., Salvadori, A., Magro, E., and Angeli, A., Involvement of a Ca++ dependent pathways in the inhibition of human natural killer (NK) cell activity by cortisol, *Immunopharmacology,* 18, 11, 1989.
31. Gatti, G., Masera, R., Carignola, E., Magro, E., Sartori, M. L., and Angeli, A., Natural killer (NK) cell activity in Cushing's syndrome, *J. Endocrinol. Invest.,* 12, 183, 1992.
32. Masera R., Gatti, G., and Bateman, A., Modulazione autocrino-paracrina dell'attività natural killer nell'uomo da parte di peptidi della famiglia delle corticostatine-defensine, *Ann. Ital. Med. Int.,* 8, 80, 1993.
33. Besedovsky, H. O. and Del Rey, A., Immune-neuroendocrine network, *Prog. Immunol.,* 6, 578, 1986.
34. Parker, L. N., Levin E. R., and Lifrak E. T., Evidence for adrenocortical adaptation to severe illness, *J. Clin. Endocrinol. Metab.,* 60, 947, 1985.
35. Daynes, R. A., Dudley, D. J., and Araneo, B. A., Regulation of murine lymphokine production *in vivo*. II. Dehydroepiandrosterone is a natural enhancer of interleukin 2 synthesis by helper T cells, *Eur. J. Immunol.,* 20, 793, 1990.
36. Baumann, H., Gauldie, J., The acute phase response, *Immunol. Today,* 15,74, 1994.
37. Garrel D. R., Corticosteroid-binding globulin during inflammation and burn injury: nutritional modulation and clinical implications, *Horm. Res.,* 45, 245, 1996.
38. Johnson, H. M., Smith, E. M., Torres, B. A., and Blalock, J. E., Regulation of the *in vitro* antibody response by neuroendocrine hormones, *Proc. Natl. Acad. Sci. U.S.A.,* 79, 4171, 1982.
39. Johnson, H. M., Torres, B. A., Smith, E. M., Dion, L. D., and Blalock, J. E., Regulation of lymphokine (γ-interferon) production by corticotropin, *J. Immunol.,* 132, 246, 1984.
40. Gatti, G., Masera, R., and Pallavicini, L., Interplay *in vitro* between ACTH, ß-endorphin and glucocorticoids in the modulation of spontaneous and lymphokine-inducible human natural killer (NK) activity, *Brain Behav. Immun.,* 7, 16, 1993.
41. Audhya, T., Brand, H., Hollander, C. S., Suppression of natural killer cytotoxicity in human peripheral T lymphocytes by corticotropin-releasing factor: an *in vitro* model of *in vivo* effects, *Clin. Res.,* 38, 595A, 1990.
42. Audhya, T., Nabriski, D., Saperstein, A., Ranson, J., and Hollander, C. S., Corticotropin-releasing factor suppresses interleukin-2 induced lymphocyte proliferation and increases cAMP production in human splenocytes, *Clin. Res.,* 38, 461A, 1991.
43. Sapolsky, R., Rivier, C., Yamamoto, G., Plotsky, P., and Vale, W., Interleukin-1 stimulates the secretion of hypothalamic corticotropin-releasing factor, *Science,* 238, 522, 1987.
44. Katsuura, G., Gottschall, P. E., Dahl, R. R., and Arimura, A., Adrenocorticotropin release induced by intracerebroventricular injection of recombinant human interleukin-1 in rats: possible involvement of prostaglandins, *Endocrinology,*122, 1773, 1988.
45. Uehara, A. M., Gottschall, P. E., Dahl, R. R., and Arimura, A., Interleukin-1 stimulates ACTH release by an indirect action which requires endogenous corticotropin releasing factor, *Endocrinology,* 21, 1580, 1987.

46. Andreis, P. G., Neri, G., Belloni, A. S., Mazzocchi,G., Kasprzak, A., and Nussdorfer, G. G., Interleukin-1ß enhances corticosterone secretion by acting directly on the rat adrenal gland, *Endocrinology,* 129, 53, 1991.
47. Denicoff, K. D., Durkin, T. M., and Lotze, M. T., The neuroendocrine effects of interleukin-2 treatment, *J. Clin. Endocrinol. Metab.,* 69, 402, 1989.
48. Naitoh, Y., Fukata, J., and Tominaga, T., Interleukin-6 stimulates the secretion of adrenocorticotropic hormone in conscious, freely-moving rats, *Biochem. Biophys. Res. Commun.,* 155, 1459, 1988.
49. Mastorakos, G., Weber, J. S., Magiakou, M. A., Gunn, H., and Chrousos, G. P., Hypothalamic-pituitary-adrenal axis activation and stimulation of systemic vasopressin secretion by recombinant interleukin 6 in humans: potential implications for the syndrome of inappropriate vasopressin secretion, *J. Clin. Endocrinol. Metab.,* 79, 934, 1994.
50. Mastorakos, G., Weber, J. S., Magiakou, M. A., Crofford, L. J., Kalogeras, K., Wilder, R., and Chrousos, G. P.,Interleukin 6 effects on the human hypothalami-pituitary-adrenal axis, in *Proc. 75th Annu. Meet. Endocrine Soc.,* Las Vegas, Abstr. 1665, 1993.
51. Bartalena, L., Hammond, G. L., Farsetti, A., Flink, I. L., and Robbins, J., Interleukin-6 inhibits corticosteroid-binding globulin synthesis by human hepatoblastoma-derived (HepG2) cells, *Endocrinology,* 133, 291, 1993.
52. D'Urso, R., Falaschi, P., Canfalone, G., Carusi, E., Proietti A., Barnaba, V., and Balsano, F., Neuroendocrine effects of recombinant α-interferon administration in humans, *Prog. Neuroendocrinimmunol.,* 4, 20, 1991.
53. Muller, H., Hiemke, C., Hammes, E., and Hess, G., Sub-acute effects of interferon-α 2 on adrenocorticotrophic hormone, cortisol, growth hormone and prolactin in humans, *Psychoneuroendocrinology,* 17, 459, 1992.
54. Holsboer, F., Stalla, G. K., Von Bardeleben, U., Miler, H., and Miller, O. A., Acute adrenocortical stimulation by recombinant γ interferon in human controls, *Life Sci.,* 42, 1, 1988.
55. Sharp, B. M., Matta, S. G., Peterson, P. K., Newton, R., Choa, C., and McCallen, K., Tumor necrosis factor-α is a potent secretagogue: comparison to interleukin-1 β, *Endocrinology,* 124, 3131, 1989.
56. Shimizu, H., Ohatni, K., Tanaka, Y., and Uehara, Y., Interleukin-6 may mediate activation of pituitary-adrenal axis by interferon-α, in *Proc. 75th Annu. Meet. Endocrine Soc.,* Las Vegas, Abstract 1666, 1993.
57. Gisslinger, H., Gilly, B., Woloszczuk, W., Mayr, W. R., Havelec, L., and Linkesch, W., Thyroid autoimmunity and hypothyroidism during long-term treatment with recombinant interferon-α, *Clin. Exp. Immunol.,* 90, 363, 1992.
58. Chrousos, G. P., The hypothalamic-pituitary-adrenal axis and immune-mediated inflammation, *N. Engl. J. Med.,* 332, 1351, 1995.
59. Tominaga, T., Fukata, J., and Naito, Y., Prostaglandin-dependent *in vitro* stimulation of adrenocortical steroidogenesis by interleukins, *Endocrinology,* 128, 526, 1991.
60. Gisslinger, H., Svoboda, T., and Templ, H., Interferon α stimulates the hypothalamic pituitary-adrenal axis *in vivo* and *in vitro,* in *Proc. 1st Int. Congr. ISNIM,* Florence, 1990, 117.
61. Kehrer, P., Turnill, D., Dayer, J. M., Muller, A. F., and Gaillard, R. C., Human recombinant interleukin-1 β and α, but not recombinant tumor necrosis factor α stimulate ACTH release from rat anterior pituitary cells *in vitro* in a prostaglandin E2 and cAMP indipendent manner, *Neuroendocrinology,* 48, 160, 1988.

62. Andreis, P. G., Neri, G., and Nussdorfer, G. G., Corticotropin-releasing hormone (CRH) directly stimulates corticosterone secretion by the adrenal gland, *Endocrinology,* 128,1198, 1991.
63. Naitoh, Y., Fukata, J., Nakaishi, S., Nakai, Y., Hirai, Y., Tamai, S., Mori, K., and Imura, H., Chronic effects of interleukin-1 on hypothalamus, pituitary and adrenal glands in rat, *Neuroendocrinology,* 51, 637, 1990.
64. Whitcomb, R. W., Linehan, W. M., Wahl, L. M., and Knazek, R. A., Monocytes stimulate cortisol production by cultured human adrenocortical cells, *J. Clin. Endocrinol. Metab.,* 66, 33, 1988.
65. Falaschi, P., Martocchia, A., Proietti, A., Pastore, R., D'Urso, R., and Barnaba, V., La Neuroendocrinoimmunologia, in *Atti dei Congressi della Società Italiana di Medicina Interna. 94° Congresso,* Edizioni L. Pozzi, Roma, 1993, 315.
66. Navarra, P., Tsagarakis, S., Faria, S. M., Rees, L. H., Besser, G. M., and Grossman, A. B., Interleukins-1 and -6 stimulate the release of corticotropin-releasing hormone-41 from rat hypothalamus *in vitro* via the eicosanoid cyclooxygenase pathway, *Endocrinology,* 128, 37, 1991.
67. Cambronero, J. C., Rivas, F. J., Borrell, J., and Guaza, C., Interleukin-2 induces corticotropin-releasing-hormone release from superfused rat hypothalami: influence of glucocorticoids, *Endocrinology,* 131, 677, 1992.
68. Blalock, J. E., The immune system as a sensory organ, *J. Immunol.* 132, 1067, 1984.
69. Demitrack, M. A., Dale, J. K., Straus, S. E., Laue, L., Listwak, S. J., Kruesi, M. J. P., Chrousos, G. P., and Gold, P. W., Evidence for impaired activation of the hypothalamic-pituitary-adrenal axis in patients with chronic fatigue syndrome, *J. Clin. Endocrinol. Metab.,* 73, 1224, 1991.
70. Chikanza, J. C., Petrou, P., Kingsley, G., Chrousos, G., Panayi, G. S., Defective hypothalamic response to immune and inflammatory stimuli in patients with rheumatoid arthritis, *Arthritis Rheum.,* 35, 1281, 1992.
71. Sibbald, W. J., Short, A., and Cohen, M. P., Variations in adrenocortical responsiveness during severe bacterial infections. Unrecognized adrenocortical insufficiency in severe bacterial infections, *Ann. Surg.,* 186, 29, 1977.
72. Chrousos, G. P., Detera-Wadleigh, S. D., and Karl, M., Syndromes of glucocorticoid resistance, *Ann. Intern. Med.,* 119, 1113, 1993.
73. Schreiber, W., Pollmacher, T., Fassbender, K., Gudwills, S., Vedder, H., Wiedemann, K., Galanos, C., and Holsboer, F., Endotoxin- and corticotropin-releasing hormone-induced release of ACTH and cortisol, *Neuroendocrinology,* 58, 123, 1993.

Index

A

Abstinence syndrome, 226–227, 241–242
Acetaminophen, 107
Acetylcholine, gonadotropin-modulatory activity, 206
Acne vulgaris, 126
Acquired immunodeficiency syndrome, *See* AIDS
Acromegalism, 191–192
ACTH, 3–4, 18, 19, 61, 122, 174, 196, *See also* Corticotropin-releasing hormone
 cytokine interactions, 5, 26–29, 179–181, 188, 225, 329–331, *See* specific cytokines
 endotoxin-induced α-MSH and, 71
 glucocorticoid resistance syndrome, 333
 immune cell expression and/or receptors, 54, 61, 179, 193, 325–327, *See* specific cell types
 immune cell function and, 174–175, 329
 α-MSH homology, 69
 thermoregulatory function, 179
 thymic expression, 189, 193
 tumor expression, 136, 137, 145
ACTH receptor, 70
Activin, 176, 206, 208
Acute-phase proteins (APP), 117–119
 glucocorticoid effects, 329
 IFN-γ and, 6
 IL-1 and, 5
 α-melanocyte-stimulating hormone (MSH) and, 21
 stress and, 125
 TNF and, 6
Acute phase response, *See* Acute-phase proteins; Inflammatory response
 cytokines and, 117–119
 stress and, 124–125
Addiction, 225–226
Adenylate cyclase, 298
Adhesion molecules, 19
Adjuvant-induced arthritis
 CRH expression, 23
 joint innervation and sensitivity, 120
 α-MSH and, 72

 prolactin antiinflammatory effects, 22
 substance P and, 121–122
Adrenal glands, 18
Adrenocorticotropin-releasing hormone, *See* ACTH
Adrenomedullin, 136, 139, 140
Adult respiratory distress syndrome (ARDS), 70
Affective disorders, *See* Depression; specific disorders
Agoraphobia, 8
AIDS, 62–65
 emerging pathogens, 62–63
 glucocorticoid receptor alterations, 333
 methionine-enkephalin treatment, 65, 77–90
 opiate abuse and, 233
AIDS dementia complex (ADC), 64
AIDS-related complex (ARC), methionine-enkephalin treatment, 77–90
Aldesleukin, 95, 101
Alfentanyl, 296
Allergic contact dermatitis, 121
Allergic encephalomyelitis, 22, 30
Allodynia, 262
Alopecia areata, 126
Analgesia, neuroimmune interactions, 263–266, *See also* Pain
Anemia
 erythropoietin therapy, 104
 treatment-associated, 95
Angiosarcoma, 135, 138, 140
Animal models (systems or general issues)
 differential HPA axis inflammatory stimulus responsiveness, 30
 IL-1 effects on immunocyte β-endorphin expression, 271–278
 nonexperimental stressors, 167
 opiate immunomodulatory effects, 234–242
 review of model stressors and effects on host defense, 165–168
 steroid-sensitive and steroid-resistant species, 164
Anorexia, 54
 cancer and, 141–142
Anorexia nervosa, 8–9, 217

Antibody response
    IL-1β effects, 165
    murine restraint stress and, 166
    neuroendocrine modulation, 165
    opioid effects, 234, 315
    TSH effects, 175
Antidepressants
    anxiety disorders and, 7
    depressed subject cytokines and, 57
    immune function and, 56
    interferon therapy and, 107
Anxiety, INF-γ- vs. IL-10-reactive subjects
        examination stress response, 40, 41,
        43, 45
Anxiety disorders, 7–8
Apoptosis, 144
Appetite
    acute phase response, 117
    cancer and, 141–142
    depression and, 52, 54
Arginine vasopressin (AVP), 19
    cytokine-induced stimulation, 329
    inflammatory cytokine interactions, 27–28
    inflammatory stress response, 332
    peripheral inflammatory response and, 26
Arthritis, experimental model, *See* Adjuvant-
        induced arthritis
Arthritis, rheumatoid, see Rheumatoid arthritis
Aspirin, 107
Asthma, 53, 121, 126
    glucocorticoid receptor alterations, 333
Astrocytoma, 144
Atherosclerosis, 124
Atopic dermatitis, 126
Autoimmune disorders and mechanisms, *See also*
        Rheumatoid arthritis; other specific
        disorders
    bromocriptine effects, 22–23
    endorphins and, 175
    interferon therapy, 221–222
    neuroendocrine (HPA axis) interactions,
        30–31
    obsessive-compulsive disorder, 7
    pregnancy and, 333
    prolactin antiinflammatory effects, 22–23
AVP, *See* Arginine vasopressin
Axon reflex, 121
AZT, 316–321

B

Bacillary parenchymal angiomatosis (BPA), 63
Bacterial infection, AIDS-related, 62–65

Bartonell henselae, 63
Basic-fibroblast growth factor (BFGF), 135, 137,
        140, 141, 246
Basophils, 18
B-cell growth factor, 175
B cells, 282
    cytokine expression, 326
    neuroendocrine hormone effects, 328
    opioid effects, 237, 239–242, 287–289, 290,
        300
    opioid expression and inflammatory pain
        control, 264
    opioid receptors, 299
    TNF-γ and HIV expression, 64–65
    TSH binding sites, 22
    TSH immunomodulatory effects, 175
7-Benzylidene-naltrexone (BNTX), 241
Bladder carcinoma, 135, 138, 140, 141
Blood-brain barrier, cytokine penetration, 62, 94,
        273
    active transport, 62, 94
    circumventricular organs, 94
    interferon, 224
    preoptic nucleus and, 179
BNTX, 241
Body weight changes
    cancer and, 141–142
    methionine-enkephalin treatment of
        AIDS-related complex, 83, 90
    restraint stress and, 166
Bombesin, 135, 136, 137, 139–140, 146, 246
Bone lesions, 142
Bradykinin, 140
Brain cancer, 135, 138, 140
Breast cancer, 2, 135, 138, 140, 144
Bromocriptine, 176, 191
    autoimmune disease treatment, 22–23
Bromodeoxyuridine, 192
Bulimia nervosa, 54

C

C_3, 55
C_4, 55
Cachectin, 181
Cachexia, 26, 73, 142
Calcitonin gene-related peptide (CGRP),
        263
CALLA, 147
Camphor, 179, 286, 287, 289, 291, 292
Campylobacter, 63
Cancer, *See also* specific cancers
    associated growth factors, 246

Index

cytokine and neuropeptide relationships, 133–147
  autocrine/paracrine tumor growth factors, 134
  clinical manifestations, 141–142
  endogenous opioids and receptors, 143–145
  receptor expression, 138–139
  specific tumor expression (tables), 135–140
cytokine therapies and indications, 95
  toxicity, *See* Central nervous system toxicity
murine rotational stress model, 166
opioid growth factor ([Met^5]-enkephalin) and, 245–256, *See* Opioid growth factor, gastrointestinal cancer and
paraneoplastic phenomena, 134, 145
stress effects on recovery, 2
Capsaicin-induced nerve damage, 24
  substance P-mediated inflammatory response and, 121, 125, 126
Cat scratch disease (CSD), 63
CBG, 330
CC16, 40–45
CD3+ cells
  AZT/met-enkephalin treatment of HIV and, 321
  methionine-enkephalin treatment of AIDS-related complex, 77, 80, 81, 84
  exogenous catecholamine effects, 164
CD4+ T-cells
  AZT/met-enkephalin treatment of HIV and, 321
  brain, 65
  exogenous catecholamine effects, 164
  glucocorticoid effects, 327–328
  methionine-enkephalin treatment of AIDS-related complex, 77, 80, 81, 84, 85
  opiate-induced immunosuppression, 234
  opioid receptors, 299
  TNF-γ and HIV expression, 64–65
CD8+ T-cells
  AZT/met-enkephalin treatment of HIV and, 321
  INF-γ- vs. IL-10-reactive subjects psychological stress response, 40–45
  methionine-enkephalin treatment of AIDS-related complex, 77, 80, 82, 84, 85
  opioid effects on cAMP production, 298
  opioid receptors, 299
  TNF-γ and HIV expression, 64–65
CD10+ cells, 147
CD25, 77, 80, 83, 84
CD38, 80, 84, 86
CD56+ NK cells, 77, 80, 84, 85

Central nervous system/immune system interactions, *See* Neuroendocrine interactions; specific hormones
Central nervous system infection (HIV), 64–65
Central nervous system toxicity, 93–108
  acute and chronic effects, 97
  colony-stimulating factors, 96, 104–105
  erythropoietin, 96, 104, 107
  IL-1, 102–103
  IL-2, 101–102
  IL-3, 106
  IL-4, 102
  IL-6, 103
  IL-11, 106
  IL-12, 102
  interferons, 97–101, 107, 222
  management of, 106–107
  mechanisms, 94–96
  stem cell factor, 105–106
  thrombopoeietin, 106
  TNF, 103
Cervical cancer, 135, 138, 140, 141
c-fos, opioid-induced expression, 145
Chagas' disease, 196–197
Chemokines, 3
Chills, cytokine-induced, 97, 101
Cholecystectomy, 6
Cholecystokinin-B (CCK-B), 135, 136, 138–140, 146
Chronic inflammation, *See* Adjuvant-induced arthritis; Autoimmune disorders and mechanisms; Rheumatoid arthritis
Chronic myelogenous leukemia, 95
Circadian rhythms
  immune cell-linked cortisol activity, 328
  peripheral blood lymphocyte populations, 163
  rheumatoid arthritis-associated plasma cytokine levels, 28
Circumventricular organs, 94
Citrulline, 210
c-jun, 21
Clara Cell Protein, 40–45
Clomipramine, 7, 57
Clostridium dificile toxin A-induced enteritis, 122
c-myc, 176, 300
Coagulation proteins, 118
Cocaine, lymphocyte function and, 284, 285, 286, 289
Codeine, 234
Cold water stress, 123, 126, 167
Colon mitosis inhibitor, 246
Colony-stimulating factors, 3, *See* specific factors
  CNS toxicity, 96, 104–105

Colorectal cancer, 135, 138, 140, 144, 146, 245–246
  [Met^5]-enkephalin and, 247–256, See Opioid growth factor, gastrointestinal cancer and
Common acute lymphocytic leukemia antigen (CALLA), 147
Complement factors, 55, 118
Concanavalin A (con A)-induced response
  opioid effects on immune cell function, 284
  trauma-associated cytokine inhibition, 6
Condylomata acuminata, 95
Coronary artery disease, 126
Corticosteroid-binding globulin (CBG), 330
Corticosteroids, 4, 119, 162–164, 327–329, See Gluocorticoids; Hypothalamic-pituitary-adrenal (HPA) axis; specific substances
Corticosterone
  animal handling stress effects, 167
  blood levels and immune cell function, 163
  cytokine interactions, 329–331
  rotational stress effects on murine plasma levels, 165
  shock stress effects, 168
Corticotropin, See ACTH
Corticotropin-releasing hormone (CRH), 3, 18, 19, 40, 46, 122, 162, 164–165, 174, 196
  acute phase response, 119
  c-fos expression induction, 145
  cytokine interactions, 5, 179–180, 265–266, 329, 331–332, See specific cytokines
    IL-1 stimulated effects, 188, 271–277
    interferon-stimulated sympathetic NK cytotoxicity modulation, 222
  immune cell binding sites, 325
  immune cell endorphin release and, 265–266, 272–277
  immune cell expression, 179, 326–327
  inflammatory stress response, 332
  nitric oxide interactions, 212
  peripheral inflammatory response and, 23–26
  receptor distribution, 165
  shock stress effects, 168
  thermoregulatory function, 179
  tumor expression, 136, 137
Cortisol, 19
  cytokine-stimulated response, 330–331
  glucocorticoid resistance syndrome, 333
  IL-1 stimulated effects, 188
  immune cell function and, 328–329
  rheumatoid arthritis patient blood levels, 332
  stress effects, 164
  thymic interactions, 189

CP96,345, 122
C-reactive protein (CRP), 40, 118–119
Critical weight hypothesis, 214
Crohn's disease, 121, 126
Cushing's syndrome, 145, 329
Cyclic AMP (cAMP)
  opioid modulation of immune cell expression, 298, 300
  TSH stimulation of, 175
Cyclic GMP, nitric oxide-stimulated gonadotropin interactions, 206, 210, 211, 213
Cyclooxygenase, 206, 209–210
Cyclosporine, 22–23
Cytokines, See specific cytokines
  CNS interactions, See Central nervous system toxicity; Neuroendocrine interactions
  stress-related psychopathology, 7–9
  synergization in CNS, 182
  therapies, See types of, 2–3
Cytokine therapies
  cancer and, See Cancer
  CNS toxicity, 93–108, See Central nervous system toxicity; specific cytokines
  selected cytokines and indications, 95
Cytomegalovirus (CMV) blastogenesis, methionine-enkephalin effects, 82, 84, 88
Cytotoxic lymphocytes, 282, See also CD8+ cells; Natural killer (NK) cell activity
  AZT/met-enkephalin treatment of HIV and, 321
  opioid effects, 235, 238–241, 284, 286

D

DADLE, 240, 298
DALEA, 297
DAMEA, 287, 288, 289, 291, 293, 294, 295, 297
DAMGO, 286, 289, 292, 293, 295, 296, 297, 300
Degenerative osteoarthritis, 333
Delayed cutaneous hypersensitivity, 22
Delta-opioid receptors, 143, 282, 283
  agonist/antagonist immunomodulatory effects, 240
  gastrointestinal expression, 247
  immune cell expression, methionine enkephalin interactions, 315
  peripheral nerve expression and inflammatory pain control, 264
Deltorphin, 240, 281, 287, 288, 289, 291, 293, 296
Dental phobia, 126
Depression, 5

Index

antidepressant effects, 56–57, *See also* Antidepressants
  cancer and, 141, 142
  cytokine function and, 7
  cytokine-induced symptoms, 99, 100
  cytokine plasma concentrations, 55
  cytokine production, 56
  elderly subjects, 56
  INF-γ- vs. IL-10-reactive subjects
    psychological stress response, 40, 41, 43, 45
  neuroendocrine interactions, 52
  neurovegetative features, 52, 54–55
  rationale for studying cytokines, 53–55
Dexamethasone-induced effects, 6, 8, 20, 25, 328, 333
Dihydromorphine, 298
Diurnal cycle
  light/dark cycle and immune function alterations, 167
  peripheral blood lymphocyte numbers, 163
Dopamine, 174
  gonadotropin-modulatory activity, 206
  NOS and prolactin interactions, 213
Dorsal horn, 261
Dorsal root ganglion, 264
DPDPE, 240, 284–291, 293–297
Drug addiction, 225
DSLET, 240, 288, 289, 290, 292
D-TIPP, 240–241
Dynorphin (dynorphin A), 143, 281, 282
  cytokine/neuropeptide analgesic effects and, 266
  lymphocyte function and, 284, 285, 289, 292, 293
  macrophage function and, 294–297
  tumor expression, 136

E

EEG, cytokine-associated effects, 96, 222–223
Eicosanoids, 65
Elderly subjects, depression and plasma cytokine levels, 56
Electroconvulsive therapy, 57
Electroencephalogram (EEG) activity, cytokine modulation of, 96, 222–223
Emotional disorders, *See* Depression; specific disorders
Emotional stress
  IL-2 expression and, 5
  immune function and, 162
Endogenous opioids, *See* Opioid peptides
Endometrial cancer, 135, 138, 144

Endorphin(s), 174, 281, 282, *See* specific types
  immune cell secretion, 54
α-Endorphin, 21
  immune cell interactions, 175
  macrophage function and, 296, 297
β-Endorphin, 3, 18, 21
  central IL-1 effects on immune cell expression, 271–278
  cytokine interactions, 180
  exogenous application for inflammatory pain control, 266–267
  gastrointestinal expression, 247
  immune cell cytokine production and, 242
  immune cell expression, 179, 272
    central IL-1 effects, 271–278
  immune cell function and, 300
  immunosuppressive properties *in vivo*, 277
  inflammatory response and, 21
  interferons and, 224
  lymphocyte function and, 284–291
  macrophage function and, 292–297
  neuropeptide-cytokine-stimulated release and inflammatory pain control, 265–266
  shock stress effects, 168
  thymic interactions, 189
  tumor expression, 135–136, 140, 143–145
  tumor growth and, 144
Endothelin-1, 136
Endotoxemia, 70
Endotoxin-induced responses, *See* Lipopolysaccharide (LPS)-induced responses
  application for assessing HPA stimulus response, 333
  α-MSH and, 71–72
Enkephalin(s), 3, 54, 281, 282, *See also* Leucine enkephalin; Methionine enkephalin
Eosinophils, 18, 121
Epidermal growth factor (EGF)
  gastrointestinal cancer and, 246
  tumor expression, 136
  receptor, 138
Epinephrine, 3, 122
  IL-12 inhibition, 6
  stress-associated cytokine stimulation, 123
  stress effects, 164
Epoeitin α, 95
Epsilon-opioid receptor, 143, 315
Erythropoietin, 3
  CNS toxicity, 96, 104, 107
  hypertensive effects, 96, 104
E-selectin, 65
Escapable shock, 168
Essential hypertension, 126
Estrogen, 206, 208

Evoked field potentials, interferon effects, 223
Examination stress, 124, 125, 166
    subject variations in INF-γ vs. IL-10 response, 39–46
Experimental autoimmune encephalomyelitis, (EAE), 221–222

F

Fatigue syndrome, 97, 98, 332
Fentanyl, 240, 281, 282, 286–292
Fever
    acute phase response, 117
    cancer and, 141, 142
    cytokine-induced, 5, 96, 97, 103, 107
    α-Melanocyte-stimulating hormone (MSH) and, 21, 70–71
    stress and, 125
Fibrinogen, 124
Fibroblast-interferon, 95
Filgrastim, 95, 104–105
Fischer rat model, 30
FITC, 298
Fluorescein isothiocyanate (FITC), 298
6–FN, 298
Follicle-stimulating hormone (FSH), 176
    cytokine interactions, 181
    hypothalamic control, 206–210, 213
    inflammatory cytokine interactions, 188–189
    leptin and, 206, 214–215
    thymic expression, 193
    thymic interactions, 189
    TSH similarity and potential leukocyte interactions, 22
Follicle-stimulating hormone-releasing factor (FSHRF), 206–208
Foot shock models, 123, 168, 275–276
Friend leukemia virus (FLV), 317–321
β-Funaltrexone (β-FNA), 286, 289, 293, 297

G

GABA-a receptor, hypothalamic LHRH neurons, 211–212
Galanin, 139
Ganglioneuroblastoma, 138
Gastric cancer, 135, 138, 140, 144
Gastrin, 135, 136, 138–140, 146, 246
Gastrin-releasing peptide (GRP), 135–137, 139–140, 146, 246
Gastrointestinal cancers, 135, 140, 147
    associated growth factors, 246

    opioid growth factor and, 247–256, *See* Opioid growth factor, gastrointestinal cancer and
Gastrointestinal opioids, 247
Generalized anxiety disorder, 8
Gentamicin, 71
Glial cytokines, 62
Glioblastoma, 144
Glucocorticoid-responsive elements (GRE), 20
Glucocorticoids, 4, 40, 53, 196, 300, 327–329
    acute phase protein expression, 124
    clinical applications, 332
    cold-water stress effects, 167
    cytokine-stimulated secretion, 331
    disease-associated receptor alterations, 333
    diurnal cycle, 163
    gene expression and, 19, 21
    immune cell binding sites, 164, 325
    immune cell function and, 163–164, 327–328
    inflammation and, 19–21
    opiate-induced immunosuppression, 234
    receptor, 20
        immune cell expression, 164, 325
    resistance syndrome, 333
Glutamic acid
    gonadotropin-modulatory activity, 206
    nitric oxide-stimulated LHRH release, 210–211
Gonadotropin, hypothalamic control of, 206–213, *See* specific hormones
Gonadotropin-releasing hormone, 206, *See also* Luteinizing hormone-releasing hormone
    tumor expression, 135
Gonadotropin-releasing hormone-associated peptide (GAP), 207
Granulocyte colony-stimulating factor (G-CSF), 95
    CNS toxicity, 96, 104–105
Granulocyte-macrophage colony-stimulating factor (GM-CSF), 95
    CNS toxicity, 105
    gonadotropin release and, 213
    hypothalamic nitric oxide-stimulated LHRH release and, 212
    opioid effects, 292
    tumor expression, 135–137, 140
    receptor, 138
Growth hormone (GH), 3, 4, 18, 61, 146, 184, *See also* Somatostatin
    cytokine interactions, 5, 181, 188–189, *See* specific cytokines
    immune cell function and, 176–177
    immune cell secretion, 179
    inflammatory response and, 22

thymic expression, 193
  receptor, 195, 196
  thymic interactions, 189, 191–192
Growth hormone releasing hormone (GHRH), 174
  cytokine interactions, 181
  immune cell secretion, 179
  nitric oxide interactions, 212
Guanylate cyclase, 206, 210
Guillan-Barre-like syndrome, 64

H

Hairy cell leukemia, 95
Handling stress, 167
Head/neck cancer, 135, 138, 144
Headaches, cytokine-induced, 96, 97, 100, 101, 103
Heat shock protein, 20
Hematopoietic growth factors, 3, *See* specific factors
Hemoglobin, nitric oxide scavenging function, 209
Hepatitis, 95
  interferon-α treatment, 330–331
Hepatocellular cancer, 135
Heroin, 239–240, 297
Herpes simplex virus, 166–167
Hepatocyte growth factor (HGF), 135, 138, 140, 141
Hippocampal cytokine receptor/cytokine expression, 326–327
HIV infection, 62–65
  CNS infection, 64–65
  erythropoietin therapy, 104
  IL-2 therapy (aldesleukin), 101–102
  immune activation effects on replication, 317
  α-MSH and, 73–74
  opioid (met-enkephalin) treatment, 316, 321
  type 1/type 2 cytokine ratio and, 318
HTLV-1, 316
Human chorionic gonadotropin
  TSH similarity and potential leukocyte interactions, 22
  tumor expression, 135, 138, 140
  tumor growth modulation, 147
Human immunodeficiency virus, *See* HIV infection
Human T-cell leukemia virus type 1 (HTLV-1), 316
Hydrogen peroxide, 283
6-Hydroxydopamine, 274
Hyperalgesia, 262
Hypertension, 96, 104, 126

Hypothalamic electrophysiology, interferon effects, 223–224, 227
Hypothalamic-pituitary-adrenal (HPA) axis, 3, 53–54, *See also* ACTH; Corticotropin-releasing hormone; Glucocorticoids; Growth hormone; Opioid peptides; Prolactin; other specific hormones
  anorexia nervosa-associated hyperactivity, 8
  autoimmune disorders, 30–31
  clinical implications, 332–333
  cytokine effects on, 26–30, 179–182, 188–189, 271, 331–332, *See* specific cytokines
  cytokine expression, 326–327
  cytokine neurotoxic mechanisms, 96, *See also* Central nervous system toxicity
  depression and, 52
  immune system modulation of, 329–332
  immune system shared signal models and receptors, 325–327
  immunomodulatory activity, 17–31, 327–329
  inflammatory cytokine effects on, 26–30
  interferon suppression, 222
  opioid effects, 282, 300, *See* Opioid peptides; specific opioids and opiates
  pregnancy and, 333
  psychological stress-induced hyperactivity, 40
  stress effects on immune function, 162–165
  stress neurobiology, 122
  thymic hormone and cytokine modulation, 188–189
  thymus physiology and, 191–192
Hypothalamic-pituitary-gonadal axis, 3, 180, 182, 205–216, *See also* Follicle-stimulating hormone; Gonadotropin-releasing hormone; Luteinizing hormone; specific related hormones
  nitric oxide function, 206, 209–213
Hypothalamic-thyroid axis, 3, 180, 182, *See* specific thyroid hormones

I

ICI 174864, 240, 241, 284, 289
Ileal cancer, 135
Immobilization stress, 124, 125, 166
Immune cells, *See* specific cells
Immune response, *See* Inflammatory response; specific immune cells and responses
  cytokines and, *See* specific cytokines
  neuroendocrine interactions, *See* Hypothalamic-pituitary-adrenal (HPA) axis; Neuroendocrine interactions; specific hormones
  stress effects, *See* Stress

Immune system, sensory function of, 62, 173, 332
Immunogenic inflammation, 116
Immunoglobulins
  endorphin immunomodulatory effects, 277
  INF-γ- vs. IL-10-reactive subjects
    psychological stress response, 40–45
  opioid effects on B cell function, 288–289
Individual stress response variability, 168–169
Indomethacin, 25, 273
Inescapable foot shock models, 123, 168, 275–276
Infection-induced inflammation, 116
  thymic-pituitary component of immune response, 196–197
Inflammatory anorexia, 54
Inflammatory bowel disease, 73, 121, 126
Inflammatory cytokines, See Interleukin-1; Interleukin-6; Tumor necrosis factor-α
Inflammatory pain, 261–266
Inflammatory response, 18–19, 115–122
  AVP and, 26
  CRH direct effects, 23–26
  cytokines and acute phase response, 117–119
  glucocorticoid effects, 19–21
  growth hormone and, 22
  immunogenic and neurogenic, 116–117
  α-MSH and, 21, 69–74
  neuroendocrine effects, 19–26
    inflammatory cytokine modulation, 26–29
  nervous innervation, 119–120
  opioid peptides and, 21
  prolactin and, 22–23
  somatostatin effects, 24–26
  stress and, 115, 122–127
  substance P and, 116, 117, 120–122
  TSH and, 21–22
Inhibin, 176, 206, 208
Insulin-like growth factor 1 (IGF-1), 146, 176
  gastrointestinal cancer and, 246
  thymus-pituitary function and, 192
Insulinoma, 138
Interferon(s), 2
  CNS effects, 221–227
    direct vs. indirect effects, 224
    EEG activity, 222–223
    evoked field potentials, 223
    neuronal activity modulation, 223–224
    opioids, 224–227
    peripheral nervous system, 222
    toxicity, 97–101, 107, 222
  opioids and, 224–227
  somnogenic effects, 223
  stress response, 52
  types, 221

Interferon-α, 221
  clinical uses, 95
  CNS effects, 221–227
    toxicity, 94, 97–99, 107
    EEG alterations, 222–223
    evoked field potentials, 223
  hepatitis treatment, 330–331
  neuroendocrine response, 330–331
  neuronal activity modulation, 223–224
  NK cell cytotoxicity and, 289
  opioids and, 224–227
  peripheral nervous system effects, 222
  somnogenic activity, 54
  therapeutic applications, 221–222
Interferon-α-2a, 95
Interferon-α-2b, 95, 98
Interferon-α-n3, 95
Interferon-β
  clinical uses, 95
  CNS toxicity, 99–100
  thermoregulatory function, 179
Interferon-β-1a, 95, 99–100
Interferon-β-1b, 95, 99–100
Interferon-γ
  cancer cachexia and, 142
  clinical/immune correlates of examination stress-induced responses, 39–46
  CNS toxicity, 100–101, 107
  cold-water stress effects, 167
  HPA axis activation, 331
  immune cell AVP receptors and, 26
  inhibin/activin effects, 176
  morphine immunosuppressive effects and, 234
  neural electrophysiology effects, 224
  neuroendocrine interactions, 175, 181, 188–189, 329
  opioids and, 224–227, 287
  shock stress effects, 168
  stress response, 6
  viral infection and methionine enkephalin-mediated immunomodulation, 316–317, 318, 319
Interferon-γ-1b, 100
Interferon-inducible protein 10, 3
Interleukin-1 (IL-1), 5, 69, 117, 271
  acute-phase reaction, 5
  blood-brain barrier penetration, 62, 273
  cancer cachexia and, 142
  central cytokine synergies, 182
  CNS toxicity, 102–103
  cold-water stress effects, 167
  HPA axis expression, 327
  hypothalamic nitric oxide-stimulated LHRH release and, 212

Index

immune cell endorphin expression and, 271–278
inflammatory pain and, 263
neuroendocrine interactions, 4–5, 18, 26–29, 53–54, 179–180, 182, 188, See also specific hormones
peripheral neuronal receptor interactions, 96
psychological stress response, 40
somnogenic activity, 54
stimulated endorphin release and inflammatory pain control, 265–266
stress response, 123–124
substance P-induced inflammatory response, 120, 122
transport into the brain, 94
Interleukin-1 receptor (IL-1R)
HPA axis, 326
pituitary expression, 180, 194, 196
Interleukin-1 receptor antagonist (IL-1ra), 2
depression and, 55
INF-γ- vs. IL-10-reactive subjects psychological stress response, 40–45
study of central IL-1 administration effects on immune cell expression of endorphin, 272
Interleukin-1α (IL-1α), 2
cerebral production, 62
CNS toxicity, 103
HIV infection and, 73–74
HPA axis activation, 329–331
HPA axis expression, 327
neureoendocrine effects, 188
tumor expression, 135, 137
Interleukin-1β (IL-1β), 2
anorexia induction, 54
antibody response and, 165
antidepressant therapy and, 57
cerebral production, 62
CNS toxicity, 103
depression and, 7, 55, 56
fever stimulation, 142
HIV and brain cell expression, 65
HPA axis activation, 329–331
HPA axis expression, 327
IL-13 and traumatic stress response, 7
leptin activation model of gonadotropin release, 216
morphine immunosuppressive effects and, 234
neureoendocrine effects, 188
panic disorder and, 8
pituitary expression, 193
post-traumatic stress disorder and, 8
stress response, 53
tumor expression, 135–140
receptors, 138–139

Interleukin-2 (IL-2), 2, 5
anorexia and, 9, 54
CNS toxicity, 94, 96, 101–102
depression and, 7, 56
emotional stress response, 5
HIV therapy, 101–102
HPA axis expression, 327
neuroendocrine interactions, 4–5, 180–181, 188–189, 328, 330–332
opiate effects, 234, 238, 239, 240, 241, 242
panic disorder and, 8
pituitary expression, 193
stress response, 52
trauma-associated inhibition, 5
viral infection and methionine enkephalin-mediated immunomodulation, 317, 318, 321
Interleukin-2 receptor (IL-2R)
antidepressant therapy and, 57
depression and, 7, 55
examination stress response, 166
generalized anxiety disorder, 8
HPA axis, 326
methionine-enkephalin treatment of AIDS-related complex, 77, 80, 83, 84
pituitary expression, 193, 195–196
Interleukin-3 (IL-3)
CNS toxicity, 106
depression and, 7, 56
glucocorticoid effects on immune cell function, 327
HPA axis expression, 327
receptor, 326
panic disorder and, 8
stress effects, 5–6
Interleukin-3 (IL-3)-like activity, antidepressant therapy and, 57
Interleukin-4 (IL-4)
acute phase response, 119
CNS toxicity, 102
neuroendocrine hormone effects, 328
opiate effects, 234, 238, 239, 240, 241
stress effects, 6
tumor expression, 136
viral infection and methionine enkephalin-mediated immunomodulation, 317, 320
Interleukin-5 (IL-5), 175
Interleukin-6 (IL-6), 2, 117
anorexia and, 9, 54
antidepressant therapy and, 57
cancer cachexia and, 142
central cytokine synergies, 182
cerebral production, 62
circadian rhythms, 28

CNS toxicity, 103
depression and, 55, 56
fever and, 142
glucocorticoid effects, 19
HPA axis expression, 327
IL-13 and traumatic stress response, 7
INF-γ- vs. IL-10-reactive subjects
    psychological stress response, 40–45
morphine immunosuppressive effects and, 234
neuroendocrine interactions, 4, 5, 6, 18, 26–29, 181, 188, 330–331, *See also* specific hormone interactions
opiate immune effects and, 234, 239
pituitary expression, 193
psychological stress response, 52–53
psychopathology and, 7
receptor
    antidepressant therapy and, 57
    depression and, 7, 55
    HPA axis expression, 194–196, 326
    tumor expression, 138–139, 141
stress-associated neuroendocrine interactions, 6
stress response, 6, 123–124
substance P-induced release, 120
transport into the brain, 94
tumor expression, 135–141
Interleukin-7, tumor expression, 136, 137
Interelukin-8 (IL-8), 3
anorexia induction, 54
glucocorticoid effects on immune cell function, 327
IL-13 and traumatic stress response, 7
tumor expression, 135–137, 140
receptors, 139
Interleukin-9 (IL-9), 2
Interleukin-10 (IL-10), 2
acute phase response, 119
AIDS and, 63
catecholamine stimulation, 6
clinical/immune correlates of examination stress-induced responses, 39–46
tumor expression, 135, 137
viral infection and methionine enkephalin-mediated immunomodulation, 317, 320
Interleukin-11 (IL-11), CNS toxicity, 106
Interleukin-12 (IL-12), 2, 6
AIDS and, 63
CNS toxicity, 102
tumor expression, 136, 137
viral infection and methionine enkephalin-mediated immunomodulation, 317, 319

Interleukin-13 (IL-13), 2
AIDS and, 63
traumatic stress response, 7
Interleukin-14 (IL-14), 2
Interleukin-15 (IL-15), 2
Intraarticular opioid effects, 266–267
Iridocyclitis, 22
Iritis, 22

K

Kaposi's myeloma, 141
Kaposi's sarcoma, 63, 95, 135, 138, 140, 147
Kaposi's sarcoma associated herpes virus (KSHV), 141
Kappa-opioid receptor, 143, 282, 283
    gastrointestinal expression, 247
    immune cell expression, methionine enkephalin interactions, 315
    peripheral nerve expression and inflammatory pain control, 264
Kit ligand, 105
Knee surgery, 266–267

L

Lamprey luteinizing hormone-releasing hormone, 206, 208
Legionella, 63
Leptin, 54
    reproductive function and, 213–217
    stimulated FSH/LH release, 206
Leucine enkephalin (LENK), 143
    macrophage function and, 292–297
    tumor expression, 135–136
Leukemia, 333
    cytokine/neuropeptide expression, 136, 139, 140
Leukemia inhibitory factor (LIF), 137, 193, 287
Leukocyte interferon, 95
Lewis rat model, 30
    animal cellular immune responses and, 167
Lipopolysaccharide (LPS)-induced responses, 116, *See also* Endotoxin-induced responses
    cytokine-HPA axis interaction model, 26
    depression and, 56
    HIV-infected subjects, 63
    Stat 3 activation, 216
β-Lipoprotein, 180, 189
Listeria-induced immunoproliferative response, 176
Locus coeruleus, 162, 223

Lung cancer, 136, 139, 140, 144, 145–146
Luteinizing hormone (LH), 61, 174
  cytokine interactions, 5, 180–181, See specific cytokines
  GM-CSF effects, 213
  gonadal control, 208
  hypothalamic control, 206–210, 213
  IL-1 stimulated effects, 188
  immunomodulatory activity, 176
  inflammatory cytokine interactions, 188
  leptin and, 206, 214–215
  receptor, tumor expression, 138
  thymic interactions, 189, 193
  TSH similarity and potential leukocyte interactions, 22
  tumor expression, 135, 138
Luteinizing hormone releasing hormone (LHRH), 174
  cytokine interactions, 180, 212
  GABA receptors, 211
  gonadal control, 208
  hypothalamic control, 206–213
  lamprey LHRH as putative FSHRH, 206, 208
  leptin and, 214–217
  nitric oxide and, 206, 209–213
  sexual behavior and, 212
  thymic interactions, 189
Lymph node cells, IL-1 effects on β-endorphin expression, 271–278
Lymph node size, methionine-enkephalin treatment of AIDS-related complex, 77, 83, 89
Lymphocyte activating factor (LAF), 164
Lymphocyte chemotactic factor (LCF), 287
Lymphocytes, 18, 282, See also Natural killer (NK) cell activity; T-cells; specific subpopulations
  AVP receptors, 26
  diurnal cycle, 163
  examination stress effects on IL-2 receptor expression, 166
  glucocorticoid effects, 327–329
  growth hormone and, 176
  INF-γ- vs. IL-10-reactive subjects psychological stress response, 40–45
  morphine/opiate immunosuppression and cytokine interactions, 234–242
  "neuroendocrine" hormone production, 61–62, 179
  neuroendocrine modulation, 4, 163–165, See specific hormones
  opioid actions, 284–289, 300–301
  opioid receptors, 298, 299
  shock stress effects, 168
  steroid receptors, 164
  stress-associated decrease in circulating population, 1
  stress-associated glucocorticoid control of, 163–164
  TSH immunomodulatory effects, 175
Lymphokines, 2, 283, See Tumor necrosis factor-β; specific interleukins
Lymphoma, 141, 136, 333

M

Macrophage activating factor (MAF), 164
Macrophage colony-stimulating factor (M-CSF)
  CNS toxicity, 105
  opioid effects, 292–293
Macrophage inflammatory protein (MIP), 3, 65, 135
Macrophage-migration inhibitory factor (MIF), 193
Macrophages, 18, 283
  acute phase response, 117
  cold-water stress effects, 167
  cytotoxic products, 283–284
    opioid effects, 293, 296–297
  "neuroendocrine" hormone production, 61–62
  neuroendocrine interactions, 4, 175, See specific hormones
  nitric oxide release, 283–284, 293, 296–297
  opioid binding sites, 298, 299
  opioid effects, 240, 241, 289–297, 300–301
    chemotaxis, 297
    inflammatory pain control, 264
    nitric oxide production, 293, 296–297
    phagocytosis, 297
    proliferation and differentiation, 289, 292–293
    respiratory burst, 293
  stress-induced cytokine release, 123–124
  substance P-induced activity, 120–121, 126
  tumoricidal activity, 297
Major depression, See Depression
Major histocompatibility complex (MHC) antigens
  ACTH effects on macrophage expression, 175
  interferons and, 2
  IFN-β effects, 99
  shock stress effects, 168
  tachyphylaxis, 97
  thymic interactions, 191
Malignant melanoma, 95
Maloney sarcoma virus, 166
Mast cell growth factor, 105
Mating behavior, 209, 211, 212
MC-1R, 70

MC-2R, 70
MC-5R, 73
Megakaryote growth and development factor (MGDF), 106
Melanocortin receptors, 70, 73
α-Melanocyte-stimulating hormone (MSH), 69–74
   ACTH homology, 69
   arthritis and, 71–72
   fever and inflammatory response, 21, 70–71
   HIV infection and, 73–74
   melanocortin receptors, 70, 73
   myocardial infarction and, 72–73
   reperfusion injury and, 72–73
   systemic inflammation, 70–71
Melanoma, 95, 136
Melanoma growth-stimulating activity, 3
Melatonin, 206
Meperidine, 107, 240, 286, 287, 288, 290, 291
Methadone, 239–240, 281, 296
Methionine enkephalin (MENK), 4, 143, 315
   adverse reactions, 84
   AIDS/ARC treatment, 65, 77–90
   cytokine/neuropeptide analgesic effects and, 266
   gastrointestinal expression, 247
   immune cell function and, 300, 315–316
   inflammatory response and, 21
   interferons and, 224
   lymphocyte function and, 284–292
   macrophage function and, 292–297
   opioid growth factor ([Met^5]-enkephalin), 246–247
      gastrointestinal cancer and, 247–256, *See* Opioid growth factor, gastrointestinal cancer and
   tumor expression, 135, 137, 140
      receptor, 140
   viral infection model and immunomodulatory cytokine response, 317–321
Methylphenidate, 107
Metoclopramide, 107
Monocytes, 18, *See also* Macrophages; Peripheral blood mononuclear cells
   opioid binding sites, 298, 299
   opioid expression and inflammatory pain control, 264
   TSH binding sites, 22
Monocyte chemotactic and activating factor, 3
Monokines, 2, *See* Tumor necrosis factor-α; specific interleukins
Morphine, 281, 282
   binding sites, interferon affinity, 225
   c-fos expression induction, 145
   chronic administration and withdrawal effects, 241–242
   dependence and tolerance, 225–226, 237–238, 241–242
   endogenous, 234
   immune cell function and, 237–239, 284–298, 300
   interferon interactions, 226
   intraarticular pain relief, 266–267
α-MSH, *See* α-Melanocyte-stimulating hormone
Mu-opioid receptor (μ), 143, 282, 283
   gastrointestinal expression, 247
   immune cell expression, methionine enkephalin interactions, 315
   peripheral nerve expression and inflammatory pain control, 264
Multiple myeloma, 141
Multiple sclerosis, 95, 99–100, 221
Muramyl dipeptide (MDP), 234
Mycobacterium avium complex, 63
Mycobacterium tuberculosis, 63
Myeloma, 137, 139, 140, 141, 333
Myocardial infarction, 72–73

N

Nadolol, 289
Naloxone, 222, 227, 277, 284, 286, 287, 289–293, 298, 316
Naltrexone, 284, 286, 289
   morphine tolerance/dependence and, 241–242
   opioid mechanisms in colon cancer, 248–249
Naltrexone-FITC, 298
Naltrindole, 286, 290
Natural killer (NK) cell activity
   antidepressant therapy and, 57
   examination stress and, 166
   interferon-mediated sympathetic modulation of, 222
   methionine-enkephalin/AZT treatment of HIV and, 321
   methionine-enkephalin treatment of AIDS-related complex, 77, 80, 82, 84, 85, 86
   neuroendocrine interactions, 4, 164, 165, 175, 329, *See* specific hormones
   opioid effects, 143, 234, 235, 238–242, 277, 289, 291–292, 300, 315
   stress-associated catecholamine effects, 164
   TSH binding sites, 22
Nb2 cells, 176
β-Neoendorphin (β-NEND), 293
Nephotic syndrome, 333
Nerve growth factor (NGF), 137, 262

Index

Nervous innervation, inflammatory processes and, 119–120
Neural endopeptidase 24.11 (NEP), 147
Neurasthenia, 97, 100
Neuroblastoma, 137, 139, 140, 144
Neuroendocrine interactions, 2, 3, 17–31, 61–62, 187, 325–333, *See also* ACTH; Corticotropin-releasing hormone; Glucocorticoids; Hypothalamic-pituitary-adrenal (HPA) axis; Opioid peptides; specific hormones, neuroendocrine axes
  autoimmune disorders, 30–31
  cellular immune system and, 174–179
  central cytokine synergies, 182
  central IL-1 effects on immunocyte endorphin expression, 271–278
  clinical implications, 332–333
  cytokine effects on the brain, 4–5, 62, 173, 179–182
  cytokine neurotoxic mechanisms, 96, *See also* Central nervous system toxicity
  cytokines and stress responses, 4–7
  peripheral pain generation, 261–263
  pregnancy, 333
  sensory function of immune system, 62, 173, 332
  stress responses, 162–165
  thermoregulatory function, 179
  thymus-pituitary axis paradigm, 187–197
Neuroendocrine tumors, 145
Neurogenic inflammation, 116–117
Neurokinin A, 120, 136, 140
Neurokinin B, 120, 136, 140
Neuromedin K, 146
Neuropathy secondary to HIV infection, 64
Neuropeptide Y, 136, 137, 216
Neurotensin, 139, 246
Neurotoxicity, *See* Central nervous system toxicity
Neutropenia, treatment-associated, 95
Neutrophils, 18
Newcastle disease virus, 4, 326
NF-kB, 21
Nitric oxide (NO)
  anterior pituitary hormone interactions, 213
  hypothalamic-level modulation of gonadotropin release, 206, 209–213
  macrophage secretions, 283–284
  opioid effects, 293, 296–297
  mating behavior and ("sexual gas"), 212
  α-MSH effect on lower bowel production, 73
Nitric oxide synthase (NOS), 284
  glucocorticoid effects, 19
  hypothalamic-level modulation of gonadotropin release, 206, 209–213
  leptin-stimulated activation, 216
Nitroprusside, 209, 211, 212
NMDA receptors, 210
NMMA, 209, 210, 213
Nociceptors, 261–263
Nonsteroidal anti-inflammatory agents, 107
Norbinaltorphimine, 293
Norepinephrine, 3, 122, 162
  central opioid effects, 300
  direct stress effects, 164
  IL-12 inhibition, 6
  nitric oxide-stimulated luteinizing hormone-releasing hormone release and, 210, 211
  gonadotropin-modulatory activity, 206

O

Ob/Ob mouse model, 213–216
Obsessive-compulsive disorder, 7–8
Octreotide, 146–147
6-OHDA, 274
OHM3295, 292
OHM3507, 289
OKT 10 (CD38), 80, 84, 86
Oncostatin M, 2, 135
Opiate abstinence syndrome, 226–227
Opiates, 233–242, *See* Morphine; Opioid peptides; Opioid receptors; specific drugs
  endogenous opiate alkaloids, 234
  immunomodulatory effects, 237–242
Opioid growth factor (OGF), 246–247, *See also* Methionine enkephalin
  gastrointestinal cancer and, 247–256
    autocrine production, 248–249
    evidence for OGF and ζ receptor presence in human colon cancer, 248
    function *in vitro*, 248–249
    modulation in xenografts, 249
    normal colon cell function and, 253
    pancreatic cancer, 253–254
    zeta receptor interactions, 246–247
Opioid peptides, 3–4, 246, 282, 300, *See also* β-Endorphin; Methionine enkephalin; Morphine; Opioid receptors; Proopiomelanocortin; specific opioids
  B cell function and, 287–289, 290
  cancer and, 143–145
  endogenous opiate alkaloids, 234
  exogenous application for inflammatory pain control, 266–267

exogenous opiates, 233–242, See Morphine; specific drugs
    immunomodulatory effects, 237–242
  gastrointestinal expression, 247
  immune cell expression and inflammatory pain control, 264
  immune cell function and, 175, 242, 281–301, See also specific cell types
  inflammatory/immune response and, 21
  interferon and, 224–227
  macrophage function and, 289–297, 300–301
    chemotaxis, 297
    cytokine production, 293
    nitric oxide production, 293, 296–297
    phagocytosis, 297
    proliferation and differentiation, 289, 292–293
    respiratory burst, 293
    tumoricidal activity, 297
  modulation of signal transduction pathways, 298, 300
  neuropeptide-cytokine-stimulated release and inflammatory pain control, 265
  NK cell activity and, 289, 291–292
  T-cell function and
    cytokine production, 286–287, 288
    cytotoxicity, 284, 286, 287
    proliferation, 284–286
  tolerance and dependence, 225–226
Opioid receptors, 21, 282, See also specific receptor types
  agonist/antagonist immunomodulatory effects, 240
  δ, 143, 240, 247, 264, 282, 283, 315
  ε, 143, 315
  gastrointestinal, 247
  immune cell expression, 298–299, 315–316
  interferon and, 222, 225, 227
  μ, 143, 247, 264, 282, 283, 315
  peripheral nerve expression and inflammatory pain control, 263–264
  tumor expression, 138–139, 143–144
  ζ, 144, 246–247, 251, 253, 254
Osteoarthritis, 23
Osteoclast activating factors (OAFs), 142
Ovarian cancer, 136, 139, 140
Oxytocin, 193, 206, 211, 212

P

Pain, neuroimmune interactions
  exogenous cytokine/neuropeptide effects, 265–266
  peripheral pain control, 263–266
  peripheral pain generation, 261–263
  sensitization, 262
Pancreatic cancer, 137, 139, 144, 246, 253–254
Panic disorder, 8
Paraganglioma, 137
Paraneoplastic phenomena, 134, 145
Paraganglia cytokine binding sites, 96
Paraventricular nucleus, 168
Penile erection, 209, 211, 212
2-N-Pentiloxy-2–phenyl-4–methyl-morpholine (PM), 284, 286, 288, 290, 293
Periaqueductal gray (PAG), opioid effects on immune function, 289, 293, 296–297, 300
Peripheral blood lymphocytes (PBLs), See also Lymphocytes
  diurnal cycle, 163
  examination stress effects on IL-2 receptor expression, 166
  INF-γ- vs. IL-10-reactive subjects psychological stress response, 40–45
  stress-associated glucocorticoid control of, 163–164
Peripheral blood mononuclear cells (PBMCs), 5
  cytokine interactions in opioid-mediated immune response, 316–317
  depression and, 56
  IL-1 effects on endorphin expression, 271–278
  opioid effects on cAMP production, 298
  psychopathology-associated cytokine alterations, 7–9
  trauma-associated IL-2 inhibition, 5
Peripheral nervous system, interferon and, 222
Peripheral neuronal receptor interactions, 96
Peripheral nociceptors, 261
Peritoneal substance P, 126
Phentolamine, 210, 211
Phospholipase A_2, 210
Phospholipase C, 298
Phytohemagglutinin (PHA)-induced responses
  cytokine interactions in opioid-mediated immune response, 316–317
  depression and, 56
  exhaustive stress and IL-2 release, 5
  methionine-enkephalin treatment of AIDS-related complex, 82, 84, 87
  opioid effects on immune cell function, 284
Pituitary, See also Hypothalamic-pituitary-adrenal (HPA) axis; Thymus-pituitary axis; specific pituitary hormones
  cytokine expression, 62, 193–195
  IL-1 receptors, 180
Pituitary hormones, See ACTH; Growth hormone; Hypothalamic-pituitary-adrenal (HPA) axis; Opioid peptides; Prolactin

Plaque-forming cell response, opioid effects, 175, 234, 289
Platelet aggregating factor, 65
Platelet-derived growth factor (PDGF), 135, 140
PM, 284, 286, 288, 290, 293
Pokeweed-induced responses, methionine-enkephalin treatment of AIDS-related complex, 77, 82, 84, 87
Poly I:C, 226, 289
Post-traumatic stress disorder (PTSD), 8
Prazosine, 210
Pregnancy, neuroendocrine-immune system interactions, 333
Preoptic nucleus, 179, 207, 224
Primary hyperalgesia, 262
Prodynorphin, 7, 143, 264
Proenkephalin, 143, 264
Prolactin, 4, 18, 174, 196
   antiinflammatory activity, 22–23
   cytokine interactions, 181, 188–189
   IL-1 stimulated effects, 188
   immune cell function and, 176–177
   immune cell secretion, 179
   leptin effects, 215
   nitric oxide and, 213
   thymic interactions, 189, 191–193, 195, 196
Prolactinomas, 191
Proopiomelanocortin (POMC), 3, 21, 143, 264, *See also* Opioid peptides; specific opioids
   cytokine modulation of gene expression, 180
   IL-2-stimulated expression, 188
   immune cell expression, 327
   interferons and, 225
   peripheral inflammation effects, 23–26
   tumor expression, 145
Prostaglandin E_2, 96
   cold-water stress effects, 167
   luteinizing hormone release and, 209
   opioid effects on macrophage function, 297
   substance P-induced release, 120–121, 122
   thermoregulatory function, 179
Prostaglandins, 116
Prostate carcinoma, 137, 139, 140, 141
Proteinase inhibitors, 118
Psoriasis, 22, 126
Psychiatric disorders, 7–9, *See also* Depression; specific disorders
   cancer and, 142
   cytokine-induced, 96, *See also* Central nervous system toxicity; specific disorders
   interferon-associated syndromes, 98–99
Psychological stress
   acute phase response, 124, 125
   cytokine response, 52–53

HPA axis hyperactivity
   immune cell function and, 166
   immune/clinical correlates of subject variations in stimulated INF-γ/IL-10 responses, 39
Psychomotor retardation, 98
Psychoneuroimmunology, 2

Q

Quinolinate, 65

R

RANTES, 3
Respiratory burst activity, 293
Reactive oxygen intermediates, 283, 293
Rectal adenocarcinomas, 247, 251
Renal cell carcinoma, 137, 139, 140, 141
Reperfusion injury, 72–73
Reproductive function, *See* specific hormones
   leptin and, 214–217
   nitric oxide and, 212
Respiratory distress syndrome, 70
Restraint stress, 6, 166
Rheumatoid arthritis, 23
   blood cortisol levels, 332
   cytokine circadian rhythms, 28
   glucocorticoid receptor alterations, 333
   α-MSH and, 72
   neuroendocrine (HPA axis) interactions, 30–31
   stress and, 126
   substance P and, 122
Rhodococcus equi, 63
Rotational stress, 123, 165–166
RU-486, 6, 20, 300

S

Salmonella, 63
Sarcoma, 137
Sargramostim, 95, 105
Secretin, 139
Seizures, erythropoietin-induced, 104
Selye, Hans, 1, 18, 161
Sensitization, 262
Sensory function of immune system, 62, 173, 332
Septic shock, 6, 70
Serotonin, gonadotropin-modulatory activity, 206
Serotoninergic neurons
   central opioid effects on immune cell function, 300

tonic control of immunocyte endorphin expression, 272, 274–275, 300
Serum amyloid A (SAA), 118–119, 124
Sexual behavior, 209, 211, 212
Sexual development, leptin and, 213–217
Scatter factor (SF), 141
Shigella, 63
Shock models of stress, 123, 168, 275–276
SIADH, 28
Signal transduction, opioid effects on immune cells, 298, 300
Single neuron electrophysiology, interferon effects, 223–224
Sleep
   acute phase response, 117
   depression and, 52, 54
   interferon effects, 223
Small-cell lung carcinoma, 136, 145–146
Somatostatin, 18, 61, 174, *See also* Growth hormone
   antitumor activity, 146
   cytokine interactions, 181
   gastrointestinal cancer and, 246
   immune cell secretion, 179
   inflammatory response and, 24–26
   nitric oxide interactions, 212
   obsessive-compulsive disorder, 7
   substance P-mediated inflammatory response and, 121
   thymic expression, 193
   tumor expression, 136, 137
      receptor, 138
Somatulin, 24
Spleen weight, opiate effects, 237, 242
Splenocytes, IL-1 effects on β-endorphin expression, 271–278
Staphylococcal enterotoxin A (SEA)-stimulated response, 21
Stat 3, 216
Steel factor, 105
Stem cell factors, 3
   CNS toxicity, 105–106
Stomach (gastric) cancer, 135, 138, 140, 144
Stress, 161
   acute phase proteins and, 124, *See* Acute-phase proteins
   beneficial, 161
   cytokines and, 4–7, 52–53
   defined, 122, 161
   fever and, 125
   immune system effects, 163–164
   individual response component, 168–169
   induction of proinflammatory cytokines, 123–124
   inflammatory disease and, 126–127

inflammatory response, 115, 122–127, *See also* Inflammatory response
   neurobiology of, 122–123
neuroendocrine/immune interactions, 3–7, 162–165, *See* Neuroendocrine interactions; specific hormones
review of immune response to model stressors, 165–168
substance P and, 125–126
thymus-pituitary connectivity, 196–197
Stress hormones, *See* ACTH; Growth hormone; Opioid peptides; Prolactin; specific hormones
Stress-related psychiatric disorders, 7–9, *See also* Depression; specific disorders
Substance K, 120, 135, 146
Substance P, 24, 61, 146
   inflammatory pain and, 263
   inflammatory response and, 116, 117, 120–122
   stress and, 125–126
   tumor expression, 135, 136
      receptor, 138, 139
SUPERFIT, 298
Superoxide production, 283, 293
Sydenham's chorea, 7
Sympathetic nervous system
   CRH immunomodulatory function and, 165
   interferon effects, 222
   opioid immunomodulatory activity and, 300
Sympathoadrenomedullary system, 3
Syncope of uncertain etiology, 106
Syncytia formation, 316
Systemic inflammation, α-MSH and, 70–71

T

Tachykinins, 146, *See also* Substance P; specific substances
   receptor, tumor expression, 138
Tachyphylaxis, 103
Tail shock, 168
T-cell growth factor (TCGF), 164
T-cells, 282–283, *See also* Lymphocytes; specific types
   glucocorticoid effects, 327–329
   intrathymic differentiation, 191, 192
   neuroendocrine modulation, 165
   opioid effects, 284–288, 300
      cytokine production, 286–287, 288
      cytotoxicity, 284, 286–287
      proliferation, 284–286
   opioid expression and inflammatory pain control, 264
   opioid receptors, 299, 315

stress-associated IL-3 inhibition, 6
TSH immunomodulatory effects, 175
Tetrahydrocannabinol, 206
T-helper cell activity, 282–283
   ACTH modulation, 174
   cytokine-associated mechanisms of Th1 and Th2 functions, 6
   opioid expression, 315
   stress-associated IL-3 inhibition, 6
   type 1 response, 234
      neuroendocrine hormone effects, 328
      pregnancy and, 333
      specific disease response, 328
      viral infection and, 316
   type 2 response, 234
      neuroendocrine hormone effects, 328, 329
      pregnancy and, 333
      specific disease response, 328
      stress-associated IL-4 stimulation, 6
      viral infection and, 316
Thermoregulation, 179
Thrombocytopenia, 106
Thrombopoietin, CNS toxicity, 106
Thromboxane B2, 121
Thymectomy, 189
Thymic atrophy restraint stress and, 166
Thymic carcinoid, 137
Thymopentin, 189
Thymopoietin, 189
Thymosin-α1, 189
Thymosin-β4, 189
Thymulin, 189, 191
   circulating levels in hypercortisolemic patients, 328
Thymus atrophy, murine rotational stress model, 166
Thymus physiology, 191
Thymus-pituitary axis, 187–197
   autocrine/paracrine control, 194–196
   similarities for cytokine and hormone production, 193–194
   stress responses, 196–197
Thymus size
   growth hormone effects, 176
   opiate effects, 237, 242
   stress-associated changes, 1
Thyroid carcinoma, 137, 139
Thyroid-stimulating hormone (TSH), 175
   cytokine interactions, 5, 18, 22, 181, 188
   immune cell binding sites, 22
   immune cell secretion, 179, 193
   inflammatory response and, 21–22
TIPP, 240–241
Tolerance, 225–226, 236–237, 241–242
Tourette's syndrome, 7

Toxicity, CNS, 93–108, *See* Central nervous system toxicity
Transcription factors, 21
Transferrin receptors, antidepressant therapy and, 57
Transforming growth factor-α (TGF-α), 137, 246
Transforming growth factor-β, 9
   gastrointestinal cancer and, 246
   HIV and brain cell expression, 65
   tumor expression, 135, 137
Transport stress, 167
Trauma-induced inflammation, 116
Traumatic stress-induced host defense deficiency, 7
Thyrotropin releasing hormone (TRH), 174, 180, 192
Tricyclic antidepressants, 7, 57
trkA, 262
Trypanosoma cruzi, 196–197
T suppressor cells, glucocorticoid effects, 329
Tuberculosis, 63, 162
Tumor necrosis factor (TNF), 70
   CNS toxicity, 96, 103
   somnogenic activity, 54
Tumor necrosis factor-α, 2, 117
   anorexia induction, 54
   cancer cachexia and, 142
   central cytokine synergies, 182
   endotoxin-induced α-MSH and, 71
   HIV infection and, 73–74
      brain cell expression, 65
      immune cell expression and, 64–65
   HPA axis activation, 331
   IL-13 and traumatic stress response, 7
   INF-γ- vs. IL-10-reactive subjects
      psychological stress response, 40–45
   neuroendocrine interactions, 18, 26–28, 181–182, 188
   opiate immune effects and, 234, 239
   pituitary expression, 193
   stress response, 6, 53, 123–124
   substance P-induced release, 120, 122
   transport into the brain, 94
   TSH inhibition, 22
   tumor expression, 135
Tumor necrosis factor-β (TNF-β), 2, 142
Tumors, *See* Cancer

U

U50,488, 281, 288, 289, 290, 294, 295, 296
U69,593, 298
Ulcerative colitis, 121, 126

V

Vascular cell adhesion molecule 1 (VCAM-1), 65
Vascular endothelial growth factor (VEGF), 135–140, 141
Vasoactive intestinal protein (VIP), 61
   receptor, tumor expression, 138, 147
   tumor expression, 140
Vasopressin, 140, *See* Arginine vasopressin
   nitric oxide interactions, 212
   thymic expression, 193
Viral infections, methionine enkephalin and immunomodulatory cytokine response, 316–321
Von Willebrand factor, 124

W

Weight loss, *See* Body weight
Withdrawal, 225
Wound healing, 118

Z

Zeta-opioid receptor (ζ), 144
   human colon cancer expression, 251, 254
   normal colon expression, 253
   opioid growth factor interaction, 246–247
Zinc, circulating levels in hypercortisolemic patients, 328